Plant Lipids

# Biological Sciences Series

A series which provides an accessible source of information at research and professional level in chosen sectors of the biological sciences.

***Series Editor:***

Professor J.A. Roberts, Plant Science Division, School of Biosciences, University of Nottingham, UK

***Titles in the series:***

**Biology of Farmed Fish**
Edited by K.D. Black and A.D. Pickering

**Stress Physiology in Animals**
Edited by P.H.M. Balm

**Seed Technology and its Biological Basis**
Edited by M. Black and J.D. Bewley

**Leaf Development and Canopy Growth**
Edited by B. Marshall and J.A. Roberts

**Environmental Impacts of Aquaculture**
Edited by K.D. Black

**Herbicides and their Mechanisms of Action**
Edited by A.H. Cobb and R.C. Kirkwood

**The Plant Cell Cycle and its Interfaces**
Edited by D. Francis

**Meristematic Tissues in Plant Growth and Development**
Edited by M.T. McManus and B.E. Veit

**Fruit Quality and its Biological Basis**
Edited by M. Knee

**Pectins and their Manipulation**
Edited by G.B. Seymour and J.P. Knox

**Wood Quality and its Biological Basis**
Edited by J.R. Barnett and G. Jeronimidis

**Plant Molecular Breeding**
Edited by H.J. Newbury

**Biogeochemistry of Marine Systems**
Edited by K.D. Black and G. Shimmield

**Programmed Cell Death in Plants**
Edited by J. Gray

**Water Use Efficiency in Plant Biology**
Edited by M.A. Bacon

**Plant Lipids-Biology, Utilisation and Manipulation**
Edited by D.J. Murphy

# Plant Lipids

## Biology, Utilisation and Manipulation

Edited by

DENIS J. MURPHY
Biotechnology Unit
School of Applied Sciences
University of Glamorgan
Cardiff, UK

Blackwell
Publishing

CRC Press

Editorial offices:
Blackwell Publishing Ltd, 9600 Garsington Road, Oxford OX4 2DQ, UK
Tel: +44 (0)1865 776868
Blackwell Publishing Asia Pty Ltd, 550 Swanston Street, Carlton, Victoria 3053, Australia
Tel: +61 (0)3 8359 1011

ISBN 1-4051-1904-7

Published in the USA and Canada (only) by
CRC Press LLC, 2000 Corporate Blvd., N.W., Boca Raton, FL 33431, USA
Orders from the USA and Canada (only) to
CRC Press LLC

USA and Canada only:
ISBN 0-8493-2361-4



First published 2005

Library of Congress Cataloging-in-Publication Data:
A catalog record for this title is available from the Library of Congress

British Library Cataloguing-in-Publication Data:
A catalogue record for this title is available from the British Library

Set in 10/12 Times New Roman
by Newgen Imaging Systems (P) Ltd.
Printed and bound in Great Britain
by MPG Books Ltd, Bodmin, Cornwall

The publisher's policy is to use permanent paper from mills that operate a sustainable forestry policy, and which has been manufactured from pulp processed using acid-free and elementary chlorine-free practices. Furthermore, the publisher ensures that the text paper and cover board used have met acceptable environmental accreditation standards.

For further information on Blackwell Publishing, visit our website:
www.blackwellpublishing.com

# Contents

# Contributors


**Dr Atanu Adhvaryu** Food and Industrial Oil Research, USDA/ARS/NCAUR, 1815 N. University Street, Peoria, Il 61604, USA

**Prof. Pierre Benveniste** Plant Molecular Biology Institute, IBMP-CNRA, 28 rue Goethe, 67084 Strasbourg, France

**Prof. Peter Dörmann** Max Planck Institute of Molecular Plant Physiology, Am Mühlenberg 1, 14476 Golm, Germany

**Dr Bjørn K. Drøbak** Cell Signalling Group, Department of Disease and Stress Biology, John Innes Centre, Norwich Research Park, Norwich NR4 7UH, UK

**Dr Sevim Z. Erhan** Food and Industrial Oil Research, USDA/ARS/NCAUR, 1815 N. University Street, Peoria, Il 61604, USA

**Prof. Dr Ivo Feussner** Department for Plant Biochemistry, Albrecht von Haller Institute for Plant Sciences, Georg August University Göttingen, Justus von Liebig Weg 11, D-37077 Göttingen, Germany

**Prof. John L. Harwood** Cardiff School of Biosciences, Biomedical Building, Museum Avenue, PO Box 911, Cardiff CF10 3US, UK

**Dr David F. Hildebrand** Department of Agronomy, N106C Agricultural Science Center North, Lexington, KY 40506–0091, USA

**Dr Reinhard Jetter** Department of Botany, University of British Columbia, 6270 University Blvd, Vancouver BC V6T 1Z4, Canada

| | |
|---|---|
| **Prof. Ljerka Künst** | Department of Botany, University of British Columbia, 6270 University Blvd, Vancouver BC V6T 1Z4, Canada |
| **Dr Charles McCracken** | Department of Agronomy, N106C Agricultural Science Center North, Lexington, KY 40506–0091, USA |
| **Prof. Denis J. Murphy** | Biotechnology Unit, School of Applied Sciences, University of Glamorgan, Treforest, Cardiff CF37 1DL, UK |
| **Dr Suryadevara S. Rao** | Department of Agronomy, N106C Agricultural Science Center North, Lexington, KY 40506–0091, USA |
| **Dr Sabine Rosahl** | Department of Stress and Developmental Biology, Institute of Plant Biochemistry, Weinberg 3, D-06120 Halle/Saale, Germany |
| **Dr A.L. Samuels** | Department of Botany, University of British Columbia, 6270 University Blvd, Vancouver BC V6T 1Z4, Canada |
| **Prof. Randall J. Weselake** | Department of Agricultural, Food and Nutritional Science, 410 Agriculture/Forestry Centre, University of Alberta, Alberta, Canada TOG 2P5 |
| **Dr Keshun Yu** | Department of Agronomy, N106C Agricultural Science Center North, Lexington, KY 40506–0091, USA |

# Preface

It is now over 30 years since publication of the seminal work by Hitchcock and Nichols on Plant Lipid Biochemistry (Academic Press, 1971). It is also over 15 years since the most recent updating of the comprehensive treatise on plant lipids edited by Stumpf and Conn (Lipids, Vol. 9 in the *Biochemistry of Plants* series, published in 1976 by Academic Press and revised in 1987). Since the publication of these two books, the field of plant lipids has changed almost beyond recognition. New research tools have revealed many surprising aspects of the dynamic nature of lipids and their participation in processes such as recognition, intra- and intercellular signalling, deterrence and defence against pathogens, membrane trafficking and protein function. This is in addition to new information on the more established roles of plant lipids as structural components of membranes and as long-term storage products. Plant lipids are also increasingly being seen as sources of a new generation of environmentally friendly, biodegradable and renewable industrial products, including biopolymers and high grade lubricants.

Over the past three decades, the proceedings of the biennial International Plant Lipid Congresses and other meetings have been published as collections of brief papers. There have also been monographs derived from meetings that concentrated on specific aspects of plant lipids, such as their biotechnological manipulation. However, no broad overview of plant lipids has been available since the two much earlier (and more biochemically-focused) works mentioned in the previous paragraph. In the present volume, researchers from major international laboratories have been brought together to provide reviews of progress in plant lipid research and its many applications. The intention is to link the various disciplines that are related to plant lipid research, in order to provide an interesting and wide-ranging perspective on this fast-moving field. There is a deliberate measure of overlap in some of the chapters where this sheds additional light on a topic from a different viewpoint or in a different context.

Chapter 1 provides a historical perspective on the study of plant lipids from its inception as a branch of alchemy in the seventeenth century to the interdisciplinary research of the post-genomic era in the twenty-first century. As new techniques become available, plant lipid research has advanced in both new and more traditional ways. Hence, the powerful tools of modern molecular genetics have contributed immensely to a far deeper understanding of lipid biochemistry and its underlying enzymology, even to the extent that some well established metabolic pathways are now in the process of being revised.

The next three chapters discuss aspects of the formation, modification and utilisation of fatty acids. Chapter 2 deals with the biosynthesis *de novo* of fatty acids from their earliest carbon precursors, through their assembly on the various types of fatty acid synthase complexes in plants, to their final release by thioesterases. The chapter also discusses recent ideas about the metabolic regulation of fatty acid formation. Chapter 3 considers the various modification reactions that fatty acids undergo after their synthesis *de novo*, and the ways in which these processes are being manipulated to create novel fatty acids that may be of interest to the biotechnology industry. Chapter 4 discusses the vast range of naturally occurring fatty acids, their glycerol esters and their many actual or potential uses as non-edible products. This chapter also provides an overview of the physical and chemical characteristics of acyl lipids.

The next two chapters deal with the major classes of macromolecular structures formed by plant lipids, namely bilayer membranes and storage bodies. Chapter 5 examines the structure, localisation and biosynthesis of the membrane glyco- and phospholipids and considers the roles of lipases in their turnover. This is followed by a discussion of the physiological role of membrane lipids in processes such as thermal tolerance and photosynthesis. Chapter 6 is a comprehensive account of the metabolism of storage lipids, including the various biosynthetic routes to triacylglycerol formation, lipid-body assembly and their subsequent degradation by lipolysis and β-oxidation. The chapter also covers physiological aspects, including environmental effects on storage lipid formation and their possible transport within plants.

Chapters 7 and 8 are concerned in different ways with some of the other types of macromolecular lipid assemblies in plants. Chapter 7 examines the roles of the various lipid-associated proteins in plants and compares these with analogous proteins in other organisms, including animals and microbes. These diverse proteins suggest new roles for some plant lipid assemblies that include membrane trafficking and dynamic interactions in physiological events such as stress responses and senescence. Chapter 8 focuses on that key plant lipid structure, the cuticle. The chapter covers the elaboration of fatty acids to form very long chain acyl derivatives and their assembly into waxes. The spatial context of this process is discussed, as is the physical nature of the various forms of crystalline wax that may be deposited under different conditions.

The final three chapters are concerned with three diverse classes of plant lipids that have numerous functions but are linked in their contributions to various aspects of signalling, either as part of normal development or during stress responses, both biotic and abiotic. Chapter 9 deals with the wide range of inositol-containing lipids, their derivatives and the proteins with which they interact. Comparisons with non-plant systems reveal both similarities and differences and highlight the many key roles of phosphoinositides in many fundamental signalling processes in plants. Chapter 10 examines the diverse class of oxygenated fatty acid derivatives, the oxylipins, which are now known to play a crucial role in the response of plants to pathogens and to more general wounding by herbivores. In particular, the formation and signalling roles of jasmonates and their interactions with other signalling pathways are described. Chapter 11 covers a vast range of lipidic compounds, namely the prenyllipids. Key members of this group are described, including quinones, carotenoids, sterols and terpenoids. The utilisation and manipulation of important prenyllipids, like carotenoids and taxol, are also discussed.

This book introduces the reader to the various aspects of plant lipids, summarises current developments and provides a resource for further study. To facilitate the latter aim, the authors have provided extensive lists of references, totalling well over two thousand non-redundant citations. It is hoped that this book will inform and stimulate the reader, as well as demonstrating the dynamic nature of plant lipid research. We also highlight some of the exciting opportunities for the application of this research in medicine and agriculture. Finally, we would point to the many new avenues that are now opening to young investigators who, hopefully, may be encouraged to pursue a career in furthering our knowledge of plant lipid science.

*Denis J. Murphy*

# 1 The study and utilisation of plant lipids: from margarine to lipid rafts

Denis J. Murphy

## 1.1 Introduction

This chapter will take a retrospective look at the study and utilisation of plant lipids, especially acyl lipids, and set current developments in a historical context. It is meant to be an outline description of the field, rather than a comprehensive history of the topic – perhaps the latter task is something that one of our readers might be inspired to undertake at some time in the future.

## 1.2 Early studies of plant lipids

Plant lipids are all around us, both in our environment and in our food. They are the second most important source of edible calories in the human diet (after carbohydrates). Indeed, for millions of years, they have also been a key source of those essential fatty acids (EFAs) that are an obligatory component of the diet of all mammals – ever since our distant animal ancestors lost the ability to introduce double bonds beyond the $\Delta_9$ position in long-chain fatty acids. Since the dawn of agriculture, over 11 000 years ago, certain plant species have been cultivated specifically for their lipid composition. The earliest olive plantations have been dated to more than 9 millennia before the present day and maize may have been domesticated in Mesoamerica as early as 10 millennia ago. Globally, there are now just 15 major crops that supply most of the human diet and five of these are high-oil crops, namely, soybeans, oil palm, maize, peanuts and coconut (Harlan, 1992). Plant oils have also been utilised by human societies for a host of non-edible applications, ranging from fuels to lighting and from cosmetics to lubricants. It was only the easy availability of plentiful, inexpensive, fossil-derived hydrocarbons (petrochemicals) in the late nineteenth century that displaced plant oils from many of their more prominent non-edible applications (oleochemicals). With the end of the era of cheap fossil hydrocarbons now in sight, probably within the present century (Murphy, 2004), plant lipids will once again be required in order to produce many of the multitudes of economically important non-food commodities that we depend upon so much.

Despite their impressive pedigree as important sources of food and raw materials for human societies, the systematic study of plant or animal lipids did not begin until after the inception of the modern scientific revolution in the seventeenth century. One of the earliest landmarks in lipidology was the publication in 1666 of the Hippocrates Chymicus by the German alchemist, Otto Tachenius. In this book, Tachenius first suggested that fats contain an acidic substance – what we now call fatty acids. He was also the first person to give a distinct definition of salt when he wrote that 'all salts are composed of two parts, of acid and alkali'. He further added that soap was the salt of an oily acid. Tachenius' statements were not accepted by the community of the time, and it was not until the French chemist Chevreul rediscovered the idea in 1816 through his laboratory work that Tachenius' definition of a salt was finally accepted (Partington, 1989).

Relatively little work of note was done on plant lipids until the nineteenth century, when many important contributions were made by French and Germanic chemists in particular. An early landmark occurred at the beginning of the nineteenth century when the Swiss chemist, Nicolas-Théodore de Saussure, demonstrated that linseed oil could condense with oxygen: this was an early hint of the existence of double bonds (de Saussure, 1804). In 1816, the renowned French chemist, Magendie, found that dogs that were fed on a diet in which the only lipid component was olive oil did not live for more than a month. Although olive oil contains about 7% linoleic acid, the high oleic/linoleic ratio would have led to marginal EFA deficiency in these animals. This result, although not appreciated at the time, flagged up the nutritional importance of some classes of polyunsaturated fatty acids (which are lacking in olive oil) and of antioxidant vitamins, including the A and E complexes.

The compositions of the major plant oils were gradually uncovered over the next few decades: with the suggestion that oleic acid and 'margaric acid' (a mixture of palmitic and stearic acids) were present as a mixture in vegetable oils (Pelouze and Boudet, 1838); the discovery of myristic acid in seeds of the Myristicaceae (Playfair, 1841); the isolation of lauric acid in the seeds of laurel, *Laurus nobilis* (Marsson, 1842); the preparation of linoleic acid from linseed (*Linum usitasissimum*) oil (Sacc, 1844) – although it was nearly a century before its structure was elucidated by Hilditch in 1939; the isolation of first known hydroxy fatty acid, ricinoleic acid (18-1-OH), from castor oil (Saalmüller, 1848); and finally, the description (Darby, 1849) and purification (Websky, 1853) of erucic acid from rapeseed oil. A link with the principal membrane lipids of animals occurred when phospholipids, often called lecithins at the time, were also shown to be present in plant seeds (Töpler, 1861), while lipases were first demonstrated in plant seeds by Muntz (1871). The chemical structure of oleic acid, including the position of the double bond, was first described by Edmed (1898) following an elegant series of oxidation steps. With this development, the stage was set for the systematic elucidation of the structures of each of the major plant fatty acids, although this task would take several more decades before it was finally achieved.

Meanwhile, the first major commercially relevant technological innovation for plant lipids came in 1869 when a French chemist called Hippolyte Mège Mouriès produced what we now know as margarine. Even in those days, plant lipid researchers had to respond to the needs of industry and government and the work of Mège Mouriès followed a call by the French emperor, Napoleon III, for an alternative to butter. Although it has been claimed that this was done to enhance the nutrition of the working classes, another (doubtless more pragmatic) motive was to provide a cheaper and more versatile source of food for the French army, which was then preparing for a conflict with the emergent state of Prussia for domination of continental Europe. As a raw material for the new product, Mège Mouriès selected a particular solid fatty acid fraction that was called margaric acid. This name derives from the lustrous pearly-white drops of the crystalline form of margaric acid that are reminiscent of pearls and which are called margarites in Greek (this is also the derivation of the name Margaret).

The earliest forms of margarine were mixtures of animal and plant fats, but the product was yet to be a great commercial success. Two technical advances tipped the balance towards plant fats in margarine and allowed it to be an effective competitor with butter. First, improved refining methods allowed the purification of a greater variety of liquid oils and solid vegetable fats that could be blended to make good spreadable margarine. Second, the process of hydrogenation, which was invented in 1901 by English chemist William Normann, allowed the large scale conversion of relatively cheap plant oils into solid fats. Not only did the hydrogenation process produce a good, inexpensive butter substitute, it also significantly reduced the amount of oxidation-prone polyunsaturates in the solid margarine, which greatly extended its shelf life and, therefore, its utility for consumers.

Like many start-up researcher/entrepreneurs of the present day, Mège Mouriès soon sold on the rights to his invention to a larger, more established company and, in 1871, the technology was acquired by the Dutch firm, Jurgens, which is now part of the Unilever group. Today, Unilever is one of the largest multinational enterprises that use plant lipid products, including margarines and detergents, as a core part of their business. Unilever scientists have also made key contributions to the study of plant lipids. For example, it was at the Unilever research labs in the United Kingdom that some of the most significant advances in lipid analysis were made in the 1960s and 1970s, as described in later sections.

The dawn of the twentieth century was appropriately marked by the demonstration of his eponymous engine by Rudolf Diesel at the International Exposition, held in Paris in 1900. The first diesel engine was powered by peanut oil, so it can be said that plant lipids played an important role in ushering in the age of the automobile (Nitske, 1965). However, plant-derived oils were almost immediately abandoned in favour of a petroleum fraction that is slightly heavier than gasoline. This fuel was named after the inventor of the engine and has been called diesel ever since. Ironically, the past decade has witnessed a partial return to plant-derived diesel fuels, normally based on methyl ester derivatives of seed oils from crops like rapeseed and sunflower. These so-called 'biodiesel' fuels have improved environmental impacts compared to regular diesel, especially in sensitive areas like inland waterways and city centres. Despite their undoubted 'green' credentials, however, biodiesel fuels remain economically questionable as petrodiesel substitutes and their use is likely to be more or less completely dependent on support from government subsidies for the foreseeable future (Murphy, 1998, 2004).

One of the earliest examples of an international honour for a plant lipid researcher came in 1915 when R.M. Willstätter was awarded the Nobel Prize for chemistry 'for his research on plant pigments, especially chlorophyll'. Willstätter had worked on the technique of chromatography late in the nineteenth century and had undertaken some particularly heroic extractions, such as the isolation of chlorophyll from several hundred kilograms of dried stinging nettles in 1905 (Willstätter, 1973).

By the early twentieth century, the results of plant lipid science were being applied widely in both industry and medicine. For example, the use of interesterification for the production of oleochemicals was patented in the United Kingdom by van Loon in 1924 (UK patent # 249 916), while in 1924, the essential role of vitamin E was described and the name was first suggested (Sure, 1924). This was followed a few years later by the discovery of the essential role of long-chain polyunsaturated fatty acids in the diet (Burr *et al.*, 1930). A further Nobel Prize followed in 1930 when R. Kuhn was honoured for his work on carotenoids and vitamins.

Plant lipid research was now about to enter the next stage of its development, which was characterised in particular by several key advances in analytical methodology and instrumentation.

## 1.3 The chemistry era – and the definition of the term 'lipid'

Most of the really important developments in advancing the chemical study of plant lipids came after World War II. However, there were a few discoveries that can be regarded as precursors for these later successes, most of which relate to the discovery and application of various forms of chromatography. Investigators like Willstätter in Germany had been experimenting with different techniques for the separation of plant lipids since the late nineteenth century, but the first public description of the chromatographic technique was given by Mikhail Tswett at a meeting in Warsaw in 1903. Tswett coined the term 'chromatography' and used the new method for the separation of plant pigments on a chalk column, although it took several more decades before the technique was taken further. In 1938, a Soviet group described the use of thin-layer chromatography (TLC) on microscope slides coated with aluminium oxide (Izmailov *et al.*, 1938). This group, based in Kharkov in the Ukraine, originally developed this early version of TLC to separate plant chemicals of potential pharmaceutical interest, the so-called galenic extracts. A further landmark came in 1940 when T.P. Hilditch, in the United Kingdom, published the seminal and much reprinted volume entitled *The chemical constitution of natural fats* (Hilditch, 1940).

It was in the 1947 edition of this book that Hilditch proposed that the term 'lipid' should be used. As Hilditch stated:

*unanimity has not yet been reached in the terminology to be adopted in classifying the various types of naturally occurring compounds in which fatty acids are present . . . even a collective title for the whole group is not completely settled.*

Prior to this, there had been several competing terms used in the English literature, including 'lipoids', which tended to be used by British investigators, and 'lipins', which was favoured by Americans. An interesting hangover from this terminology is the current use of the term 'oxylipin' for oxygenated lipid derivatives, as employed in Chapter 10 of the present volume. The term 'lipid' had originally been proposed by Sperry (1926), but was not widely adopted until Hilditch's suggestion in 1947.

Lipids were originally defined according to chemical criteria, namely, as fatty acids and their derivatives. However, it has become more common to define lipids on the basis of their physical properties, namely, as oily, fatty or waxy organic compounds, which while insoluble in water are readily soluble in organic solvents. Obviously, exceptions can be found to the latter definition, e.g. it would not apply to some of the more polar monoacylglycerols or fatty acid salts. Even no less an authority than Bill Christie admits that, 'there is no widely-accepted definition' of a lipid (Christie, 2003). Historically, the term 'fats' was used for naturally occurring 'triglycerides', whether solid or liquid (these are now correctly called triacylglycerols – although, alas, this seems to have escaped the attention of some in the field). Lipids can also encompass a host of other naturally occurring non-polar compounds that do not contain acyl groups. These include a host of prenyl derivatives including carotenoids, tocopherols, terpenes and quinones as well as phytylated pyrrole derivatives, such as the chlorophylls, and numerous families of fatty acid derivatives, some of them relatively polar, including eicosanoids, oxylipins and volatiles like hexenal and pentenone (see Chapter 4 for more on lipid definitions).

Despite the international conflicts of the 1940s, there were some notable advances in lipid analysis during this decade. Martin and Synge (1941) first described partition chromatography at a meeting of the Biochemical Society in the then war-torn city of London. Both investigators subsequently shared the 1952 Nobel Prize for Chemistry 'for the invention of partition chromatography'. Many lipid analyses at this stage employed large amounts of starting material, as described above in the case of Willstätter and his stinging nettles. This was necessary, either because the lipids of interest were such minor components and/or because the relatively crude detection systems that were then available required large quantities of purified lipid. Therefore, the development of an initial purification step involving high-capacity column chromatography was of great use. Early columns used alumina and could separate non-saponifiable lipid mixtures into several fractions (Swain, 1948), although silicic acid columns were soon developed and used to separate relatively complex mixtures of neutral lipids and phospholipids (Borgström, 1952). Reversed-phase partition chromatography on Kieselguhr was first used to separate C12–C18 fatty acids by

Howard and Martin (1950), followed shortly thereafter by the first application of TLC to lipids – in this case, terpenes from citrus juices, which were studied by Kirchner's group at the USDA Agriculture, Fruit and Vegetable Laboratory in Southern California (Kirchner, 1951). TLC was developed further by Stahl and colleagues at Saarbrücken in Germany in the mid- to late 1950s and by the end of the decade, he and other workers, such as Weicker (1959) in Germany and Mangold (1959) at the Hormel Institute in Austin, Minnesota, had developed silica-based TLC into a highly precise and accurate technique that could be used for both preparative and analytical separations. The utility of TLC was extended further by the usage of impregnated silica or other stationary phases. For plant lipid researchers, one of the most useful methods is argentation-TLC, whereby silver nitrate is added to the silica. This allows for the separation of long-chain fatty acyl groups on the basis of their unsaturation, as well as chain length, and is especially valuable for analysis of plant acyl lipids because of their high degree of unsaturation.

Meanwhile, in the early 1950s, A.J.P. Martin, who had jointly invented partition chromatography a decade earlier, teamed up with another young chemist, A.T. James, in London. This duo presented a paper announcing the extension of partition chromatography to include a gas as the mobile phase at a Biochemical Society meeting in 1950. This marked the birth of gas–liquid chromatography or GLC, and the landmark paper describing the technique appeared 18 months later (James and Martin, 1952). Part of the impetus for developing GLC came from a colleague of James and Martin, who was keen to find an alternative to paper chromatography for the effective resolution of fatty acid mixtures. Since then, gas chromatography (GC), as it became known, has developed rapidly, particularly during the 1960s, to provide both a preparative and an analytical tool for lipid, and especially fatty acid, analysis. The technique has now been applied in almost every area of analytical and biochemical research. Since the mid-1990s, hybrid techniques have been developed that have added a further dimension to GC analyses. Probably the most effective of these is the use a mass spectrometer as a detector rather than straightforward thermal conductivity or flame ionisation detection systems. By employing a mass spectrometer in tandem with GC (called GC–MS), one can effectively do a two-dimensional separation and analysis of a lipid mixture. A further refinement is to add a second mass spectrometer detector in series (called GC–MS–MS), which enables the simultaneous separation and identification, based on comparison with existing libraries of spectra stored on the computer, of thousands of compounds in a mixture. This approach is the basis of 'metabolomics', as discussed in Section 1.6.

Numerous additional physical and chemical techniques were applied to the analysis of plant lipids from the late 1950s and beyond. Examples include the many forms of spectroscopy, such as ultraviolet, infrared, Raman and nuclear magnetic resonance; X-ray crystallography; hydrogenation; oxidation; and the various staining reagents that ranged from iodine vapour and flourescein sprays to charring with concentrated mineral acids. Not all of these methods can be described here, but their use in the 1960s and 1970s has been comprehensively reviewed elsewhere

(Hitchcock and Nichols, 1971; Gurr and James, 1971–1991; Gunstone, 1976, 1976). Other techniques used for plant lipid analysis include the various forms of calorimetry, such as differential scanning calorimetry and differential thermal analysis that have been especially valuable in the study of the phase behaviour of lipids, both in pure form and as mixtures *in situ*. There are several more recent manuals relating to the chemical analysis of lipids, which describe some of the contemporary techniques that are available (Christie, 1992–2003, 2003; Gunstone *et al.*, 1994; Grob and Barry, 1995; Baugh, 1997; McDonald and Mossoba, 1997; Hamilton, 1998).

## 1.4 The biochemistry era

Prior to the 1950s, most studies of plant lipids tended to focus on the chemical analysis of the lipids themselves, i.e. their extraction, separation, purification and the description of their structures. One example of such work is the first isolation of the key intermediate, phosphatidic acid (1,2-diacyl-*sn*-glycerol-3-phosphate), from cabbage leaves by Chibnall and Channon (1927). Unlike microbial researchers, plant investigators at this stage did not yet have access to powerful genetic tools, which would assist in the elucidation of the complexities of lipid metabolism *in vivo* (but see the next section for details of an exception to this, namely rapeseed/canola studies in the 1950s–1970s). This meant that much of the early work on lipid biochemistry relied greatly on inferences from microbial and animal systems. This situation began to change in the early 1950s when several technical advances encouraged a new generation of researchers to tackle the formidable challenges posed by lipid metabolism in plants. As described above, the progress in analytical techniques greatly facilitated the separation and identification of acyl lipids in particular. In addition, one of the more useful spin-offs from the nuclear weapons programmes, which were initiated in the 1940s, was the availability of radioactive isotopes, such as $^{14}C$, $^{3}H$, $^{32}P$ and $^{35}S$. This opened the door to the synthesis of radiolabelled tracers like [$^{14}C$]acetate or [$^{14}C$]$CO_2$. Such tracers were especially powerful tools as they could be used to follow the pathways of metabolites in either *in vitro*, cell-free systems or with *in vivo* systems that might consist of a whole organism or a detached organ or tissue such as a leaf disc.

The foundations of plant lipid biochemistry, as we now know it, were laid in the 1950s and 1960s when a relatively small number of groups began the detailed description of the main processes of fatty acid and acyl lipid metabolism. This period saw the emergence of a series of plant lipid biochemistry labs in Europe and North America, with the University of California (UC) making a particularly prominent contribution at its various campuses. These investigators included Beevers at UC Santa Cruz, Mudd at UC Riverside and, most prominently, Stumpf at UC Berkeley and then UC Davis. In 1958, the former university agricultural station at Davis, a small town about an hour's drive inland from the San Francisco Bay Area, was established as an independent campus of the UC system. It was on this new campus that Paul Stumpf (together with Eric Conn and Lloyd Ingraham) established the Department of Biochemistry and Biophysics and set up what became the world's most influential plant lipid biochemistry lab over the next few decades. As early as 1952, Stumpf's group was using radiolabelled compounds to investigate fatty acid biosynthesis in slices from peanut cotyledons and other plant tissues. Among other achievements, they found that acetate was the most effective exogenous fatty acid precursor (Newcombe and Stumpf, 1952). In 1955, the structure of coenzyme A (CoA) was elucidated (Baddiley, 1955), and by 1962 Stumpf's group had discovered the fatty acyl thiokinase (now called acyl-CoA synthetase) (Barron and Stumpf, 1962). Throughout the 1960s, Stumpf's group went on to elucidate the mechanism of fatty acid biosynthesis in plants with their description of thioesterases (1965), acetyl CoA carboxylases (1966) and the purification of acyl-binding protein, or ACP (1968). By the end of the 1960s, the outline and many of the details of plant fatty acid biosynthesis *de novo* were elucidated.

The various modification reactions of plant fatty acids, including desaturation and hydroxylation, were also characterised in the 1960s. The aerobic pathway of fatty acid desaturation, originally described for yeast and involving an acyl-ACP substrate (Bloomfield and Bloch, 1960) was soon shown to also apply to plants (Mudd and Stumpf, 1961; James, 1963; Stumpf and James, 1963). A.T. James (who had developed the first GC system with Martin in the early 1950s – see Section 1.2) went on to establish a particularly influential lipid biochemistry group at the Unilever Research Labs in Colworth, United Kingdom, where he was joined by workers such as Gurr, Nichols, Morris and Hitchcock. During the 1960s–1970s, this team made impressive contributions in the field of plant lipids as well as doing important work with microbial and animal systems. Although desaturation from stearate to oleate was relatively easily demonstrated, it was more challenging to elucidate the higher desaturations required to produce polyunsaturates. One problem was that whereas isolated chloroplasts could synthesise fatty acids as far as oleate, further desaturations required the presence of additional cytosolic fractions (i.e. microsomes). It was the Unilever group of Nichols, Gurr and colleagues who eventually showed that these higher desaturations occurred on oxygen ester substrates, and especially on phosphatidylcholine (Nichols *et al.*, 1967; Gurr *et al.*, 1969). This group, together with the labs of both Gunstone and Stumpf, also laid the foundation for elucidation of other fatty acids modifications in plants, including hydroxylation, conjugated desaturation and *trans* desaturation (reviewed in Hitchcock and Nichols, 1971).

The assembly of acyl groups onto glycerol backbones to produce acylglycerols was first characterised in *E. coli* and is known as the Kennedy pathway. Stumpf, Kates and co-workers showed that the same mechanism of stepwise acylation applied to higher plants (Barron and Stumpf, 1962; Sastry and Kates, 1966). It was already known that, unlike animals, most membranes in plant photosynthetic tissues were composed primarily of glycolipids – principally mono- and di-galactolipids. Galactolipids had been originally discovered in wheat flour by Carter *et al.* (1956).

However, it still came as a surprise when, in 1959, Benson at the Scripps Institute of Oceanography near San Diego, California, discovered an entirely new type of acyl lipid that contained sulfur, namely, sulfoquinovosyldiacylglycerol (Benson, 1959, 1963). Originally discovered in the microalga Chlorella, these sulfolipids were subsequently shown to be present in the thylakoid membranes of all lower and higher plants. The assembly of galactolipids, from UDP-galactose and diacylglycerol precursors, was demonstrated using isolated chloroplasts and [$^{14}$C]galactose in a series of papers by Mudd and colleagues (Ongun and Mudd, 1968; reviewed in Leech and Murphy, 1976). Benson (1963) suggested that sulfolipids were synthesised by an analogous mechanism involving a nucleoside diphosphate sulfoquinovose precursor that had already been detected in extracts of Chlorella cells. The only prominent phospholipid in plant photosynthetic membranes is phosphatidylglycerol and early studies of its synthesis were undertaken by the groups of Benson in the United States of America, Kates in Canada and Douce in France (Benson and Maruo, 1958; Benson and Miyano, 1961; Douce *et al.*, 1966; Sastry and Kates, 1966). The characterisation of the other major phospholipids, which are mostly extraplastidial in higher plants, was started in the 1960s in the labs of Benson, Kates and Nichols (reviewed in Hitchcock and Nichols, 1971). During the 1970s, workers such as Galliard also made key contributions in the early characterisation of the various forms of lipid oxidation reactions that are now known to play such important roles in processes ranging from flavour generation to hormonal signalling in plants.

Regarding lipid catabolism, Stumpf's lab once again played an important role in showing that β-oxidation of fatty acids occurred in plants (Stumpf and Barber, 1956). The groups of both Beevers and Stumpf then showed that β-oxidation takes place in a newly discovered organelle, the glyoxysome (Cooper and Beevers, 1969; Hutton and Stumpf, 1969). Over the next two decades, Beevers' group went on to characterise the details of plant β-oxidation and other aspects of glyoxysomal metabolism, including the glyoxylate cycle and generation of sucrose from catabolised fatty acids. Plant lipases acting under non-physiological (acid) conditions had been discovered previously, but the first neutral lipase was described by Yamada (1957). This enzyme was shown to be located on lipid bodies in germinating seedlings (Ching, 1968).

Despite all of these advances in the 1960s, plant lipid biochemistry at this stage still remained something of a Cinderella discipline, compared to the much larger and better-resourced animal and microbial groups. The emergence of plant lipids as an area of research with its own distinctive identity and unique personalities came about largely thanks to an initiative by a group of British workers in the early 1970s. The initial landmark was the publication in 1971 of the first textbook for plant researchers, *Plant Lipid Biochemistry* (Hitchcock and Nichols, 1971). This book has been an invaluable resource for many of us and my own much-thumbed copy still occupies a privileged and readily accessible place right above my office desk. However, in the early 1970s, there was still no real regular forum for plant lipid researchers to present and discuss their work, except as a relatively marginal part of meetings dominated by animal and microbial investigators. As Galliard and Mercer stated in 1975:

*Those engaged in research on plant lipids have suffered in the past at biochemical meetings in which colleagues working with animal or microbial systems have dominated the proceedings. One of us has resorted to using only the term Solanum tuberosum when referring to the humble potato in the hope that the non-phytochemical majority might think that this was the name of a micro-organism and thus be acceptable to the biochemical establishment.*

This grim situation led to the organisation by Galliard, Gurr, Mercer and Rhodes of an international symposium on the 'Chemistry and Biochemistry of Plant Lipids', which was held at the Food Research Institute in Norwich, United Kingdom in 1974. Unwittingly, the organisers of this meeting set in train the establishment of a whole series of biennial plant lipid symposia that continue to this day (see Table 1.1 for a full list of these meetings).

The Norwich meeting was deemed so successful by the participants that four German colleagues, Lichtenthaler, Heinz, Mangold and Tevini, organised a second symposium in Karlsruhe in 1976. The subsequent progress of plant lipid biochemistry from the 1970s to the present day is best documented in the series of proceedings that have been published from each of the biennial International Congresses that have been held up to the present day. These volumes are listed in Table 1.1, together with books published from several additional notable meetings where plant lipids were a major theme. In the 1990s, a National Plant Lipid Cooperative (NPLC) was established in the United States of America to further research, collaboration and information sharing within the community. Despite its name, NPLC soon became an international resource and a link to the expanding series of databases relating to lipid studies. The principal initiative of NPLC was to establish a new series of plant lipid meetings that were held in the United States of America in the intervening years between the biennial International Congresses. No formal proceedings of these meetings, held near Lake Tahoe, California, have been published although abstracts may be available via the NPLC web site (http://www.msu.edu/user/ohlrogge). Rather than attempting to summarise the immense amount of work in plant lipid biochemistry over the past three decades, I will select just a few achievements and try to link these with the subject of the following section, i.e. the molecular genetics revolution of the 1990s. Readers who are interested in further details of progress in lipid biochemistry and physiology during this period are recommended to consult the proceedings of the various meetings and other reference works as listed in Tables 1.1 and 1.2.

Two notable milestones that occurred in Stumpf's lab shortly before his retirement in the early 1980s, were the purification of the separated components of the fatty acid synthase complex (Stumpf and Shimakata, 1983) and the first purification of a plant acyl desaturase (McKeon and Stumpf, 1982). The availability of these and

**Table 1.1** Thirty years of Plant Lipid Congresses, 1974–2004

*List of International Plant Lipid Congresses and principal organisers*

1. 1974, Norwich, UK, Terry Galliard
2. 1976, Karlsruhe, Germany, Hartmut Lichtenthaler
3. 1978, Sweden, Stockholm, Connie Liljenberg
4. 1980, Paris, France, Paul Mazliak
5. 1982, Groningen, The Netherlands, Jan Wintermans
6. 1984, Neuchatel, Switzerland, Paul-Andre Siegenthaler
7. 1986, Davis, California, USA, Paul Stumpf
8. 1988, Budapest, Hungary, Peter Biacs
9. 1990, Wye, UK, Peter Quinn
10. 1992, Djerba, Tunisia, Abelkader Cherif
11. 1994, Paris, France, Jean-Claude Kader
12. 1996, Toronto, Canada, John Williams
13. 1998, Seville, Spain, Juan Sanchez
14. 2000, Cardiff, UK, John Harwood
15. 2002, Okazaki, Japan, Norio Murata
16. 2004, Budapest, Hungary, Peter Biacs

*Published proceedings from the International Plant Lipid Congresses**

Galliard, T. and Mercer, E.J., eds (1975) *Recent Advances in the Chemistry and Biochemistry of Plant Lipids*, Academic Press, London

Tevini, M. and Lichtenthaler, H.K., eds (1977) *Lipids and Lipid-Polymers in Higher Plants*, Springer Verlag, Berlin

Appelqvist, L.A. and Liljenberg, C., eds (1979) *Biochemistry and Physiology of Plant Lipids*, Elsevier, Amsterdam

Mazliak, P., Benveniste, P., Costes, C. and Douce, R., eds (1981) *Biogenesis and Function of Plant Lipids*, Elsevier, Amsterdam

Wintermans, J.F.G.M. and Kuiper, P.J.C. (1982) *Biochemistry and Metabolism of Plant Lipids*, Elsevier, Amsterdam

Siegenthaler, P.A. and Eichenberger, W., eds (1984) *Structure, Function and Metabolism of Plant Lipids*, Elsevier, Amsterdam

Stumpf, P.K., Mudd, J.B. and Nes, W.D., eds (1987) *Metabolism, Structure and Function of Plant Lipids*, Plenum, New York

Biacs, P.A., Gruiz, K. and Kremmer, T., eds (1989) *Biological Role of Plant Lipids*, Plenum, New York

Quinn, P.J. and Harwood, J.L., eds (1991) *Plant Lipid Biochemistry, Structure and Utilization*, Portland Press, London

Cherif, A., Daoud, D., Marzouk, B., Smaoui, A. and Zarrouk. M., eds (1992) *The Metabolism, Structure and Utilization of Plant Lipids*, Centre National Pédagogique, Tunis

Kader, J.C. and Mazlaik, P., eds (1995) *Plant Lipid Metabolism*, Kluwer, Dordrecht

Williams, J.P., Khan, M.U. and Lem, N.W., eds (1997) *Physiology, Biochemistry and Molecular Biology of Plant Lipids*, Kluwer, Dordrecht

**Table 1.1** (continued)

Sanchez, J., Cerda-Olmedo, E. and Martinez-Force, E., eds (1998) *Advances in the Biochemistry and Physiology of Plant Lipids*, Universidad de Sevilla, Secretariado de Publicationes

Harwood, J.L. and Quinn, P.J. (2001) *Recent Advances in the Biochemistry of Plant Lipids*, Portland Press, London

Murata, N., Yamada, M., Nishida, I. *et al.* eds (2003) *Advanced Research on Plant Lipids*, Kluwer, Dordrecht

* Dates refer to publication, which is sometimes in the year after the meeting in question.

**Table 1.2** Books on Plant Lipids

*Other conference proceedings*

Thompson, W.W., Mudd, J.B. and Gibbs, M., eds (1983) *Biosynthesis and Function of Plant Lipids*, American Society of Plant Physiologist, Baltimore, USA (6th Ann. Symp. Bot. Riverside, California, 1983)

Murata, N. and Somerville, C., eds (1993) *Biochemistry and Molecular Biology and Storage of Plant Lipids*, American Society of Plant Physiologist, Rockville, USA (US/Japan Binational Seminar, Kona, Hawaii, 1992)

Harwood, J.L. ed (1998) *Plant Lipid Biosynthesis, Fundamentals and Agricultural Applications*, Cambridge University Press, Cambridge (Soc. Exp. Biol. conference, Canterbury, UK, 1997)

*Other books related to plant lipids*

Gunstone, F.D., Harwood, J.L. and Padley, F.B., eds (1994) *The Lipid Handbook*, 2nd edn, Chapman and Hall, London (This is one of the premiere reference works dealing with lipid composition and analysis.)

Gurr, M.I. and James, A.T. (1971) *Lipid Biochemistry: An Introduction*, Chapman and Hall, London

(Early editions contain many descriptions of experimental techniques and examples applied to plant lipids. The latest is the 5th edition and is written by Gurr, M.I., Harwood, J.L. and Frayn, K.N. (2002) Blackwell Science, Oxford.)

Harwood, J.L. and Russell, N.J. (1984) *Lipids in Plants and Microbes*, George Allen and Unwin, London

Hitchcock, C. and Nichols, B.W. (1971) Plant Lipid Biochemistry: The Biochemistry of Fatty Acids and Acyl Lipids with Particular Reference to Higher Plants and Algae

Stumpf, P.K. ed (1980) *Lipids: Structure and Function*, Vol 4, The biochemistry of plants, a comprehensive treatise, Stumpf PK and Conn eds in chief, Academic Press, New York

Moore, T.S. (1993) *Lipid Metabolism in Plants*, CRC Press, Boca Raton

other purified proteins allowed for the initial steps in obtaining amino acid sequence information and/or the preparation of antibody probes, which would eventually lead to the cloning of the corresponding genes. The work on fatty acid synthase was taken further in the 1990s by Slabas *et al.* in Durham, United Kingdom, who refined the purification of the fatty acid synthase components and then collaborated with Rice *et al.* in Sheffield, United Kingdom to use X-ray crystallography to determine the first high-resolution structures of the proteins. These and other structural studies have provided valuable information for attempts at enzyme engineering.

Another key advance was the description of the dual localisation of lipid metabolism, with separate but linked pathways in the plastids and the cytosol, as reviewed by two of the major contributors to the hypothesis in Roughan and Slack (1982). It was during the investigation of this hypothesis that Browse *et al.* (1989) first reported the use of Arabidopsis genetics in the elucidation of plant lipid metabolism. The use of a combination of genetics and biochemistry have now given us significant insights into the complexities of galactolipid, sulfolipid and phospholipid metabolism as outlined in the succeeding chapters of this book.

In addition to their role in cell membranes, lipids are important storage products, especially in plants producing oilseeds and in oily fruits such as avocado, olive and oil palm. Some of the earliest work to characterise lipid-body formation in seeds was carried out by Frey-Wyssling *et al.* (1963) in Zurich, by several US labs, including Jacks *et al.* (1967), Sorokin (1967), Totten (Mollenhauer and Totten, 1971), Stumpf (Harwood *et al.*, 1972) and also at the Unilever labs in the United Kingdom (Gurr *et al.*, 1974; Smith, 1974). This was followed by many notable contributions, especially that of Slack *et al.* (1980) who first isolated the lipid-body proteins now known as oleosins. The first oleosin gene was isolated by Huang's group in 1987 (Vance and Huang, 1987) and since then oleosins and lipid storage have also been studied by several other labs including those of Murphy (1989, 2001), Herman (1987), Moloney (Holbrook *et al.*, 1991) and Napier (Thoyts *et al.*, 1995).

Among the many examples of the synergy between the academic progress of plant lipid science and its commercial application, are the studies of the grass-specific herbicides that target acetyl-CoA carboxylases (see Chapter 2 for more details). These are some of the most effective herbicides that are used to control monocot weeds, both in agricultural crops and in domestic gardens. The use of these herbicides as research tools allowed several groups, including Lichtenthaler and Harwood, to characterise the different structures and mechanisms of acetyl-CoA carboxylation in monocots and dicots (reviewed in Harwood, 1988, 1996). This was regarded as a key and potentially rate-limiting step in fatty acid biosynthesis, the manipulation of which might result in the creation of higher yielding oil crops. The rate limiting nature of acetyl-CoA carboxylase in fatty acid biosynthesis was demonstrated in 1994, as described by Page *et al.* (1994). In turn, the better understanding of the biochemical basis of these herbicides has assisted in the design of improved forms that can selectively and effectively target weeds, while leaving non-target plants unaffected.

One of the most frustrating aspects of lipid biochemistry is trying to work with non-polar, insoluble substrates; labile, membrane-bound enzymes; and time-consuming, discontinuous assays that often involve radiolabelled precursors and toxic, volatile reagents. This made for slow progress in studying many key enzymes, including most acyltransferases and desaturases, which are integral membrane proteins. Two groups that made especially noteworthy contributions in the purification of such recalcitrant proteins, right through until the late 1990s, were Heinz *et al.* in Hamburg and the Calgene labs in Davis, California. To some extent, technical advances have been of assistance here, especially the availability of new types of milder and more versatile detergents. But it is still mostly a combination of hard, routine biochemical work and a good bit of imagination, e.g. in assay design and detergent use, that has now made many membrane associated enzymes accessible for study for the first time.

## 1.5 The molecular genetics revolution

From the mid-1980s, a quiet revolution began to have an ever-increasing impact on plant lipid research. In 1985, the first paper was published in which mutants of *Arabidopsis thaliana* were used to elucidate the role of a plant lipid – in this case, the chloroplast-specific fatty acid, $\Delta_3$ *trans* hexadecenoic acid (Browse *et al.*, 1985). However, it was at the Lipid Congress in Budapest in July 1988 that many of us became aware of the real power of 'the weed', as *A. thaliana* soon became known. Before this time, the field was very much dominated by biochemists, with a few chemists and physiologists for good measure, but very few of us were truly aware of the power of genetics. In their paper to the Budapest Congress, Browse *et al.* showed how Arabidopsis plants with mutations in various steps of lipid metabolism could be used to modify and improve upon the existing scheme for lipid biosynthesis in leaves (Browse *et al.*, 1989). It was even more exciting to learn about other emerging developments that would soon make Arabidopsis the key plant for both molecular genetic and biochemical investigations of plant lipids.

*Arabidopsis thaliana*, also called thale cress, is a small annual plant from the Brassicaceae family, the same family as rapeseed and the many brassica vegetable crops. Arabidopsis was first described by Johannes Thal (hence, thaliana) in the Harz mountains in the sixteenth century. Arabidopsis mutants were described as early as 1873, when Alexander Braun wrote about – what we would now call – an AGAMOUS plant (a floral mutation), which he had found in a field in Berlin (Braun, 1873). But the real foundations of Arabidopsis studies were laid by Laibach who, while still a PhD student in Bonn, noted that the plant had only five small chromosomes (Laibach, 1907). In the 1940s, Laibach's group developed Arabidopsis further as a genetic model that could potentially serve as the Drosophila of the plant world (Laibach, 1943). Therefore, at the same time as Martin and Synge were developing chemical methods for lipid analysis under difficult conditions in war-ravaged London, Laibach's group was developing biological methods under equally testing conditions in wartime Germany. A flavour of these challenges is given by the story of one of Laibach's students, called Erna Reinholz. Her project involved developing X-ray mutagenesis to create additional variation in Arabidopsis (some important lipid-related genes have been identified from such *mutants*). The manuscript of her work was seized by the occupying forces of the US military in 1945. They were looking for anything that could possibly relate to the attempted development of nuclear weapons in Germany and noticed that her work related to radiation. Luckily for plant science, the Reinholz PhD thesis was published in 1947 by the Joint Intelligence Objectives Agency of the Occupying Powers, who cryptically described it as 'an unclassified captured document', doubtless not realising the biological significance of her work (Reinholz, 1947).

Further progress was made in the 1950s and 1960s in the creating of informative mutant collections of Arabidopsis and the first two International Arabidopsis Congresses were held in 1965 and 1976. At this stage, Arabidopsis research shared a Cinderella status with plant lipid research and in both cases, the effective solution was to organise regular, focused international meetings. It was at this time that George Rédei formally proposed the use of Arabidopsis as a general research tool to elucidate processes in plant physiology, and by implication plant biochemistry (Rédei, 1975). This stimulated many geneticists, and some notable future cloners of lipid-related genes, to explore the Arabidopsis model (Meyerowitz, 2001). The development of recombinant DNA technology in the late 1970s was soon taken up by plant scientists and the full potential of Arabidopsis as a molecular genetic tool was outlined in a paper in *Science* in 1985 (Meyerowitz and Pruitt, 1985). The first Arabidopsis gene sequences were published in 1986 and other necessary tools were developed in the late 1980s and beyond, including efficient transformation with exogenous DNA, DNA-marker-based mapping and map-based cloning all of which culminated in a project to sequence the entire genome. The latter became an epic international endeavour that finally bore fruit in the publication of the first draft of the genome sequence in *Nature* in December 2000. However, long before this, the use of tools like mutagenesis, gene cloning and EST (expressed sequence tags) libraries had transformed our knowledge of lipid metabolism in Arabidopsis, as well as spurring on new developments in biochemistry and cell biology of plant lipids in general.

Luckily for plant lipid research, one of the labs at the forefront of Arabidopsis molecular genetics, that of Somerville at Michigan State University (later moving to Stanford in California), was also keenly interested in lipid metabolism. Many workers from this lab went on to establish their own lipid labs, including some of the authors of the present volume. As outlined above, the process began in the mid- to late 1980s with the use of Arabidopsis *mutants* to investigate the pathways of fatty acid biosynthesis in leaves. By the early 1990s, many genes encoding lipid-related enzymes had been isolated using EST libraries or by screening other expression libraries with antibody or oligonucleotide probes based on partial sequence information. This allowed many groups to leapfrog over the technical hurdles associated with the purification of often-recalcitrant enzymes from plants. Instead, they were now able to express the corresponding genes in a suitable organism and use affinity tags (e.g. polyhistidine) to obtain milligram quantities of virtually pure protein. This could then be used for structural studies and/or for kinetic studies of enzyme mechanism and function.

The molecular genetics revolution also spawned a huge interest in the possibilities regarding the manipulation of plant lipid metabolism for commercial purposes. Several new biotech companies were established with the goal of using the new technology for the production of transgenic oil crops having improved lipid profiles, either for edible or industrial uses. The interest in 'designer oil crops' has been well documented elsewhere (Murphy, 1994, 1999, 2003) but it is worth mentioning one small biotech company that had a particularly important influence on plant lipid research, which was out of all proportion to its modest size. Calgene Inc. was established in the early 1980s by a group of researchers from the neighbouring UC Davis campus. Calgene produced several innovative (albeit commercially unsuccessful) products, including the FlavrSavr™ tomato (Martineau, 2000), but was best known for its work on the manipulation of oil content in crops. To achieve this, the Calgene team made an impressive series of advances both in lipid biochemistry and in molecular genetics. These included the purification and characterisation of enzymes involved in short-chain and long-chain fatty acid biosynthesis, stearate desaturation and wax ester biosynthesis and the production of transgenic plants in which these activities had been manipulated in order to alter oil composition (Knutzon and Knauf, 1998). In retrospect, the most notable achievement of the Calgene group (the company was acquired by Monsanto in 1998 and no longer has a separate identity) has been their contribution to basic lipid research rather than to the commercialisation of transgenic crops. During the early 1990s, there were numerous other small companies interested in lipid manipulation and several of the larger biotech companies, such as Dupont, Pioneer and Monsanto, also established plant lipid research groups. This decade was characterised by an increasing interaction of university and commercial groups, both in basic and applied research into plant lipids. However, as noted in the previous section, plant lipid researchers have worked closely with industry for many decades, ever since Mège Mouriès was persuaded to develop plant lipids (margarine) for military applications and then to sell the technology to an overseas company, way back in 1871.

Lest we think that large industrial groups and sophisticated molecular biology are the best way to improve oil crops, it is instructive to think back to what was achieved in a few decades by a small publicly funded group in Canada. Although Arabidopsis is regarded as the key organism for plant genetics, there had been a foretaste of the power of genetics in facilitating the study and manipulation of plant lipids way back in the early 1960s. Indeed, this earlier work gave rise to an entirely new food crop and a multi-billion dollar worldwide industry that still eclipses in value all the collective achievements of modern crop genetic engineering. This refers to the work of Keith Downey and colleagues at the Agriculture Canada labs in Saskatchewan, Canada. Downey's genetic analysis of rapeseed plants, *Brassica napus*, indicated that erucic acid (C22:1, $\Delta_{13}$) was formed by the sequential elongation of oleic acid via eicosenoic acid. The pathway was confirmed soon afterwards by monitoring acetate incorporation patterns in immature rapeseed pods. The genetic analysis showed that this elongation pathway was controlled by only a few genes and was thus amenable to manipulation (Downey and Craig, 1964). Downey's group went on to select and breed naturally occurring mutant varieties of rapeseed in which this fatty acid elongation pathway was almost totally abolished.

Part of the reason for their success was the innovative use of the 'half-seed' technique to screen the many hundreds of potentially mutated seeds that had to be examined in order to identify suitable low-erucic variants. The half-seed method involves the careful removal from a dry seed of the whole or part of a cotyledon without damaging the radicle or the other cotyledon. Downey's group then analysed the fatty acid content of the cut portion of the cotyledon by the still relatively new method of GC. Meanwhile, the remainder of the seed was germinated and went on to produce a normal adult rapeseed plant. Just to illustrate the power of the technological advance represented by GC, we should consider its logistics compared to earlier methods of fatty acid analysis. What had previously required many hundreds of seeds, many grams of seed oil and several weeks of analysis could now be achieved by collecting part of a single 5 mg seed that could be analysed for fatty acid content in just a few hours. This enabled the team to analyse huge numbers of individual seeds and thereby to obtain genetically pure low-erucic varieties as part of a modestly funded but immensely effective research programme.

The result of abolishing the elongation pathway in rapeseed was spectacular. What had previously been a rather low-value break crop with a high content of anti-nutritional (according to data then available, although this has now been challenged, e.g. see Green and Innis, 2000) erucic acid was suddenly turned into a premium quality, high oleic, edible oil crop that was promptly rechristened 'canola'. Canola is now a mainstay of Canadian prairie agriculture and has been adopted enthusiastically in parts of the United States of America. Similar zero-erucic varieties have been adopted worldwide and the annual value of this relatively new oilseed crop is in excess of $ 6 billion. As a coda to this story, advances in molecular genetics in the late 1990s have now allowed us to discover the exact nature of the mutations that Downey *et al.* selected in order to create the low-erucic acid phenotype of modern rapeseed/canola. The plants contain single point mutations in two genes encoding isoforms of β-ketoacyl CoA synthase. This enzyme is part of the fatty acid elongase complex now known to mediate the formation of erucic acid from oleate (Fourmann *et al.*, 1998). The achievement of such a dramatic result, namely, a new high-value oil crop, from such a small change, namely, the alteration of just two nucleotides in a genome of over 400 million bases, demonstrates the power of genetics as applied to plant breeding. Such a result would be the envy of any latter-day biotechnologist and is a useful reminder that genetic engineering is not the only way to achieve the precise manipulation of a crop.

## 1.6 New frontiers – cell biology and the 'omics

During the 1990s, much attention was focused on the various worldwide genome sequencing programmes, as exemplified for plants by Arabidopsis. The acquisition and analysis of whole-genome sequence data became known as genomics. By the early years of the twenty-first century, the complete genomes of several hundred organisms had been sequenced and deposited in public databases, such as NCBI-BLAST (http://www.ncbi.nlm.nih.gov/BLAST). Extensive sequence data from major crops like maize, rice and soya are also available for research purposes from various company databases. Today, terabytes of such data can be accessed by researchers and the major bottleneck is no longer the acquisition of information, but rather its analysis – hence, the growth of bioinformatics, or the analysis of biologically relevant data.

Despite the plethora of sequence data, it is now apparent that even a full inventory of all of the genes in an organism does not necessarily lead to an understanding of its biochemistry, much less its cell biology or physiology. Although we have known the complete Arabidopsis genome sequence for several years, at least one-third of the genes are of completely unknown function, while many others are only assigned putative function on the basis of sequence similarities with other organisms. Therefore, we are now moving into what has been termed the 'post-genomics' era of research. As applied to plant lipid studies, this has included the establishment of inventories of all known lipid-related genes and their expression patterns in several key tissues or stages of development. The most complete public database of this nature is hosted by the Ohlrogge group at Michigan State University (http://www.canr.msu.edu/lgc/index.html). This group has also made impressive contributions to describing the pattern of expression of over 100 lipid-related genes in the major plant tissues, such as seeds, roots, flowers and leaves. A further list of contemporary web-based resources for plant lipid researchers is given in Table 1.3. One example of the challenges thrown up by the sequencing of the Arabidopsis genome is the fact that there are almost 400 putative genes with sequence similarities to a single class of lipid-related genes, namely, the glycosylhydrolases/glycosidases (Henrissat and Bairoch, 1996; see also Chapter 5). At present, we know little about the biochemical function of the enzymes encoded by all of these genes. While this task may seem a daunting challenge, it does underline the fact that far from making biochemists redundant, genomics has opened up many new opportunities for them.

In addition to genomics, we now have access to further tools for the large-scale investigation of the function of lipids in whole plants, in individual tissues, in different developmental stages and under different environmental conditions. Such tools include other 'omics, including proteomics, metabolomics, transcriptomics and even lipidomics. The term 'lipidomics' emerged into the literature in 2003 and has been defined as 'the full characterization of lipid molecular species and of their biological roles with respect to the expression of proteins involved in lipid metabolism and function, including gene regulation' (Lagarde *et al.*, 2003). To a

**Table 1.3** Links to databases and other resources for plant lipids (much of this information comes from the Lipid Analysis Unit site, as listed below)

*General lipids*

Cyberlipid – a web site containing much useful information on lipid chemistry, biochemistry and analysis (http://www.cyberlipid.org/)

Lipid Nomenclature – this is the IUPAC guide (http://www.chem.qmw.ac.uk/iupac/lipid)

Conjugated Linoleic acid – Wisconsin Food Research Institute (http://www.wisc.edu/fri/clarefs.htm). These pages give a comprehensive list of references to papers dealing with CLA

Compilation of trivial names of fatty acids (by R.O. Adlof and F.D. Gunstone) (http://www.aocs.org/member/division/analytic/fanames.htm)

Lipidat. – a relational database of thermodynamic and associated information on lipid mesophase and crystal polymorphic transitions, including lipid molecular structures (glycero- and sphingolipids) (http://www.lipidat.chemistry.ohio-state.edu/)

Lipid Bank for Web – a database of information on lipid structures and properties with thousands of references (http://lipid.bio.m.u-tokyo.ac.jp/)

Lipid Analysis Unit – this site is a general information resource on lipids supported by the Lipid Analysis Unit at the Scottish Crop Research Institute (http://www.lipid.co.uk)

*Plant lipids*

On-line chemical database for new seed crops produced by the New Crop Research Unit at NCAUR, Peoria, IL, USA (http://www.ncaur.usda.gov/nc/ncdb/search.html-ssi/) (see Abbot *et al.* (1997) *Journal of the American Chemical Society*, **74**, 723–726 and correction on p. 1181 for instructions) – chromatographic, physical chemical and spectroscopic information on oil seeds

Similar database to the previous provided by BAGKF (Institute for Chemistry and Physics of Lipids), Munster, Germany – SOFA (Seed Oil Fatty Acids) (www.bagkf.de/sofa) (see Aitzetmuller *et al.* (2003) *Eur. J. Lipid Sci. Technol.*, **105**, 92–103)

A catalogue of genes for plant lipid biosynthesis at Michigan State University (http://www.canr.msu.edu/lgc/index.html)

NPLC (National Plant Lipid Cooperative) (http://www.msu.edu/user/ohlrogge/)

A further source of links to web-based lipid information, includes: NPLC Directory of Plant Lipid Scientists

The NPLC Electronic Mailing List

The NPLC Database of Plant Lipid Literature

The Plant Lipid Home Page – (http://blue.butler.edu/~kschmid/lipids.html)

Maintained by Katherine M. Schmid this page contains many excellent lipid-related links.

The Malaysian Palm Oil Board (MPOB) – web site devoted to all aspects of palm oil-biology, technology, food and non-food uses, and commercial matters (http://www.mpob.gov.my)

**Food- and industry-related**

*Loders Croklaan* is part of Unilever Ltd. The site covers science, technology, and nutrition related to lipids in general and to Unilever's products in particular (www.croklaan.com)

**Table 1.3** (continued)

*ITERG*. French Research Institute dealing with oils and fats research and technology (www.iterg.com)
*Natural of Norway*. Manufacturers of conjugated linoleic acid for health food and other applications (www.natural.no).
The web site of the *British Nutrition Foundation* carries information on lipids in addition to other food components (www.nutrition.org.uk)
*CTVO* – a European group devoted to the chemical and technological utilization of vegetable oils (www.danet.de/fnr/ctvo)
*IENICA* – an Interactive European Network for industrial crops and their applications(www.csl.gov.uk/ienica/)
*ACTIN* – a UK group devoted to non-food uses of oils and fats (www.actin.co.uk)
European web site for the *American Soybean Association* (www.asa-europe.org)
*International Food Science and Technology* – contains information on various food problems, including those involving lipids (www.ifst.org)
*Oil World* is a German company producing data on a weekly basis for oilseeds, oils and fats and oilmeals and covers production, imports, exports and disappearance. Information is based on different commodity oils and fats and is presented on the basis of individual countries (www.oilworld.de).
*FFA Sciences* is a company manufacturing probes to measure free fatty acid levels in oils and clinical samples (www.ffasciences.com)
*Britannia Foods* has some articles 'By Invitation Only' of interest to lipid technologists mainly (www.britanniafood.com)
*Peter Lapinskas* – consultant to the Oils and Fats Industry – some interesting data on unusual seed oils (www.lapinskas.com)

**Lipidomics-related web sites**

http://www.ksu.edu/lipid/lipidomics
http://hcc.musc.edu/research/shared_resources/lipidomics.cfm
http://medschool.mc.vanderbilt.edu/brownlab/comlip.html
http://www.wissenschaft-online.de/gbm/homepage/abstract_detail.php?artikel_id=265
http://www1.elsevier.com/gej-ng/29/50/lipids/119/47/26/article.pdf

great extent, this is simply repackaging a variety of biochemical and molecular genetic approaches under a new name, when they are applied specifically to lipids. Proteomics is the analysis of the entire protein complement of a cell(s), tissue(s) or organism, often sampled at a particular developmental stage or environmental state. Metabolomics is similar except that it is the analysis of metabolites rather than proteins, while transcriptomics looks at the subset of genes transcribed in a given organism/tissue/cell. An extremely powerful technique used in transcriptomics is DNA-microarrays, which allows us to determine the mRNA expression level of practically every gene of an organism. For more on such *in silico* techniques, see http://www.bio-itworld.com/bio/microarrays.

The power of the various forms of 'omics lies in their ability to acquire comparative information on all the expressed genes, proteins, metabolites etc. that are present in a biological system and then to do the same analysis on the same system under different developmental or environmental conditions. These analyses are already showing that lipids may be involved in hitherto unexpected areas, including acting as components in plant responses to fungal pathogens and salt stress (Kirsch *et al.*, 1997; Xiao *et al.*, 2001). Plant lipids can also function as signalling molecules and mediators in a wide range of physiological processes as described in Chapters 9, 10 and 11.

As we discover more about the details of plant lipid metabolism and their various physiological roles, there is a further frontier for study that has been less well-explored but may now become amenable following recent technical advances. This area relates to the cellular context of plant lipids, either as discrete molecules or more frequently as macromolecular assemblies, such as bilayer membranes or lipid bodies. We already know that plant lipid assemblies are much more dynamic than was thought until quite recently (Murphy, 2001) and new tools are becoming ever more available to researchers that will enable them to study the dynamic behaviour of lipids in real time. Hitherto, most cell biological investigations of lipid and membrane processes have relied upon static imaging systems, such as transmission, freeze-etch and freeze fracture electron microscopy. These techniques have been used effectively in combination with labelled probes in immunocytochemical studies, e.g. the use of gold labelling to study protein distribution in cells.

In the last decade, however, there have been significant developments of improved high-resolution imaging systems, such as laser-assisted confocal microscopy employing double-photon and, since the late 1990s, multi-photon imaging. This allows the investigator to analyse whole cell images with great resolution and with a large depth of field, rather than relatively thin slices of small sectors of cells. Along with the improved detector systems has come the development of powerful new reporter systems, such as novel fluorophores and immunofluorescent tags. Fluorophores and antibody tags no longer need be large proteins like green fluorescent protein (GFP) or β-glucuronidase (GUS), but can be much smaller molecules that can be conveniently conjugated to proteins with little or no impact on their behaviour. Fluorophores of different but overlapping absorbances can be designed to assay whether two labelled proteins are adjacent to one another. If the proteins are within a few Å of each other, the resultant fluorescence resonance energy transfer, or FRET, means that excitation of one fluorophore will result in an emission from the other fluorophore. This allows us to study the interactions of proteins in membrane in very fine detail. Moreover, such studies can also be done in real time with living cells. Thanks to highly sensitive detection methods like multi-photon imaging, the movement of labelled lipids and proteins can be studied as it happens. It is these and similar developments that are beginning to give us a window onto the dynamic behaviour of lipids within cells, as discussed in Chapter 7.

## 1.7 Conclusions and future prospects

In this chapter, I have tried to present a broad outline of how far the study of plant lipids has come since Tachenius first hinted at the existence of fatty acids almost 350 years ago. As with most empirical scientific disciplines, progress in our knowledge has been very much dependent on the development of new or improved techniques. We now have access to an enviable set of investigative tools at every level – from the finest chemical analysis and automated genome wide screening to live whole cell imaging. The study of plant lipids has emerged in the past 25 years from being something of a secondary discipline to one that is at the frontiers of basic research as well as of considerable potential commercial utility. For example, the first desaturase genes to be characterised and manipulated were from plants, as were the first lipid-body proteins. However, there is still much that plant lipid researchers can learn from colleagues working in other systems (and vice versa). A good example of the crossover between the various types of model organisms can be seen in the latest edition of the *Lipid Biochemistry* textbook by Gurr *et al.* (2002).

In the chapters of this book, there are many other references to this crossover of knowledge with non-plant workers. Such comparative approaches have already considerably enriched our knowledge of lipid metabolism and are now increasingly impinging on cell biology studies as well. One current example is the realisation that lipid rafts and other elements of intracellular trafficking mechanisms, possibly including caveolin-like structures, are probably present in plant membranes, as well as other higher eukaryotes. It is therefore important that plant researchers make themselves aware of current developments in other fields and attend more general lipid meetings with their non-plant colleagues. This does not mean abandoning the valuable plant lipid meetings, although we could perhaps invite more non-plant speakers in selected areas. However, following the advances of the past 25 years, one would hope that plant lipidologists are now self-confident enough to participate in the mainstream meetings without resorting to Terry Galliard's desperate subterfuge of pretending that we are working on bacteria!


## Acknowledgements

I am indebted to Mike Gurr and John Harwood for their valuable comments and suggestions on the text of this chapter. I am also grateful to Martin Rumsby for awakening an interest in lipids when I was an undergraduate at the University of York in 1971–1974 and to Rachel Leech for her supervision of my DPhil thesis on plant lipid metabolism at the same institution. Finally, I thank Paul Stumpf for accepting me as a Fulbright postdoctoral scholar at the University of California, Davis from 1977–1980, during which time I was privileged to work alongside many notable plant lipid scientists.

## References

Baddiley, J. (1955) The structure of coenzyme A. *Advances in Enzymology*, **16**, 1–21.

Barron, E.J. and Stumpf, P.K. (1962) Fat metabolism in higher plants. XIX. The biosynthesis of triglycerides by avocado-mesocarp enzymes. *Biochimica et Biophysica Acta*, **60**, 329–337.

Baugh, P.J. (1997) *Gas Chromatography: A Practical Approach*, IRL Press, Oxford.

Benson, A.A. (1963) The plant sulfolipid. *Advances in Lipid Research*, **1**, 387.

Benson, A.A. and Maruo, B. (1958) Plant phospholipids. I. Identification of the phosphatidyl glycerols. *Biochimica et Biophysica Acta*, **27**, 189–195.

Benson, A.A. and Miyano, M. (1961) *Biochemical Journal*, The phosphatidylglycerol and sulfolipid of plants: asymmetry of the glycerol moiety **81**, 31P.

Benson, A.A., Daniel, H and Wiser, R. (1959) A sulfolipid in plants. *Proceedings of the National Academy of Sciences USA*, **45**, 1582–1587.

Bloomfield, D.K. and Bloch, K. (1960) The formation of delta 9-unsaturated fatty acids. *Journal of Biological Chemistry*, **235**, 337–345.

Borgström, B. (1952) Investigation on lipid separation methods. Separation of phospholipids from neutral fat and fatty acids, *Acta Physiologica Scandinavica*, **25**, 101–110.

Braun, A. (1873) *Freunde zu Berlin*, p. 75.

Browse, J.A., McCourt, P.J. and Somerville, C.R. (1985) A mutant of Arabidopsis lacking a chloroplast specific lipid. *Science*, **227**, 763–765.

Browse, J.A., Kunst, L., Hugly, S. and Somerville, C.R. (1989) Modifications to the two pathway scheme of lipid metabolism based on studies of Arabidopsis mutants, in *Biological Role of Plant Lipids* (eds P.A. Biacs, K. Gruiz and T. Kremmer), Plenum, New York, pp. 21–24.

Burr, G.O. and Burr, M.M. (1930) On the nature and role of the fatty acids essential in nutrition. *Journal of Biological Chemistry*, **86**, 587–621.

Carter, H.E., McCluer, R.H. and Slifer, E.D. (1956) Lipids of wheat flour. I. Characterization of galactosylglycerol components. *Journal of the American Chemical Society*, **78**, 3735–3738.

Chibnall, A.C. and Channon, H.J. (1927) The ether-soluble substances of cabbage leaf cytoplasm. II. Calcium salts of glyceridephosphoric acids. *Biochemical Journal*, **21**, 233–246.

Ching, T.M. (1968) Neutral lipase activities on lipid bodies in germinating seedlings. *Lipids*, **3**, 482–488.

Christie, W.W. (2003) What is a lipid? Available online: http://www.lipid.co.uk/infores/Lipids/whatlip/file.pdf

Christie, W.W. (2003) *Lipid Analysis, Isolation, Separation, Identification and Structural Analysis of Lipids*, 3rd edn, The Oily Press, Bridgewater, UK.

Christie, W.W., ed. (1992–2003) *Advances in Lipid Methodology*, vols 1–5, The Oily Press, Bridgewater, UK.

Cooper, T.G. and Beevers, H. (1969) β-Oxidation in glyoxysomes from castor bean endosperm. *Journal of Biological Chemistry*, 244, 3514–3520.

Darby, S. (1849) *Annales*, **69**, 1.

de Saussure, N.T. (1804) Recherches chimiques sur la végétation, Paris.

Douce, R., Foure, M. and Marechal, J. (1966) *Comptes Rendues de l'Academie de Sciences de Paris* **262**, 1549–1552.

Downey, R.K. and Craig, B.M. (1964) Genetic control for fatty acid biosynthesis in rapeseed (*Brassica napus* L.). *Journal of the American Chemical Society*, **41**, 475–478.

Edmed, F.G. (1898) The structure of oleic acid. *Journal of Chemical Society*, **73**, 627.

Fourmann, M., Barret, P., Renard, M., Pelletier, G., Delourme, R. and Brunel, D. (1998) The two genes homologous to Arabidopsis FAE1 co-segregate with the two loci governing erucic acid content in Brassica napus. *Theoretical and Applied Genetics*, **96**, 852–858.

Frey-Wyssling, A., Grieshaber, E. and Muhlethaler, K. (1963) Origin of spherosomes in plant cells. *Journal of Ultrastructural Research*, **8**, 506–516.

Galliard, T. and Mercer, E.J., eds (1975) *Recent Advances in the Chemistry and Biochemistry of Plant Lipids*, Academic Press, London.

Green, T.J. and Innis, S.M. (2000) Low erucic acid canola oil does not induce heart triglyceride accumulation in neonatal pigs fed formula. *Lipids*, **35**, 607–612.

Grob, R.L. and Barry, E.F. (1995) *Modern Practice of Gas Chromatography*, 3rd edn, Wiley, New York.

Gunstone. F.D., McLaughlan, J., Scrimgeour, C.M. and Watson, A.P. (1976) Improved procedures for the isolation of pure oleic, linoleic, and linolenic acids or their methyl esters from natural sources. *Journal of the Science of Food and Agriculture*, **27**, 675–680.

Gunstone, F.D., Pollard, M.R., Scrimgeour, C.M. and Vedanayagam, H.S. (1977) Fatty acids. Part 50. 13C nuclear magnetic resonance studies of olefinic fatty acids and esters. *Chemistry and Physics of Lipids*, **18**, 115–129.

Gunstone, F.D., Harwood, J.L. and Padley, F.B., eds (1994) *The Lipid Handbook*, Chapman and Hall, London.

Gurr, M.I., Blades, J., Appleby, R.S., Smith, C.G., Robinson, M.P. and Nichols, B.W. (1974) Studies on seed-oil triglycerides. Triglyceride biosynthesis and storage in whole seeds and oil bodies of *Crambe abyssinica. European Journal of Biochemistry*, **43**, 281–290.

Gurr, M.I., Harwood, J.L. and Frayn, K.N. (2002) *Lipid Biochemistry*, 5th edn, Blackwell Science, Oxford.

Gurr, M.I. and James, A.T. (1971–1980) *Lipid Biochemistry*, 1st–3rd edns, Gurr MI and Harwood JL, 4th edn, Chapman and Hall, London.

Gurr, M.I., Robinson, M.P. and James, A.T. (1969) The mechanism of formation of polyunsaturated fatty acids by photosynthetic tissue. The tight coupling of oleate desaturation with phospholipid synthesis in *Chlorella vulgaris. European Journal of Biochemistry*, **9**, 70–78.

Hamilton, R. (1998) *Lipid Analysis of Oils and Fats*, Kluwer, Dordrecht.

Harlan, J.R. (1992) *Crops and Man*, 2nd edn, American Society of Agronomy, Madison, WI.

Harwood, J.L. (1988) The site of action of some selective graminaceous herbicides is identified as acetyl-CoA carboxylase. *Trends in Biochemical Sciences*, **13**, 330–331.

Harwood, J.L. (1996) Recent advances in the biosynthesis of plant fatty acids. *Biochimica et Biophysica Acta*, **1301**, 7–56.

Harwood, J.L., Sodja, A., Stumpf, P.K. and Spurr, A.R. (1972) On the origin of oil droplets in maturing castor bean seeds, *Ricinus communis. Lipids*, **6**, 851–854.

Henrissat, B. and Bairoch, A. (1996) Updating the sequence-based classification of glycosyl hydrolases. *Biochemical Journal*, **316**, 695–696.

Herman, E.M. (1987) Immunogold-localization and synthesis of an oil-body membrane protein in developing soybean seeds. *Planta*, **172,** 336–345.

Hilditch, T.P. (1940) *The Chemical Constitution of Natural Fats*, 1st edn, Chapman and Hall, London.

Hilditch, T.P. *et al.* (1939) *Journal of Society for Chemistry in Industry*, **58**, 233.

Hitchcock, C. and Nichols, B.W. (1971) *Plant Lipid Biochemistry*, Academic Press, London and New York.

Holbrook, L.A., van Rooijen, G.J.H., Wilen, R.W. and Moloney, M.M. (1991) Oil body proteins in microspore-derived embryos of *Brassica napus* L. cv Reston. Hormonal, osmotic, and developmental regulation of synthesis. *Plant Physiology*, **97**, 1051–1058.

Howard, G.A. and Martin, A.J.P. (1950) *Biochemical Journal*, **46**, 532–538.

Hutton, D. and Stumpf, P.K. (1969) Microbodies in higher plants. *Plant Physiology*, **44**, 508–516.

Izmailov, N.A. and Shraiber, M.S. (1938) *Farmatsiya* (Mosk.) **3**, 1.

Jacks, T.J., Yatsu, L.Y. and Altschul, A.M. (1967) Isolation and characterization of peanut spherosomes. *Plant Physiology*, **42**, 585–597.

James, A.T. (1963) The biosynthesis of long-chain saturated and unsaturated fatty acids in isolated plant leaves. *Biochimica et Biophysica Acta*, **70**, 9–19.

James, A.T. and Martin, R.M.L. (1952) Gas–liquid partition chromatography: the separation and micro-estimation of ammonia and the methylamines. *Biochemical Journal*, **50**, 679–690.

Kirchner, J.G. (1951) Use of silica thin layers platos for separation of terpenes. *Analytical Chemistry*, **23**, 420–425.

Kirsch, C., Hahlbrock, K. and Somssich, I.E. (1997) Rapid and transient induction of a parsley microsomal delta 12 fatty acid desaturase mRNA by fungal elicitor. *Plant Physiology*, **115**, 283–289.

Knutzon, D. and Knauf, V. (1998) Manipulating seed oils for polyunsaturated fatty acid content, in *Plant Lipid Biosynthesis* (ed. J.L. Harwood), Cambridge University Press, Cambridge, pp. 27–304.
Lagarde, N., Geloen, A., Record, M. *et al.* (2003) Lipidomics is emerging. *Biochimica et Biophysica Acta*, **1634**, 61.
Laibach, F. (1907) *Botanische Centralblat Beihefte (I)*, **22**, 191–210.
Laibach, F. (1943) *Arabidopsis thaliana* (L.) Heynh. als object fur genetische und entwicklung-sphysiologische untersuchungen. *Bot Archiv*, 44, 439–455.
Leech, R.M. and Murphy, D.J. (1976) The cooperative function of chloroplasts in the biosynthesis of small molecules, in *The Intact Chloroplast* (ed. J. Barber), Elsevier, Amsterdam.
Mangold, H.K. (1959) Zur analyze von lipiden mit hilfe der radioreagenz-methoden. *Fette Seifen Anstrichm.*, **61**, 877–881.
Martineau, B. (2000) *First Fruit – The Creation of the Flavr Savr$^{TM}$ Tomato and the Birth of Biotech Food*, McGraw Hill, New York.
Marsson, T. (1842) *Annales* **41**, 329.
Martin, A.J.P. and Synge, R.L.M. (1941) A new form of chromatogram employing two liquid phases.1. A theory of chromatography. 2. Application of the microdetermination of the higher monoaminoacids in proteins. *Biochemical Journal*, **35**, 1358–1368.
McDonald, R.E. and Mossoba, M.M., eds (1997) *New Techniques and Applications in Lipid Analysis*, Amer. Oil Chem. Soc. Champaign, IL.
McKeon, T.A. and Stumpf, P.K. (1982) Purification and characterization of the stearoyl-acyl carrier protein desaturase and the acyl-acyl carrier protein thioesterase from maturing seeds of safflower. *Journal of Biological Chemistry*, **257**, 12141–12147.
Meyerowitz, E.M. (2001) Prehistory and history of Arabidopsis. *Plant Physiology*, **125**, 15–19.
Meyerowitz, E.M. and Pruitt, R.E. (1985) *Arabidopsis thaliana* and Plant molecular genetics, *Science*, **229**, 1214–1218.
Mollenhauer, H.H. and Totten, C. (1971) Studies on seeds. 3. Isolation and structure of lipid-containing vesicles. *Journal of Cell Biology*, **48**, 533–541.
Mudd, J.B. and Stumpf, P.K. (1961) Fat metabolism in higher plants. XIV. Factors affecting the synthesis of oleic acid by particulate preparations from avocado mesocarp. *Journal of Biological Chemistry*, **236**, 2602–2609.
Muntz (1871) *Ann. Chem. Liebigs.*, **22**, 472.
Murphy, D.J., ed. (1994) *Designer Oil Crops*, VCH Press, Weinheim, pp. 604–698.
Murphy, D.J. (1998) *Biodiesel and its Prospects*, NRC-PBI Bulletin, Saskatoon, Canada, May, 8–10.
Murphy, D.J. (1999) Production of novel oils in plants. *Current Opinion in Biotechnology.*, **10**, 175–180.
Murphy, D.J. (2001) Biogenesis and functions of lipid bodies in animals, plants and microorganisms. *Progress in Lipid Research*, **40,** 325–438.
Murphy, D.J. (2003) Agricultural biotechnology and oil crops – current uncertainties and future potential. *Applied Biotechnology, Food Science and Policy*, **1**, 25–38.
Murphy, D.J. (2004) Biotechnology, its impact and future prospects, in *From Molecular to Global Photosynthesis* (eds J. Barber and M. Archer), Imperial College Press, London.
Murphy, D.J. and Cummins, I. (1989) Purification and immunogold localisation of the major oil-body membrane protein of oilseed rape. *Plant Sci.*, **60**, 47–54.
Murphy, D.J., Cummins, I. and Kang, A.S. (1989) Synthesis of the major oil-body membrane protein in developing rapeseed (*Brassica napus*) embryos. Integration with storage-lipid and storage-protein synthesis and implications for the mechanism of oil-body formation. *Biochemical Journal*, **258**, 285–293.
Newcombe, E.H. and Stumpf, P.K. (1952) Fatty acid biosynthesis in peanut cotyledons, in *Phosphorus Metabolism* (eds W.D. McElroy and B. Glass), Johns Hopkins Press, Baltimore, pp. 291–300.
Nichols, B.W., James, A.T. and Breuer, J. (1967) Interrelationships between fatty acid biosynthesis and acyl-lipid synthesis in *Chlorella vulgaris*. *Biochemical Journal*, **104**, 486–496.

Nitske, W.R. and Wilson, C.M. (1965) *Rudolf Diesel, Pioneer of the Age of Power*, Univ. Oklahoma Press.
Ongun, A. and Mudd, J.B. (1968) Biosynthesis of galactolipids in plants. *Journal of Biological Chemistry*, **248**, 1558–1566.
Page, R.A., Okada, S. and Harwood, J.L. (1994) Acetyl-CoA carboxylase exerts strong flux control over lipid synthesis in plants. *Biochimica et Biophysica Acta*, **1210**, 369–372.
Partington, J.R. (1989) *A Short History of Chemistry*, 3rd ed, Dover Publications Inc, New York, pp. 60, 221–222.
Pelouze, M. and Boudet, E. (1838) *J. Pharm.*, **24**, 3.
Playfair, L. (1842) Fatty acids of oils in Myristicaceae, *Annales*, **37**, 152.
Rédei, G.P. (1975) Arabidopsis as a genetic tool. *Annual Review of Genetics*, **9**, 111–1127.
Reinholz, E. (1947) Field Inf Agency Technical Report No. **1006**, 1–70.
Roughan, P.G. and Slack, C.R. (1982) Cellular organisation of glycerolipid metabolism. *Ann. Rev. Plant Physiology*, **33**, 97–132.
Saalmüller, L. (1848) *Annales*, **64**, 108.
Sacc, F. (1844) *Annales*, **51**, 213.
Sastry, P.S. and Kates, M. (1966) Biosynthesis of lipids in plants. II. Incorporation of glycerophosphate-32-P into phosphatides by cell-free preparations from spinach leaves. *Canadian Journal of Biochemistry*, **44**, 459–467.
Slack, C.R., Bertaud, W.S., Shaw, B.D., Holland, R., Browse, J. and Wright, H. (1980) Some studies on the composition and surface properties of oil bodies from the seed cotyledons of safflower (*Carthamus tinctorius*) and linseed (*Linum usitatissimum*). *Biochemical Journal*, **190**, 551–61.
Smith, C.G. (1974) The ultrastructural development of spherosomes and oil bodies in the developing embryo of *Crambe abyssinica*. *Planta*, **119**, 125–142.
Sorokin, H. (1967) The spherosomes and the reserve fat in plant cells. *American Journal of Botany*, **54**, 1008–1016.
Sperry, W.M. (1926) Lipid excretion III. Further studies of the quantitative relations of fecal lipids. *Journal of Biological Chemistry*, **68**, 357–383.
Stumpf, P.K. and Barber, G.A. (1956) Detection of β-oxidation of fatty acids in plant extracts, *Plant Physiology*, **31**, 304–308.
Stumpf, P.K. and James, A.T. (1963) The biosynthesis of long-chain fatty acids by lettuce chloroplast preparations. *Biochimica et Biophysica Acta*, **70**, 20–32.
Stumpf, P.K. and Shimakata, T. (1983) Molecular structure and functions of plant fatty acid synthetase enzymes, in *Biosynthesis and Function of Plant Lipids* (eds W.W. Thompson, J.B. Mudd and M. Gibbs), *American Society of Plant Physiologist*, Baltimore, USA, pp. 1–15.
Sure, B. (1924) Dietary requirements for reproduction. II. The existence of a specific vitamin for reproduction. *Journal of Biological Chemistry*, **58**, 693–697.
Swain, L.A. (1948) Lipid separations in alumina columns, *Canadian Journal of Chemistry*, **32**, 533.
Tachenius, O. (1666) *Hippocrates Chymicus*, Paris.
Thoyts, P.J., Millichip, M.I., Stobart, A.K., Griffiths, W.T., Shewry, P.R. and Napier, J.A. (1995) Expression and *in vitro* targeting of a sunflower oleosin. *Plant Molecular Biology*, **29**, 403–410.
Töpler, H. (1861) *Landw. Vers. Sta.*, **3**, 85.
Vance, V.B. and Huang, A.H.C. (1987) The major protein from lipid bodies of maize. Characterization and structure based on cDNA cloning. *Journal of Biological Chemistry*, **262**, 11275–11279.
Websky, F. (1853) *Journal fur Praktische Chemise*, **58**, 449.
Weicker, H. (1959) Dünnschicht-Chromato graphie. *Klinische Wochenschrift*, **37**, 763.
Willstätter, R. (1973) *Aus mein Leben; Von Arbeit, Musse und Freunden*, 2nd edn, Verlag Chimie, Weinheim, Germany, p. 156.
Xiao, F., Tang, X. and Zhou, J.M. (2001) Expression of 35S::Pto globally activates defense-related genes in tomato plants. *Plant Physiology*, **126**, 1637–1645.
Yamada, M. (1957) *Sci. Pap. Coll. Gen. Educ. Tokyo*, **7**, 97–99.

# 2 Fatty acid biosynthesis

John L. Harwood

## 2.1 Introduction

Most of the important lipids in plant cells are acyl lipids. The main storage lipid is triacylglycerol, while the main membrane lipids (depending on the membrane source) are glycosylglycerides or phosphoglycerides. Therefore, fatty acid synthesis as the virtually exclusive source of the fatty acid components of these lipids is vital for tissue health and plant survival. Moreover, acyl lipids are also major components of the surface coverings (waxes, cutin, suberin: Kolattukudy, 1980) while acyl lipids, or fatty acids derived from them, give rise to most of the lipid signalling molecules.

Thus, it is clear that fatty acid biosynthesis is one of the primary pathways of metabolism. In 1967, the first component protein of fatty acid synthase was purified (Simoni *et al.*, 1967) and there has been a continuous advance in our knowledge of the enzymology and protein biochemistry of fatty acid synthesis ever since. Over the last 15 years, the techniques of molecular biology have been a significant boost to our search for further details of the various processes involved. In addition, molecular biology has been utilised to produce transgenic plants, which have given us insights (sometimes unexpected) into fatty acid synthesis on the one hand or to develop crops that have, in some cases, been commercially successful.

This chapter deals with the *de novo* formation of long-chain fatty acids. The following chapter will deal with the further modification (desaturation, elongation, etc.) that can occur to the 16 or 18C products of fatty acid synthase. The most recent comprehensive review that deals with the subject is that of Harwood (1996) although sections of Williams *et al.* (1997), Sanchez *et al.* (1998), Harwood (1998), Harwood and Quinn (2000) and Murata *et al.* (2003) cover specific aspects. In addition, older books (Hitchcock and Nichols, 1971; Stumpf, 1980, 1987) or reviews (Roughan and Slack, 1982; Harwood, 1988; Browse and Somerville, 1991) remain classic sources of information, much of which is still very relevant today. Two more recent reviews that provide important information about regulation are Ohlrogge and Jaworski (1997) and Voelker and Kinney (2001).

## 2.2 Carbon supply for fatty acid formation

Except for the brief period when storage material in seeds provides carbon for fatty acid synthesis during germination, photosynthesis produces the precursor molecules required. Since acetyl-CoA can be regarded as one key precursor, much early work focused on how acetyl-CoA was provided for fatty acid synthesis. Two main pathways were suggested – a direct supply of plastid acetyl-CoA by pyruvate dehydrogenase/decarboxylase or indirectly using mitochondrial pyruvate dehydrogenase. In the latter case, the mitochondrial acetyl-CoA is hydrolysed, acetate then moves to the plastid where regeneration of acetyl-CoA can take place by acetyl-CoA synthase. Early work on these pathways was summarised by Liedvogel (1987) and by Harwood (1988).

Work by two groups has recently shed further light on carbon supply for fatty acid synthesis. Bao *et al.* (2000) looked at the question in several different plant leaves using stable isotope ($^{13}C$) methods. They did this by measuring the *in vivo* rate of fatty acid formation in both the light and the dark. As expected, the dark rates were much lower (<5% ) than those in the light. These experiments allowed them to estimate fatty acid degradation/turnover as about 4% of total leaf fatty acids per day. They then measured the acetate pool and, by using pulse or continuous [$^{14}C$]carbon dioxide labelling of fatty acids concluded that the bulk pool of acetate was not involved in fatty acid synthesis. By implication, the authors concluded that direct utilisation of newly fixed carbon via plastid pyruvate dehydrogenase was the most likely route although they could not absolutely eliminate the possibility that a small pool of acetate, metabolically distinct from the bulk pool, was being used (Bao *et al.*, 2000). Further work by the same group using stable isotope labelling methods with developing oilseed rape embryos concluded that plastid glycolysis provided at least 90% of precursors for plastid acetyl-CoA (Schwender and Ohlrogge, 2002).

In a second study, Ke *et al.* (2000) considered carbon supply for lipid synthesis in developing Arabidopsis seeds. They pointed out that, although four sources of carbon have been proposed for plastid acetyl-CoA production, a function for ATP-citrate lyase (Ratledge *et al.*, 1997) or for acetylcarnitine (Masterson *et al.* 1990) are both uncertain at the present time. Thus, ATP-citrate lyase was in the wrong compartment (the cytosol) according to Kaethner and ap Rees (1985) while Roughan *et al.* (1993) challenged a role for acetylcarnitine. This left plastid pyruvate dehydrogenase or mitochondrially generated acetate as the two alternative sources. Ke *et al.* (2000) isolated cDNA clones for plastidic pyruvate dehydrogenase and for acetyl-CoA synthetase. Accumulation of pyruvate dehydrogenase mRNA mimicked that of mRNAs for the heteromeric plastid acetyl-CoA carboxylase (ACCase) and, also, correlated strongly with lipid accumulation within the developing embryo. In contrast, the level of mRNA for acetyl-CoA synthetase did not correlate with lipid accumulation either in time or by tissue distribution (Ke *et al.*, 2000).

In conclusion, there is now considerable evidence that plastid pyruvate dehydrogenase is the major source of acetyl-CoA for fatty acid synthesis. It seems unlikely that alternative supply routes are normally significant although, of course, there may be certain circumstances where this is the case.

Rawsthorne and co-workers have, for a number of years, been examining the interrelationships between carbohydrate and lipid metabolism in a number of developing seeds. Their work has been recently reviewed (Rawsthorne, 2002). For most of their systems, carbon transport into the plastid was critical but there were differences between different systems and also during various developmental stages (see below). In developing castor bean endosperm, malate-dependent fatty acid synthesis was much higher than that driven by pyruvate (Eastmond *et al.*, 1997). This finding agreed with previous work (Miernyk and Dennis, 1983; Smith *et al.*, 1992). A malate/inorganic phosphate exchange translocator was shown to be involved for the import of carbon (Eastmond *et al.*, 1997).

Glucose 6-phosphate can enter the (oilseed rape embryo) plastid via a specific transporter and contribute to starch and fatty acid synthesis as well as being used for the oxidative pentose phosphate pathway. At early developmental stages, glucose 6-phosphate was a much better source of carbon for fatty acid synthesis than exogenous pyruvate. However, at the late cotyledon stage, when oil accumulation was rapid, pyruvate was preferred. Under these conditions, glucose 6-phosphate was mainly (60%) used for the oxidative pentose phosphate pathway (Eastmond and Rawsthorne, 2000), which could be a major factor in regulating fatty acid formation through NADPH supply (as demonstrated in developing soybean embryos: Pike *et al.*, 2003). In addition, because acyl-CoA esters inhibit both the plastidial adenylate transporter as well as the glucose 6-phosphate transporter (Fox *et al.*, 2000, 2001; Johnson *et al.*, 2000) it has been suggested that the latter could be important for the regulation of oil accumulation. Indeed, transgenic rape and Arabidopsis plants with the glucose 6-phosphate transporter down-regulated by RNAi showed reduced oil accumulation (Nield *et al.*, 2003). A summary of carbon partitioning in developing oilseed rape is shown in Figure 2.1.

Other sources of carbon for fatty acid synthesis in developing oilseed rape have been investigated (Hill and Rawsthorne, 2000) and a potential role for the phosphoenolpyruvate transporter, particularly at late stages of development, has been suggested (Kubis and Rawsthorne, 2000). These matters and other aspects of carbon supply under different circumstances are reviewed in Rawsthorne (2002).

## 2.3 Acetyl-CoA carboxylase

Acetyl-CoA carboxylase (ACCase: acetyl-coenzyme A: bicarbonate ligase (ATP), EC 6.4.1.2) catalyses the first committed step in fatty acid synthesis. ACCase is a soluble Class 1 biotin-containing enzyme (Knowles, 1989) that catalyses the ATP-dependent formation of malonyl-CoA from bicarbonate and acetyl-CoA. This malonyl-CoA is used for *de novo* synthesis of fatty acids, which is located inside plastids. In addition, malonyl-CoA is needed for elongation of fatty acids on the endoplasmic reticulum (Harwood, 1988; Nikolau *et al.*, 2003). Furthermore, the cytosol is the site of synthesis for various secondary metabolites, such as flavonoids and anthocyanins (Hahlbrock, 1981), which also requires malonyl-CoA.

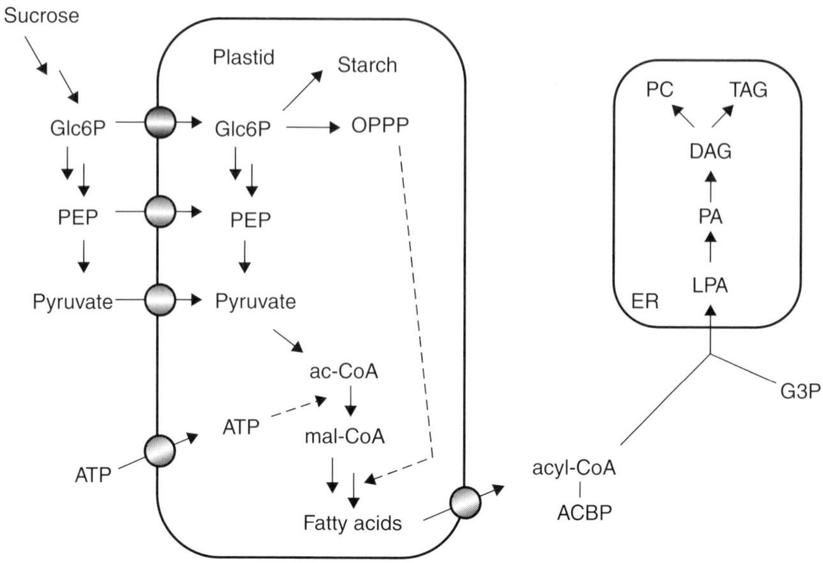


**Figure 2.1** An overview of the current understanding of metabolism associated with fatty acid synthesis and plastidial carbon partitioning in the developing oilseed rape embryo based on metabolic studies. The compartmental nature of the metabolism from sucrose to lipids involving the cytosol, plastids and the endoplasmic reticulum is illustrated, as is the interaction between the oxidative pentose phosphate pathway and fatty acid synthesis (dashed arrow). Transporters on the plastid envelope are shown as shaded circles. Transporters for which genes have been cloned and their proteins functionally characterised are indicated by darkened shading. Double arrows show multi-step pathways. ac-CoA; acetyl-CoA; mal-CoA; malonyl-CoA; OPPP; oxidative pentose phosphate pathway. Reproduced from Rawsthorne (2002) with kind permission of the author and Elsevier.

As expected from such requirements, two isoforms of ACCase are found in different plants. These isoforms have distinct properties and account for ACCase activity, respectively, in the plastid and non-plastid (presumed cytosolic) compartments (Egli *et al.*, 1993; Alban *et al.*, 1994; Herbert *et al.*, 1996).

### *2.3.1 Structure of ACCase*

The overall reaction of ACCase is catalysed in two main steps (Figure 2.2) that need physically distinct active sites (Knowles, 1989). The first partial reaction is catalysed by biotin carboxylase and uses ATP to drive the carboxylation of the biotin prosthetic group of biotin carboxyl carrier protein (BCCP). A carboxyphosphate intermediate is probably used with bicarbonate as the source of carbon.

The second partial reaction is that of the carboxyltransferase. The carboxyl group is transferred from BCCP to acetyl-CoA to yield malonyl-CoA. This reaction, in contrast to the biotin carboxylase partial reaction, is unique to each biotin-containing carboxylase and, hence, determines specificity. Even so, there are analogies in

A $\text{BCCP-biotin} + HCO_3^- + \text{ATP} \underset{\textit{biotin carboxylase}}{\overset{Mg^{++}}{\rightleftharpoons}} \text{BCCP-biotin-}CO_2 + \text{ADP} + \text{Pi}$

B $\text{BCCP-biotin-}CO_2 + \text{Acetyl-CoA} \underset{\textit{carboxyl transferase}}{\rightleftharpoons} \text{BCCP-biotin} + \text{Malonyl-CoA}$

**Figure 2.2** The overall acetyl-CoA carboxylase reaction.

protein structure and function to other enzymes with multiple active sites for acetyl modification, such as pyruvate dehydrogenase (Knowles, 1989).

An analogous ACCase to the plant multisubunit form is, of course, that of *Escherichia coli*. Like the chloroplast multisubunit ACCase, that from *E. coli* has four subunits, two of which (CT-α, CT-β) belong to the carboxyltransferase (Cronan and Waldrop, 2002). The BCCPs of *E. coli* and soybean align with 39% identical residues; the biotin carboxylase proteins have 54% identity and the CT-β (plastid-encoded in plants) proteins have 42% identity. For the CT-α subunit of the carboxyltransferase, the soybean protein has a long *C*-terminal extension. This extension is of unknown function but often contains repeated sequences suggesting a possible structural role (Cronan and Waldrop, 2002). It can be deleted without loss of carboxyltransferase activity (Kozaki *et al.*, 2001).

Although there is considerable sequence identity between soybean and *E. coli* multisubunit ACCase sequences, it should be noted that most plants have multiple genes (even in Arabidopsis there are two genes for BCCP: Mekhedor *et al.*, 2000) and these show considerable heterogeneity and size. Nevertheless, despite their complexity, plant ACCase proteins contain all those sequences considered essential for *E. coli* ACCase activity (Cronan and Waldrop, 2002). This gives one confidence that the latest advances in our understanding of the reaction mechanism (see Cronan and Waldrop, 2002) can be extrapolated, with some confidence, from *E. coli* to higher plants.

Figure 2.2 summarises the overall ACCase reaction and Figure 2.3 shows the probable mechanism of the reaction as deduced for the maize multifunctional plastidic ACCase (Herbert *et al.*, 1996). In this case, evidence was obtained for an ordered Ter-Ter mechanism. Distinctions between this mechanism (where a quaternary complex is formed between the enzyme and the three substrates) and other possibilities are discussed by Harwood (1996) where relevant references are also cited.

One feature of all plant ACCases that have been studied thus far is that the Km value for the bicarbonate is significantly higher than that for ATP and acetyl-CoA. The latter values are usually around $10^{-4}$ M whereas the Km for bicarbonate is $10^{-3}$ M or more (Harwood, 1988). This is true even for ACCases that have unusual features (Price *et al.*, 2003b).

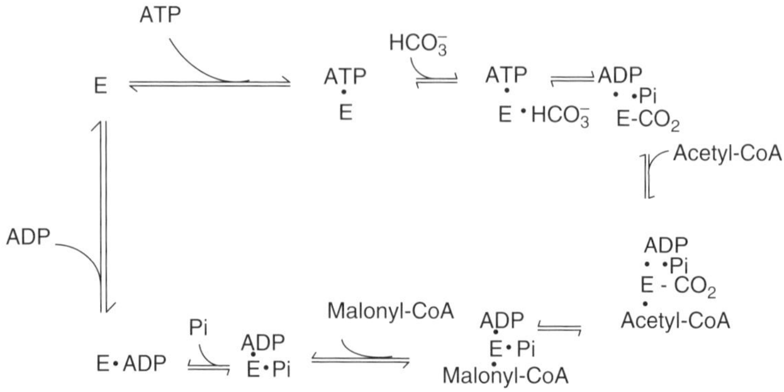


**Figure 2.3** Depiction of a probable ordered Ter-Ter mechanism for maize acetyl-CoA carboxylase (see Herbert *et al*., 1996).

As mentioned in the introduction to Section 2.3, there are (at least) two isoforms for ACCase in plant cells. In addition, for ACCase in general, there are two main structures that are possible (Gurr *et al*., 2002). For bacteria, such as *E. coli*, the enzyme is a multisubunit complex with four separate proteins (Cronan and Waldrop, 2002). This dissociable structure is often referred to as a 'prokaryotic form' even though it can occur in eukaryotic organisms. The alternative main structure is a multifunctional protein where enzymes catalysing the individual partial reactions and the BCCP are domains within the same large molecular mass protein. Such an arrangement is also found in the mammalian ACCases (Gurr *et al*., 2002).

For many years, the nature of the plant ACCase structure was a subject of some controversy. Indeed it is perhaps ironic that a major impetus to solving the problem was the commercial exploitation of the sensitivity of certain plant ACCases to selective herbicides (Section 2.3.3). It is now clear that plants can contain both types of structure. For the 'cytosolic' form of ACCase, all plants (and eukaryotic algae), which have been examined, contain a multifunctional protein of 200–240 kDa. For the plastid isoform, most plants contain a multi-enzyme (multisubunit) complex. This includes indirect evidence from Roesler *et al.* (1996) which suggests that a multisubunit complex may be found in Liliaceae. Exceptions within the dicotyledons appear to be (at least some) members of the Geraniaceae (Christopher and Holtum, 2000) and within the monocotyledons, members of the Graminaceae (grasses) (Lichtenthaler *et al*., 1989). These comments are based on available evidence and may prove to be over-simplifications in the future, when more information becomes available.

The first substantial purification of a multisubunit complex form of ACCase was from pea (Alban *et al*., 1994). This work clearly confirmed the conclusions for a 'prokaryotic form' of the enzyme in dicotyledons made much earlier with spinach

(Kannangara and Stumpf, 1972). Other studies on the chequered history of ACCase purification from different plants and the conflicting results, which were sometimes obtained, are summarised in Harwood (1988, 1996).

### *2.3.2 Properties of different isoforms of ACCase*

The two isoforms of ACCase, which were separated from pea, allowed the first direct experiments of their comparable properties that had been made in plants (Alban *et al.*, 1994). Detailed enzyme kinetics of the plastid multienzyme complex and the cytosolic multifunctional form were reported. The multifunctional form, which was slightly sensitive to graminicide herbicides, could carboxylate propionyl-CoA at significant rates (20% that for acetyl-CoA) whereas the multisubunit complex could not. The high molecular mass pea cytosolic isoform was also shown to contain two graminicide binding sites when in its native form (actually a homodimer). The two binding sites for graminicides seem to be a general feature of all multifunctional forms, which have been examined so far (see Price *et al.*, 2003b). In contrast, the ability of ACCase to carboxylate propionyl-CoA, which was suggested to be relevant to graminicide sensitivity (Dehaye *et al.*, 1994), does not now seem to be so (Price *et al.*, 2003b).

Two multifunctional forms of ACCase were purified substantially from maize leaves (Egli *et al.*, 1993; Herbert *et al.*, 1996). The minor isoform appeared to represent about 15% of the total activity and had the smaller molecular mass (219 kDa on SDS-PAGE). It is presumed to be cytosolic although that has never been confirmed experimentally. The major isoform (227 kDa by SDS-PAGE) is present in the plastids. Both isoforms can carboxylate propionyl-CoA at around 30% of the rate for acetyl-CoA (Herbert *et al.*, 1996) and they have a similar reaction mechanism (Harwood, 1996). Despite their similar sizes and the above properties, antibodies against the major (227 kDa) isoform only reacted weakly against the 219 kDa isoform (Egli *et al.*, 1993). Moreover, the major plastid isoform was inhibited much better by acyl-CoAs than the minor enzyme (Herbert *et al.*, 1996). This could be significant physiologically because the minor isoform might be expected to encounter significant levels of acyl-CoAs in the cytosol.

General methods for the purification of ACCase have been discussed. More recent methods, as utilised for multifunctional ACCases from maize leaves (Herbert *et al.*, 1996) have been applied successfully to a variety of grasses (Price *et al.*, 2003b). In addition, it is possible to use cell cultures as a convenient source of enzyme, when the original plants have very fine leaves and/or grow non-uniformly (Price *et al.*, 2003a).

### *2.3.3 Herbicides acting on ACCase*

Two commercially important groups of selective herbicides have ACCase as their primary target. These are the aryloxyphenoxypropionic acids (AOPPs) and the cyclohexanediones (CHDs) (Figure 2.4). These two classes of herbicide accounted, in

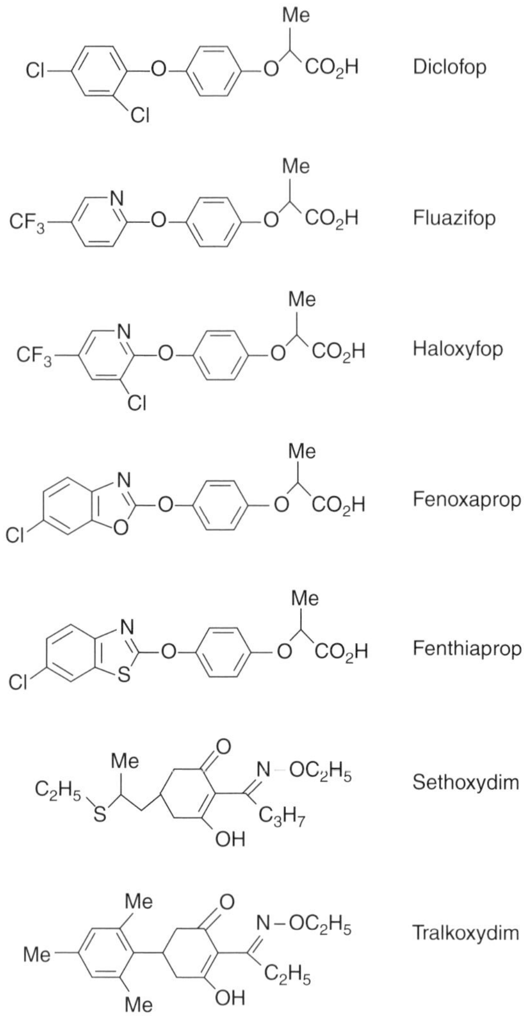


**Figure 2.4** Structures of some important graminicides belonging to the aryloxyphenoxypropionate ('fops') and cyclohexanedione ('dims') classes, which inhibit acetyl-CoA carboxylase.

1998, for about 10% of the global herbicide market (Harwood, 1999). Although the mechanism of action of the AOPPs and CHDs was unclear at first, a variety of evidence suggested that ACCase could be the target site (Burton *et al.*, 1987; Focke and Lichtenthaler, 1987; Rendina and Felts, 1988; Secor and Cseke, 1988). Unambiguous data showed this to be the case (Walker *et al.*, 1988) and further experiments showed that the carboxyltransferase partial reaction was the site of action (Rendina *et al.*, 1989).

Despite their very different structures, the AOPP and CHD herbicides show double inhibition on kinetic analysis. This means that the binding of a member of one herbicide class prevents binding of a herbicide of a second class (Rendina *et al.*, 1989). For maize isoforms, where the extra-plastid isoform was partly sensitive, the inhibition of activity could be fitted to a simple hyperbolic curve. In contrast, the plastid (insensitive) isoform showed co-operativity. Thus, binding of a molecule of herbicide to one subunit of the native dimer of ACCase affected binding of the second herbicide molecule (Herbert *et al.*, 1996). This agreed with previous data for crude ACCase fractions from resistant and sensitive *Lolicum multiflorum* (Evenson *et al.*, 1994). Similarly, when ACCase activity from the inherently resistant grass *Poa annua* was examined, co-operativity for herbicide binding was noted (Herbert *et al.*, 1995). Thus, co-operativity of graminicide binding seemed to be one property that distinguished insensitive ACCase from sensitive enzymes. As a supplement, it had also often been found that the insensitivity of ACCase preparations to graminicides appeared to be correlated also with a smaller activity towards propionyl-CoA (Herbert *et al.*, 1995).

In order to see which of the above properties might be consistently correlated with sensitivity of ACCase to graminicides, Price *et al.* (2003b) purified isoforms of the enzyme from several grasses with either acquired or inherent resistance to the herbicides. They found no correlation between the ability of isoforms to carboxylate propionyl-CoA and their sensitivity to graminicides. However, purified forms of graminicide-insensitive ACCase were characterized by herbicide-binding co-operativity whereas, in contrast, sensitive forms of the enzymes were not. These experiments, together with previous data, showed that herbicide binding co-operativity is the only kinetic property that differentiates naturally occurring or selected insensitive ACCases from the typical sensitive forms usually found in grasses (Price *et al.*, 2003b).

Apart from a few grass species that can metabolise particular graminicides (e.g. wheat and diclofop: Zimmerlin and Durst, 1992), resistance in the whole plant is almost invariably correlated with an insensitive plastid isoform of ACCase (Devine, 1997). Some grasses such as *P. annua* or species of *Festuca* show inherent resistance to at least some AOPPs and CHDs. In addition, acquired resistance is becoming an increasing problem for agriculture (Devine, 1997; Harwood, 1999). Examination of several of the resistant biotypes has shown that acquired resistance is associated with the presence of insensitive forms of the enzyme (Devine, 1997). Several of the graminicide-resistant biotypes have been characterised and reduced ACCase sensitivity shown to be an important resistance mechanism (e.g. Gronwald *et al.*, 1992; Tardif *et al.*, 1993; Price *et al.*, 2004). Inheritance studies have shown that ACCase resistance to both types of graminicide is governed by a single dominant or semi-dominant nuclear gene and that resistance to different herbicides is encoded at the same gene locus (see, e.g. Murray *et al.*, 1996; Tardif *et al.*, 1996).

Work with chimeric genes containing wheat plastid ACCase that complemented the yeast *ACCI* mutation, located the herbicide-sensitivity section in a 400 amino acid fragment (Nikolskaya *et al.*, 1999). By comparison with other sequences this fragment was identified as part of a carboxyltransferase domain, in agreement with the biochemical data of Rendina *et al.* (1989). Further work identified an isoleucine to leucine substitution as being critical for herbicide interaction, although other residues may play a role in defining binding specificity (Zagnitko *et al.*, 2001). The isoleucine to leucine substitution has also been show in acquired-resistance biotypes of blackgrass (Brown *et al.*, 2002).

### *2.3.4 Genes coding for ACCase*

It is, perhaps, ironic that the first substantial sequence information for plant ACCases came for the high-molecular-mass multifunctional form. Thus, primary structures of multifunctional isoforms were deduced from full-length nucleotide sequences for the alga *Cyclotella cryptica* (Roessler and Ohlrogge, 1993), wheat (Gornicki *et al.*, 1994) and alfalfa (Shorrosh *et al.*, 1994).

The multienzyme form of ACCase, as exemplified by the plastid enzyme from dicotyledons, contains four proteins. In the early 1990s, a single biotinylated polypeptide was detected in peas (Baldet *et al.*, 1993; Sasaki *et al.*, 1993; Alban *et al.*, 1994) and assumed to be the biotinylated subunit of BCCP. At this time, a pea homologue of the *E. coli* gene that encodes the β-subunit of the carboxyltransferase was studied and biochemical evidence for its identity provided (Sasaki *et al.*, 1993).

A tobacco cDNA which encoded a chloroplast peptide with good sequence homology (52–65%) to the biotin carboxylases of ACCases from prokaryotes was studied by Shorrosh *et al.* (1995). Further proof of its identity was provided by Roesler *et al.* (1996) and Shorrosh *et al.* (1996). In addition, antibodies to the castor bean biotin carboxylase cross-reacted to proteins in a number of dicotyledons and in the monocotyledons, day lily, iris and onion. Interestingly, there was no cross-reactivity of the antibodies detected in members of the Gramineae (barley and switchgrass).

In view of the strong evidence that ACCase activity is important in controlling the flux of carbon to fatty acid synthesis in leaves (Post-Beittenmiller *et al.*, 1992; Page *et al.*, 1994), it was of interest to modify expression of the component proteins of this enzyme. When biotin carboxylase expression levels were changed in tobacco leaves, plants with less than 25% of the wild-type biotin carboxylase levels showed severely retarded growth under low light. Furthermore, such plants also had a 26% lower fatty acid content compared to wild-type (Shintani *et al.*, 1997). Of interest also (see later in this chapter) was that manipulation of biotin carboxylase expression did not change that of BCCP.

For the model plant, *Arabidopsis thaliana*, the BCCP gene was identified and characterised by Nikolau's group (Choi *et al.*, 1995; Ke *et al.*, 1997), while two labs characterised the *CAC2* gene, which encodes the biotin carboxylase (Bao *et al.*, 1997; Sun *et al.*, 1997). Later the *CAC3* gene, which encodes the α-subunit of carboxyltransferase, and the plastidic *accD* gene, which encodes the β-subunit, were isolated and characterised (Ke *et al.*, 2000). These studies allowed further experiments on the coordination of expression of the constituent proteins of the multienzyme form of ACCase. It was found that all four subunit mRNAs accumulated at a constant molar ratio, implying that the nuclear and plastid genomes were able to communicate (Ke *et al.*, 2000). This contrasted with the differing protein levels for BCCP and biotin carboxylase as reported in transgenic tobacco (Shintani *et al.*, 1997) and also to Arabidopsis protein levels in transgenic plants with different amounts of BCCP2 expression (Thelen and Ohlrogge, 2002a). Comparison of the $\alpha$-subunit of Arabidopsis carboxyltransferase with that from *E. coli* showed that the plant protein had an additional *C*-terminal domain that was suggested to have a structural function (Ke *et al.*, 2000; see Cronan and Waldrop, 2002).

In Arabidopsis, three subunits of multisubunit ACCase are encoded by single genes (Mekhedor *et al.*, 2000) whereas two cDNAs encoding BCCP have been found. The BCCP isoform 1 is constitutively expressed and was particularly abundant in developing leaves and seeds (Thelen *et al.*, 2000, 2001). In contrast, the BCCP isoform 2 was predominantly expressed in flowers and developing seeds, particularly 8–12 days after flowering. As expected, antisense-mediated down regulation of BCCP 2 (reduced by an average of 38%) resulted in a reduction (average 7%) in the fatty acid content of seeds (Thelen and Ohlrogge, 2002a). What was surprising was that over-expression of BCCP 2 also resulted in a reduction in seed fatty acid content. In addition, reduced seed setting was often seen and there was a 65% reduction in ACCase activity. These results were attributed to assembly of non-biotinylated ACCase complexes, which were, of course, not active (Thelen and Ohlrogge, 2002a). However, the data in general were consistent with an important role for ACCase in regulating fatty acid (and lipid) synthesis (Harwood, 1996).

The single subunit, multifunctional form of ACCase in dicotyledons is (as mentioned before) viewed as a cytoplasmic enzyme. However, in oilseed rape, this structural isoform is also found in the plastids, in addition to the multisubunit form of ACCase (Roesler *et al.*, 1997; Schulte *et al.*, 1997). In antisense lines of oilseed rape, activity of the multifunctional isoform (as measured via its associated propionyl-CoA carboxylase) was reduced about 50% and this was associated with wrinkled seeds and a lower lipid content (65% and 86% wild-type in two homozygous lines) (Sellwood *et al.*, 2000). Again this provides evidence for an important role for ACCase in controlling carbon flux to lipids.

Of course, the multisubunit form of ACCase must be able to associate easily. In fact, in studies with the two subunits of the carboxyltransferase, a tight association had already been shown in immunoprecipitation experiments (Ke *et al.*, 2000). Recently, it has been shown that the whole multisubunit ACCase is strongly associated with the chloroplast envelope through non-ionic interactions to the carboxyltransferase units (Thelen and Ohlrogge, 2002b). In lysed pea chloroplasts, at least 60% of the ACCase subunits were associated with the membrane fraction. The association seemed to be through Triton-X100-resistant domains (possibly similar to the lipid rafts in animal plasma membranes that may also be present in plants

(Peskan *et al.*, 2000)) and it was proposed that an unidentified integral membrane protein was involved (Thelen and Ohlrogge, 2002b). The association of ACCase with membranes was, in fact, first suggested over 30 years ago (Kannangara and Stumpf, 1972) but, because of variable amounts of association found by others (see Thelen and Ohlrogge, 2002b), were not considered seriously. Such an association might well have an important functional significance because of the suggestion for substrate channelling in plastid fatty acid synthesis (Roughan and Ohlrogge, 1996).

### *2.3.5 Regulation of ACCase*

Acetyl-CoA carboxylase catalyses the first committed step in fatty acid synthesis. Therefore, intuitively, it would be appear to be a good enzyme to exert strong flux control over the pathway. Early evidence for this hypothesis was mainly circumstantial but was substantial, as summarised by Harwood (1988). Two pieces of research gave firm evidence. First, Post-Beittenmiller and associates carefully measured the pools of acyl-thioesters following substantial changes in fatty acid formation. These experiments, using light-exposure of leaves, showed that, while levels of malonyl-thioesters were now detectable on shifting from dark to light, the pool size of acetyl-CoA remained stable (Post-Beittenmiller *et al.*, 1992). Further stimulation of fatty acid synthesis with Triton-X100 caused the acetyl-CoA pool to decline. These data provided *in situ* experimental evidence for a regulatory role for ACCase in fatty acid synthesis (see Harwood, 1996; Ohlrogge and Jaworski, 1997).

A quantitative measure of the importance of an individual enzyme in a metabolic pathway can be made using metabolic control theory. For a metabolic pathway, control is shared by all the various steps in a pathway and there is no such thing as a single 'rate-limiting' enzyme (Fell, 1997). The importance of an individual enzyme is measured (as its flux control coefficient) by manipulating its activity and measuring the effect on the overall pathway. The availability of specific inhibitors for ACCase in grasses (i.e. the graminicides) allowed us to measure flux control coefficients. This showed (Table 2.1) that ACCase exerted 45–61% of the total control of flux to lipids, under the conditions of the experiments (Page *et al.*, 1994),

**Table 2.1** Apparent flux control coefficients for grass leaf acetyl-CoA carboxylases during lipid synthesis in the light (Page *et al.*, 1994).

| Inhibitor | Plant used | |
|---|---|---|
| | Barley | Maize |
| Fluazifop | $0.61 \pm 0.05$ | $0.59 \pm 0.16$ |
| Sethoxydim | $0.54 \pm 0.17$ | $0.45 \pm 0.13$ |

*Note:* These data show that 45–61% of the total control of a pathway utilising at least 21 enzyme resides at the level of acetyl-CoA carboxylase. See Page *et al.* (1994) for experimental details and the theory behind the calculations.

i.e. it was a very important enzyme in the pathway. These and other experiments are discussed further by Ohlrogge and Jaworski (1997).

Since ACCase is so important in controlling fatty acid (and lipid) formation in, at least, some circumstances, one can immediately ask; how does this regulation occur? A number of mechanisms have been proposed. These include changes in the stromal properties (pH, ATP conc., etc.), in acyl-thioester concentrations and in CoA levels. The experiments generating such data have been discussed in detail (Harwood, 1988, 1996; Ohlrogge and Jaworski, 1997).

More recently, further evidence has accumulated for post-translational regulation of ACCase. Reversible phosphorylation associated with high ATP levels (Savage and Ohlrogge, 1999) and redox modulation (Sasaki *et al.*, 1997) would both be caused indirectly by light-stimulated photosynthesis (Nikolau and Hawke, 1984). Another form of regulation would be by feedback regulation with long-chain acyl-CoAs (such as can occur in mammals: see Gurr *et al.*, 2002). Extra-chloroplastic fatty acyl-CoAs have been shown to decrease fatty acid synthesis significantly (Shintani and Ohlrogge, 1995) and the recent demonstrated association of ACCase with the plastid envelope (Thelen and Ohlrogge, 2002b) may provide a mechanism. Such an association might allow the demand for cytosolic fatty acids to be perceived.

## 2.4 Fatty acid synthase

Together with ACCase, fatty acid synthase (FAS) is the second major enzyme complex involved in *de novo* fatty acid formation. The plant FAS is a type II dissociable multiprotein complex, like the *E. coli* system but unlike that of animals, i.e. the individual proteins that make up FAS can be isolated and their function demonstrated separately. The overall reactions involved in fatty acid biosynthesis are depicted in Figure 2.5. Two acyl transferases are shown. Malonyl-CoA: ACP transacylase is involved in generating the all-important source of 2C units, malonyl-ACP. This intermediate is used by all of the condensing reactions for *de novo* synthesis. A second acyl transferase is acetyl-CoA: ACP transacylase but its physiological relevance is in some doubt since it was discovered that the first condensation reaction used acetyl-CoA directly, rather than acetyl-ACP (see Harwood, 1996). There are three different β-ketoacyl-ACP synthases (KASs; condensing enzymes). The first one used is KAS III and is equivalent to the FabH protein of *E. coli*. It condenses acetyl-CoA with malonyl-ACP to give a 4C keto-intermediate. Successive reduction, dehydration and a second reduction then produce a 4C fatty acid, butyrate, which is attached to ACP. Although KAS III can probably catalyse further condensation reactions, it is considered that KAS I (equivalent to FabB in *E. coli*) is used to produce 6–16C fatty acids. The final condensation for producing stearate, uses KAS II (equivalent to FabF in *E. coli*).

The remaining enzymes of FAS, are β-ketoacyl-ACP reductase, β-hydroxylacyl-ACP dehydrase and enoyl-ACP reductase. Although preparations of the two

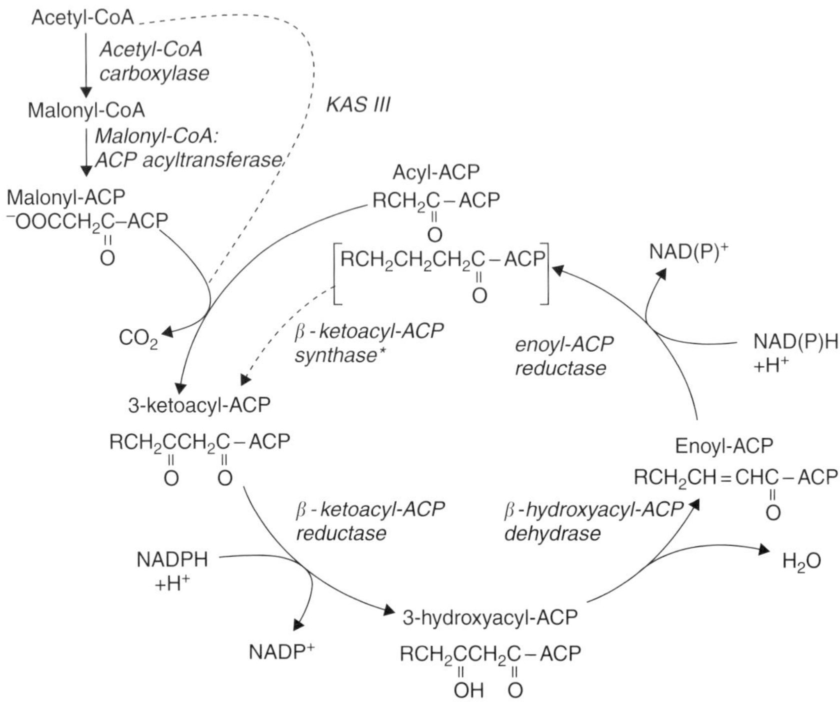


**Figure 2.5** The overall reaction of fatty acid synthase. * The first condensation reaction is catalysed by β-ketoacyl-ACP synthase III (KAS III) that uses acetyl-CoA and malonyl-ACP substrates. The next six condensations are catalysed by KAS I and the final reaction between palmitoyl-ACP and malonyl-ACP uses KAS II.

reductases from different plants show varying nucleotide selectivity, there seems to be only one gene for each in the Arabidopsis genome (Mekhedor *et al.*, 2000). By contrast, two genes seem to be present for the β-hydroxyacyl-ACP dehydrase.

Completion of *de novo* fatty acid synthesis is made in one of three ways. Either the product of acyl-ACP is hydrolysed by a thioesterase, the palmitate or stearate products are desaturated or acyl-ACPs are used directly for complex lipid formation using plastid acyltransferases. The action of the thioesterases is particularly important because, through their action, non-esterified fatty acids are made available for conversion to acyl-CoAs. This conversion destines the acyl chains for assembly into complex lipids in the extra-plastid compartment (see Figure 2.11).

### 2.4.1 *Acyl carrier protein*

Acyl carrier protein (ACP) was the first protein of plant lipid metabolism to be purified to homogeneity (Simoni *et al.*, 1967) and also the first such gene to be cloned (Ohlrogge *et al.*, 1991). Like its counterpart in *E. coli*, it is a small molecular mass protein with many charged residues. Early evidence from protein purification suggested that isoforms were present (Harwood, 1996) and ten independent cDNA clones were isolated from an oilseed rape embryo library (Safford *et al.*, 1988). Further work showed that at least three different mature coding sequences could be demonstrated. Likewise, in barley, three ACPs were found in early work (Hoj and Svendsen, 1984) and cDNA clones isolated (Hansen, 1987). This research has been thoroughly reviewed (Ohlrogge, 1987; Harwood, 1996).

One feature of the search for isoforms of ACP was the discovery of their presence in mitochondria, in addition to their expected location in plastids. The ACP in *Neurospora* (Brody and Mikolajczyk, 1988) was shown to be capable of supporting *de novo* fatty acid synthesis (Mikolajczyk and Brody, 1990; see Section 2.4.5). It was suggested that a mitochondrial FAS system might be necessary for the formation of mitochondrially-located lipids, such as the specific component, cardiolipin. In partial support of this suggestion, was the incorporation of radioactivity from [2-$^{14}$C]malonate into myristoyl moieties of cardiolipin (Brody *et al.*, 1990). There were also various suggestions that ACP was needed for different functions in connection with the electron chain (see Shintani and Ohlrogge, 1994). The most accepted suggestion (Section 2.4.5) is now that ACP (and fatty acid synthesis) is needed for the production of mitochondrial lipoic acid (Wada *et al.*, 1997).

So far as the Arabidopsis genome is concerned, three genes for mitochondrial ACP and five for plastidial ACP have been deduced (Mekhedov *et al.*, 2000). Based on protein analysis ACP 1, ACP 2 and ACP 3 (all plastid forms) have been identified in all tissues examined (Hlousek-Radojcic *et al.*, 1992) whereas ACP 4 is expressed mainly in Arabidopsis leaves (see Bonaventure and Ohlrogge, 2002). The latter was the most abundant isoform in leaves and its expression was induced several-fold by light. In contrast, mRNA levels for ACP 2 and ACP 3 were unchanged by light. Consistent with the quantitative importance of ACP 4 in leaves, transgenic Arabidopsis (where this isoform was reduced severely) showed a bleached phenotype, a lowered photosynthetic efficiency and a 22–42% reduction in leaf lipids on a fresh weight basis (Branen *et al.*, 2003).

Bonaventure and Ohlrogge (2002) identified untranslated regions for the ACP mRNAs, which seemed important during expression. For ACP 4, a GATA-like motif was found in the upstream region, which was similar to other genes that underwent light-induced induction. For ACP 1 and ACP 2, 5′-UTR regions favoured expression in seeds and roots, respectively. There were also two boxes of conserved regions in the UTR sequences and when these were deleted for ACP 1, there was 10–20-fold less gene expression in transgenic plants.

Other aspects of ACP, including gene structure and expression, properties of acyl-ACPs made with different ACP isoforms and the generation of holo-ACP from apo-ACP are covered in Harwood (1996).

### *2.4.2 Condensing enzymes*

As discussed in the introductory section, there are three different condensing enzymes (KAS I, KAS II, KAS III) that are found generally in plants. In *Cuphea*, a fourth condensing enzyme, termed KAS IV, has also been identified as a medium-chain specific enzyme (Dehesh *et al.*, 1998; Slabaugh *et al.*, 1998).

KAS I extends the fatty acid chain length from 4C to 16C during six rounds of 2C addition. Classically, the enzyme was distinguished from KAS II by its relative insensitivity to arsenite (Harwood and Stumpf, 1971). Later, KAS I was shown to be more sensitive to cerulenin than KAS II (see Jaworski, 1987). KAS I has been purified from several plants and its properties studied (Harwood, 1996). The enzyme from barley chloroplasts was purified as a series of dimers with αα, αβ and ββ arrangements (Siggard-Anderson *et al.*, 1991). Probes generated from sequence information from the β-subunit were used to identify a cDNA with an open reading frame coding for a protein of 462 residues plus a 35-amino acid transit peptide. This showed considerable sequence homology to the *fabB*-encoded KAS I from *E. coli*. The gene encoding the β2 isozyme of barley KAS I has been characterised in some detail (Kauppinen, 1992) and sequences have been reported for other plant KAS I-encoding cDNAs (see Harwood, 1996).

The similarity of the plant KAS I with the *E. coli* condensing enzyme has allowed crystal structures of the latter (Olsen *et al.*, 1999) to be utilised in order to deduce reaction mechanisms (Olsen *et al.*, 2001). The new carbon–carbon bonds are created in a Claisen condensation performed in three steps (Figure 2.6). KAS I was also shown to belong to the thiolase-fold group of enzymes. The KAS I dimer was not changed by formation of complexes but there was an asymmetric binding of substrates. Two histidines were needed for decarboxylation; one donates

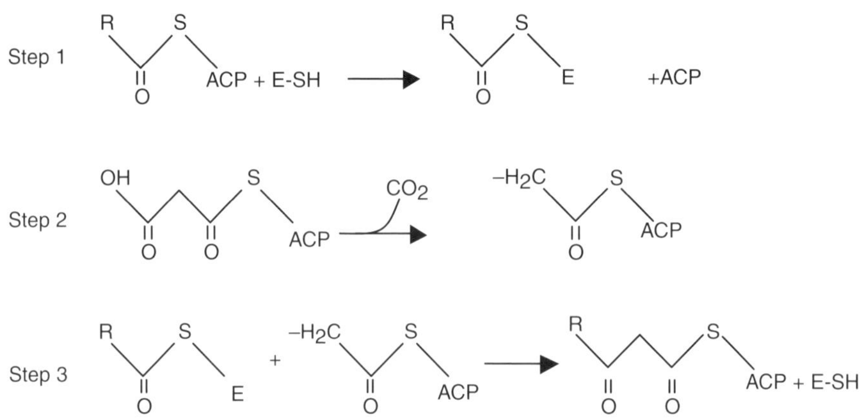


**Figure 2.6** The three-step mechanism characterising the carboxylating Claisen condensing enzymes (see Olsen *et al.*, 2001).

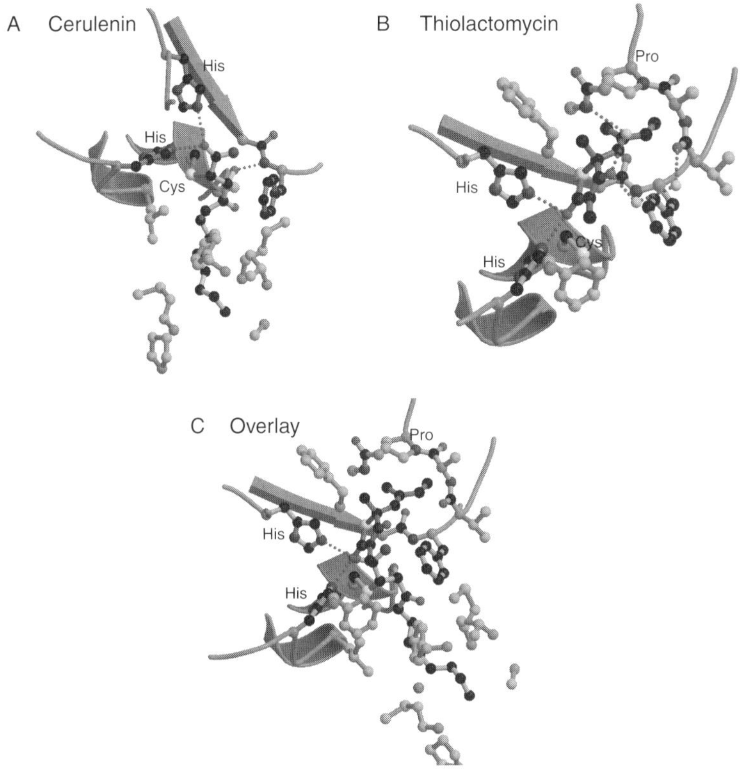


**Figure 2.7** Mechanism of cerulenin and thiolactomycin inhibition of the fatty acid synthase condensing enzyme FabB of *E. coli*. A. Cerulenin located in the active site of FabB. O1 of cerulenin is covalently attached to the active site Cys163. O2 hydrogen bonds to His198 and His333, while O3 occupies the oxyanion hole formed by the backbone amides of Phe392 and Cys163. B. Thiolactomycin in the active site of FabB. O3 of thiolactomycin hydrogen bonds to His298 and His333. The binding is stabilized by the intercalation of the isoprenoid tail in the space between Pro272 and the peptide bond between Gly391 and Phe392. C. Superimposition of cerulenin and thiolactomycin in the FabB active site. Cerulenin binds to the acyl-enzyme intermediate half of the active site, while thiolactomycin binds to the malonyl-ACP side. Only O2 of cerulenin and O1 of thiolactomycin overlap hydrogen bonded to the His dyad. Reproduced from Heath *et al.* (2001) with kind permission from the authors and Elsevier. This figure is reproduced in colour in the colour plate section.

a hydrogen bond to the malonyl-thioester oxo group and the other abstracts a proton from the leaving group (Olsen *et al.*, 2001). The reaction mechanism and the binding/inhibition by cerulenin or thiolactomycin (Figure 2.7) are discussed further by Heath *et al.* (2001) where a full discussion of other inhibitors of Type II fatty acid synthases will be found also.

KAS II, which catalyses the elongation of palmitate to stearate and, hence, serves to regulate the ratio of 18C/16C products is sensitive to arsenite (Harwood and

Stumpf, 1971) and less so to cerulenin (Jaworski, 1987). Indeed, KAS II expression has been modified to change levels of palmitate (see Harwood, 1996) and an Arabidopsis mutant *fab1* has been identified with a reduced ability to elongate palmitate (Wu *et al.*, 1994). The *E. coli* KAS II equivalent enzyme (FabB) was expressed and crystallised (Huang *et al.*, 1998). It has very similar features to the KAS I enzyme. Each has a Cys-His-His active site triad and are proposed to have similar reaction mechanisms (Olsen *et al.*, 2001). Moreover, the acyl binding pockets of KAS I and KAS II are so similar that they alone do not seem able to provide a basis for the quite different substrate specificities.

In Arabidopsis, 69% of the palmitate is elongated to stearate (Browse *et al.*, 1986), resulting in a final ratio for 18C/16C acids of 2.4 in wild-type (Kunst *et al.*, 1988). The Arabidopsis *fab 1* mutant referred to above and which is partly deficient in KAS II activity (Wu *et al.*, 1994) has been studied with complementation experiments. *Fab 1* plants were transformed with a cDNA from *Brassica napus* encoding the KAS II enzyme and complementation resulted in a restoration of leaf fatty acid composition to near wild-type levels. The mutant phenotype tissue damage and/or death at low temperature were also eliminated. The *fab 1* mutation was shown to be a single nucleotide change that resulted in a Leu 337 Phe substitution. An equivalent substitution in *E. coli* KAS II caused a 10-fold decrease in substrate affinity and severely impaired its condensation activity (Carlsson *et al.*, 2002).

Plant KAS enzymes were also used recently to try to complement *E. coli mutants*, like CY244, that were deficient in fatty acid biosynthesis at 42°C (Rasmussen *et al.*, 2003). Unfortunately, although expression of *E. coli* KAS enzymes could complement the *E. coli* CY244 mutant, plant cDNAs did not successfully complement the same mutant, probably because of solubility problems during their expression (Rasmussen *et al.*, 2003).

An unusual seed-specific KAS enzyme has been found in coriander. Coriander seeds accumulate petroselenic acid (*cis* Δ6-18:1), which is made, rather surprisingly, via Δ4-desaturation followed by elongation of 16:1 to 18:1 (Cahoon and Ohlrogge, 1994). A divergent form of KAS I was isolated from a cDNA library prepared from developing coriander seeds but no KAS II-type cDNAs could be detected. Multiple alignment of the derived amino acid sequences revealed that, although the divergent KAS was closely related to KAS I, its active site motif showed several substitutions to those found in either KAS I or KAS II (Mekhedor *et al.*, 2001). The divergent KAS was specifically expressed in coriander seeds but not in leaves. In agreement that the divergent KAS was a KAS I- and not a KAS II-type enzyme, the elongation of Δ4-16:1 to Δ6-18:1 in seed extracts was strongly inhibited by cerulenin (Mekhedor *et al.*, 2001).

Another unusual KAS has been reported in leek sheath epidermis. This enzyme, in contrast to that located in leek parenchyma and lamina epidermis tissues, was able to synthesise products of up to 22C. It had all the properties of plant KAS II enzymes (moderately sensitive to cerulenin, used acyl-ACP substrates) except for its substrate selectivity. It was suggested that the sheath KAS II could play a particular role in wax biosynthesis (Domergue and Post-Beittenmiller, 2000; see also Chapter 8 on plant cuticular waxes).

Voelker and Kinney (2001) have pointed out some difficulties with the plant KAS nomenclature at the present time. They point out that the subunits of KAS I and KAS II dimers are encoded by two closely related plastid genes, named *KAS A* and *KAS B*. They suggest that KAS I is a homodimer with a subunit mass of 46 kDa that is encoded by the *KAS B* gene. The KAS II enzyme may be a heterodimer containing a subunit encoded by the *KAS B* gene and another (of 50 kDa) encoded by the *KAS A* gene. These assignments should be regarded as tentative at present but, nevertheless, are a good working model.

KAS III, also known as short-chain condensing enzyme, catalyses the initial condensation reaction of plant FAS (Figure 2.5). The enzyme, which uses acetyl-CoA rather than acetyl-ACP as its primer, was reported initially in *E. coli* FAS (Jackowski and Rock, 1987). Jaworski *et al.* (1989) detected the enzyme in spinach and it has been found in all plants examined (Walsh *et al.*, 1990). The enzyme has been purified from spinach (Clough *et al.*, 1992), avocado (Gulliver and Slabas, 1994) and pea (Jones *et al.*, 2003). When the corresponding gene was cloned from spinach (Jaworski *et al.*, 1994), its derived protein sequence was found to be similar to other KAS IIIs but distinct from the KAS I and KAS II enzymes. The KAS III is insensitive to cerulenin (which was how it was discovered: Jaworski and Rock, 1987) and is the least sensitive of the condensing enzyme isoforms to thiolactomycin and its derivatives (Jones *et al.*, 1995b, 2003).

A crystal structure for *E. coli* KAS III was deduced by Qui *et al.* (1999) and Davies *et al.* (2000). The active site of the enzyme, in contrast to KAS I and KAS II is characterised by a Cys-His-Asn triad. In fact, the plant KAS IIIs all show good conservation of sequence around these three residues. In addition, another stretch of about 20 amino acids was almost completely conserved in plant KAS IIIs. By deriving a structure for the pea KAS III (Figure 2.8) using crystal structures from other KAS enzymes, Jones *et al.* (2003) suggested that this 20-residue motif fitted over the active site. The motif contains segments that have been suggested to be involved in substrate binding and in catalysis.

Studies carried out by Jaworski *et al.* (1989) showed that KAS III activity matched total fatty acid synthesis. This was of interest because KAS III preparations contain acetyl-CoA: ACP acyltransferase (ACAT) activity (Jaworski and Rock, 1987; Clough *et al.*, 1992; Jones *et al.*, 2003) and this latter reaction was suggested to be limiting for fatty acid synthesis (Shimakata and Stumpf, 1982). However, with the discovery that KAS III used acetyl-CoA for condensation, the ACAT reaction became strictly redundant. Nevertheless, an ACAT has been purified from *E. coli* (Lowe and Rhodes, 1988) and ACATs have been separated from KAS III in avocado Gulliver and Slabas (1994) and pea (Jones *et al.*, 2003). It may be that ACAT allows the generation of acetyl-ACP which could be (poorly) used by KAS I and, hence, generate plasticity in the fatty acid synthase system.

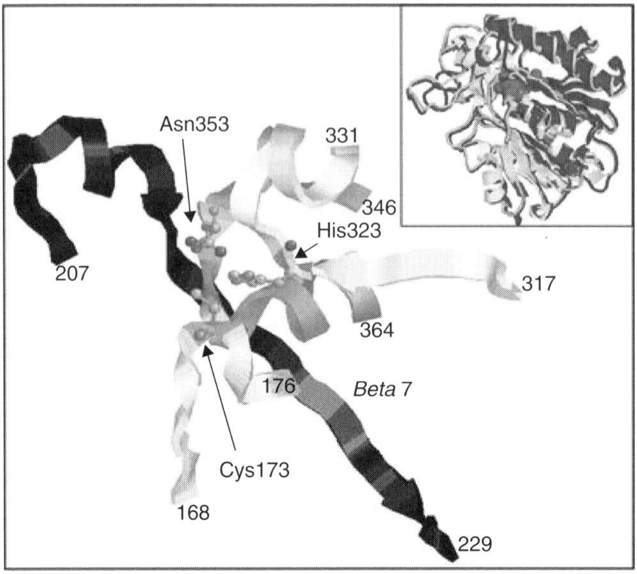


**Figure 2.8** The active-site region of the pea KAS III. This shows the three key residues and their amino acid sequences, which are completely conserved among the known plant KAS IIIs. The sequence $Val^{207}$ to $Ala^{229}$ in the pea sequence, which is almost completely conserved in other plant KAS IIIs is also shown lying across the active site. The model is based on the three-dimensional structure of the *E. coli* KAS III (Davies *et al.*, 2000) and shows the preferred conformations of a α-helix, β-sheet and β-turns. Amino acid numbering is derived from the translated pea transcript accession number CAC08184. For details of the sequences shown, see Jones *et al.* (2003). Reproduced from Jones *et al.* (2003) with kind permission from the authors and Springer-Verlag. This figure is reproduced in colour in the colour plate section.

Expression of cDNAs encoding KAS III showed that the enzyme was ubiquitous (Dehesh *et al.*, 2001) as had been deduced from activity measurements (Walsh *et al.*, 1990). Overexpression of the enzyme resulted in changes in palmitate synthesis, indicating a universal role in plant fatty acid synthesis, regardless of the species or the tissue examined (Dehesh *et al.*, 2001). Moreover, a change induced in fatty acid products was also found when *B. napus* was modified by the introduction of the *E. coli fabH* gene, which encodes a KAS III-equivalent enzyme (Verwoert *et al.*, 1995).

The lack of build-up of intermediates during fatty acid synthesis suggests that the condensing enzymes may be important for controlling overall flux. However, experiments to demonstrate this for KAS III have not been very convincing. Thus, if KAS III is overexpressed up to 40-fold, butyryl-ACP builds up but there is no increase in total fatty acid synthesis. The build-up in butyryl-ACP suggested that KAS I had now become limiting (Ohlrogge and Jaworski, 1997). Nevertheless, as discussed earlier, the condensing enzyme, KAS II, clearly has an important role in ensuring adequate 18C fatty acid supply and, therefore, influences fatty acid quality.

### *2.4.3 The other component enzymes of FAS*

After condensation, the β-ketoacyl-ACP intermediate is reduced. Although isoforms of β-ketoacyl-ACP reductase, which differ in their nucleotide selectivity, have been reported (see Harwood, 1996), it is usually regarded as a NADPH enzyme. The enzyme has been isolated and purified from several plant tissues and the monomer from *B. napus* has a molecular mass of 28 kDa (Slabas *et al.*, 1992). However, it is thought to function as a tetramer *in vivo*. cDNAs have been isolated from a number of plants and show 55–81% identity for their derived sequences (see Harwood, 1996). The Arabidopsis library shows only one gene for β-ketoacyl-ACP reductase so far identified (Mekhedov *et al.*, 2000).

The NADP-dependent β-ketoacyl-ACP reductase from oilseed rape was crystallised (Fisher *et al.*, 2000a) and features of its substrate binding and catalysis deduced (Fisher *et al.*, 2000b). The structure revealed a homotetramer (see above) in which each subunit had a classic dinucleotide-binding fold. In keeping with the sequence similarity of the β-ketoacyl-ACP reductase with the enoyl-ACP reductase of FAS (Figure 2.9) and the short-chain alcohol dehydrogenase (SDR) family, a catalytic triad of Ser-Tyr-Lys was found. Modelling of the substrate into the active site indicated the need for conformational changes in the enzyme (Fisher *et al.*, 2000b).

The *B. napus* β-ketoacyl-ACP reductase was down-regulated using antisense in order to study effects on seed lipid accumulation or leaf membrane biosynthesis (O'Hara *et al.*, 2000). As part of the study, these workers developed a new assay and examined the relationship between the activity of the reductase, leaf expansion and overall fatty acid synthesis (O'Hara *et al.*, 2001). Only during the rapid phase of leaf expansion (days 4–7 after emergence) was it thought that β-ketoacyl-ACP reductase could contribute to any significant effect on total fatty acid synthesis rates. At other times, the major control of flux appeared to reside elsewhere in the pathway. Nevertheless, seeds gathered from antisensed *Brassica* lines showed some distinct morphologies and a reduced lipid content (O'Hara *et al.*, 2000).

The next enzyme in the cycle of 2C additions (Figure 2.5) is the β-hydroxyacyl-ACP dehydrase. This enzyme was purified from spinach leaves by Shimakata and Stumpf (1982). It is encoded by two genes in Arabidopsis (Mekhedor *et al.*, 2000) in contrast to the two reductases of FAS, each of which is encoded by a single gene. The dehydrase is specific for the D(-) stereoisomer of the substrate.

The second reductase of the FAS cycle is enoyl-ACP reductase (Figure 2.5). As discussed by Harwood (1988, 1996), enoyl-ACP reductase can show absolute nucleotide selectivity (NADH-requiring) or be able to utilise NADPH, depending on the plant source. The best-studied enzyme is that from *B. napus* where a purified preparation was shown, like the β-ketoacyl-ACP reductase, to be functional as a tetramer (Slabas *et al.*, 1986). The subunit molecular mass was 35 kDa (Slabas *et al.*, 1990). Covalent modification of arginine residues with phenylglyoxal resulted in inhibition of enoyl reductase activity and this has been studied in some detail

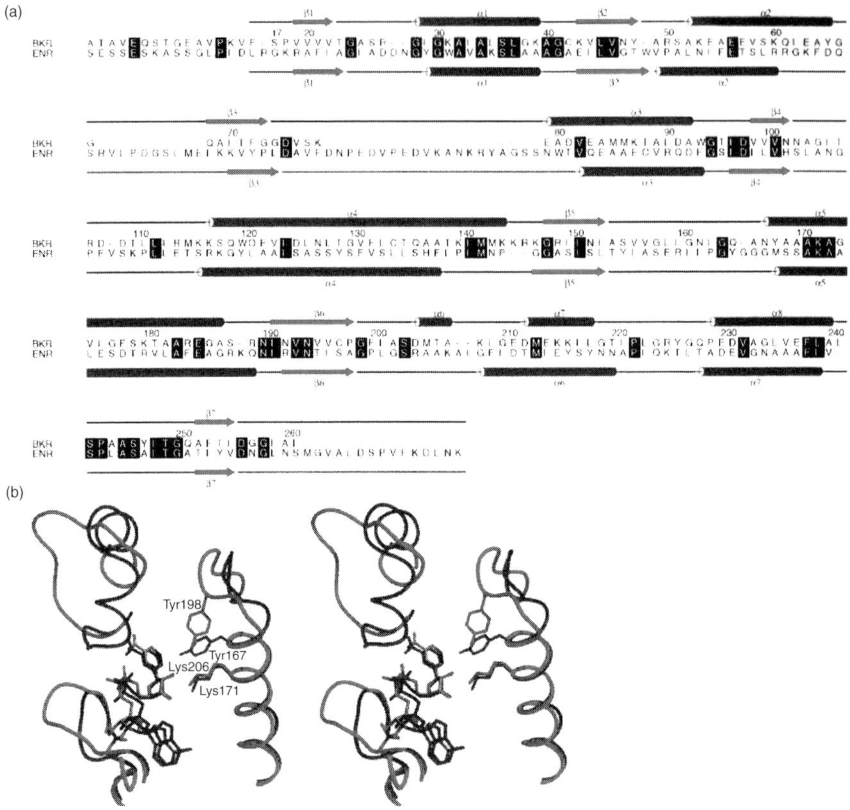


**Figure 2.9** Sequence and structural alignments of *Brassica napus* β-ketoacyl-ACP reductase (BKR) and enoyl-ACP reductase (ENR). A. The aligned sequences of *B. napus* BKR and ENR with the conserved residues shown in reverse typeface. The secondary structure elements of BKR and ENR are shown above and below the alignment, respectively; α helices are blue and β strands red. B. Stereo view superimposition of the Cα backbones of the active sites of *B. napus* BKR (green) and ENR (red). The catalytic tyrosine and lysine residues and bound nucleotides are shown in stick format. Reproduced from Fisher *et al.* (2000b) with kind permission from the authors and Elsevier. This figure is reproduced in colour in the colour plate section.

(Cottingham *et al.*, 1989). A cDNA was cloned by screening a seed-specific cDNA library with an antibody raised against the purified rapeseed protein (Kater *et al.*, 1991). Examination of the temporal expression during seed development showed an increase in steady state mRNA levels prior to lipid storage synthesis. Although *B. napus* is an allotetraploid and contains two pairs of enoyl reductase-encoding genes, neither of these is expressed specifically in leaves or seeds (Fawcett *et al.*, 1994). In other plants like *Petunia hybrida* or *A. thaliana*, there is a single gene for enoyl reductase while *Nicotiana tabacum* (which like rape is an allotetraploid) contains two (de Boer *et al.*, 1998). As enoyl reductase can be regarded as a housekeeping gene, increase in its activity during times of high fatty acid demand (e.g. during seed maturation) requires careful modification of its expression levels. Because of this, a detailed study has been made of flanking regions and the positions and sequences of introns and exons in different enoyl reductase genes (de Boer *et al.*, 1998).

The mechanism of reaction of enoyl reductase has been studied using crotonyl-ACP as substrate and MALDI-TOF analysis. The reaction was shown to form a compulsory-ordered ternary complex with NADH binding before binding of the acyl substrate. The authors also noted that the use of chemically synthesised crotonyl-ACP could complicate interpretation/structural aspects of the enoyl reductase–crotonyl-ACP complex because of additional crotonylations of the ACP in such cases (Fawcett *et al.*, 2000).

Enoyl reductase has been identified as a target for a metabolite of isoniazid, an important drug for the treatment of tuberculosis. It should also be noted that isoniazid may (also) act on β-ketoacyl-ACP synthase and, according to Mdluli *et al.* (1998), this may be its primary mode of action. Accordingly, there has been an interest in crystallising enoyl reductase from Type II dissociable FAS enzymes in order to study structure–function relationships. The enoyl reductase from *E. coli* was crystallised and interactions with thienodiazoborine or benzodiazeborine studied (Baldock *et al.*, 1996). Further structural studies down to 2.1-Å resolution were able to reveal features of the nucleotide binding and conserved tyrosine and lysine residues important for catalytic activity. In addition, a hydrophobic region on the protein surface, adjacent to the bound nicotinamide ring was thought to be the binding site for the fatty acyl substrate (Baldock *et al.*, 1998).

The widely used biocide, triclosan, has been suggested to block lipid synthesis in *E. coli* by specifically inhibiting its enoyl reductase (McMurray *et al.*, 1998). To provide further evidence that triclosan can act as an enoyl reductase inhibitor, Levy *et al.* (1999) crystallised the enzyme in a complex with triclosan and $NAD^+$. Further work on the crystal structure, including the use of crystals of *B. napus* enoyl reductase, revealed how triclosan could act as a site-directed picomolar inhibitor of the enzyme by mimicking its substrate. Elements of both the protein and the nucleotide cofactor appeared to play important roles in triclosan recognition (Roujeinikova *et al.*, 1999).

Another compound that can inhibit plant enoyl reductase is diflufenican. This herbicide has its primary mode of action against carotenoid formation but also affects *de novo* fatty acid formation (Ashton *et al.*, 1992). The mechanism appeared to involve the enoyl reductase reaction and, in avocado, both NADH- and NADPH-driven enoyl reductions were inhibited (Ashton *et al.*, 1994). In contrast, diflufenican has no effect on purified *E. coli* enoyl reductase (Heath *et al.*, 2001).

### *2.4.4 Termination of FAS*

In most plants, oleate and palmitate are the major products of fatty acid synthesis in plastids. These products are released from their ACP thioesters by the activity of thioesterases. The activity of acyl-ACP thioesterases has two functions. First, it terminates synthesis by releasing free fatty acids. The latter are then exported outside plastids as acyl-CoAs (Harwood, 1996). So, a second function of the thioesterases is connected with the export of acyl chains because this destines fatty acids for the 'eukaryotic pathway' of lipid metabolism. In contrast, if acyl-ACPs are used directly by the plastid acyltransferases then a 'prokaryotic pathway' is followed (Browse and Somerville, 1991; Harwood, 1996).

Acyl-ACP thioesterases were first purified from soya bean seeds (Kinney *et al.*, 1990) and oilseed rape (Hellyer and Slabas, 1990). Once enzyme characteristics were assessed, it was clear that there were two different classes of thioesterases (Jones *et al.*, 1995; Voelker *et al.*, 1997), which differed in both amino acid sequence and in substrate specificity. The FATA class has high activity with oleoyl-ACP but much lower activity with saturated substrates. The FATB class has highest activity with saturated acyl-ACPs but also has good activity with oleoyl-ACP and other unsaturated substrates (Doermann *et al.*, 1995; Voelker *et al.*, 1997; Salas and Ohlrogge, 2002). In the Arabidopsis genome, there are two genes for *FATA* and one for *FATB* (Mekhedor *et al.*, 2000).

The FATB-like enzymes can be subdivided into two functional groups according to the chain length of their preferred substrate. Most plants contain FATB enzymes with a selectivity for acyl-ACPs in the range 14–18C, but a few plants (which produce medium chain acids in their seed oils) contain FATB enzymes, which use substrates in the range 6–14C. While virtually all tissues show good FATA activity and this can be considered a housekeeping gene (Gibson *et al.*, 1994), the activity of FATB is usually more variable. Nevertheless, both genes are expressed at good levels in different tissues including seeds of *Cuphea hookeriana* (Doermann *et al.*, 1993), *A. thaliana* (Doermann *et al.*, 1995) and *Garcinia mangostana* (Hawkins and Kridl, 1998). This result was also found in developing sunflower seeds, where at 15 days after flowering, both FATA- and FATB-like enzymes were found. In contrast, later in the period of maturation (30 days after flowering), only a single enzyme could be detected by bisubstrate assays (Martinez-Force *et al.*, 2000). Interestingly, in a sunflower mutant that produced a higher content of palmitate, a modified FATA-like thioesterase was found which had better activity with palmitoyl-ACP than usual (Martinez-Force *et al.*, 1999, 2000). It would be interesting to re-examine plants like *Theobroma cacao*, where general thioesterase activity produces a large amount of stearate as well as palmitate hydrolysis (Griffiths *et al.*, 1993), to see whether modified FATA and/or FATB enzymes are present.

An antisense and overexpression study of FATB in Arabidopsis showed that this thioesterase was involved, at least in part, in the production of saturated fatty acids for flowers and, in agreement with the results above, in seeds (Doermann *et al.*, 2000). Similarly, down-regulation of *FATB* expression in soybean caused a reduction in the seed content of palmitate (Wilson *et al.*, 2001; Buhr *et al.*, 2002). Further work with a disrupted *FATB* gene in Arabidopsis demonstrated an essential role for saturated fatty acids in plant growth. The palmitate content was reduced by about half in a variety of tissues and this was coincident with a slower growth rate (50% less fresh weight at 4 weeks) and seeds with a low viability and altered morphology. Moreover, an adequate saturated fatty acid supply, normally provided through FATB activity, seemed to be important for normal wax synthesis, which was impaired in *mutants* (Bonaventure *et al.*, 2003).

These data with *FATB mutants* show that, even though the thioesterase makes a minor contribution to the total flux of carbon for leaf glycerolipid formation in Arabidopsis (Figure 2.10), this small input is essential for normal plant growth and maturation (Bonaventure *et al.*, 2003).

Since medium-chain fatty acids are a valuable renewable resource, attempts have been made by researchers to produce high-yielding annual crops using genetic transformation. Eccleston transformed *B. napus* using a medium-chain acyl-ACP thioesterase isolated from *Umbellularia californica* (California bay). Although laurate accumulated (7%) in seeds, none was detectable in leaves despite very high active levels of the medium-chain thioesterase. If the medium-chain thioesterase was expressed in seeds using a napin promoter, then up to 60% of the total seed fatty acids were laurate. At the same time β-oxidation was increased and this was believed to be responsible for the limited accumulation of seed laurate when the constitutive CaMV 35S promoter was used. In the high-laurate seeds, enzymes of the fatty acid synthase (and ACP) were increased, perhaps to compensate for the lauric acid lost through β-oxidation (Eccleston *et al.*, 1996; Eccleston and Ohlrogge, 1998).

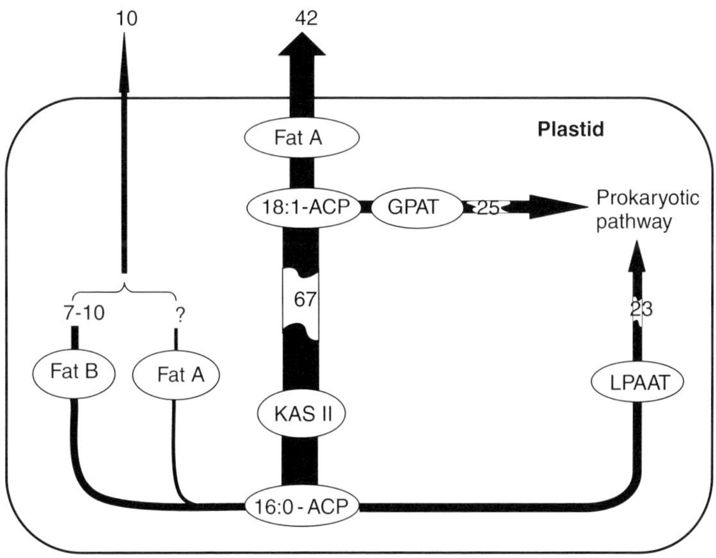


**Figure 2.10** Predicted fluxes of 16C and 18C fatty acids into membrane leaf glycerolipids in Arabidopsis (redrawn from Bonaventure *et al.*, 2003).

### 2.4.5 *Mitochondrial FAS*

Following the discovery of ACPs in the mitochondria of plants (Chuman and Brody, 1989; Shintani and Ohlrogge, 1994), as well as a variety of other organisms, the function of these proteins has been an enigma. In addition to their role in fatty acid synthesis, ACPs also function in a number of other metabolic reactions (see Wada *et al.*, 1997). In mitochondria from several different plants, ACP has been identified as a component of the NADH: ubiquinone oxidoreductase complex (Complex 1) and an acylated form of mitochondrial ACP is found associated with the peripheral part of the complex (Schulte *et al.*, 1994). In *Neurospora,* mitochondrial ACP was reported to be involved in *de novo* fatty acid synthesis (Mikolajczyk and Brody, 1990) and, certainly, mitochondrial ACP from Arabidopsis was able to function as a cofactor for chloroplast fatty acid synthase *in vitro* (Shintani and Ohlrogge, 1994).

Further work, using mitochondria isolated from pea leaves, showed that $^{14}C$-malonate was incorporated into fatty acids and acylated-ACP was detected. Malonate was a much better precursor than acetate, suggesting that ACCase activity may have been lacking. Cerulenin inhibited labelling of fatty acids, in keeping with the presence and use of a typical KAS I enzyme in the fatty acid synthase. Over 30% of the label incorporated was found associated with H protein that contains lipoic acid. Together, the data showed that ACP was involved in *de novo* fatty acid formation by plant mitochondria and that a major role for this pathway was to make lipoic acid precursors (Wada *et al.*, 1997).

## 2.5 Regulation of fatty acid formation

Before discussing the regulation of fatty acid synthesis, a mention must be made of the possible existence of a metabolon, which contains many or all of the enzymes for fatty acid formation. The concept of metabolons has received increasing credence over the last couple of decades and is now accepted as plausible organisation framework for a number of (soluble) metabolic pathways. This includes the β-oxidation of fatty acids where there is much evidence in favour of a metabolon although its existence has not been conclusively proven (Eaton, 2002). In 1996, Roughan and Ohlrogge published evidence that the chloroplast enzymes of fatty acid synthesis were organised into a multienzyme assembly that channelled acetate into long-chain acids (Roughan and Ohlrogge, 1996). Such a situation would need not only specific interaction and organisation of the relevant proteins *in vivo* but also careful control of expression of their genes and turnover of the enzymes to maintain their relative activities optimally.

Perhaps in keeping with the idea that synthesis and, hence, the activity of enzymes involved in fatty acid formation needs to be carefully regulated and changes coordinated well is the fact that expression of many such enzymes occurs at constant molar ratios (O'Hara *et al.*, 2002). Thus, the biotin carboxylase of ACCase, β-ketoacyl-ACP reductase, enoyl-ACP reductase and acyl-ACP thioesterase (and stearoyl-ACP desaturase) accumulated in a coordinate manner during *B. napus* embryogenesis. Their mRNAs were present in a constant stoichiometric ratio as judged by dilution series on Northern gels. Furthermore, in a study of ACCase, the various components (including the plastid-encoded carboxyltransferase component) were co-ordinately expressed at a constant molar ratio (Ke *et al.*, 2000). In addition, use of yeast two-hybrid system has revealed direct interactions between enoyl-ACP reductase, stearoyl-ACP desaturase and acyl-ACP thioesterase in oilseed rape (Honeyman and Fawcett, 2000). Perhaps, surprisingly, there were no interactions detected between ACP and other FAS proteins. All these data suggest that a fatty acid synthesis metabolon may exist, although it has still to be demonstrated convincingly.

The overall topic of the regulation of fatty acid synthesis has been covered in a review by Ohlrogge and Jaworski (1997). *In vitro* data from a number of different types of experiments have implied that ACCase could be important – at least in leaves. This conclusion has been confirmed by measurement of acyl-thioester levels under different rates of fatty acid formation as discussed in Section 2.3.5. Further, measurement of flux control coefficients for two different ACCases showed that this enzyme exerted strong flux control over fatty acid synthesis (Table 2.1; Page *et al.*, 1994).

Apart from ACCase, Ohlrogge and Jaworski (1997) suggest that the condensing enzymes of FAS may be important. The lack of significant levels of accumulating acyl intermediates tend to rule out the other components of FAS as being significant. Since KAS III catalyses the first reaction of the FAS cycle, logic would suggest that it might be important. However, if all three KAS enzymes show similar activities (as they often do) then the effect of manipulating one would be small. Indeed when KAS III was overexpressed (in Arabidopsis and tobacco), fatty acid synthesis rates remained the same and butyryl-ACP accumulated, suggesting that KAS I was now limiting (Tai and Jaworski, 1995).

When lipid accumulation was examined in cultures from important oil crops, flux control analysis revealed that fatty acid synthesis was more important than lipid assembly (Ramli *et al.*, 2002a,b) (Figure 2.11). This was in keeping with the previous observation in developing *Cuphea* and other embryos that exogenous fatty acids could stimulate lipid accumulation (Bafor *et al.*, 1991; Bao and Ohlrogge, 1999). Dissection of fatty acid synthesis by the use of an inhibitor of FAS indicated that, in oil palm callus tissue, FAS exerted about half of the total control over fatty acid formation (Ramli *et al.*, unpublished data). Further analysis of the *de novo* fatty acid synthesis reactions by the use of inhibitors is limited by the lack of effective and specific chemicals. However, transgenic plants are available in several cases and may be able to shed further light on these questions.

Several other factors can also influence the overall rates of fatty acid production. Feedback inhibition by exogenous fatty acids can be easily demonstrated (Shintani and Ohlrogge, 1995; Ramli *et al.*, 2002b). Although acyl-ACPs can reduce KAS III activity (Bruck *et al.*, 1996) *in vitro*, the reproducibility of this observation and

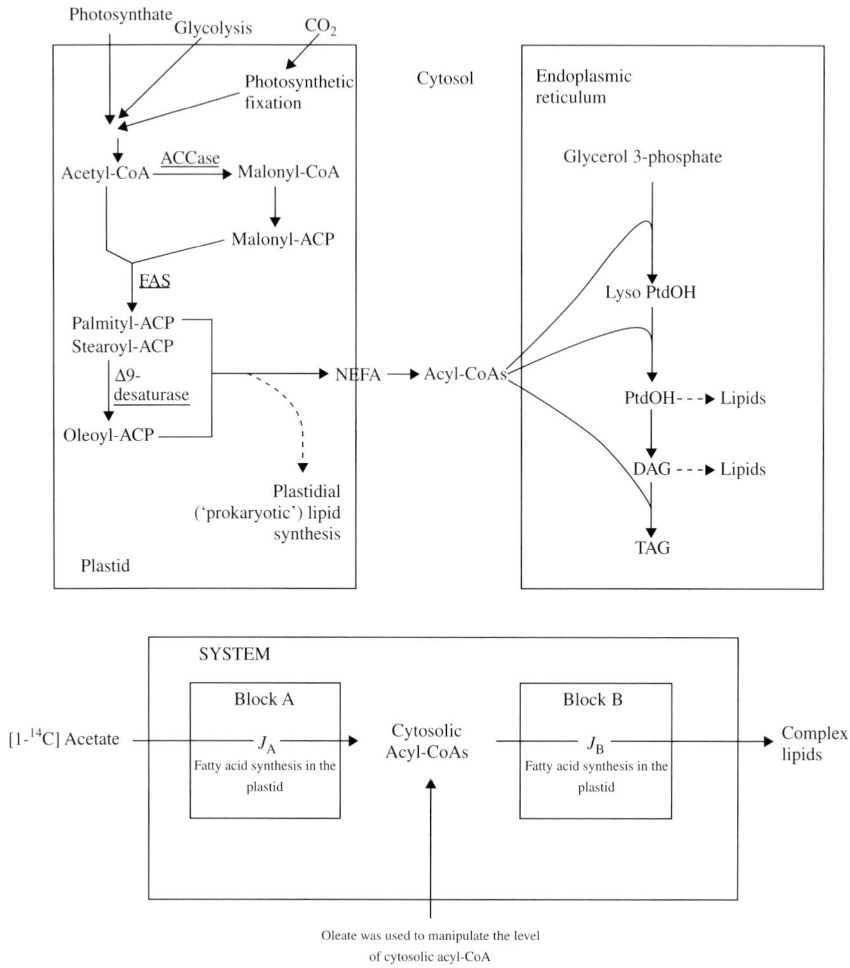


**Figure 2.11** The experimental system used to measure block control coefficients for lipid accumulation in callus cultures from oil crops. The upper part of the figure shows a simplified depiction of plant oil crop lipid biosynthesis with fatty acid formation in the plastid and lipid assembly outside. The acyl-ACPs produced in the plastid are either used directly for plastidic lipid production or hydrolysed and exported to the cytosolic compartment as acyl-CoAs. The latter are used by the acyltransferase of the Kennedy pathway on the endoplasmic reticulum. Abbreviations: ACCase, acetyl-CoA carboxylase; ACP, acyl carrier protein; FAS, fatty acid synthase. The lower half of the figure shows the pathway reactions for lipid biosynthesis grouped together to allow the conceptually simplified system used for control analysis. Carbon flux from substrate pool acetate, enters Block A to produce fatty acids. These fatty acids are then exported to the cytosol as acyl-CoA esters, which then serve as substrates for the Kennedy-pathway enzymes during TAG synthesis on the endoplasmic reticulum. Therefore, the two blocks of reactions are connected via a single pool of cytosolic acyl-CoA (as the system's intermediate). For the purpose of measuring kinetics of each block (i.e. $J_A$ or $J_B$) independently, [1-$^{14}$C]acetate was used as carbon precursor for fatty acid synthesis in the plastid, while complex lipid assembly was followed conveniently with [U-$^{14}$C]glycerol, which proved to be selective for endoplasmic reticulum activities. Reproduced from Ramli *et al.* (2002b) with kind permission from the authors and Portland Press.

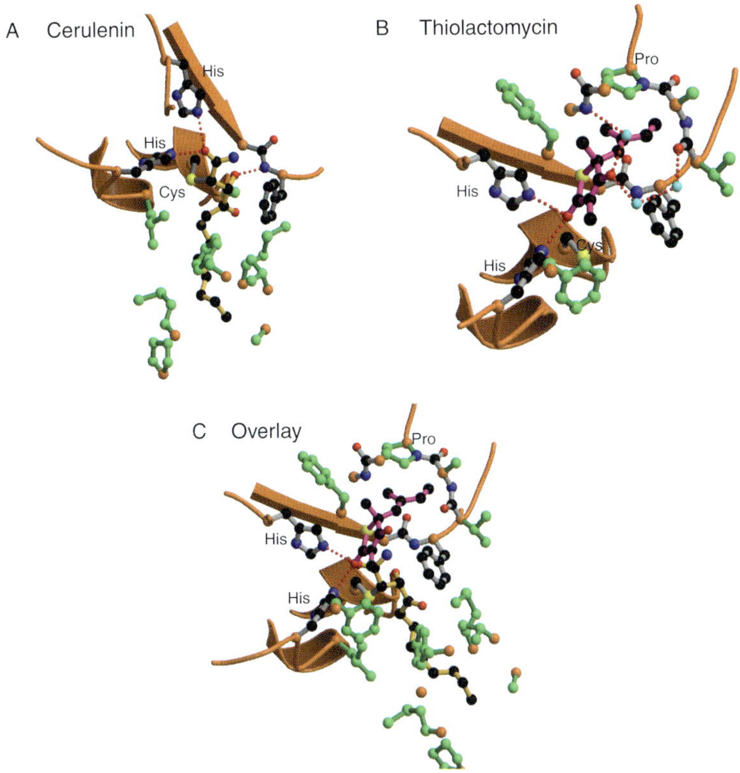


**Figure 2.7** Mechanism of cerulenin and thiolactomycin inhibition of the fatty acid synthase condensing enzyme FabB of *E. coli*. A. Cerulenin, shown with orange bonds, located in the active site of FabB. O1 of cerulenin is covalently attached to the active site Cys163. O2 hydrogen bonds to His198 and His333, while O3 occupies the oxyanion hole formed by the backbone amides of Phe392 and Cys163. B. Thiolactomycin, shown with magenta bonds, in the active site of FabB. O3 of thiolactomycin hydrogen bonds to His298 and His333. The binding is stabilized by the intercalation of the isoprenoid tail in the space between Pro272 and the peptide bond between Gly391 and Phe392. C. Superimposition of cerulenin and thiolactomycin in the FabB active site. Cerulenin binds to the acyl-enzyme intermediate half of the active site, while thiolactomycin binds to the malonyl-ACP side. Only O2 of cerulenin and O1 of thiolactomycin overlap hydrogen bonded to the His dyad. Reproduced from Heath *et al.* (2001) with kind permission from the authors and Elsevier.

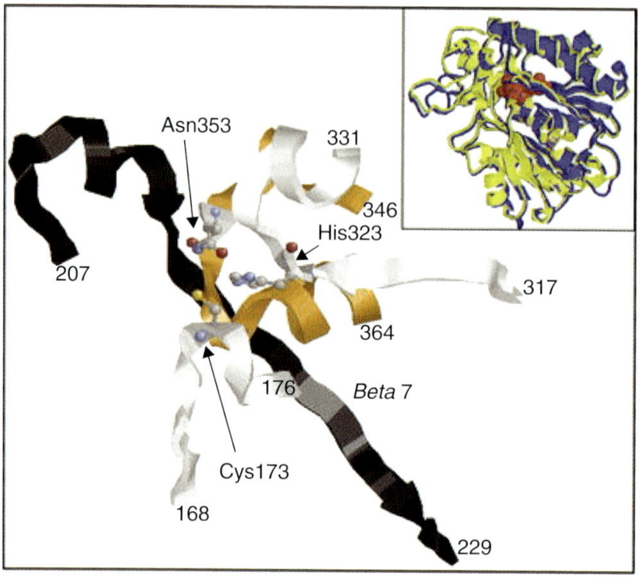


**Figure 2.8** The active-site region of the pea KAS III. This shows the three key residues and their amino acid sequences, which are completely conserved among the known plant KAS IIIs. The sequence $Val^{207}$ to $Ala^{229}$ in the pea sequence, which is almost completely conserved in other plant KAS IIIs is also shown lying across the active site. The model is based on the three-dimensional structure of the *E. coli* KAS III (Davies *et al*., 2000) and shows the preferred conformations of a α-helix, β-sheet and β-turns. Amino acid numbering is derived from the translated pea transcript accession number CAC08184. For details of the sequences shown, see Jones *et al*. (2003). Reproduced from Jones *et al*. (2003) with kind permission from the authors and Springer-Verlag.

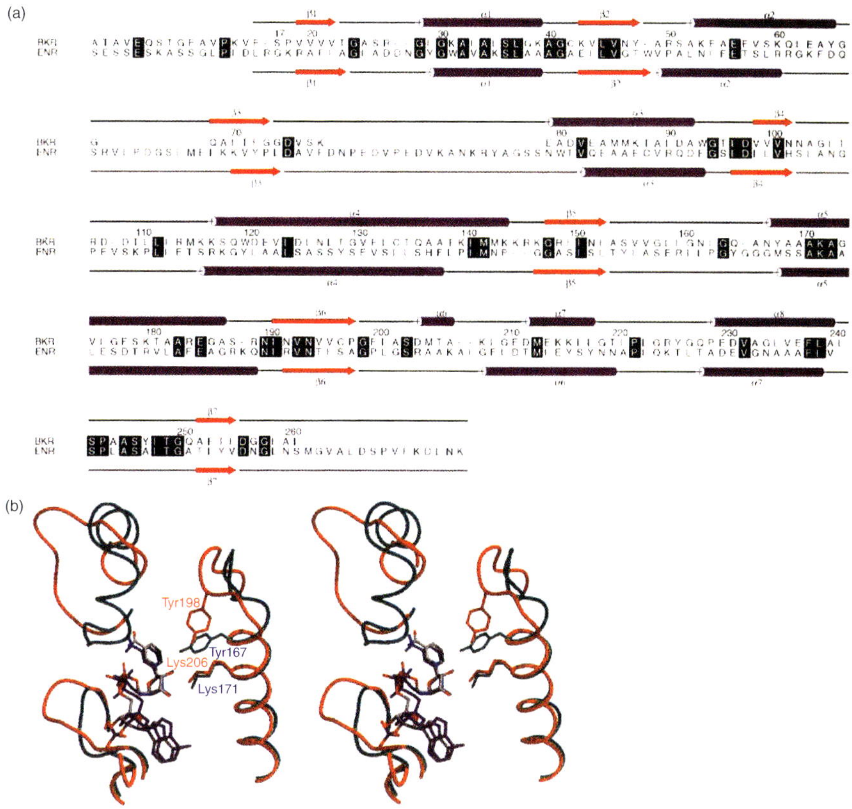


**Figure 2.9** Sequence and structural alignments of *Brassica napus* β-ketoacyl-ACP reductase (BKR) and enoyl-ACP reductase (ENR). A. The aligned sequences of *B. napus* BKR and ENR with the conserved residues shown in reverse typeface. The secondary structure elements of BKR and ENR are shown above and below the alignment, respectively; α helices are blue and β strands red. B. Stereo view superimposition of the Cα backbones of the active sites of *B. napus* BKR (green) and ENR (red). The catalytic tyrosine and lysine residues and bound nucleotides are shown in stick format. Reproduced from Fisher *et al.* (2000b) with kind permission from the authors and Elsevier.

its physiological relevance has been questioned (Ohlrogge and Jaworski, 1997). In fact, the latter reviewers conclude that the mechanism of feedback regulation (by fatty acids or acyl-ACPs) is unknown at present.

The other side of the coin is whether the supply of carbon precursors and cofactors can limit fatty acid synthesis and lipid production. Measurements of acetyl-CoA levels would suggest that this is not usually a limiting factor and, furthermore, supply of ATP and NAD(P)H is generally adequate for plant growth and, by implication, for primary pathways of metabolism (Ohlrogge and Jaworski, 1997). Nevertheless, there are examples from transgenic plants where carbon can be diverted away from lipid synthesis by overexpression of enzyme(s) in alternative pathways (Ohlrogge and Jaworski 1997). Under such conditions, the supply of carbon would, indeed, limit lipid accumulation.

Chloroplasts are able to generate NADPH and ATP during photosynthesis and these are thought normally to be adequate for fatty acid synthesis. Moreover, in the light, ACCase activity is increased by redox activation, as discussed earlier. In heterotrophic tissues, however, the plastids need to import these cofactors or to generate them intraplastidically through carbohydrate catabolism or metabolite shuttles (Rawsthorne, 2002). Thus, ATP has been shown to be imported (via ATP/ADP exchange translocators) during fatty acid synthesis in a variety of isolated plastids. Moreover, in rapeseed, exogenous ATP supports the highest rates of fatty acid synthesis, which shows no light dependency (Eastmond and Rawsthorne, 1998). It should be noted, however, that interpretation of *in vitro* experiments and their extrapolation to *in vivo* situations may be complicated (Rawsthorne, 2002).

A similar situation to that noted for ATP can also be applied to the source of reductant. Several studies with heterotrophic tissues have shown no requirement for exogenous reduced nucleotides whereas the use of an acetate substrate needs exogenous NAD(P)H for maximum fatty acid synthesis (e.g. Smith *et al.*, 1992). These authors have explained the difference in ATP requirement when using various substrates by the fact that metabolism of some substrates can lead to the production of cofactors. For example, the intraplastidial conversion of malate to acetyl-CoA also forms NADH and NADPH. In plastids from rapeseed embryos (one of the best-researched tissues), ATP-dependent incorporation of acetate into fatty acids can occur in the dark, implying that reduced nucleotides can be generated adequately inside the plastids under these conditions. Various possible pathways for the formation of reduced nucleotides are discussed by Rawsthorne (2002).

The end products of lipid synthesis in plants are, of course, the lipids and not their fatty acid constituents. Evidence has already been cited above which imply that, in oil crops, the rate of supply of fatty acids exerts an important control over lipid accumulation (e.g. Ramli *et al.*, 2002b). Additional evidence comes from lower plants like the alga *Chlamydomonas*, where exogenous lipids stimulate triacylglycerol formation, showing that the synthesis of triacylglycerols is not limited by the activity of triacylglycerol synthesis enzymes (Grenier *et al.*, 1991). Likewise, a careful comparison of oleaginous with non-oleaginous yeasts, led to the conclusion that the differences in lipid accumulation were due to the rates of fatty acid production rather than its utilization (Botham and Ratledge, 1979).

In some agreement with the above conclusions, transgenic plants expressing high levels of acyl-ACP thioesterases often showed an increased ability to accumulate lipid (see Ohlrogge and Jaworski, 1997). Under these circumstances, it can be argued that the thioesterase serves two purposes. First, by hydrolysing acyl-ACPs it removes end-product inhibition over fatty acid synthesis. Second, by releasing an increased supply of fatty acids, it allows the latter limitation to be overcome and more lipid to be accumulated. However, significant increases in rapeseed or Arabidopsis oil content were also achieved when a yeast lysophosphatidic acid acyltransferase was expressed in developing seeds. Similar increases were not achieved by manipulating plant LPAATs, suggesting that the effect of a yeast LPAAT could be due to its lack of regulation in the plant host (see Ohlrogge and Jaworski, 1997). Nevertheless, these data imply that, under some circumstances, lipid assembly (i.e. control of fatty acid synthesis by demand or sink) may play a significant role in regulating fatty acid biosynthetic rates.

## 2.6 Biotechnological aspects

With our increasing knowledge of the individual enzymes involved in fatty acid biosynthesis and their genes, opportunities for creating commercial transgenic crops have been considered (e.g. Murphy, 1994; Ohlrogge, 1994). So far as saturated fatty acid production is concerned, there has only been one 'success' to date and that is the manipulation of canola varieties of oilseed rape to produce medium-chain fatty acids. This was described in Section 2.4.4. Thus, in summary, the gene for a medium-chain acyl-ACP thioesterase was isolated from a plant, which normally produces medium-chain triacylglycerols in its seed oil (California bay) and inserted into *B. napus*. Where an appropriate seed-specific promoter (napin) was used, substantial (up to 60 mole%) amounts were found in seeds.

However, there are other goals for biotechnology, which include increasing overall oil yields (Bright and Hawkes, 1998) and the proportions of different saturated fatty acids (Browse *et al.*, 1998). Although most of the genetic manipulation that has been carried out in research labs has involved anti-sensing individual genes and, hence, has tended to reveal more about essentiality or function rather than possible future crops, important information has been obtained. Thus, we now have considerable information about multigene families and about promoters controlling activity of relevant genes (c.f. Ohlrogge and Jaworski, 1997). The application of fundamentals of plant biochemistry and molecular biology to agricultural aspects is covered in a book by Harwood (1998), and the application of biotechnology to oil crops in a recent article by Murphy (2003).

Of course, fundamental science may be able to help the commercial sector in other ways. For example, elucidation of the mechanism of action of graminicides was due to experiments designed to understand molecular mechanisms of the ACCases. And, indeed, the presence of weeds with inherent or acquired resistance has also been explained by biochemical and molecular biological experiments (see Section 2.3.3). Thus, as concluded by two experts (Bright and Hawkes, 1998), industry needs an excellent science base. This is not only to provide highly trained scientists but also to give answers to current questions and problems. Moreover, because we often do not really know what questions we need to answer, we need to research on a broad front so that we cover as many possibilities as possible. Thus, progress in industry, as well as in pure science, will only be achieved when we have sufficient data to answer confidently the 'big' questions. Perhaps the scientists amongst us should bear that in mind and not just do research to suit our own prejudices. In fact, to recall Albert Camus' comment in *The Fall* (though meant in a very different context), 'You know what charm is: a way of getting the answer yes without having asked any clear question'. If only getting the right answer were so easy!

## References


Alban, C., Baldet, P. and Douce, R. (1994) Localisation and characterisation of two structurally different forms of acetyl-CoA carboxylase in young pea leaves, of which one is sensitive to aryloxylphenoxypropionate herbicides. *The Biochemical Journal*, **300**, 557–565.

Ashton, I.A., Abulnaja, K.O., Pallett, K.E., Cole, D.J. and Harwood, J.L. (1994) The mechanism of inhibition of fatty-acid synthase by the herbicide diflufenican. *Phytochemistry*, **35**, 587–590.

Bafor, M.E., Jonsson, L., Stobart, A.K. and Stymme, S. (1991) Regulation of triacylglycerol biosynthesis in embryos and microsomal preparations for developing castor-bean endosperm. *The Biochemical Journal*, **280**, 502–514.

Baldet, P., Alban, C. and Douce, R. (1993) Localization of free and bound biotin in cells from green pea leaves. *Archives of Biochemistry and Biophysics*, **303**, 67–73.

Baldock, C., Rafferty, J.B., Sedelnikova, S.E., Baker, P.J., Stuitje, A.R., Slabas, A.R., Hawkes, T.R. and Rice, D.W. (1996) A mechanism of drug action revealed by structural studies of enoyl reductase. *Science*, **274**, 2107–2110.

Baldock, C., Rafferty, J.B., Stuitje, A.R., Slabas, A.R. and Rice, D.W. (1998) The X-ray structure of *Escherichia coli* enoyl reductase with bound NAD+ at 21 Å resolution. *Journal of Molecular Biology*, **284**, 1529–1546.

Bao, X. and Ohlrogge, J.B. (1999) Supply of fatty acid is one limiting factor in the accumulation of triacylglycerol in developing embryos. *Plant Physiology*, **120**, 1057–1062.

Bao, X., Focke, M., Pollard, M. and Ohlrogge, J.B. (2000) Understanding *in vivo* carbon precursor supply for fatty acid synthesis in leaf tissue. *The Plant Journal*, **22**, 39–50.

Bao, X., Shorrosh, B.S. and Ohlrogge, J.B. (1997) Isolation and characterisation of an *Arabidopsis* biotin carboxylase gene and its promoter. *Plant Molecular Biology*, **35**, 539–550.

Bonaventure, G. and Ohlrogge, J.B. (2002) Differential regulation of mRNA levels of acyl carrier protein isoforms in *Arabidopsis. Plant Physiology*, **128**, 223–235.

Bonaventure, G., Salas, J.J., Pollard, M.R. and Ohlrogge, J.B. (2003) Disruption of the *FAT B* gene in *Arabidopsis* demonstrates an essential role of saturated fatty acids in plant growth. *The Plant Cell*, **15**, 1020–1033.

Botham, P.A. and Ratledge, C. (1979) A biochemical explanation for lipid accumulation in *Candida* 107 and other oleaginous micro-organisms. *Journal of General Microbiology*, **114**, 361–375.

Branen, J.K., Kwon, M. and Engeseth, N.J. (2003) Expression of antisense ACP-4 reduces the lipid content in *Arabidopsis* leaf tissue, in *Advanced Research on Plant Lipids* (ed. N. Murata, M. Yamada, I. Nishida, H. Okuyama, J. Sekiya and H. Wada), Kluwer, Dordrecht, pp. 73–76.

Bright, S.W.J. and Hawkes, T.R. (1998) The future: industry requirements from advances in plant lipid research, in *Plant Lipid Biosynthesis: Fundamentals and Agricultural Applications* (ed. J.L. Harwood). Cambridge University Press, Cambridge, UK, pp. 351–363.

Brody, S. and Mikolajczyk, S. (1988) *Neurospora* mitochondria contain an acyl carrier protein. *European Journal of Biochemistry*, **173**, 353–359.

Brody, S., Mikolajczyk, S. and Chapman, L. (1990) Studies on *de novo* fatty acid synthesis in mitochondria, in *Plant Lipid Biochemistry, Structure and Utilization* (ed. P.J. Quinn and J.L. Harwood), Portland Press, London, pp. 117–119.

Brown, A.C., Moss, S.R., Wilson, Z.A. and Field, L.M. (2002) An isoleucine to leucine substitution in the ACCase of *Alopecurus myosuroides* is associated with resistance to the herbicide sethoxydim. *Pesticide Biochemistry and Physiology*, **72**, 160–168.

Browse, J. and Somerville, C.R. (1991) Glycerolipid synthesis: biochemistry and regulation. *Annual Review of Plant Physiology and Molecular Biology*, **42**, 467–506.

Browse, J., Spychalla, J., Okuley, J. and Lightner, J. (1998) Altering the fatty acid composition of vegetable oils, in *Plant Lipid Biosynthesis: Fundamentals and Agricultural Applications* (ed. J.L. Harwood). Cambridge University Press, Cambridge, UK, pp. 131–153.

Browse, J.A., Warwick, N., Somerville, C.R. and Slack, C.R. (1986) Fluxes through the prokaryotic and eukaryotic pathways of lipid synthesis in the 16:3 plant *Arabidopsis thaliana. The Biochemical Journal*, **235**, 25–31.

Bruck, F.M., Brummel, H., Schuch, R. and Spener, F. (1996) In-vitro evidence for feed-back regulation of β-ketoacyl carrier protein synthase III in medium-chain fatty acid biosynthesis. *Planta*, **198**, 271–278.

Buhr, T., Sato, S., Ebrahim, F. *et al.* (2002). Ribozyme termination of RNA transcripts down-regulate seed fatty acid genes in transgenic soybean. *The Plant Journal*, **30**, 155–163.

Burton, J.D., Gronwald, J.W., Somers, D.A., Connelly, J.A., Gengenbach, B.G. and Wyse, D.L. (1987) Inhibition of plant ACCase by the herbicides sethoxydim and haloxyfop. *Biochemical and Biophysical Research Communications*, **148**, 1039–1044.

Cahoon, E.B. and Ohlrogge, J.B. (1994) Metabolic evidence for the involvement of a delta(4)-palmitoyl-acyl carrier protein desaturase in petroselinic acid synthesis in coriander endosperm and transgenic tobacco cells. *Plant Physiology*, **104**, 827–837.

Carlsson, A.S., LaBrie, S.T., Kinney, A.J., von Wettstein-Knowles, P. and Browse, J. (2002) A KAS 2 cDNA complements the phenotypes of the *Arabidopsis* fab 1 mutant that differs in a single residue bordering the substrate binding pocket. *The Plant Journal*, **29**, 761–770.

Choi, J.K., Yu, F., Wurtele, E.S. and Nikolau, B.J. (1995) Molecular cloning and characterisation of the cDNA coding for the biotin-containing subunit of the chloroplastic acetyl-CoA carboxylase. *Plant Physiology*, **109**, 619–625.

Christopher, J.T. and Holtum, J.A.M. (2000) Dicotyledons lacking the multisubunit form of the herbicide-target enzyme acetyl-CoA carboxylase may be restricted to the family Geraniaceae. *Australian Journal of Plant Physiology*, **27**, 845–850.

Chuman, L. and Brody, S. (1989) Acyl carrier protein is present in the mitochondrion of plants and eukaryotic microorganisms. *European Journal of Biochemistry*, **184**, 643–649.

Clough, R.C., Matthis, A., Barnum, S.R. and Jaworski, J.G. (1992) Purification and characterisation of 3-ketoacyl-acyl carrier protein synthase III from spinach. *Journal of Biological Chemistry*, **267**, 20992–20998.

Cottingham, I.R., Austin, A. and Slabas, A.R. (1989) Inhibition and covalent modification of rape seed enoyl-ACP reductase by phenylglyoxal. *Biochimica et Biophysica Acta*, **995**, 273–278.

Cronan, J.E. Jr. and Waldrop, G.L. (2002) Multi-subunit acetyl-CoA carboxylases. *Progress in Lipid Research*, **41**, 407–435.

Davies, C., Heath, R.J., White, S.W. and Rock, C.O. (2000) The 1.8 A crystal structure and active-site architecture of β-ketoacyl-acyl carrier protein synthase III (Fab H) from *Escherichia coli. Structure*, **8**, 185–195.

deBoer, G.J., Kater, M.M., Fawcett, T., Slabas, A.R., Nijkamp, H.J.J. and Stuitje, A.R. (1998) The NADH-specific enoyl-ACP reductase: characterization of a housekeeping gene involved in storage lipid synthesis in seeds of Arabidopsis and other plant species. *Plant Physiology and Biochemistry*, **36**, 473–486.

Dehesh, K., Edwards, P., Fillatti, J., Slabaugh, M. and Byrne, J. (1998) KAS IV: a 3-ketoacyl-ACP synthase from *Cuphea* sp. is a medium chain specific condensing enzyme. *The Plant Journal*, **15**, 383–390.
Dehesh, K., Tai, H., Edwards, P., Byrne, J. and Jaworski, J.G. (2001) Overexpression of 3-ketoacyl-acyl carrier protein synthase III in plants reduces the rate of lipid synthesis. *Plant Physiology*, **125**, 1103–1114.
Devine, M.D. (1997) Mechanisms of resistance to ACCase inhibitors: a review. *Pesticide Science*, **51**, 259–264.
Doermann, P., Spencer, F. and Ohlrogge, J.B. (1993) Characterization of two acyl-ACP thioesterases from developing *Cuphea* seeds specific for medium-chain and oleoyl-ACP. *Planta*, **189**, 425–432.
Doermann, P., Voelker, T.A. and Ohlrogge, J.B. (1995) Cloning and expression in *E.coli* of a novel thioesterase from *Arabidopsis thaliana* for long-chain acyl-ACPs. *Archives of Biochemistry and Biophysics*, **316**, 612–618.
Doermann, P., Voelker, T.A. and Ohlrogge, J.B. (2000) Accumulation of palmitate in Arabidopsis mediated by the acyl-ACP thioesterase FATB1. *Plant Physiology*, **123**, 637–644.
Domergue, F. and Post-Beittenmiller, D. (2000) Biochemical characterisation of 3-ketoacyl-ACP synthase II from leek epidermis. *Biochemical Society Transactions*, **28**, 610–613.
Eastmond, P.J. and Rawsthorne, S. (1998) Comparison of the metabolic properties of plastids isolated from developing leaves or embryos of *Brassica napus*. *Journal of Experimental Botany*, **49**, 1105–1111.
Eastmond, P.J. and Rawsthorne, S. (2000) Coordinate changes in carbon partitioning and plastidial metabolism during the development of oilseed rape embryos. *Plant Physiology*, **122**, 767–774.
Eastmond, P.J., Dennis, D.T. and Rawsthorne, S. (1997) Evidence that a malate/inorganic phosphate exchange translocator imports carbon across the leucoplast envelope for fatty acid synthesis in developing castor bean endosperm. *Plant Physiology*, **114**, 851–856.
Eaton, S. (2002) Control of mitochondrial β-oxidation flux. *Progress in Lipid Research*, **41**, 197–239.
Eccleston, V.S. and Ohlrogge, J.B. (1998) Expression of lauroyl-ACP thioesterase in *Brassica napus* seeds induces pathways for both fatty acid oxidation and biosynthesis and implies a set point for triacylglycerol accumulation. *The Plant Cell*, **10**, 613–621.
Eccleston, V.S., Cranmer, A.M., Voelker, T.A. and Ohlrogge, J.B. (1996) Medium-chain fatty acid biosynthesis and utilization in *Brassica napus* plants expressing lauroyl-ACP thioesterase. *Planta*, **198**, 46–53.
Egli, M.A., Gengenbach, B.G., Gronwald, J.W., Somers, D.A. and Wyse, D.L. (1993) Purification and characterisation of acetyl-CoA carboxylase from diclofop-resistant and susceptible *Lolium multiflorum*. *Plant Physiology*, **105**, 671–680.
Evenson, K.J., Gronwald, J.W. and Wyse, D.L. (1994) Purification and characterisation of ACCase from diclofop-resistant and -susceptible *Lolium multiflorum*. *Plant Physiology*, **105**, 671–680.
Fawcett, T., Copse, C.L., Simon, W.J. and Slabas, A.R. (2000) Kinetic mechanism of NADH-enoyl-ACP reductase from *Brassica napus*. *Federation of European Biochemical Societies Letters*, **484**, 65–68.
Fawcett, T., Simon, W.J., Swinhoe, R. Shanklin, J., Nishida, I., Christie, W.W. and Slabas, A.R. (1994) Expression of mRNA and steady-state levels of protein isoforms of enoyl-ACP reductase from *Brassica napus*. *Plant Molecular Biology*, **26**, 155–163.
Fell, D.A. (1997) *Understanding the Control of Metabolism*, Portland Press, London.
Fisher, M., Kroon, J.T.M., Martindale, W., Stuitje, A.R., Slabas, A.R. and Rafferty, J.B. (2000b) The X-ray structure of *Brassica napus* β-ketoacyl carrier protein reductase and its implications for substrate binding and catalysis. *Structure*, **8**, 339–347.
Fisher, M., Sedelnikova, S.E., Martindale, W. *et al.* (2000a) Crystallization of the NADP-dependent β-ketoacyl-acyl carrier protein reductase from *Brassica napus*. *Acta Crystallographica*, **D56**, 86–88.
Focke, M. and Lichtenthaler, H.K. (1987) Inhibition of the ACCase of barley chloroplasts by cycloxydim and sethoxydim. *Zeitschrift für Naturforschung*, **42c**, 1361–1363.

Fox, S.R., Rawsthorne, S. and Hills, M.J. (2000) Role of acyl-CoAs and acyl-CoA binding protein in regulation of carbon supply for fatty acid biosynthesis. *Biochemical Society Transactions*, **28**, 672–674.

Fox, S.R., Rawsthorne, S. and Hills, M.J. (2001) Fatty acid synthesis in pea root plastids is inhibited by the action of long-chain acyl-coenzyme as on metabolite transporters. *Plant Physiology*, **126**, 1259–1265.

Gibson, S., Falcone, D.L., Browse, J. and Somerville, C.R. (1994) Use of transgenic plants and mutants to study the regulation and function of lipid composition. *Plant Cell and Environment*, **17**, 627–637.

Gornicki, P., Podkowinski, J., Scappino, I.A., Di Maio, J., Ward, E. and Haselkorn, R. (1994) Wheat acetyl-CoA carboxylase cDNA and protein structure. *Proceedings of the National Academy of Sciences USA*, **91**, 6860–6864.

Grenier, G., Guyon, D., Roche, O., Dubertret, G. and Tremolieres, A. (1991) Modification of the membrane fatty acid composition of *Chlamydomonas reinhardtii* cultured in the presence of liposomes. *Plant Physiology and Biochemistry*, **29**, 429–440.

Griffiths, G., Walsh, M.C. and Harwood, J.L. (1993). Acyl-thioesterase activity in developing seeds of cocoa. *Phytochemistry*, **32**, 1403–1405.

Gronwald, J.W., Eberlain, C.V., Betts, K.J., Baerg, R.J., Ehlke, N.J. and Wyse D.L. (1992) Mechanism of diclofop resistance in an Italian ryegrass biotype. *Pesticide Biochemistry and Physiology*, **44**, 128–139.

Gulliver, B.S. and Slabas, A.R. (1994) Acetoacyl-acyl carrier protein synthase from avocado – its purification, characterisation and clear resolution from acetyl CoA-ACP transacylase. *Plant Molecular Biology*, **25**, 179–191.

Gurr, M.I., Harwood, J.L. and Frayn, K.N. (2002) *Lipid Biochemistry*, 5th ed. Blackwell Science, Oxford.

Hahlbrock, K. (1981) Flavonoids, in *The Biochemistry of Plants* (eds. P.K. Stumpf and E.E. Conn) Vol. 7, Academic Press, New-York, pp. 425–456.

Hansen, L. (1987) Three cDNA clones for barley leaf acyl carrier proteins I and III. *Carlsberg Research Communications*, **52**, 381–392.

Harwood, J.L. (1988) Fatty acid metabolism. *Annual Review of Plant Physiology and Molecular Biology*, **39**, 101–138.

Harwood, J.L. (1996) Recent advances in the biosynthesis of plant fatty acids. *Biochimica et Biophysica Acta*, **1301**, 7–56.

Harwood, J.L., ed. (1998) *Plant Lipid Biosynthesis: Fundamentals and Agricultural Applications*, Cambridge University Press, Cambridge, UK.

Harwood, J.L. (1999) Graminicides which inhibit lipid synthesis. *Pesticide Outlook*, **10**, 154–158.

Harwood, J.L. and Quinn, P.J., eds (2000) *Recent Advances in the Biochemistry of Plant Lipids*, Portland Press, London.

Harwood, J.L. and Stumpf, P.K. (1971) Control of fatty acid synthesis in germinating seeds. *Archives of Biochemistry and Biophysics*, **142**, 281–291.

Hawkins, D.J. and Kridl, J. (1998) Characterisation of acyl-ACP thioesterases of mangosteen seed and high levels of stearate production in transgenic canola. *The Plant Journal*, 13, 743–752.

Heath, R.J., White, S.W. and Rock, C.O. (2001) Lipid biosynthesis as a target for antibacterial agents. *Progress in Lipid Research*, **40**, 467–497.

Hellyer, A. and Slabas, A.R. (1990) Acyl-[acyl-carrier protein] thioesterase from oil seed rape: purification and characterisation, in *Plant Lipid Biochemistry, Structure and Utilization* (ed. P.J. Quinn and J.L. Harwood), Portland Press, London, pp. 157–159.

Herbert, D., Harwood, J.L., Cole, D.J. and Pallett, K.E. (1995) Characteristics of aryloxyphenoxypropionate herbicide interactions with ACCases of different graminicide sensitivities. *Proceedings of the Brighton Crop Protection Conference – Weeds.* British Crop Protection Council, Farnham, UK, pp. 387–392.

Herbert, D., Price, L.J., Alban, C., Dehaye, L., Job, D., Cole, D.J., Pallett, K.E. and Harwood, J.L. (1996) Kinetic studies on two isoforms of acetyl-CoA carboxylase from maize leaves. *The Biochemical Journal*, **318**, 997–1006.

Hill, L.M. and Rawsthorne, S. (2000) Carbon supply for storage by-product synthesis in developing seeds of oilseed rape. *Biochemical Society Transactions*, **28**, 667–669.

Hitchcock, C. and Nichols, B.W. (1971) *Plant Lipid Biochemistry*, Academic Press, London, UK.
Hlousek-Radojcic, A., Post-Beittenmiller, D. and Ohlrogge, J.B. (1992) Expression of constitutive and tissue specific acyl carrier protein isoforms in *Arabidopsis. Plant Physiology*, **98**, 206–214.
Hoj, P.B. and Svendsen, I.B. (1984) Barley chloroplasts contain two acyl carrier proteins coded for by different genes. *Carlsberg Research Communications*, **49**, 483–492.
Honeyman, G. and Fawcett, T. (2000) Protein interactions of fatty acid synthase II, in *Recent Advances in the Biochemistry of Plant Lipids* (eds. J.L. Harwood and P.J. Quinn), Portland Press, London, pp. 615–616.
Huang, W., Jia, J., Edwards, P., Dehesh, K., Schneider, G. and Lindquist, Y. (1998) Crystal structure of the ketoacyl-acyl carrier protein syntase II from *E.coli* reveals the molecular architecture of condensing enzymes. *EMBO Journal*, **17**, 1183–1191.
Jackowski, S. and Rock, C.O. (1987) Acetoacetyl-acyl carrier protein synthase, a potential regulator of fatty acid biosynthesis in bacteria. *Journal of Biological Chemistry*, **262**, 7927–7931.
Jaworski, J.G. (1987) Biosynthesis of monoenoic and polyenoic fatty acids, in *The Biochemistry of Plants*, Vol. 9 (eds P.K. Stumpf and E.E. Conn), Academic Press, New York, pp. 159–174.
Jaworski, J.G., Clough, R.C. and Barnum, S.R. (1989) A cerulenin-insensitive short-chain 3-ketoacyl acyl carrier protein synthase in *Spinacea oleracea. Plant Physiology*, **90**, 41–44.
Jaworski, J.G., Tai, H., Ohlrogge, J.B. and Post-Beittenmiller, D. (1994) The initial reactions of fatty acid biosynthesis in plants. *Progress in Lipid Research*, **33**, 47–54.
Johnson, P.E., Fox, S.R., Hills, M.J. and Rawsthorne, S. (2000) Inhibition by long-chain acyl-CoAs of glucose 6-phosphate metabolism in plastids isolated from developing embryos of oilseed rape. *The Biochemical Journal*, **348**, 145–150.
Jones, A., Davies, H.M. and Volker, T.A. (1995a) Palmitoyl-ACP thioesterase and the evolutionary origin of plant acyl-ACP thioesterases. *Plant Cell*, **7**, 359–371.
Jones, A.L., Dancer, J.E. and Harwood, J.L. (1995b) Effects of thiolactomycin on fatty acid synthesis in pea. *Phytochemistry*, **39**, 511–514.
Jones, A.L., Gane, A.M., Herbert, D., Willey, D.L., Rutter, A.J., Kille, P., Dancer, J.E. and Harwood, J.L.(2003) β-ketoacyl-acyl carrier protein synthase III from pea: properties, inhibition by a novel thiolactomycin analogue and isolation of a cDNA encoding the enzyme. *Planta*, **216**, 752–761.
Jones, A.L., Herbert, D., Rutter, A.J., Dancer, J.E. and Harwood, J.L. (2000) Novel inhibitors of the condensing enzymes of the Type II fatty acid synthase of pea. *The Biochemical Journal*, **347**, 205–209.
Kaethner, T.M. and ap Rees, T. (1985) Intracellular location of ATP citrate lyase in leaves of *Pisum sativum. Planta*, **163**, 290–294.
Kannangara, C.G. and Stumpf, P.K. (1972) Prokaryotic type acetyl-CoA carboxylase in spinach chloroplasts. *Archives of Biochemistry and Biophysics*, **152**, 83–91.
Kater, M.M., Koningstein, G.M., Nijkamp, H.J.J. and Stuitje, A.R. (1991) cDNA cloning and expression of *Brassica napus* acyl-acyl carrier protein reductase in *Escherichia coli. Plant Molecular Biology*, **17**, 895–909.
Kauppinen, S. (1992) Structure and expression of the *Kas 12* gene encoding a β-ketoacyl-ACP synthase 1 isozyme from barley. *Journal of Biological Chemistry*, **267**, 23999–24006.
Ke, J., Behal, R.H., Back, S.L., Nikolau, B.J., Wurtele, E.S. and Oliver, D.J. (2000) The role of pyruvate dehydrogenase and acetyl-CoA synthetase in fatty acid synthesis in developing *Arabidopsis* seeds. *Plant Physiology*, **123**, 497–508.
Ke, J., Choi, J-K., Smith, M., Horner, H.T., Nikolau, B.J. and Wurtele, E.S. (1997) Structure of the *CAC1* gene and *in situ* characterization of its expression: the *Arabidopsis thaliana* gene coding for the biotin-containing subunit of the plastidic acetyl-CoA carboxylase. *Plant Physiology*, **113**, 357–365.
Ke, J., Wen, T-N, Nikolau, B.J. and Wurtele, E.S. (2000) Coordinate regulation of the nuclear and plastidic genes coding for the subunits of the heteromeric acetyl-CoA carboxylase. *Plant Physiology*, **122**, 1057–1071.
Kinney, A.J., Hirtz, W.D. and Yadav, N.S. (1990) Stearoyl-ACP desaturase and a β-ketoacyl-ACP synthase from developing soybean, in *Plant Lipid Biochemistry, Structure and Utilization* (eds P.J. Quinn and J.L. Harwood), Portland Press, London, pp. 126–128.
Knowles, J.R. (1989) The mechanism of biotin-dependent enzymes. *Annual Review of Biochemistry*, **58**, 195–221.

Kolattukudy, P.G. (1980) Cutin, suberin and waxes, in *The Biochemistry of Plants*, Vol. 4 (ed. P.K. Stumpf), Academic Press, New York, pp. 571–645.

Kozaki, A., Mayumi, K. and Sasaki, Y. (2001) Thiol-disulfide exchange between nuclear-encoded and chloroplast-encoded subunits of pea acetyl-CoA carboxylase. *Journal of Biological Chemistry*, **276**, 39919–39925.

Kubis, S.E. and Rawsthorne, S. (2000) The role of plastidial transporters in developing embryos of oilseed rape for fatty acid synthesis. *Biochemical Society Transactions*, **28**, 665–666.

Kunst, L., Browse, J. and Somerville, C. (1988) Altered regulation of lipid biosynthesis in a mutant of *Arabidopsis* deficient in chloroplast glycerol phosphate acyltransferase activity. *Proceedings of the National Academy of Sciences USA*, **85**, 4143–4147.

Levy, C.W., Roujeinikova, A., Sedelnikova, S. *et al.* (1999) Molecular basis of triclosan activity. *Nature*, **398**, 383–384.

Lichtenthaler, H.K., Kobek, K. and Focke, M. (1989) Differences in sensitivity and tolerance of monocotyledonous and dicotyledonous plants towards inhibitors of acetyl-coenzyme A carboxylase. *Proceedings of the Brighton Crop Protection Conference – Weeds*. British Crop Protection Council, Farnham UK, pp. 173–182.

Liedvogel, B. (1987) Lipid precursors in plant cells: the problem of acetyl-CoA generation for plastid fatty acid synthesis, in *The Metabolism, Structure and Function of Plant Lipids* (ed. P.K. Stumpf, J.B. Mudd and W.D. Nes), Plenum, New York, pp. 509–512.

Lowe, P.N. and Rhodes, S. (1988) Purification and characterisation of [acyl carrier protein]acetyltransacylase from *Escherichia coli*. *The Biochemical Journal*, **250**, 789–796.

Martinez-Force, E., Alvarez-Ortega, R. and Garces, R. (1999) Enzymatic characterization of high palmitic acid sunflower mutants. *Planta*, **207**, 533–538.

Martinez-Force, E., Cantisan, S., Serrano-Vega, M.J. and Garces, R. (2000) Acyl-acyl carrier protein thioesterase activity from sunflower seeds. *Planta*, **211**, 673–678.

Masterson, C., Wood, C. and Thomas, D.R. (1990) L-Acetylcarnitine, a substrate for chloroplast fatty acid synthesis. *Plant Cell Environment*, **13**, 755–765.

McMurray, L.M., Oethinger, M. and Levy, S.B. (1998) Triclosan targets lipid synthesis. *Nature*, **394**, 531–532.

Mdluli, K., Slayden, R.A., Zhu, Y.Q., Ramaswamy, S., Pan, X., Mead, D., Crane, D.D., Musser, J.M. and Barry, C.E. (1998) Inhibition of a *Mycobacterium tuberculosis* β-ketoacyl ACP synthase by isoniazid. *Science*, **280**, 1607–1610.

Mekhedov, S., Cahoon, E.B. and Ohlrogge, J.B. (2001) An unusual seed-specific 3-ketoacyl-ACP synthase associated with the biosynthesis of petroselenic acid in coriander. *Plant Molecular Biology*, **47**, 507–518.

Mekhedov, S., de llarduya, O.M. and Ohlrogge, J.B. (2000) Toward a functional catalog of the plant genome. A survey of genes for lipid biosynthesis. *Plant Physiology*, **122**, 389–401.

Miernyk, J.A. and Dennis, D.T. (1983) The incorporation of glycolytic intermediates into lipids by plastids from developing endosperm of castor oil seeds. *Journal of Experimental Botany*, **34**, 712–718.

Mikolajczyk, S. and Brody, S. (1990) *De novo* fatty acid synthesis mediated by acyl carrier protein in *Neurospora* mitochondria. *European Journal of Biochemistry*, **187**, 431–437.

Murata, M., Yamada, M., Nishida, I., Okuyama, H., Sekiya, J. and Wada, H., eds (2003). *Advanced Research on Plant Lipids*, Kluwer, Dordrecht.

Murphy, D.J., ed. (1994) *Designer Oil Crops*, VCH, Weinheim.

Murphy, D.J. (2003) Agricultural biotechnology and oil crops – current uncertainties and future potential. *Applied Biotechnology and Food Science Policy*, **1**,25–38.

Murray, B.G., Brule-Babel, A.L. and Morrison, I.N. (1996) Two distinct alleles encode for acetyl-CoA carboxylase resistance in wild oat. *Weed Science*, **44**, 476–481.

Nield, P.C., Rawsthorne, S., Flugge, U-I. and Hills, M.J. (2003) The role of the plastidial glucose 6-phosphate transporter in the control of fatty acid biosynthesis of developing embryos of oilseed rape, in *Advanced Research on Plant Lipids* (ed. N. Murata, M. Yamada, I. Nishida, H. Okuyama, J. Sekiya and H. Wada), Kluwer, Dordrecht, pp. 81–84.

Nikolau, B.J., Ohlrogge, J.B. and Wurtele, E.S. (2003) Plant biotin-containing carboxylases. *Archives of Biochemistry and Biophysics*, **414**, 211–222.

Nikolskaya, T., Zagnitko, O., Tevzadze, G., Haselkorn, R. and Gornicki P. (1999) Herbicide sensitivity determinant of wheat plastid ACCase is located in a 400-amino acid fragment of the carboxyltransferase domain. *Proceedings of the National Academy of Sciences USA*, **96**, 14647–14651.

O'Hara, P., Slabas, A.R. and Fawcett, T. (2000) Modulation of fatty acid biosynthesis by β-keto reductase expression. *Biochemical Society Transactions*, **28**, 613–615.

O'Hara, P., Slabas, A.R. and Fawcett, T. (2001) Fatty acid synthesis in developing leaves of *Brassica napus* in relation to leaf growth and changes in activity of 3-oxoacyl-ACP reductase. *Federation of European Biochemical Societies Letters*, **488**, 18–22.

O'Hara, P., Slabas, A.R. and Fawcett, T. (2002) Fatty acid and lipid biosynthetic genes are expressed at constant molar ratios but different absolute levels during embryogenesis. *Plant Physiology*, **129**, 310–320.

Ohlrogge, J.B. (1987) Biochemistry of plant acyl carrier proteins, in *The Biochemistry of Plants*, Vol. 9 (eds P.K. Stumpf and E.E. Conn), Academic Press, New York, pp. 137–157.

Ohlrogge, J.B. (1994) Design of new plant products: engineering of fatty acid metabolism. *Plant Physiology*, **104**, 185–190.

Ohlrogge, J.B. and Jaworski, J.G. (1997) Regulation of fatty acid synthesis. *Annual Review of Plant Physiology and Plant Molecular Biology*, **48**, 109–136.

Ohlrogge, J.B., Browse, J. and Somerville, C.R. (1991) The genetics of plant lipids. *Biochimica et Biophysica Acta*, **1082**, 1–26.

Olsen, J.G., Kadziola, A., von Wettstein-Knowles, P., Siggaard-Anderson, M., Lindquist, Y. and Larsen, S. (1999). The X-ray crystal structure of β-ketoacyl[acyl carrier protein] synthase I. *Federation of European Biochemical Societies Letters*, **460**, 46–52.

Olsen, J.G., Kadziola, A., von Wettstein-Knowles, P., Siggaard-Anderson, M. and Larsen, S. (2001) Structures of β-ketoacyl-acyl carrier protein synthase 1 complexed with fatty acid elucidate its catalytic machinery. *Structure*, **9**, 233–243.

Page, R.A., Okada, S. and Harwood, J.L. (1994) Acetyl-CoA carboxylase exerts strong flux control over lipid synthesis in plants. *Biochimica et Biophysica Acta*, **1210**, 369–372.

Peskan, T., Westermann, M. and Oelmuller, R. (2000) Identification of low-density Triton-X-100-insoluble plasma membrane microdomains in higher plants. *European Journal of Biochemistry*, **267**, 6989–6995.

Pike, M.J., Everett, C.J. and Rawsthorne, S. (2003) Carbon metabolism in developing soybean embryos, in *Advanced Research on Plant Lipids* (ed. N. Murata, M. Yamada, I. Nishida, H. Okuyama, J. Sekiya and H. Wada) Kluwer, Dordrecht, pp. 69–71.

Post-Beittenmiller, D., Roughan, P.G. and Ohlrogge, J.B. (1992) Regulation of plant fatty acid biosynthesis. Analysis of acyl-CoA and acyl-acyl carrier protein substrate pools in spinach and pea chloroplasts. *Plant Physiology*, **100**, 923–930.

Price, L.J., Herbert, D., Cole, D.J. and Harwood, J.L. (2003a) Use of plant cell cultures to study graminicide effects on lipid metabolism. *Phytochemistry* **63**, 533–541.

Price, L.J., Herbert, D., Moss, S.R., Cole, D.J. and Harwood, J.L. (2003b) Graminicide insensitivity correlates with herbicide-binding co-operativity on acetyl-CoA carboxylase isoforms. *The Biochemical Journal*, **375**, 415–423.

Price, L.J., Moss, S.R., Cole, D.J. and Harwood, J.L. (2004) Graminicide resistance in a blackgrass (*Alopecurus myosuroides*) population correlates with insensitivity of acetyl-CoA carboxylase. *Plant Cell and Environment* **27**, 15–27.

Qui, X.Y., Janson, C.A., Konstantinidis, A.K. *et al.* (1999) Crystal structure of β-ketoacyl-acyl carrier protein synthase III. *Journal of Biological Chemistry*, **274**, 36465–36471.

Ramli, U.S., Baker, D.S., Quant, P.A. and Harwood, J.L. (2002a) Control mechanisms operating for lipid biosynthesis differ in oil-palm (*Elaeis guineensis* Jacq.) and olive (*Olea europaea* L.) callus cultures. *The Biochemical Journal*, **364**, 385–391.

Ramli, U.S., Baker, D.S., Quant, P.A. and Harwood, J.L. (2002b) Control analysis of lipid biosynthesis in tissue cultures from oil crops shows that flux control is shared between fatty acid synthesis and lipid assembly. *The Biochemical Journal*, **364**, 393–401.

Rasmussen, A.V., Yasuno, R. and von Wettstein-Knowles, P. (2003) Characterising KAS enzymes using the *E.coli* double mutant strain CY244, in *Advanced Research on Plant Lipids* (eds. N. Murata, Y. Yamada, I. Nishida, H. Okuyama, J. Sekiya and H. Wada) Kluwer, Dordrecht, pp. 65–68.

Ratledge, C., Bowater, M.D.V. and Taylor, P.N. (1997) Correlation of ATP/citrate lyase activity with lipid accumulation in developing seeds of *Brassica napus*. *Lipids*, **32**, 7–12.

Rawsthorne, S. (2002) Carbon flux and fatty acid synthesis in plants. *Progress in Lipid Research*, **41**, 182–196.

Rendina, A.R. and Felts, J.M. (1988) Cyclohexanedione herbicides are selective and potent inhibitors of ACCase from grasses. *Plant Physiology*, **86**, 983–986.

Rendina, A.R., Beaudoin, J.D., Craig-Kennard, A.C. and Breen, M.K. (1989) Kinetics of inhibition of ACCase by the aryloxyphenoxypropionate and cyclohexanedione graminicides. *Proceedings of the Brighton Crop Protection Conference – Weeds*. British Crop Protection Council, Farnham, UK, pp. 163–172.

Roesler, K., Shintani, D., Savage, L., Boddupalli, S. and Ohlrogge, J.B. (1997) Targetting of the *Arabidopsis* homomeric acetyl-CoA carboxylase to *Brassica napus* seed plastids. *Plant Physiology*, **113**, 75–81.

Roesler, K.R., Savage, L.J., Shintani, D.K., Shorrosh, B.S. and Ohlrogge, J.B. (1996) Co-purification, co-immunoprecipitation and coordinate expression of acetyl-CoA carboxylase activity, biotin carboxylase and biotin carboxyl carrier protein of higher plants. *Planta*, **198**, 517–525.

Roessler, P.B. and Ohlrogge, J.B. (1993) Cloning and characterization of the gene that encodes acetyl-CoA carboxylase in the alga *Cyclotella cryptica*. *Journal of Biological Chemistry*, **268**, 19254–19259.

Roughan, P.G. and Ohlrogge, J.B. (1996) Evidence that isolated chloroplasts contain an integrated lipid-synthesizing assembly that channels acetate into long-chain fatty acids. *Plant Physiology*, **110**, 1239–1247.

Roughan, P.G. and Slack, C.R. (1982) Cellular organisation of glycerolipid metabolism. *Annual Review of Plant Physiology*, **33**, 97–132.

Roughan, P.G., Post-Beittenmiller, D., Ohlrogge, J.B. and Browse, J. (1993) Is acetylcarnitine a substrate for fatty acid synthesis in plants? *Plant Physiology*, **101**, 1157–1162.

Roujeinikova, A., Levy, C.W., Rowsell, S. *et al.* (1999) Crystallographic analysis of triclosan bound to enoyl reductase. *Journal of Molecular Biology*, **294**, 527–535.

Safford, R., Windust, J.H.C., Lucas, C., De Silva, J., James, C.M., Hellyer, A., Smith, C.G., Slabas, A.R. and Hughes, S.G. (1988) Plastid-localized seed acyl-carrier protein of *Brassica napus* is encoded by a distinct, nuclear multigene family. *European Journal of Biochemistry*, **174**, 287–295.

Salas, J.J. and Ohlrogge, J.B. (2002). Characterization of substrate specificity of plant FatA and FatB acyl-ACP thioesterases. *Archives of Biochemistry and Biophysics*, **403**, 25–34.

Sanchez, J., Cerda-Olmedo, E. and Martinez-Force, E., ed. (1998). *Advances in Plant Lipid Research*, Universidad de Sevilla, Seville.

Sasaki, Y., Hakamada, K., Suama, Y., Nagano, Y., Furusawa, I. and Matsuno, R. (1993) Chloroplast-encoded protein as a subunit of acetyl-CoA carboxylase in pea plant. *Journal of Biological Chemistry*, **268**, 25118–25123.

Sasaki, Y., Kozaki, A. and Hatano, M. (1997) Link between light and fatty acid synthesis: thioredoxin-linked reductive activation of plastidic acetyl-CoA carboxylase. *Proceedings of the National Academy of Sciences USA*, **94**, 11096–11101.

Savage, L.J. and Ohlrogge, J.B. (1999) Phosphorylation of pea chloroplast acetyl-CoA carboxylase. *The Plant Journal*, **18**, 521–527.

Schulte, U., Fecke, W., Krull, C. *et al.* (1994) *In vivo* dissection of the mitochondrial respiratory NADH: ubiquinone oxidoreductase (complex 1). *Biochimica et Biophysica Acta*, **1187**, 121–124.

Schulte, W., Schell, J. and Topfer, R. (1994) A gene encoding acetyl-CoA carboxylase from *Brassica napus*. *Plant Physiology*, **106**, 793–794.

Schulte, W., Topfer, R., Stracke, R., Schell, J. and Martini, N. (1997) Multifunctional acetyl-CoA carboxylase from *Brassica napus* is encoded by a multi-gene family. Indication for plastidic localization of at least one isoform. *Proceedings of the National Academy of Sciences USA*, **94**, 3465–3470.

Schwender, J. and Ohlrogge, J.B. (2002) Probing *in vivo* metabolism by stable isotope labelling of storage lipids and proteins in developing *Brassica napus* embryos. *Plant Physiology*, **130**, 347–361.

Secor, J. and Cseke, C. (1988) Inhibition of ACCase activity by haloxyfop and tralkoxydim. *Plant Physiology*, **86**, 10–12.

Sellwood, C., Slabas, A.R. and Rawsthorne, S. (2000) Effects of manipulating expression of acetyl-CoA carboxylase 1 in *Brassica napus* embryos. *Biochemical Society Transactions*, **28**, 598–600.

Shimakata, T. and Stumpf, P.K. (1982) Purification and characterisations of β-ketoacyl-ACP reductase, β-hydroxyacyl-ACP dehydrase and enoyl-ACP reductase from *Spinacia oleracea* leaves. *Archives of Biochemistry and Biophysics*, **218**, 77–91.

Shintani, D., Roesler, K., Shorrosh, B., Savage, L. and Ohlrogge, J.B. (1997) Antisense expression and overexpression of biotin carboxylase in tobacco leaves. *Plant Physiology*, **114**, 881–886.

Shintani, D.K. and Ohlrogge, J.B. (1994) The characterisation of a mitochondria acyl carrier protein isoform isolated from *Arabidopsis thaliana. Plant Physiology*, **104**, 1221–1229.

Shintani, D.K. and Ohlrogge, J.B. (1995) Feedback inhibition of fatty acid synthesis in tobacco suspension cells. *The Plant Journal*, **7**, 577–587.

Shorrosh, B.S., Dixon, R.A. and Ohlrogge, J.B. (1994) Molecular cloning, characterisation and elicitation of acetyl-CoA carboxylase from alfalfa. *Proceedings of the National Academy of Sciences USA*, **91**, 4323–4327.

Shorrosh, B.S., Roesler, K.R., Shintani, D., van de Loo, F.J. and Ohlrogge, J.B. (1995) Structural analysis, plastid localization and expression of the biotin carboxylase subunit of acetyl-CoA carboxylase from tobacco. *Plant Physiology*, **108**, 805–812.

Shorrosh, B.S., Savage, L.J., Soll, J. and Ohlrogge, J.B. (1996) The pea chloroplast membrane-associated protein, IEP96, is a subunit of acetyl-CoA carboxylase. *The Plant Journal*, **10**, 261–268.

Siggaard-Andersen, M., Kauppinen, S. and von Wettstein-Knowles, P. (1991) Primary structure of a cerulenin-binding β-ketoacyl [acyl-carrier protein] synthase from barley chloroplasts. *Proceedings of the National Academy of Sciences, U.S.A.*, **88**, 4114–4118.

Simoni, R.D., Criddle, R.S. and Stumpf, P.K. (1967) Purification and properties of plant and bacterial acyl carrier proteins. *Journal of Biological Chemistry*, **242**, 573–581.

Slabas, A.R., Chase, D., Nishida, I., Murata, N., Sidebottom, C., Safford, R., Sheldon, P.S., Kekwick, R.G.O., Hardie, D.G. and MacKintosh, R.W. (1992) Molecular-cloning of higher-plant 3-oxoacyl-(acyl carrier protein) reductase - sequence identities with the nodg-gene product of the nitrogen-fixing soil bacterium Rhizobium-meliloti. *The Biochemical Journal*, **283**, 321–326.

Slabas, A.R., Cottingham, I.R., Austin, A., Hellyer, A., Safford, R. and Smith, C.G. (1990) Immunological detection of NADH-specific enoyl-ACP reductase from rape seed (*Brassica napus*). *Biochimica et Biophysica Acta*, **1039**, 181–188.

Slabas, A.R., Sidebottom, C.R., Hellyer, A., Kessel, R.M.J. and Tombs, M.P. (1986) Induction, purification and characterisation of NADH-specific enoyl acyl carrier protein reductase from developing seeds of oil seed rape. *Biochimica et Biophysica Acta*, **877**, 271–280.

Slabaugh, M., Leonard, J. and Knapp, S. (1998) Condensing enzymes from *Cuphea wrightii* associated with medium chain fatty acid biosynthesis. *The Plant Journal*, 13, 611–620.

Smith, R.G., Gauthier, D.A., Dennis, D.T. and Turpin, D.H. (1992) Malate- and pyruvate-dependent fatty acid synthesis in leucoplasts from developing castor endosperm. *Plant Physiology*, **98**, 1233–1238.

Stumpf P.K., ed. (1980) *The Biochemistry of Plants*, Vol. 4, Academic Press, New York.

Stumpf, P.K., ed. (1987) *The Biochemistry of Plants*, Vol. 9, Academic Press, New York.

Sun, J., Ke, J., Johnson, J.L., Nikolau, B.J. and Wurtele, E.S. (1997) Biochemical and molecular biological characterisation of *CAC2*: the *Arabidopsis thaliana* gene coding for the biotin carboxylase subunit of plastidic acetyl-coenzyme A carboxylase. *Plant Physiology*, **115**, 1371–1383.

Tai, H. and Jaworski, J.G. (1995) Expression of cDNA encoding 3-ketoacyl-acyl carrier protein synthase III (KAS III) in tobacco and *Arabidopsis. Presented at the Biochemistry and Molecular Biology of Plant Fatty Acids and Glycerolipids Symposium*, South Lake Tahoe, California, USA.

Tardif, F.J., Holtum, J.A.M. and Powles, S.B. (1993) Occurrence of a herbicide-resistant ACCase mutant in annual rye grass selected by sethoxydim. *Planta*, **190**, 176–181.

Tardif, F.J., Preston, C., Holtum, J.A.M. and Powles, S.B. (1996) Resistance to ACCase-inhibiting herbicides endowed by a single major gene encoding a resistance target site in a biotype of *Lolium rigidum. Australian Journal of Plant Physiology*, **23**, 15–23.

Thelen, J.J. and Ohlrogge, J.B. (2002a) Both antisense and sense expression of biotin carboxyl carrier protein isoform 2 inactivates the plastid acetyl-CoA carboxylase in *Arabidopsis thaliana. The Plant Journal*, **32**, 419–431.

Thelen, J.J. and Ohlrogge, J.B. (2002b) The multisubunit acetyl-CoA carboxylase from pea chloroplasts is associated with envelope membranes in a non-ionic manner through the carboxyl-transferase subunit. *Archives of Biochemistry and Biophysics*, **400**, 245–257.

Thelen, J.J., Mekhedov, S. and Ohlrogge, J.B. (2000) Biotin carboxyl carrier protein isoforms in Brassicaceae oilseeds. *Bichemical Society Transactions*, **28**, 595–598.

Thelen, J.J., Mekhedov, S. and Ohlrogge, J.B. (2001) Brassicaceae express multiple isoforms of biotin carboxyl carrier protein in a tissue-specific manner. *Plant Physiology*, **125**, 2016–2028.

Verwoert, I.I.G.S., Linden, K. Hvd., Walsh, M.C., Nijkamp, H.J.J. and Stuitje, A.R. (1995). Modification of *Brassica napus* seed oil by expression of the *Escherichia coli* fab H gene, encoding 3-ketoacyl-acyl carrier protein synthase III. *Plant Molecular Biology*, **27**, 875–886.

Voelker, T. and Kinney, A.J. (2001) Variations in the biosynthesis of seed-storage lipids. *Annual Review of Plant Physiology and Plant Molecular Biology*, **52**, 355–361.

Voelker, T.A., Jones, A., Cranmer, A.M., Davies, H.M. and Knutzon, D.S. (1997) Broad-range and binary-range acyl-ACP thioesterases suggest an alternative mechanism for medium-chain production in seeds. *Plant Physiology*, **114**, 669–677.

Wada, H., Shintani, D. and Ohlrogge, J. (1997) Why do mitochondria synthesise fatty acids? Evidence for involvement in lipoic acid production. *Proceedings of the National Academy of Sciences USA*, **94**, 1591–1596.

Walker, K.A., Ridley, S.H., Lewis, T. and Harwood, J.L. (1988) Fluazifop, a grass-selective herbicide which inhibits acetyl-CoA carboxylase in sensitive plant species. *The Biochemical Journal*, **254**, 307–310.

Walsh, M.C., Klopfenstein, W.E. and Harwood, J.L. (1990) The short chain condensing enzyme has a widespread occurrence in the fatty acid synthetases from higher plants. *Phytochemistry*, **29**, 3797–3799.

Williams, J.P., Khan, M.U. and Lem, N.W., eds (1997) *Physiology, Biochemistry and Molecular Biology of Plant Lipids*, Kluwer, Dordrecht.

Wilson, R.F., Marquardt, T.C., Novitzky, W.P. *et al.* (2001). Metabolic mechanisms associated with alleles governing the 16:0 concentration of soybean oil. *Journal of the American Oil Chemists Society*, **78**, 335–340.

Wu, J., James, J.W., Dooner, H.K. and Browse, J. (1994) A mutant of Arabidopsis deficient in the elongation of palmitic acid. *Plant Physiology*, **106**, 143–150.

Zagnitko, O., Jelinska, J., Tevzadze, G., Haselkorn, E. and Gornicki, P. (2001) An isoleucine/leucine residue in the carboxyltransferase domain of ACCase is critical for interaction with aryloxyphenoxypropionate and cyclohexanedione inhibitors. *Proceedings of the National Academy of Sciences USA*, **98**, 6617–6622.

Zimmerlin, A. and Durst, F. (1992) Aryl hydroxylation of the herbicide diclofop by a wheat cytochrome P-450 monooxygenase: substrate specificity and physiological activity. *Plant Physiology*, **100**, 874–881.

# 3 Fatty acid manipulation

David F. Hildebrand, Keshun Yu, Charles McCracken
and Suryadevara S. Rao

## 3.1 Introduction

The major oilseed crops currently supply close to 100 million metric tons of oil worldwide (Hildebrand and Yu, 2003; Mielke, 2003). About two-thirds of this oil is from soybean, palm and canola (genetically improved rapeseed). The major fatty acids from the world oil supply are palmitic, linoleic and oleic acids. Many unusual fatty acids can accumulate in seed oils as their major components, or in substantial amounts sometimes comprising more than 90% of the seed oil (Hildebrand and Yu, 2003) (see below and Table 3.1). Unusual fatty acids include variations in carbon chain length and degree of unsaturation. Most fatty acids have double bonds in the *cis* configuration. However, *trans* double bonds are also found. There are also different positional isomers, conjugated unsaturated, acetylenates, hydroxy, epoxy and keto fatty acids. Cyclopropenoids, cyclopentenoids and even fluoro fatty acids add more diversity to the fatty acid species. To increase utilization of some of these unusual fatty acids, scientists have been trying to domesticate several unusual fatty acid producing oilseeds. Unfortunately, domestication of most of these unusual oilseeds has a diversity of problems, such as climatic requirements, low yield, harvesting, weed control and other management factors. Thus, the most promising route to development of such renewable resources is via oilseed genetic engineering. This oilseed engineering is predicated on the identification and cloning of key genes involved in the biosynthesis and accumulation of unusual fatty acids. In this chapter, we will cover the cloning and characterization of Δ9 and Δ12 desaturase genes and the various homologs that apparently evolved from them. The enzymes encoded by these genes are involved in unusual fatty acid synthesis and the accumulation of these compounds in triacylglycerols.

Higher plants possess a number of different unsaturated fatty acids (Hilditch and Williams, 1964), of which the principal ones have double bonds at the 9, 12 and 15 (ω3) positions of 18-carbon acyl chains (Browse and Somerville, 1991). Plants have two main types of fatty acid desaturases, which use glycerolipids and acyl carrier proteins (ACP) as substrates unlike most animal and fungal desaturases. Interestingly, all the plant membrane-bound desaturases prefer glycerolipids (Pugh and Kates, 1977) to acyl-CoA as substrates. Fatty acid desaturases are, therefore,

**Table 3.1** Some unusual fatty acids that accumulate in seed oils and examples of seeds with high levels.

| Common name | Chemical name | High accumulator | % unusual fatty acid | Reference |
|---|---|---|---|---|
| Alchornoic | 14-epoxy,*cis*-11-eicosenoic | *Alchornea cordifolia* | 50 | Kleiman *et al.*, 1977 |
| Axillarenic | 11,13-dihydroxytetracos-*trans*-9-enoic | *Baliospermum axillare* | 3 | Husain *et al.*, 1980 |
| Calendic | *trans*-8,*trans*-10,*cis*-12-octadecatrienoic | *Calendula officinalis* | 63 | Meier zu Beerentrup and Roebblen, 1987 |
| Catalpic | *trans*-9,*trans*-11,*cis*-13-octadecatrienoic | *Catalpa bignonioides* | | Joh *et al.*, 1995 |
| Dimorphecolic | 9-hydroxy-*trans*-10-*trans*-12-octadecadienoic | *Dimorphotheca pluvialis* | 60 | Muuse *et al.*, 1994 |
| Coronaric | 9-epoxy,*cis*-12-octadecenoic | *Chrysanthemum coronarium* | | Smith *et al.*, 1959 |
| Crepenynic | 12acetylenic,*cis*-9-octadecenoic | *Crepis alpina* | 74 | Meier zu Beerentrup and Roebblen, 1987 |
| Eleostearic | *cis*-9,*trans*-11,*trans*-13-octadecatrienoic | *Aleurites fordii* | 80 | Dyer *et al.*, 2003 |
| Epoxystearic | 9-epoxy-octadecanoic | *Tragopogon porrifolius* | 3 | Chisholm and Hopkins, 1959 |
| Isanolic | 8-hydroxyoctadec-17-ene-9,11-di-ynoic | *Ongokea gore* | | Uzzan, 1961 |
| Isoricinoleic | 9-hydroxy-12-*cis*-octadecenoic | *Wrightia coccinea* | 76 | Siddiqi *et al.*, 1980 |
| Licanic | 4-oxo-*cis*-9,*trans*-11,*trans*-13-octadecatrienoic | *Licania rigida* | 78 | Paschke *et al.*, 1953 |
| Lesquerolic | 14-hydroxy-*cis*-11-eicosenoic | *Lesquerella fendleri* | 55 | Carlson *et al.*, 1990 |
| Parinaric | *cis*-9,*trans*-11,*trans*-13,*cis*-15-octadecatetraenoic | *Parinarium laurunum* | 54 | Takagi, 1965 |
| Punicic | *cis*-9,*trans*-11,*cis*-13-octadecatrienoic | *Punicia granatum* | 86 | Wang *et al.*, 1998 |
| Phloionolic | 9,10,18-trihydroxyoctadecanoic | *Chamaepeuce afra* | 9 | Mikolajczak and Smith, 1967 |
| Ricinoleic | 12-hydroxy-9-*cis*-octadecenoic | *Ricinus communis* | 88 | Borch-Jensen *et al.*, 1997 |
| Vernolic | 12-epoxy,*cis*-9-octadecenoic | *Vernonia galamensis* | 80 | Ayorinde *et al.*, 1990 |

*Note:* t, trans; h, hydroxy; e, epoxy; a, acetylene; o, oxo; me, methylene with double bond, m, methylene.

mainly classified into three types: acyl-lipid, acyl-ACP and acyl-CoA desaturases (Avelange-Macherel *et al.*, 1995).

## 3.2 The soluble Δ9 desaturases

Plant stearoyl-ACP desaturase is unusual in that it is the only soluble Δ9 desaturase identified to date (Shanklin and Somerville, 1991). These authors purified the stearoyl-ACP desaturase from avocado mesocarp and used mono-specific polyclonal antibodies to isolate full-length cDNA clones from castor seed and cucumber. There was no detectable identity between the deduced amino acid sequences of the castor Δ9 desaturase from rat liver or yeast Δ9 desaturases and the Δ9 acyl-lipid desaturase and Δ12 desaturase from *Synechocystis*, suggesting that Δ9 desaturases of higher plants are structurally different from the Δ9-acyl-CoA desaturases of yeast and mammals and are evolved independently (Shanklin and Somerville, 1991). The soluble Δ9 desaturases are localized in the plastids implying they might have close links to their cyanobacterial homologs – the obvious phylogenetic source of the chloroplasts. Interestingly, when expressed in *Escherichia coli*, Δ9 acyl-lipid desaturases from two cyanobacteria, *Anabaena variabilis* and *Synechocystis sp. PCC 6803* desaturated stearic acid to yield oleic acid at the C-1 positions of phosphatidylethanolamine and phosphatidylglycerol (Sakamoto *et al.*, 1994). The deduced amino acid sequences of the Δ9 acyl-lipid desaturases are similar in part to those of stearoyl-CoA desaturases of the rat, the mouse and *Saccharomyces cerevisiae*, but not to those of acyl-(acyl-carrier-protein) desaturases of higher plants (Sakamoto *et al.*, 1994).

Thompson *et al.* deduced the amino acid sequence of purified safflower stearoyl-ACP desaturase and designed primers based on this sequence to clone a full length cDNA (Thompson *et al.*, 1991). However, the enzyme was active only when spinach ferredoxin (Fd) was added to the *E. coli* extract when the gene was expressed in *E. coli*. This led them to the conclusion that *E. coli* Fd functions poorly as an electron donor for plant desaturases. The *N*-terminal regions of all the three cDNAs of castor, cucumber and safflower have chloroplast transit peptide sequences indicating that the soluble stearoyl-ACP desaturase is localized in the plastids of higher plants. An NCBI BLAST search with the *Brassica* stearoyl-ACP desaturase reveals 17 plant stearoyl-ACP desaturases. Although none of these were from Arabidopsis, the complete expressed sequence tag (EST) analysis available in the Arabidopsis lipid gene database suggests that seven stearoyl-ACP desaturases are present in Arabidopsis, out of which, two isoforms are preferentially expressed in seeds (Beisson *et al.*, 2003). The Arabidopsis database search led to the identification of more interesting acyl-CoA desaturases (Figure 3.1), which will be explained later (Mekhedov *et al.*, 2001).

The only prokaryotic sequences coding for proteins similar to acyl-ACP desaturases but of unknown function are found in different strains of *Mycobacterium* and

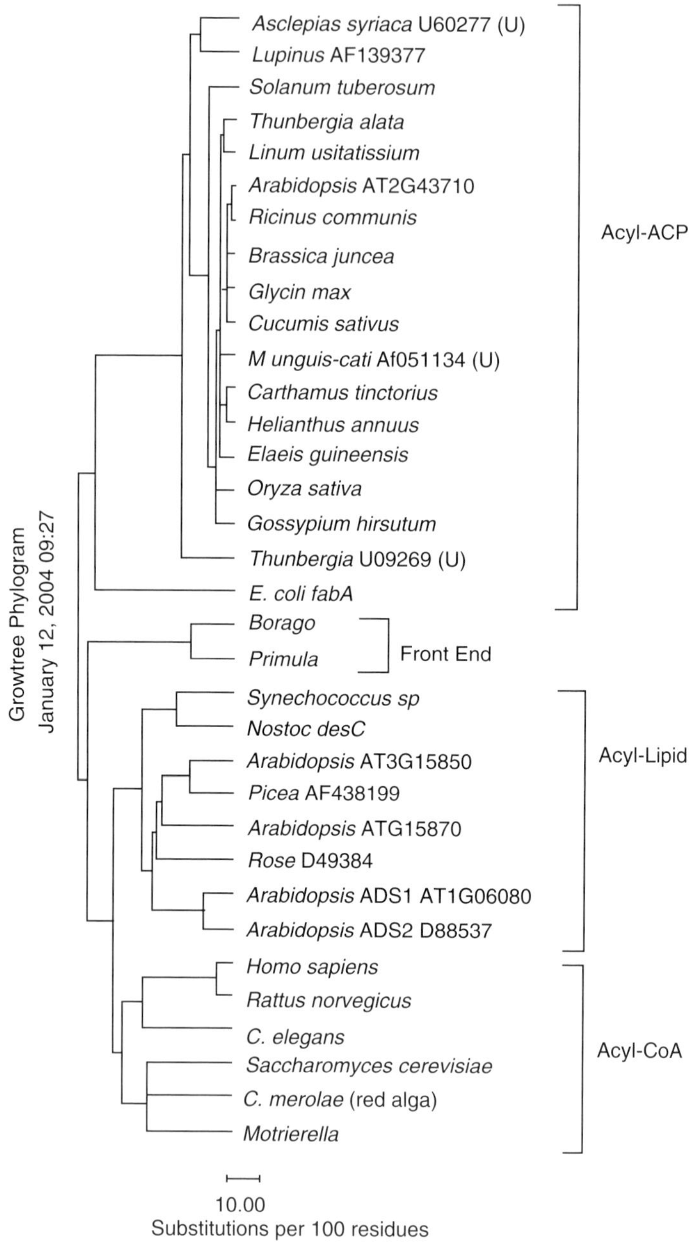


**Figure 3.1** A phylogram showing sequence relationships at the amino acid level of the Δ9 desaturases from different species. Unusual ACP-desaturases are marked with the letter U. Only one Arabidopsis Δ9 desaturase is an ACP-desaturase and the rest of the family members are acyl-lipid desaturases. Numbers following some of the entries are GenBank accession numbers.

*Streptomyces*. The relevance of this observation is presently restricted to the fact that such sequences do exist in prokaryotic organisms (Sperling *et al.*, 2003). In addition to stearoyl/palmitoyl-ACP desaturase, several species of plants express various soluble acyl-ACP desaturase activities capable of producing a diverse assortment of fatty acids (in excess of 200) with alterations in chain length, double-bond position and oxygenated functional groups (Kishore and Somerville, 1993). These include a Δ4-palmitoyl (16:0)-ACP desaturase of *Coriandrum sativum* (coriander) seed (Cahoon *et al.*, 1992), Δ6-16:0-ACP desaturase of *Thunbergia alata* (black-eyed Susan vine) seed (Cahoon *et al.*, 1994) and a Δ9-myristoyl (14:0)-ACP desaturase of *Pelargonium xhortorum* (geranium) trichomes (Schultz *et al.*, 1996). As their names indicate, these enzymes introduce double bonds into different positions (Δ4, Δ6 or Δ9) of fatty acids of different chain lengths (14, 16 or 18 carbon atoms) (Cahoon *et al.*, 1994; Suh *et al.*, 2002). These enzymes share more than or equal to 60% amino acid sequence identity with the Δ9-18:0-ACP desaturase (Cahoon and Shanklin, 2000). The seed oil in *Macadamia integrifolia* contains about 30% palmitoleic acid [16:1(Δ9)] (Gummeson *et al.*, 2000).

Characterization of the Δ9-18:0-ACP desaturase demonstrated that Fd-reducing systems, dependent on either NADPH/ferredoxin $NADP^+$ reductase (FNR) or on photoreduction of Fd supplied with chloroplast lamellae, could provide electrons to the desaturase reaction (Jacobson *et al.*, 1974). Fd has been most extensively studied in photosynthetic tissue where this protein serves as the major electron carrier from photosystem I to produce NADPH. Five Fd isoproteins were isolated from tomato and these isoproteins showed differential temporal and spatial accumulation patterns (Wada *et al.*, 1989; Aoki and Wada, 1996). Studies with Fd and its influence on biochemical reactions have revealed that distinct isoforms have the potential to function differently in separate reactions (Suh *et al.*, 2002). After the discovery of the acyl-ACP desaturase reaction and its preference for Fd as electron donor, essentially all studies of acyl-ACP desaturases in plants have used spinach Fd as the cofactor for *in vitro* studies. However, Suh *et al.*, used combinations of *in vitro* acyl-ACP desaturase assays and (Aach *et al.*, 1997) malonyl-coenzyme A labeling studies to determine that heterotrophic Fd isoforms support up to 20-fold higher unusual acyl-ACP desaturase activity in coriander (*C. sativum*), *T. alata* and garden geranium (*P. xhortorum*) when compared with photosynthetic Fd isoforms. Fd forms from the same species and tissues as the unusual acyl-ACP desaturases also increase activity of the ubiquitous Δ9-18:0-ACP desaturase, 1.5- to 3.0-fold, in both seed and leaf extracts.

### *3.2.1 Engineering chain length specificity of soluble Δ9 desaturases*

Cahoon *et al.* generated a series of chimeric enzymes by combining portions of Δ6-16:0-ACP desaturase and Δ9-18:0-ACP desaturase of *T. alata* to understand the structural basis of difference in these two activities (Cahoon *et al.*, 1997). The activities of the resulting chimeric enzymes led to the identification of 30 amino acids corresponding to residues 178–207 of the Δ9-18:0-ACP desaturase that contained determinants of both chain length and double bond positional specification. Two of these chimeras, chimeric 1 and chimeric 2, were generated by the introduction of amino acid residues 178–207 and 178–202 from Δ9-18:0-ACP desaturase in place of the corresponding portion of Δ6-16:0-ACP desaturase. The resulting chimeric 1 and chimeric 2 proteins have activities similar to Δ9-18:0-ACP desaturase, with chimeric 1 having specific activity with 18:0-ACP twice that of 16:0-ACP and chimeric 2 having 15-fold more specificity toward 18:0. Out of the 30 amino acids, 9 amino acid residues are different in Δ9-18:0 and Δ6-16:0-ACP desaturases. Using site-directed mutagenesis of these amino acids, independently or in combination, activities similar to chimeric 1 and chimeric 2 were created by these workers. The determination of the castor stearoyl-ACP desaturase structure (Lindqvist *et al.*, 1996) has provided a logical framework for these activity changes.

The soluble Δ9-desaturase enzyme contains a di-iron center, with ligands from four of the alpha helices of the helix bundle. The iron ions are bound in a highly symmetric environment, with one of the irons forming interactions with the side chains of E196 and H232 and the second iron with the side chains of E105 and H146. Adjacent to the iron atoms is a deep and narrow channel that likely represents the binding pocket for the stearate (18:0) portion of the substrate. The channel imposes a bent conformation on the fatty acid chain between carbon atoms 9 and 10, the site of desaturation. This conformation corresponds to the *cis* configuration of the oleoyl (18:1-Δ9)-ACP product and means that the double bond insertion site is located close to the di-iron center. In this model, the architecture of the fatty acid binding channel in relation to the catalytic iron cluster affects both substrate specificity and the position at which double bonds are introduced. With regard to substrate specificity of acyl-ACP desaturases, the geometry of the lower portion of the binding pocket places severe constraints on the length of the acyl chain that can be accommodated beyond the point where the double bond is to be introduced. As such, residues at the bottom of the substrate channel are likely the most critical for determining fatty acid chain-length specificity. This explains the functional properties of chimeric 2 and mutant A188G/Y189F, which display increased Δ6 desaturase activity with 18:0-ACP compared with the wild-type Δ6-16:0-ACP desaturase. Based on the model of the active site, substitution of alanine-188 for a smaller glycine and tyrosine-189 for phenylalanine extends the cavity at the bottom of the active site enough to accommodate the two additional carbon atoms at the methyl end of 18:0-ACP. As a result, chimeric 2 and mutant A188G/Y189F are able to desaturate 18:0-ACP at the Δ9 with a rate comparable to that observed with 16:0-ACP at the Δ6 position (Cahoon *et al.*, 1997).

Comparison of the predicted amino acid sequence of the cat's claw (that accumulates nearly 80% palmitoleic acid (16:1Δ9) plus *cis*-vaccenic acid (18:1Δ11) in its seed oil) Δ9-18:0-ACP desaturase enzyme with that of the castor enzyme (Cahoon *et al.*, 1998) suggested that a single amino acid substitution (L118W) might account in large part for the differences in substrate specificity between the two desaturases.

By converting leucine-118 to tryptophan in the mature castor Δ9-18:0-ACP desaturase they were able to increase the relative specificity of this enzyme for 16:0-ACP by 80-fold. As the amino acid sequences of all of the acyl-ACP desaturases are highly homologous and colinear over most of their length, these researchers targeted the eight residues located near the bottom of the substrate-binding pocket that help to determine the chain-length specificity of the enzymes (Cahoon *et al.*, 1997). The first of these residues, M114 of the mature castor protein, corresponds to M167 of the cat's claw sequence and M133 of the milkweed sequence (which has seeds also enriched in palmitoleic (16:1Δ9) and *cis*-vaccenic (18:1Δ11) acids). Out of the eight residues, three are altered in the milkweed enzyme relative to the castor sequence (L115I, T117R and P179T), but only one change is observed in the cat's claw protein L118W. Because of its bulky side chain, tryptophan 118 can be predicted to reduce the depth of the substrate pocket and thus favor the binding of 16:0-ACP in comparison to 18:0-ACP. To test this hypothesis Cahoon and Shanklin (2000) engineered the L118W variant of the Δ9-18:0-ACP castor desaturase. Its activity with 16:0-ACP was 115% relative to its activity with 18:0-ACP. Similarly, when the ORF of G188L mutant was linked to the plastid transit peptide of the coriander Δ4-16:0-ACP desaturase and expressed in *Arabidopsis thaliana* under napin promoter single seeds from transgenic $T_1$ plants accumulated 16:1Δ9, 18:1Δ11 and 20:1Δ13 to amounts as high as 13% of the total fatty acids of the seed oil. In contrast, these fatty acids accounted for 3–4% of the total fatty acids in seeds of wild-type Arabidopsis transformed with the expression vector alone or with the wild-type castorΔ9-18:0-ACP desaturase.

In most cases, the production level of unusual fatty acid structures in transgenic plants has been substantially below the level found in the species from which genes for its biosynthesis were derived (Schultz *et al.*, 1996; Cahoon *et al.*, 1992; Cahoon, 2000). Although no single factor was identified as limiting unusual monoene products in transgenic plants, the work done by Schultz *et al.* (1996) on three distinct acyl-ACP desaturases from coriander, *T. alata* and geranium suggests that accumulation of unusual monoenoic fatty acids may be limited by the fact that novel desaturases do not fold properly into active enzymes when expressed in non-native environments. Alternatively, it may be that other yet-to-be-discovered accessory proteins are needed to provide maximal activity of the variant acyl-ACP desaturases. The low activity of the novel desaturases in transgenic systems may mean that protein engineering of the stearoyl-ACP desaturase will be a more effective strategy toward production of novel monoenes in oilseed crops (Cahoon *et al.*, 1997). In contrast, the work by Whittle and Shanklin (2001) showed that combinatorial mutagenesis produces superior mutant enzymes with greater efficiency. The point mutations generated earlier where the G188L modification resulted in a mutant castor stearoyl-ACP desaturase specific to palmitic acid turned out to have lower turnover rates. These reduced turnover rates with 16 and 14 carbon substrates have been correlated with a loss of binding energy corresponding to the dissolvation of two or four fewer methylene groups (Haas and Fox, 1999). This led Whittle and Shanklin (2001) to follow combinatorial saturation mutagenesis of wild-type castor enzyme. These experiments resulted in the identification of two key substitutions (T117R/G188L) that produced a mutant Δ9-16:0-ACP desaturase with a turnover rate approaching that of the wild-type. Kinetic analysis revealed changes in $K_m$ and $K_{cat}$ that resulted in an 82-fold improvement in specific activity for a 16-carbon substrate compared with the wild-type enzyme. Improved substrate orientation apparently compensated for loss of binding energy that results from the loss of dissolvation energy for 16-carbon substrates.

A bacterial selection system capable of distinguishing between soluble desaturase enzymes with different chain length specificities (Cahoon and Shanklin, 2000) has been successfully used for the identification of these mutant enzymes. Soluble desaturases specific for 18-carbon chain length are unable to complement MH13, an unsaturated fatty acid auxotroph of *E. coli* that is defective in the anaerobic dehydratase/isomerase pathway of double bond incorporation into fatty acids, presumably due to the lack of available 18:0-ACP substrate. However, desaturase variants capable of recognizing substrates of 16- or 14-carbon chain length are able to complement the unsaturated fatty acid auxotrophy.

### *3.2.2 Stearoyl-CoA desaturases*

In animal and fungal cells, monounsaturated fatty acids are synthesized via an aerobic process from saturated fatty acid precursors by a microsomal, membrane-bound, three-component enzyme system involving cytochrome b5, NADH-dependent cytochrome b5 reductase and the Δ9 fatty acid desaturase. Isolation and characterization of fatty acid desaturase enzymes has proved difficult due to their highly hydrophobic nature and tight association with membranes. Although fatty acid desaturation was first described using the yeast Δ9 desaturase system, only animal Δ9 enzymes have been successfully purified to homogeneity (Strittmatter *et al.*, 1974; Prasad and Joshi, 1979). The Δ9 desaturase from rat liver has been most extensively characterized. It is a protein consisting of 358 amino acids of which 62% are hydrophobic (Thiede *et al.*, 1986). Examination of deduced amino acid sequences for the membrane desaturases from mammals, fungi, insects, higher plants and cyanobacteria has revealed three regions of conserved primary sequence containing HX(3 or 4)H, HX(2 or 3)HH and HX(2 or 3)HH. This motif is also present in the bacterial membrane enzymes alkane hydroxylase (omega-hydroxylase) and xylene mono-oxygenase. Hydropathy analyses indicate that these enzymes contain up to three long hydrophobic domains, which is long enough to span the membrane bilayer twice. The conserved His-containing regions have a consistent positioning with respect to these potential membrane-spanning domains. Taken together, these observations suggest that the membrane fatty acid desaturases and hydrocarbon hydroxylases have a related protein fold, possibly arising from a common ancestral origin. Using parsimony analysis software, which takes into account the number of mutations required to transform one amino acid sequence into another, Shanklin *et al.* (1994) concluded that membrane desaturases and hydroxylases represent a gene family that has evolved from a prototype most closely resembling the rat Δ9 desaturase gene.

The Δ9-desaturase gene (OLE1) has been isolated from a number of yeasts and filamentous fungi and all have similar structures (Stukey *et al.*, 1990; Gargano *et al.*, 1995; Meesters and Eggink, 1996; Anamnart *et al.*, 1997). The substrate for the initial Δ9-desaturation of 18:0 to oleic acid (18:1) in fungi and animals is stearoyl-CoA but some of the subsequent desaturation steps, including Δ12- and Δ6-desaturation, occur using the fatty acyl chains of phospholipid molecules (Jackson *et al.*, 1998). The open reading frame of the *Saccharomyces cerevisiae* Ole1 aligns with and shows 36% identity and 60% similarity to the rat liver stearoyl-CoA desaturase protein. This comparison disclosed three short regions of high consecutive amino acid identity (>70% ) including a 11 of 12 perfect residue match (Stukey *et al.*, 1990). Expression of Ole1 is highly regulated in several organisms, including *S. cerevisiae*, and this control is exerted both at the transcriptional and post-transcriptional level (Choi *et al.*, 1996; Gonzalez and Martin, 1996). The Ole1 genes of *Mortirella alpina* (Wongwathanarat *et al.*, 1999), *Phalaris angusta* and *Yarrowia lipolytica* (Anamnart *et al.*, 1997) also showed transcriptional regulation in response to supplementation with Δ9-unsaturated fatty acids although no such repression was observed for *Kluyveromyces thermotolerans* or for *Saccharomyces kluveri* (Kajiwara, 2002). Similarly, fatty acid analysis showed that the ratio of palmitoleic acid to oleic acid was lower in *S. kluveri* (Kajiwara, 2002) and *M. alpina* (Wongwathanarat *et al.*, 1999) compared to *S. cerevisiae*.

Known fungal fatty acid desaturases are all endoplasmic reticulum (ER) membrane-bound enzymes that have their active site on the cytoplasmic face of the ER. The active site comprises three histidine-rich boxes, normally containing eight essential histidine residues, which fold to form the di-iron binding site in the native protein (Shanklin *et al.*, 1994). Cytochrome b5 is used as the electron donor and in the majority of cases the desaturase is a protein fusion with a cytochrome b5 domain attached at the *C* terminus. In the yeast Δ9-desaturase (OLE1), the cytochrome b5 domain exists as a 113-amino acid *C*-terminal fusion. Expression of OLE1 encoded on the yeast expression plasmid, rescued yeast double *mutants* that lacked both the OLE1 and microsomal cytochrome b5 genes (Sakuradani *et al.*, 1999). In addition, yeast cells in which the *C*-terminal-fused b5 domain of OLE1 is deleted remain fatty acid auxotrophic even in the presence of endogenous yeast cytochrome b5. These findings demonstrate that cytochrome b5 is not able to act in *trans* to complement the loss of the fusion domain of the yeast Δ9-desaturase (Mitchell and Martin, 1995). Therefore, the fusion domain plays an essential role in the desaturase reaction of this enzyme. In case of the rat liver Δ9 desaturase, however, no such fusion is present. So far one stearoyl-CoA Δ9 desaturase from the plant kingdom with a carboxy terminal cytochrome b5 has been reported from the red alga, *Cyanidioschyzon merolae* (Itoh *et al.*, 1998). The sequence of the *C. merolae* desaturase gene was well conserved compared to those of other species, except at the amino- and carboxyl-termini. The histidine motifs, which are associated with the di-iron-oxo moieties of desaturase (HRXXXH (residues 130–135), HRXHH (167–171) and HNXHH (304–308)), are highly conserved.

Until now, a plant Δ9 fatty acid desaturase gene homologous to animal and fungal Δ9 acyl-CoA desaturase has only been reported in rose petals (Fukuchi-Mizutani *et al.*, 1995). In fact, the rose Δ9 desaturase is more closely related to the cyanobacterial Δ9 desaturases, which are classified as acyl-lipid desaturases. Such sequences are also reported from Arabidopsis and *Picea* (Mekhedov *et al.*, 2001). When a phylogram of the Δ9 desaturases from different organisms is constructed, only one of the sequences from Arabidopsis is the acyl-ACP type and the rest of the sequences are more closely grouped with acyl-lipid Δ9 desaturases from cyanobacteria. This leads us to think that one group of Δ9 desaturases from the plants are evolved from the cyanobacteria but acyl-lipid Δ9 desaturases may be the precursors of acyl-ACP desaturases. Interestingly, the Δ9 desaturase from red alga (Itoh *et al.*, 1998) is grouped more closely to the fungal and animal desaturases (Figure 3.1).

A search of an Arabidopsis database led to the identification of an interesting acyl-CoA desaturase gene family that is most similar to animal and fungal acyl-CoA desaturases. There are 10 independent ESTs in Arabidopsis; 4 in tomato and 2 in Brassica, but this sequence class appears to be completely missing from rice. At this time, the precise function of these desaturase-like genes remains unknown. However, further analysis suggests that members of this family encode acyl-lipid desaturases, which most likely include palmitate-specific monogalactosyldiacylglycerol desaturase (FAD5). In animals and yeast, this class of desaturase genes encodes palmitoyl and stearoyl Δ9 desaturases, which use acyl-CoA (or possibly acyl-lipids) as substrates (Thiede *et al.*, 1986; Stukey *et al.*, 1989; Stukey *et al.*, 1990). The most obvious candidates for these reactions would be the desaturation of palmitoyl at the Δ7 position on monogalactosyldiacylglycerol or at the Δ3 *trans* position on phosphatidylglycerol. Both these reactions are associated with the Arabidopsis mutations *fad4* (Δ3) and *fad5* (Δ7) (Kunst *et al.*, 1989; Browse, 1985).

## 3.3 Front-end desaturases

A few plant species (such as borage and evening primrose) accumulate the $C_{18}$ polyunsaturate, γ-linolenic acid (GLA; 18:3 n-6), as a result of the Δ6-desaturation of linoleic acid (18:2, n-6) (Stymne and Stobart, 1986). Δ6-Desaturase is physically linked to its electron donor in the form of an *N*-terminal extension, distinguishing it from all other previously characterized plant microsomal desaturases (Napier *et al.*, 1997). This led Napier *et al.* (1997) to hypothesize that the presence of this *N*-terminal cytochrome b5 domain might enable these enzymes to perform carboxy-end directed desaturation (i.e. desaturation between a pre-existing double bond and the carboxyl-end of a fatty acid) as opposed to the much more prevalent methyl-directed desaturation found in plants, thus, calling them front-end desaturases. Δ6-Desaturases were also identified from the moss *Physcomitrella patens*, mice and humans, with more recent studies identifying orthologs from *Pythium irregulare*, *Echium* and *M. alpina* (Napier *et al.*, 2003) and Primula species, *Primula farinosa* L. (syn *Aleuritia farinosa* L.) and *Primula vialii Franchet* (syn. *Muscarioides vialii Franchet*) (Napier *et al.*, 2003). All these enzymes contained an *N*-terminal cytochrome b5 domain. The presence of a non-consensual 'histidine box' motif (in which the first histidine of the third histidine box is substituted with a glutamine) is present in all functionally characterized Δ6-desaturases. When this glutamine residue was mutated to a histidine artificially recreating the consensus motif, it failed to display any Δ6-desaturase activity leading the authors to conclude that the presence of glutamine in the third histidine box is essential (Sayanova *et al.*, 2001).

## 3.4 Δ12 Desaturase-like enzymes and their use in the modification of fatty acid residues

### *3.4.1 Structures and functions*

Fatty acid desaturases in plants introduce *cis*-double bonds into acyl chains with regio-specificity and adjacent double bonds are normally methylene-interrupted. However, some plants have evolved new functions for their desaturases. These new enzymes are termed 'desaturase-like' due to their similarities to desaturases. Desaturase-like enzymes create new functional groups for fatty acids, such as conjugated double bonds, hydroxy, epoxy and acetylenic groups (Kinney, 2001; Voelker and Kinney, 2001; Hildebrand and Yu, 2003). The resulting 'novel' fatty acids are just a small portion out of hundreds of diverse fatty acid species known to be produced in the plant kingdom. Most of the desaturase-like enzymes identified to date are encoded by diverged *Fad2* genes. Such genes include hydroxylases from *Ricinus communis* and *Lesquerella fendleri*, epoxygenases from *Vernonia galamensis* and *Crepis palaestina*, acetylenase from *Crepis alpina*, and conjugases from *Mormordica charantia*, *Impatiens balsamica* and *Calendula officinalis* (Hildebrand and Yu, 2003).

Similar to desaturation, the reactions for hydroxylation, epoxidation and acetylenation of fatty acids also first involve the enzymatic removal of a hydrogen atom from a methylene group in an acyl chain (Shanklin and Cahoon, 1998). Because the breaking of the C–H bond of a methylene group requires approximately 98 kcal/mol, an activated oxygen and a metal cofactor are needed for initiation of the reactions (Shanklin and Cahoon, 1998). A short electron transport chain is also required for these reactions (Shanklin and Cahoon, 1998). In the ER, the electron transport consists of cytochrome b5 and cytochrome b5 reductase and the electron donor is NADH (Spatz and Strittmatter, 1971; Dailey and Strittmatter, 1979;

Hackett and Strittmatter, 1984). Most desaturase-like enzymes identified thus far are ER-associated, non-heme di-iron proteins with a consensus motif composed of histidines (HX3-4H, HX2-3HH, (H/Q)X2HH) that are in equivalent hydrophilic domains separated by equivalent hydrophobic domains and positioned on the cytoplasmic face of the membrane (Shanklin and Cahoon, 1998). Site-directed mutagenesis studies revealed that all the histidines are essential for catalysis (Shanklin *et al.*, 1994).

To explain the reaction mechanism, two classes of active sites are thought to exist, one favoring desaturation-like reactions and the other favoring hydroxylation-like reactions (Shanklin and Cahoon, 1998). When a substrate is saturated at the target position, desaturation or hydroxylation occurs but when a substrate has a double bond at the target position, a triple bond or epoxy group is formed (Shanklin and Cahoon, 1998). Because desaturase and hydroxylase functionality can be interconverted by substitutions of a few amino acids to alter the active site geometry (Broun *et al.*, 1997, 1998; Voelker and Kinney, 2001), there probably is a common activated intermediate for desaturation and hydroxylation and, apparently, also for acetylenation and epoxidation (Shanklin and Cahoon, 1998). It is hypothesized that the intermediate results in hydrogen abstraction (Que and Dong, 1996; Wallar and Lipscomb, 1996; Shu *et al.*, 1997). At this stage, subtle changes in the active site geometry would favor either a second hydrogen abstraction leading to desaturation or oxygen rebound from the di-iron center to the acyl chain leading to hydroxylation (Shanklin and Cahoon, 1998).

### *3.4.2 Substrates and products*

In addition to being a substrate for the biosynthesis of polyunsaturated fatty acids, such as linoleate and α-linolenate, phosphatidylcholine has been shown to be a substrate for the biosynthesis of unusual fatty acids such as hydroxy (Moreau and Stumpf, 1981; Bafor *et al.*, 1991), epoxy (Bafor *et al.*, 1993), acetylenic (Lee *et al.*, 1998), conjugated (Liu *et al.*, 1997) and trans-desaturated (Cahoon *et al.*, 2001; Qiu *et al.*, 2001; Jaworski and Cahoon, 2003) fatty acids, where the acyl-group esterified at the *sn*-2 position is the target for the reactions (Hildebrand *et al.*, 2001). Most of the enzymes involved in these reactions are found to be divergent *Fad2* desaturase-like enzymes, such as oleate Δ12 hydroxylase (Loo FN and CR, 1995), linoleate Δ12 epoxygenase (Lee *et al.*, 1998), linoleate Δ12 acetylenase (Lee *et al.*, 1998), conjugases (Cahoon *et al.*, 1999; Dyer *et al.*, 2002; Hornung *et al.*, 2002) and trans-desaturases (Cahoon *et al.*, 2001; Jaworski and Cahoon, 2003) although a cytochrome P450-epoxygenase from *Euphorbia* is responsible for the epoxidation of *sn*-2 linoleoyl-PC (Hildebrand *et al.*, 2001; Cahoon *et al.*, 2002). As well as typically acting on Δ12 positions of the acyl groups, some desaturases can also act on an existing Δ9 double bond (Cahoon *et al.*, 2001; Qiu *et al.*, 2001; Jaworski and Cahoon, 2003). Although only one reaction product results from most of these enzymes, a desaturase/hydroxylase bifunctional enzyme from *Lesquerella* has been identified (Broun *et al.*, 1998; Voelker and Kinney, 2001).

### *3.4.3 Gene isolation, characterization and testing*

There are hundreds of fatty acid species identified in the oils of exotic oilseed species (Badami and Patil, 1981), some of which are potentially valuable. Examples include conjugated, hydroxy, epoxy, acetylenic, keto, ethylenic, cyclopropanoid and fluoro fatty acids (Kinney, 2001). Over the past decade, tremendous effort has focused on isolation of the genes that are responsible for the synthesis of these fatty acids. One of the successful approaches to isolating these genes has been the use of high-throughput screening of cDNA libraries, or automated EST sequencing to identify the genes by their sequence homology to known genes (Kinney *et al.*, 1998; Cahoon *et al.*, 1999). Genes encoding enzymes typically account for 0.1–1.0% of the total cDNAs in the libraries prepared from developing seeds (van de Loo *et al.*, 1995; Cahoon *et al.*, 1999, 2002; Bao *et al.*, 2002). Additionally, cloned genes can be used to design degenerate primers based on the conserved regions for cloning of related genes (Kinney, 2001). This method has led to the identification of *Vernonia* (Hitz, 1998), *Stokesia* (Hildebrand, Rao *et al.*, 2001), and *Crepis* (Lee *et al.*, 1998) epoxygenases, *Ricinus* (van de Loo *et al.*, 1995) and *Lesquerella* (Broun *et al.*, 1997) hydroxylases, acetylenases (Lee *et al.*, 1998) and *Mormordica charanti*, *Impatiens balsamica* and *C. officinalis* conjugases (Cahoon *et al.*, 2001; Kinney, 2001). As a result of this genes are diverged from *Fad2* desaturases, their sequence homology has aided in their isolation and identification (Kinney, 2001). Analysis of the phylogenetic relationships of acetylenase, conjugase, epoxygenase, hydroxylase and desaturase amino acid sequences indicates that acetylenases, conjugases, epoxygenases and hydroxylases group semi-randomly among $\Delta 12$ desaturase sequences (Figure 3.2) indicating that these enzymes arose independently many times from preexisting desaturases during plant evolution. Some $\omega 3$ desaturases also are highly homologous to $\Delta 12$ desaturases (Figure 3.2) indicating that $\omega 3$ desaturases may also have evolved from $\Delta 12$ desaturases. Once the potential gene was identified, it was expressed in plants or in microorganisms to test its functionality (Kinney, 2001). For example, expression of *Mormordica*, *Impatiens* and *Calendula* conjugases in soybean and/or tobacco had led to the formation of the conjugated fatty acids $\alpha$-eleostearic acid (18:3, $\Delta 9$-*cis*, 11*trans*, 13*trans*), parinaric acid (18:4, $\Delta 9$*cis*, 11*trans*, 13*trans*, 15*cis*) and calendic acid (18:3, $\Delta 8$*trans*, 10*trans*, 12*cis*) (Kinney, 2001).

The *Vernonia* epoxygenase cDNA (Hitz, 1998) is quite different from a partial sequence of a *V. galamensis* epoxygenase-like gene (Lenman *et al.*, 1998), which suggests possible isozymes of *Vernonia* epoxygenase (Hildebrand *et al.*, 2001). The amino acid sequence of *Stokesia laevis* epoxygenase is more closely related to that of *V. galamensis* than *Crepis palaestina* (submitted for publication). Of the three desaturase-like epoxygenase genes cloned in the *Asteraceae* family, the *Vernonia*

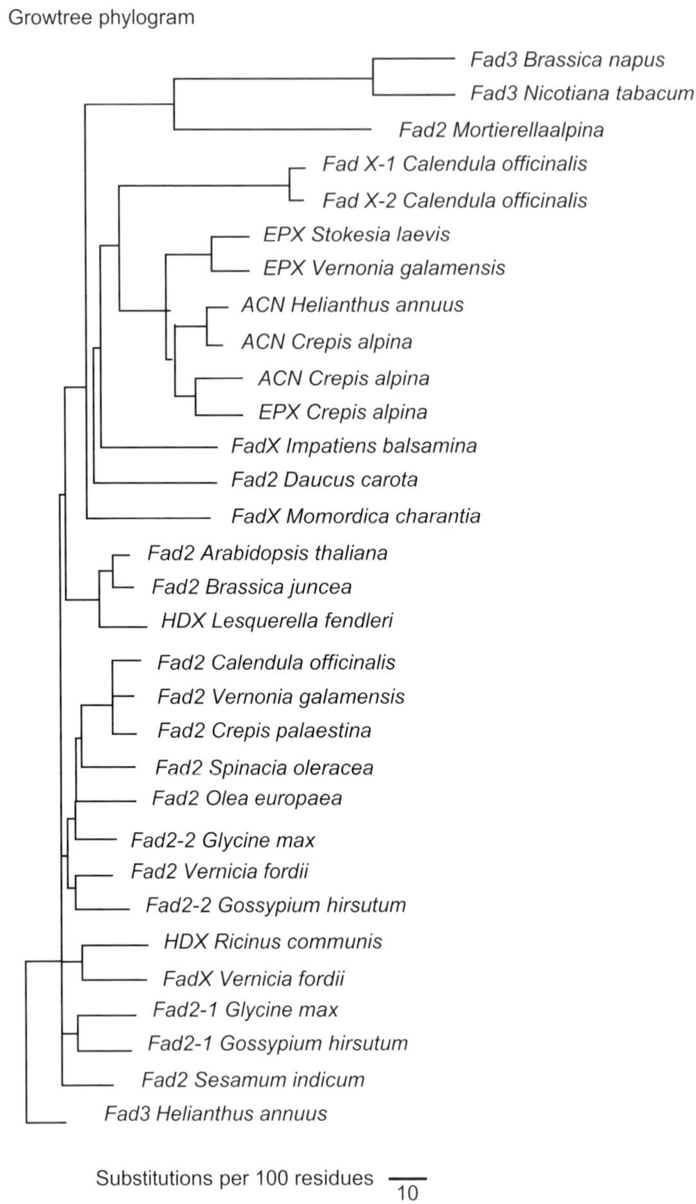


**Figure 3.2** Evolutionary relationships of Δ12 desaturases, acetylenases, conjugases, epoxygenases, hydroxylases and some related ω3 desaturase protein sequences. This Growtree phylogram used the Kimura distance correction method and neighbor joining tree construction method. ACN, acetylenase; EPX, epoxygenase; HDX, hydroxylase; FadX, conjugase.

and *Stokesia* genes are more closely related than that from *Crepis* (Hildebrand *et al.*, 2001).

### *3.4.4 Rational gene design*

The castor Δ9-18:0-ACP desaturase can be changed to a Δ9-16:0-ACP desaturase by replacing leucine 118 and proline 179 at the bottom of the binding pocket of the enzyme with the two bulkier residues – phenylalanine and isoleucine (Shanklin and Cahoon, 1998). Conversely, Δ6-16:0-ACP desaturase can be changed to a bifunctional Δ9-18:0/16:0-ACP desaturase by replacing alanine 188 and tyrosine 189 to the two smaller amino acids – glycine and phenylalanine (Shanklin and Cahoon, 1998). These substitutions change the available space at the lower portion of the binding pocket to accommodate different acyl chain lengths such that the requirement of the reaction substrate changes (Shanklin and Cahoon, 1998). A Δ6-16:0-ACP desaturase was converted to a Δ9-18:0-ACP desaturase by replacing five amino acid residues (Cahoon *et al.*, 1997). *A. thaliana* oleate desaturase (FAD2) and *L. fendleri* hydroxylase/desaturase can be interconverted by exchange of seven amino acids at four key locations (Broun *et al.*, 1998).

These studies suggest that new enzymes may be designed based on the crystallographic and primary structural data of the encoded enzymes in such a way that the encoded new enzymes will put functional groups at exactly the positions of acyl chains that are considered to be valuable (Cahoon *et al.*, 1997; Shanklin and Cahoon, 1998; Kinney, 2001). This technique of rationale design of enzyme function as well as directed evolution (Minshull and Stemmer, 1999) will accelerate the process of obtaining new enzymes for the production of useful novel fatty acids such as hydroxylases, acetylenases and epoxygenases, which may no longer be constrained by the availability in nature (Kinney, 2001). Determination of crystal structures of the desaturase-like enzymes will greatly help our understanding of the reaction mechanisms governing novel fatty acid synthesis (Shanklin and Cahoon, 1998).

## 3.5 Segregation of novel fatty acids from membrane lipids

The acyl groups in plant structural membrane glycerolipids are almost exclusively 16-carbon and 18-carbon fatty acids, with up to three methylene-interrupted double bonds (Millar *et al.*, 2000). However, the acyl groups in seed oils of many plant species have great diversity, some of which have different chain length, double bonds in unusual positions, functional groups such as conjugated double bonds, hydroxy, epoxy, acetylenic, cyclic and halogen, and are considered as novel fatty acids that in some plants can accumulate up to 90% of the total fatty acid content in the seed oil (Badami and Patil, 1981; van de Loo *et al.*, 1993; Millar *et al.*, 2000). The novel fatty acids are largely excluded from membrane lipids in the seeds and other parts of the plant, which is probably because the special physical and chemical properties of the novel fatty acids can perturb the integrity of the membrane structure such that they are deleterious to the cell (Millar *et al.*, 2000). Novel fatty acids, such as very long and medium chains and oxygenated fatty acids, are found only in very small amounts in phospholipids of many seed tissues (Bafor *et al.*, 1990, 1991; Sreenivas and Sastry, 1994; Stahl *et al.*, 1995; Wiberg and Bafor, 1995; Seither *et al.*, 1996, 1997; Wiberg *et al.*, 1997).

It is not clear why seed oils have such diverse fatty acid constituents other than some speculation on it being a defense mechanism (Millar *et al.*, 2000; Hildebrand *et al.*, 2001). It may be a result of random variation during the plants' evolutionary process (Voelker and Kinney, 2001). Although the novel fatty acids may not have selective advantages, they apparently are not a problem for the plants because they are mostly sequestered into oil bodies that have no structural functions (Millar *et al.*, 2000). However, the segregation of novel fatty acids from membrane lipids is very important because it likely is essential for generating transgenic crop plants aiming at accumulation of high levels of unusual fatty acids such as epoxy, hydroxy, acetylenic and conjugated fatty acids, which represent valuable feedstocks for the chemical industry (Stymne *et al.*, 1998; Millar *et al.*, 2000).

So far the mechanisms for plants to accomplish segregation and tight segregation of these unusual fatty acids are not completely understood (Millar *et al.*, 2000). Two hypotheses have been considered, one of which is compartmentation of membrane and storage lipid synthesis although there is not a lot of evidence to support it. The other is the selective accumulation of the novel fatty acids in triacylglycerol (TAG) after synthesis, which has more supporting evidence (Millar *et al.*, 2000). In addition, different novel fatty acids may be different in terms of segregation mechanisms and strictness (Millar *et al.*, 2000). For example, petroselinic acid (18:1$\Delta$6) is synthesized by a diverged acyl-ACP desaturase in the plastid but ricinoleic acid is synthesized by a diverged *Fad2* in the ER. It is conceivable that they may involve different channeling pathways for their segregation from membrane lipids. Also, 18:1$\Delta$6 has only one double bond position difference from common acyl groups while ricinoleic acid has a polar hydroxy group. Hence, 18:1$\Delta$6 may be more compatible with membrane lipids and not as strictly segregated from the membrane lipids as ricinoleic acid (Millar *et al.*, 2000).

### *3.5.1 Compartmentation of storage and membrane lipid synthesis*

This hypothesis proposes that membrane and storage lipids may be assembled by distinct subsets of enzymes spatially separated from each other (Lacey and Hills, 1996; Cahoon and Ohlrogge, 1994; Vogel and Browse, 1996). Evidence for this hypothesis is the different localization of enzymes in the ER for membrane and storage lipid synthesis from sucrose density-gradient centrifugation of microsomes of developing zygotic embryos of rapeseed (Lacey and Hills, 1996). Other evidence is from the analysis of petroselinic acid accumulation in coriander that suggests a distinct phosphatidylcholine pool for TAG synthesis (Cahoon and Ohlrogge, 1994).

However, further studies, such as using immunolocalization of enzymes, involved for the two classes of lipid synthesis is needed to corroborate this hypothesis (Millar *et al.*, 2000). Fatty acid channeling to membrane and storage lipids is also discussed in Chapters 5 and 6.

## 3.6 Selective accumulation of novel fatty acids in oil bodies

The prokaryotic and eukaryotic membrane lipid synthesis pathways both start from the sequential acylation of glycerol-3-phosphate (G3P) to produce phosphatidic acid. The latter is either activated to CDP-DAG for phosphatidyl glycerol and phosphatidyl serine synthesis or converted to diacylglycerol (DAG) for phosphatidyl choline, phosphatidyl ethanolamine and glycolipid synthesis. In developing seeds, DAG is further acylated to produce TAG (Figure 3.3). The production of storage and membrane lipids shares common precursors (DAG, acyl-CoA) and localization (in the ER) and occurs simultaneously in the cells of the developing seeds (Ohlrogge

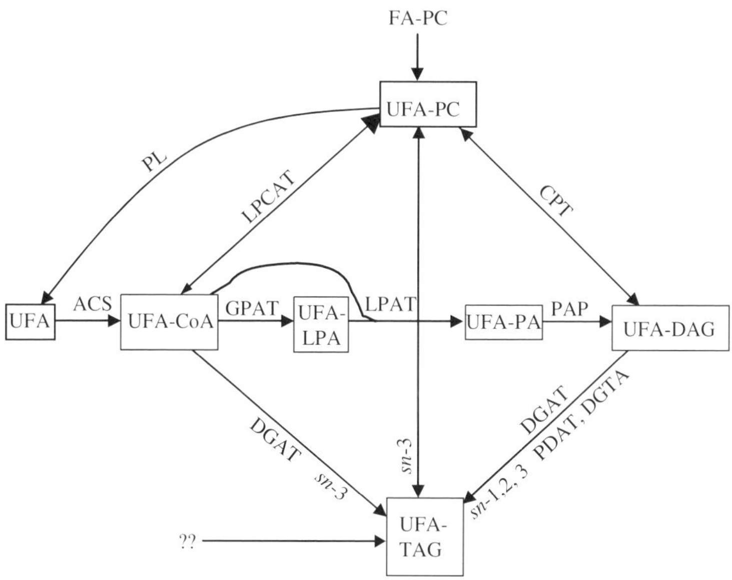


**Figure 3.3** Schematic pathways for unusual fatty acid (UFA) incorporation into triacylglycerol (TAG) after synthesis in phosphatidylcholine (PC). The *sn* numbers on a pathway indicate the possible positions of vernoloyl groups in TAG. LPA, lysophosphatidic acid; PA, phosphatidate; PDAT, phospholipid:diacylglycerol acyltransferase; CPT, cholinephosphotransferase; DGAT, acyl-CoA:diacylglycerol acyltransferase; DGTA, diacylglycerol transacylase; LPCAT, lysophosphatidylcholine acyltransferase; GPAT, glycerol-3-phosphate acyltransferase; LPAT, lysophosphatidic acid acyltransferase; PAP, phosphatidate phosphatase; PL, phospholipase; ACS, acyl-CoA synthetase.

and Browse, 1995; Millar *et al.*, 2000; Voelker and Kinney, 2001). The segregation of novel fatty acids may be at least partly due to enzyme specificities involved in the synthesis of these two lipid classes. More specifically, the phospholipid or glycolipid biosynthetic enzymes may not use phosphatidic acid and DAG molecular species containing novel fatty acids efficiently but the TAG synthetic enzymes may selectively utilize novel fatty acid containing substrates. Novel fatty acids that are already in the membrane may be quickly removed or otherwise degraded by β-oxidation (Millar *et al.*, 2000).

### *3.6.1 Medium-chain fatty acids*

Specific thioesterases remove newly formed medium-chain acyl groups from ACP and prevent them from being used by the prokaryotic pathway of glycerolipid synthesis because the G3P acylating enzymes require an acyl-ACP substrate (Millar *et al.*, 2000). In the ER, the partitioning of the medium-chain acyl groups between membrane and storage lipid might be achieved by acyltransferases. In *Cuphea*, G3P acyltransferase can use both medium-chain and common fatty acids (Ichihara, 1984; Bafor *et al.*, 1990). However, lysophosphatidic acid acyltransferase from *Cuphea* preferentially uses medium-chain fatty acids (Knutzon *et al.*, 1995; Frentzen, 1998). In the final step of TAG synthesis, DAG acyltransferase (DGAT) preferentially uses 10:0/10:0-diacylglycerol species and, almost exclusively, uses medium-chain acyl-CoAs (Bafor *et al.*, 1990; Wiberg *et al.*, 1994). Thus, medium-chain fatty acids are efficiently incorporated into TAG in the developing seeds of *Cuphea* (Millar *et al.*, 2000). In contrast, cholinephosphate acyltransferase (CPT) was not found to be involved in the selective accumulation of medium-chain fatty acid in TAG (Vogel and Browse, 1996; Voelker and Kinney, 2001). However, selective removal of laurate from the *sn*-2 position of phosphatidylcholine (PC) was found, which might have been due to a specialized phospholipase $A_2$ or a specialized phospholipid:diacylglycerol acyltransferase, PDAT (Stahl *et al.*, 1995; Voelker and Kinney, 2001).

Medium-chain fatty acids are not completely excluded from the PC of medium-chain fatty acid producers. For example, 7% laurate was found in the PC of *Actinodaphne hookeri* developing seeds that contain 95% laurate in their oil at maturity (Sreenivas and Sastry, 1994; Voelker and Kinney, 2001). Similarly, 1–2% laurate and 2% myristate were found in the PC of developing palm endosperm that contains 55% laurate and 15% myristate in its TAG at maturity (Wiberg and Bafor, 1995; Voelker and Kinney, 2001).

Studies of novel fatty acid production by transgenic plants have provided new insight on the segregation of the novel fatty acid from membrane lipids. For example, rapeseeds that are expressing a California bay tree (*Umbellularia californica*) medium-chain fatty acid thioesterase (*FatB1*), using the seed-specific Napin promoter, can produce high amounts of laurate. In some of these transgenic lines, although only 6 mol% laurate was found in PC of the membrane lipids of mature seeds that had 55 mol% of laurate in the seed oil, the laurate levels in PC of the membrane lipids of developing seeds could be as high as 46 mol% instead of the 1–4 mol% found in natural laurate producers (Wiberg *et al.*, 1997; Stymne *et al.*, 1998). This study demonstrated the relative ineffectiveness of novel fatty acid segregation from membrane lipids in transgenic producers compared to natural producers (Stymne *et al.*, 1998). In addition, transgenic oilseed rape that accumulates 60 mol% lauric acid has high activities of β-oxidation and glyoxylate enzymes, which might have been triggered by the accumulation of laurate in membrane lipids and this may, in turn, upregulate fatty acid biosynthesis to form a futile cycle (Eccleston and Ohlrogge, 1998; Ruuska *et al.*, 2002). A transgenic stearic acid producing rapeseed was also found to accumulate stearic acid in the membrane lipids other than in storage lipids in the developing seeds (Knutzon *et al.*, 1992). High stearate in membranes can reduce membrane fluidity and result in relatively poor germination rates (Thompson and Li, 1997). Interestingly, another similar transgenic rapeseed produced using a different transgene and also accumulating high stearate does not have low germinating rates (Facciotti *et al.*, 1999; Wiberg *et al.*, 2000; Voelker and Kinney, 2001).

When *FatB1* is expressed in oilseed rape under the control of the 35*S* promoter, medium-chain fatty acids can only be found in seeds even though higher *FatB1* activity levels are found by *in vitro* analysis isolated chloroplasts of the vegetative tissues (Eccleston and Ohlrogge, 1998). The medium-chain fatty acids are probably produced in the leaves and then rapidly broken down because medium-chain fatty acid targeted enzymes of both the β-oxidation pathway and the glyoxylate cycle pathways are induced in the leaves of the transgenic plants (Eccleston *et al.*, 1996; Eccleston and Ohlrogge, 1998). Secretion of medium-chain fatty acids by *E. coli mutants* blocked in β-oxidation expressing *FatB1* further supports the above hypothesis (Voelker and Davies, 1994; Millar *et al.*, 2000).

### *3.6.2 Very-long-chain fatty acids*

Very-long-chain fatty acids (VLCFAs) are also largely excluded from the membrane lipids in their natural producers. For example, only 3 mol% of erucic acid is found in the phospholipids of *Crambe abysinnica* that contains more than 60% of erucic acid in the seed oil (Gurr *et al.*, 1972; Voelker and Kinney, 2001). Microsomal radioactive feeding studies have shown that VLCFA CoA was excluded but oleoyl-CoA was accepted for phospholipid synthesis (Battey and Ohlrogge, 1988; Sun *et al.*, 1988; Taylor *et al.*, 1991; Voelker and Kinney, 2001). However, unlike medium-chain fatty acids producing rapeseeds, transgenic *Arabidopsis* expressing a fatty acid elongase 1 (*FAE1*) under the control of the 35S promoter accumulated up to 30 mol% of VLCFAs in the phospholipids of the vegetative tissues, which caused changes in plant morphology and organellar structures with the severity of these changes strongly correlated with the levels of VLCFA accumulation (Millar *et al.*, 1998; Voelker and Kinney, 2001). In contrast, the VLCFAs were excluded from membrane lipids in the developing seeds of the transgenic Arabidopsis. The inefficient catabolism of VLCFAs in the non-seed tissues of the transgenic Arabidopsis may have resulted in the accumulation of VLCFAs in these tissues (Millar *et al.*, 2000; Voelker and Kinney, 2001). It is speculated that VLCFAs may play a role in the membrane bilayer so that they are not catabolized based on the presence of C24–C26 VLCFAs in sphingolipids (Millar *et al.*, 2000). One possibility is that VLCFAs may be required for the formation of highly curved membrane structures, which may mimic the structural role of sphingolipids (Schneiter and Kohlwein, 1997; Millar *et al.*, 2000). The yeast Slc mutant, which lacks sphingolipids synthesis, can survive without sphingolipids by making novel glycerolipids containing C26 fatty acids strongly supports this hypothesis (Lester *et al.*, 1993). This yeast gene supports VLCFA accumulation in plants also (Zou *et al.*, 1997).

### *3.6.3 Novel monoenoic fatty acids*

Novel monoenoic fatty acids, such as petroselinic acid (18:1 Δ6) are synthesized while esterified to ACP and selectively hydrolyzed by specific ACP thioesterases. Free monoenoic fatty acids are moved out of plastids to the ER for triacylglycerol synthesis (Ohlrogge, 1994; Millar *et al.*, 2000). In developing coriander and *Thunbergia* seeds, petroselinic acid was found to cycle through PC most likely at the *sn*-2 position before entering TAG (Cahoon and Ohlrogge, 1994). PDAT might be involved for the selective transfer of the petroselinic acid from PC to TAG (Dahlqvist *et al.*, 1998; Schultz and Ohlrogge, 2000, 2001; Voelker and Kinney, 2001). Transgenic rapeseed producing petroselinic acid can induce selective catabolism of petroselinic acid by β-oxidation and the glyoxylate cycle (Fairbairn *et al.*, 1998–1999).

### *3.6.4 Novel fatty acids produced by diverged Fad2 enzymes*

PC is the substrate for novel fatty acid synthesis by many diverged *Fad2* enzymes as well as by cytochrome P450-type enzymes (Bafor *et al.*, 1993; Voelker and Kinney, 2001; Cahoon *et al.*, 2002). As with other unusual fatty acids, these are rare in membrane lipids of either developing or mature seeds in their natural producers (Cahoon and Ohlrogge, 1994; Padley *et al.*, 1994; Voelker and Kinney, 2001). For example, vernolic acid in *Euphorbia lagascae* and ricinoleic acid in castor are maximally only a few percent in the membrane lipids (Dahlqvist *et al.*, 1998). After these novel fatty acids are synthesized at the *sn*-2 position of PC, they may be selectively hydrolyzed by specialized phospholipase $A_2$ (PLA2) to produce free fatty acids (Stahl *et al.*, 1995). Surprisingly, microsomes from developing seeds of plants that do not produce ricinoleic acid were also found to have phospholipase activity specific for ricinoleic acid (Bafor *et al.*, 1991). Thus, PLA2s were thought to be used by plants to remove oxygenated fatty acids from membrane lipids as similarly postulated in animal cells (van Kuijk *et al.*, 1987; Bafor *et al.*, 1991; Banas *et al.*, 1992; Stahl *et al.*, 1995; Weselake, 2001).

The free novel fatty acids may be esterified to CoA by acyl-CoA synthetase and follow the Kennedy pathway to produce TAG as directed by the selective acyltransferases, namely, GPAT, LPAAT and DAGAT. In addition, diacylglycerol transacylase (DAGTA) and PDAT may also be selective for the novel fatty acid transfer into TAG (Stobart *et al.*, 1997; Dahlqvist *et al.*, 1998; Stymne *et al.*, 1998). For example, *sn*-2-ricinolate-PC is more efficiently incorporated into TAG than *sn*-2-oleate-PC by PDAT (Millar *et al.*, 2000). CPT has been speculated to be involved in the exclusion of novel fatty acids from membrane lipids but this was not found to be the case, at least, with decanoic, ricinoleic and erucic acids (Vogel and Browse, 1996; Schultz, 2001).

Mature seeds of transgenic tobacco and Arabidopsis plants expressing a *Ricinus* fatty acid hydroxylase (*FAH12*) gene under the control of the 35S promoter accumulated 17 mol% of ricinoleic acid in TAG but only 1 mol% in phospholipids while no ricinoleic acid was found in non-seed tissues although *FAH12* transcript is present at high levels in non-seed tissues and, *in vitro* assays have shown hydroxylase activity in roots and leaves (van de Loo *et al.*, 1995; Broun *et al.*, 1997; Broun and Somerville, 1997; Broun *et al.*, 1998; Voelker 2001). However, α-eleostearic acid produced by soybean embryos expressing a *Mormordica* conjugase were equally distributed between PC and TAG (Cahoon *et al.*, 1999; Voelker and Kinney, 2001). One limitation to the accumulation of these unusual fatty acids in lipids appears to be due to their catabolism via a futile cycle of β-oxidation (Moire *et al.*, 2004).

### *3.6.5 Gene specific promoters for tissue specific novel fatty acid accumulation*

To avoid accumulation of novel fatty acids in non-seed tissues of transgenic plants, 'seed specific' gene promoters such as oleosin, β-conglycinin and phaseolin promoters are commonly used. However, the oleosin promoter has been found to direct gene expression in non-seed tissues, such as developing roots of rapeseed (Murphy *et al.*, 2001). The possible physiological effects of such 'side' expression have not been determined (Murphy, 2002).

## 3.7 Structures and occurrences of hydroxy, conjugated and epoxy fatty acids in plant seed oils

There are more than 210 fatty acids that can be classified into 18 different structural classes (van de Loo *et al.*, 1993; Matthaus *et al.*, 2003). The occurrence of these fatty acids in relation to the taxonomy of plants varies considerably. They may occur only in certain species of plants in a particular family and genus or can be widespread, occurring in many unrelated species across several families. Of these fatty acids, the epoxy, hydroxyl and conjugated fatty acids are known to be of industrial significance (Jaworski and Cahoon, 2003). As an example, vernolic acid (an epoxy fatty acid) can be used for creating resins and coatings, ricinoleic acid (a hydroxy fatty acid) is useful as a plasticizer as well as in cosmetics, and eleostearic acid from tung oil (a conjugated fatty acid) may be used in varnishes and enamels (Murphy, 2002) (Table 3.1).

This section will deal with the chemical structures and occurrences of hydroxy, conjugated and epoxy fatty acids found in plant seed oils.

### *3.7.1 Hydroxy fatty acids*

Of the fatty acids that have a hydroxy group, ricinoleic acid 12-hydroxy-9-*cis*-octadecenoic acid is best known. It is 12-hydroxy-9-*cis*-octadecenoic acid. Castor oil from castor (*Ricinus communis*) seeds (family Euphorbiaceae), can be as much as 87.7% ricinoleic acid (Borch-Jensen *et al.*, 1997). Ricinoleic acid is also found in other plant oils, including *Anogiessus latifolia* (Combretaceae), at 26.6% (Hosamani, 1994); *Hiptage madablota* (family Malpighiaceae), at 70% (Badami and Kudari, 1970); and *Capparis zeylanica* (Capparaceae), at 30% (Daulatabad *et al.*, 1991). Ricinoleic acid has been found in many other plants as well; at least 10 families have been found with species that have ricinoleic acid (van de Loo *et al.*, 1993). The cDNA for the 12-hydroxylase from *R. communis* has been cloned and expressed in tobacco and Arabidopsis (Broun and Somerville, 1997).

Isoricinoleic acid (9-hydroxy-12-*cis*-octadecenoic acid) is found at 76.1% in *Wrightia coccinea* (Apocynaceae) seed oil (Siddiqi *et al.*, 1980). It is also found in *Celsia coromandeliana* (Scrophulariaceae) seed oil at 22% (Hosamani and Pattanashettar, 2000) and *Nerium oleander* (Apocynaceae) seed oil at 12% (Gummeson *et al.*, 2000).

Lesquerolic acid was first identified in *Lesquerella lasiocarpa*, family Brassicaceae, where it is found at 40–45% of the fatty acids. The structure of lesquerolic acid was found to be 14-hydroxy-*cis*-11-eicosenoic acid (Smith *et al.*, 1961). It has also been found in *L. fendleri* at 55% (Carlson *et al.*, 1990), as well as in the species *Physaria floribunda* and *Heliophilia amplexicaulis*, both of the Brassicaceae (Hayes *et al.*, 1995). In *Lesquerella*, it was found that the hydroxylase acts on oleic acid (18:1) to produce 14-hydroxy-oleic acid, which is then elongated to lesquerolic acid (20:1-0H) (Reed *et al.*, 1997). The hydroxylase for lesquerolic acid has also been cloned and expressed in Arabidopsis (Thomaeus *et al.*, 2001) as well as an elongase that is specific for 18-carbon hydroxy fatty acids (Moon *et al.*, 2001). Other hydroxy fatty acids were also found in the genus *Lesquerella*, including densipolic acid (12-hydroxy-*cis*-9,*cis*-15-octadecadienoic acid) in *Lesquerella densipila* (Smith *et al.*, 1962) and auricolic acid (14-hydroxy-*cis*-11,*cis*-17-eicosadienoic acid) in *Lesquerella auriculata* (Kleiman *et al.*, 1972).

Some hydroxy fatty acids have conjugated double bonds. Among the more well-known is dimorphecolic acid, 9-hyroxy-*trans*-10-*trans*-12-octadecadienoic acid, first isolated from *Dimorphotheca aurantiaca* seed oil (Smith *et al.*, 1960). It is found at 60% of the fatty acids in *Dimorphotheca pluvialis*, family Asteracae

(Muuse *et al.*, 1994). Other hydroxy fatty acids including a conjugated triene or diene structure include kamolenic acid (18-hydroxy-9-11-13-octadecatrienoic acid) from *Mallotus phillippinensis* (Gupta *et al.*, 1953) and coriolic acid (13-hydroxy-*cis*-9,*trans*-11-octadecadienoic acid) from *Coriaria napalensis* (Tallent *et al.*, 1966).

Many other hydroxy acids, often isolated as a small percentage of seed oil fatty acids, have been found in addition to those above, including isanolic acid [8-hydroxyoctadec-17-ene-9,11-di-ynoic acid from boleko (*Ongokea gore*)] (de Vries, 1957; Uzzan, 1961) with a conjugated alkyne group, the dihydroxy axillarenic acid (11,13-dihydroxytetracos-*trans*-9-enoic acid) isolated at 2.8% of the fatty acids of *Baliospermum axillare* (Husain *et al.*, 1980), and the trihydroxy fatty acid phloionolic acid (9,10,18-trihydroxyoctadecanoic acid) at 9% in *Chamaepeuce afra* (Mikolajczak and Smith, 1967).

### *3.7.2 Conjugated fatty acids*

Among the conjugated fatty acids that occur in plants, the best known is probably eleostearic acid from the tung plant, *Aleurites fordii* (Euphorbiaceae), where it occurs as 80% of the fatty acids (Dyer *et al.*, 2003). Eleostearic acid is *cis*-9,*trans*-11,*trans*-13-octadecatrienoic acid and is responsible for the industrial importance of tung oil as a good drying oil. The tung conjugase that makes eleostearic acid from linoleic acid is a desaturase analog and has been cloned and expressed in yeast (Dyer *et al.*, 2002). Other plants in the genus Aleurites are also significant sources of eleostearic acid such as *Aleurites montana* (67% eleostearic acid) (Radunz *et al.*, 1998) and *Aluerites trisperma* (38% eleostearic acid) (Weismann, 1976).

There are many other sources of eleostearic acid. One of the most important alternate sources of eleostearic acid is the Chinese bitter gourd, *Momordica charantia*, a crop plant throughout Asia. Seeds of this plant are up to 68% eleostearic acid (Chang *et al.*, 1996). Other plants that have been found to be sources of eleostearic acid include *Ricinodendron heudelotti* (up to 56% eleostearic acid) (Kapseu and Tchiegang, 1995), *Fevillea cordifolia* (31% eleostearic acid) (Achenbach *et al.*, 1992) and *Parinari montana* (36% eleostearic acid) (Spitzer *et al.*, 1992).

Calendic acid (*trans*-8,*trans*-10,*cis*-12-octadecatrienoic acid) is found throughout the genus *Calendula* of the Asteraceae (Chisholm and Hopkins, 1967). *C. officinalis* has been investigated as a new oil crop and it has a fatty acid profile of 62.8% calendic acid (Meier zu Beerentrup and Roebbelen, 1987). The enzyme responsible for the formation of calendic acid from linoleic acid has been isolated and cloned into yeast (Fritsche *et al.*, 1999; Qiu *et al.*, 2001) and somatic soybean embyros (Cahoon *et al.*, 2001). This enzyme (called a conjugase) acts on the double bond at position 9 in linoleic acid to make the conjugated *trans*, $\Delta 8$, $\Delta 10$ double bonds (Cahoon *et al.*, 2001).

Punicic acid was first isolated from pomegranate seed (Toyama and Tsuchiya, 1935). Punicic acid (*cis*-9,*trans*-11,*cis*-13-octadecatrienoic acid) occurs in pomegranate seeds (*Punicia granatum*) at 86% (Hildebrand *et al.*, 1998). Punicic acid also occurs in several species of *Trichosanthes*, up to 52% in *Trichosanthes nervifolia* (Datta, 1987; Lakshminarayana *et al.*, 1988; Joh *et al.*, 1995). The fatty acid in the *Trichosanthes* was not recognized as punicic acid in early reports and was called trichosanic acid (Crombie and Jacklin, 1957). Other plants that have punicic acid include *Mormodica balsamina* at 50% (Gaydou *et al.*, 1987), *Diplocyclos palmatus* (38.2% ) (Gowrikumar *et al.*, 1981) and *Apodanthera undualta* (30% by weight) (Bemis *et al.*, 1967).

Genes encoding conjugases have been cloned from both pomegranate and *Trichosanthes kirilowii* and expressed in yeast and Arabidopsis (Hornung *et al.*, 2002; Iwabuchi *et al.*, 2003). The enzyme was Δ12 desaturase-like and acted on the 12 position double bond in linoleic acid, creating the *trans*-11,*cis*-13 bonds of the conjugated triene (Hornung *et al.*, 2002). Interestingly, the conjugase was also found to be bifunctional, and exhibited Δ12 desaturase activity (Iwabuchi *et al.*, 2003).

Parinaric acid (*cis*-9,*trans*-11,*trans*-13,*cis*-15-octadecatetraenoic acid) was first isolated from the seed of *Parinarium laurinum* at a level of 53.5% (Tsujimoto, 1936; Takagi, 1965). Parinaric acid is also found in several species of *Impatiens*, including *Impatiens edgeworthii* (48%) (Bagby *et al.*, 1966), *I. balsamina* (29.1%) (Sarkar and Chakrabarty, 1956), *I. capensis* (42%) and *I. pallida* (30%) (Nozzolillo *et al.*, 1986). Parinaric acid also occurs in *Sebastiana brasiliensis* (39%) (Spitzer *et al.*, 1996). It is used as a fluorescent probe in membrane research (Sklar *et al.*, 1976).

Licanic acid (4-oxo-*cis*-9,*trans*-11,*trans*-13-octadecatrienoic acid) is a ketone isomer of eleostearic acid and was first isolated from oiticica oil, from the plant *Licania rigida* at 78.2% (McKinney and Jamieson, 1936; Machado, 1938; Paschke *et al.*, 1953). Couepic acid (from the seed of *Couepia grandiflora*) was found to be identical to licanic acid (van Loon and Steger, 1931; Kappelmeier, 1936). Licanic acid has been used as a drying oil in varnishes and paints (Bush, 1951). Other plants with licanic acid include *Licana mollis*, *Licana venosa* (Bennett *et al.*, 1950), *Licana arborea* (73%) (Rose and Jamieson, 1943) and *Couepia longipendula* (21.8% of the fatty acids) (Spitzer *et al.*, 1991).

Catalpic acid (*trans*-9,*trans*-11,*cis*-13-octadecetrienoic acid) is an isomer of eleostearic acid isolated first from *Catalpa bignonioides* (Chollet, 1946; Joh *et al.*, 1995). A conjugase has been cloned from *Catalpa ovata* but it is not yet reported whether it produces catalpic or puncic acid (Murase and Imamura, 2002).

### *3.7.3 Epoxy fatty acids*

Of the several epoxy acids found in the seed oil of plants, vernolic acid (12-epoxy,*cis*-9-octadecenoic acid) is best known. It was first elucidated in *Vernonia anthelmintica* seed (Gunstone, 1954). Vernolic acid occurs in *V. anthelmintica* at 65–75% (Vioque and Maza, 1971), *V. galamensis* (79–81%) (Ayorinde *et al.*, 1990) and *V. volkameriaefolia* (63.5%) (Siddiqi *et al.*, 1984). It has also been found in *S. laevis* (70%) (Gunn and White, 1974), several *Crepis* spp. (18–68%) (Earle *et al.*,

1966), *Erlangea tomentosa* (52%) (Phillips *et al.*, 1969), *Centratherum ritchiei* (30.1%) (Ahmad *et al.*, 1989), *E. lagascae* (57%) (Kleiman *et al.*, 1965) and *Bernardia pulchella* (91%) (Spitzer *et al.*, 1996). Of these, *E. lagascae* and *V. galamensis* have been investigated for possible crop use (Pascual-Villalobos *et al.*, 1992; Singh and Kaul, 1993; Callan and Kennedy, 1995). Many other plants also have lower but detectable vernolic acid levels. These include *Kigelia pinnata* (22.3%) (Afaque *et al.*, 1987), *Cassia corymbosa* (9.2%) (Hosamani and Sattigeri, 2003), *Ficus benghalensis* (8.2%) (Hosamani and Pattanashettar, 2003), *Xeranthemum annuum* (2%) (Powell *et al.*, 1967), *Piper nigrum* (7.7%) (Daulatabad *et al.*, 1995) and *Plectranthus mollis* (6.1%) (Daulatabad *et al.*, 1989), among many others.

A gene encoding Δ12 epoxygenase was first isolated from a specie of *Crepis*, *C. palaestina*, and was found to be a Δ12 desaturase-like enzyme. This gene was expressed in Arabidopsis and yeast (Lee *et al.*, 1998). Epoxygenase genes from *S. laevis* and *V. galamensis* have also been cloned (Hitz, 1998; Hildebrand *et al.*, 2001; Green *et al.*, 2002). The epoxygenase gene from *E. lagascae* was isolated and found to be a cytochrome P450 (Cahoon *et al.*, 2002).

Coronaric acid (9-epoxy-*cis*-12-octadecenoic acid) was first isolated from *Chrysanthemum coronarium* (Smith *et al.*, 1959). Many other plants since then have been reported to have coronaric acid, including *X. anuum* (8%) (Powell *et al.*, 1967), *Lactuca sativa* (16.9%) (Ansari *et al.*, 1987) and several *Acacia* species (3.5–6.8%) (Jain and Iqbal, 1989). Interestingly, sunflower seed showed the presence of coronaric acid (2%) after prolonged storage (Mikolajczak *et al.*, 1968). With *Cichorium intybus*, after prolonged storage, the total oxygenated fatty acids increased from 1% to 17% . The oxygenated fatty acids were found to consist of conjugated hydroxy fatty acids and coronaric acid (Spencer *et al.*, 1973).

Little work has been done with the other epoxy fatty acids found in nature. These acids include 9-epoxystearic acid from *Tragopogon porrifolius* (3%) (Chisholm and Hopkins, 1959), alchornoic acid (14-epoxy-*cis*-11-eicosenoic acid) from *Alchornea cordifolia* (50%) (Kleiman *et al.*, 1977). Other epoxy fatty acids, such as 15-epoxy-*cis*-9,*cis*-12-octadecadienoic acid occur in small amounts in a small number of species. These epoxy fatty acids (like coronaric and epoxystearic acid) may be formed during seed storage (Earle, 1970).

## Acknowledgements

Much thanks to Erin Yost, Eric Hudson, Danny Walls, Hirotada Fukushige and Susan Leopold for useful suggestions and help in putting this chapter together.

## References

Aach, H., Hornig, F. and Heise, K.P. (1997). Distribution of lipid radioactivity after fractionation of $^{14}$C-labeled zygotic rape embryos. *Journal of Plant Physiology*, **151**(3), 323–328.

Achenbach, H., Hefter-Buebl, U., Waibel, R., Constenla, M.A. and Hahn, H. (1992). Gas chromatographic analysis of the seed fat of *Fevillea cordifolia* (Cucurbitaceae) from Central America. *Fett Wissenschaft Technologie – Fat Science Technology*, **94**(8), 294–297.

Afaque, S., Siddiqui, M.M., Ahmad, I., Siddiqi, M.S. and Osman, S.M. (1987). *Kigelia pinnata* seed oil – a moderate source of vernolic acid. *Fett Wissenschaft Technologie – Fat Science Technology*, **89**(11), 433–435.

Ahmad, R., Ahmad, I. and Osman, S.M. (1989). *Centratherum ritchiei* seed oil – characterization of vernolic acid.*Fett Wissenschaft Technologie – Fat Science Technology*, **91**(12), 488–490.

Anamnart, S., Tomita, T., Fukui, F. *et al.* (1997). The P-OLE1 gene of *Pichia angusta* encodes a D9-fatty acid desaturase and complements the ole1 mutation of *Saccharomyces cerevisiae*. *Gene*, **184**, 299–306.

Ansari, M.H., Ahmad, S., Ahmad, F., Ahmad, M. and Osman, S.M. (1987). Co-occurrence of coronaric and vernolic acids in Compositae seed oils. *Fett Wissenschaft Technologie – Fat Science Technology*, **89**(3), 116–118.

Aoki, K. and Wada, K. (1996). Temporal and spatial distribution of ferredoxin isoproteins in tomato fruit. *Plant Physiology*, **112**, 651–657.

Avelange-Macherel, M.H., Macherel, D., Wada, H. and Murata, N. (1995). Site-directed mutagenesis of histidine residues in the 112 acyl-lipid desaturase of Synechocystis. *FEBS Letters*, **361**, 111–114.

Ayorinde, F.O., Butler, B.D. and Clayton, M.T. (1990). *Vernonia galamensis*: a rich source of epoxy acid. *Journal of the American Oil Chemists' Society*, **67**(11), 844–845.

Badami, R.C. and Kudari, S.M. (1970). Analysis of *Hiptage madablota* seed oil. *Journal of the Science of Food and Agriculture*, **21**(5), 248–249.

Badami, R.C. and Patil, K.B. 1981. Structure and occurence of unusual fatty acids in minor seed oils. *Progress in Lipid Research*, **19**, 119–153.

Bafor, M., Jonsson, L., Stobart, A.K. and Stymne, S. (1990) Regulation of triacylglycerol biosynthesis in embryos and microsomal preparations from the developing seeds of *Cuphea lanceolata*. *Biochemical Journal*, **272**, 31–38.

Bafor, M., Smith, M.A., Jonsson, L., Stobart, K. and Stymne, S. (1991) Ricinoleic acid biosynthesis and triacylglycerol assembly in microsomal preparations from developing castor-bean (*Ricinus communis*). *Biochemical Journal*, **280**, 507–514.

Bafor, M., Smith, M.A., Jonsson, L., Stobart, K. and Stymne, S. (1993) Biosynthesis of vernoleate (*cis*-12-epoxyoctadeca-*cis*-9-enoate) in microsomal preparations from developing endosperm of *Euphorbia lagascae*. *Archives of Biochemistry and Biophysics*, **303**(1), 145–151.

Bagby, M.O., Smith, C.R.J. and Wolff, I.A. (1966) Stereochemistry of a-parinaric acid from *Impatiens edgeworthii* seed oil. *Lipids*, **1**(4), 263–267.

Banas, A., Johansson, I. and Stymne, S. (1992). Plant microsomal phospholipases exhibit preference for phosphatidylcholine with oxygenated acyl groups. *Plant Science*, **84**, 137–144.

Bao, X., Katz, S., Pollard, M. and Ohlrogge, J. (2002) Carbocyclic fatty acids in plants: biochemical and molecular genetic characterization of cyclopropane fatty acid synthesis of *Sterculia foetida*. *Plant Physiology*, **99**, 7172–7177.

Battey, J.F. and Ohlrogge, J.B. (1988) A comparison of the metabolic fate of fatty acids of different chain lengths in developing oilseeds. *Plant Physiology*, **90**, 835–840.

Beisson, F., Koo, A.J.K., Ruuska, S. *et al.* (2003) Arabidopsis genes involved in acyl lipid metabolism. A census of the candidates, a study of the distribution of expressed sequence tags in organs, and a web-based database. *Plant Physiology*, **132**(2), 681–697.

Bemis, W.P., Moran, M.D., Berry, J.W. and Deutschman, A.J., Jr. (1967). Composition of *Apodanthera undulata* oil. *Canadian Journal of Chemistry*, **45**(21), 2637.

Bennett, H., Brown, E. and Islip, H.T. (1950) *Licania mollis* and *Licania venosa* fruits from British Guyana. *Colonial Plant and Animal Products*, **1**, 232–236.

Borch-Jensen, C., Jensen, B., Mathiasen, K. and Mollerup, J. (1997) Analysis of seed oil from *Ricinus communis* and *Dimorphotheca pluvialis* by gas and supercritical fluid chromatography. *Journal of the American Oil Chemists' Society*, **74**(3), 277–284.

Broun, P. and Somerville, C. (1997) Accumulation of ricinoleic, lesquerolic, and densipolic acids in seeds of transgenic Arabidopsis plants that express a fatty acyl hydroxylase cDNA from castor been. *Plant Physiology*, **113**(3), 933–943.

Broun, P., *et al.* (1997). Expression of castor and *Lesquerella fendleri* oleate 12-hydroxylases in transgenic plants, in *Physiology, Biochemistry and Molecular Biology of Plant Lipids* (eds J.P. Williams, M.U. Khan and N.W. Lem). Dordrecht, The Netherland, Kluwer, pp. 342–344.

Broun, P., Boddupalli, S. and Somerville, C. (1998) A bifunctional oleate 12-hydroxylase: desaturase from *Lesquerella fendleri*. *Plant Journal*, **13**(2), 201–210.

Broun, P., Shanklin, J., Whittle, E. and Somerville, C. (1997). Switching between desaturation and hydroxylation of oleate by manipulating the key residues of a hydroxylase and a desaturase. *Biochemistry and Molecular Biology of Plant Fatty Acids Glycerolipids Symposium*, Lake Tahoe, California.

Broun, P., Shanklin, J., Whittle, E. and Somerville, C. (1998). Catalytic plasticity of fatty acid modification enzymes underlying chemical diversity of plant lipids. *Science*, **282**, 1315–1317.

Browse, J. and Somerville, C. (1991). Glycerolipid synthesis: biochemistry and regulation. *Annual Review of Plant Physiology and Plant Molecular Biology*, **42**, 467–506.

Browse, J., McCourt, P.J. and Somerville, C.R. (1985). A mutant of Arabidopsis lacking a chloroplast-specific lipid. *Science*, **227**, 763–765.

Bush, W.A. (1951) Oiticica oils. *Paint, Oil and Chemical Review*, **114**(22), 14, 16–17.

Cahoon, E.B. and Shanklin, J. (2000) Substrate-dependent mutant complementation to select fatty acid desaturase variants for metabolic engineering of plant seed oils. *Proceedings of the National Academy of Sciences USA*, **97**, 12350–12355.

Cahoon, E.B. and Ohlrogge, J.B. (1994). Apparent role of phosphatidylcholine in the metabolism of petroselinic acid in developing Umbelliferae endosperm. *Plant Physiology*, **104**(3), 845–855.

Cahoon, E.B., Carlson, T.J., Ripp, K.G. *et al.* (1999). Biosynthetic origin of conjugated double bonds: production of fatty acid components of high-value drying oils in transgenic soybean embryos. *Proceedings of the National Academy of Sciences USA*, **96**(22), 12935–12940.

Cahoon, E.B., Cranmer, A.M., Shanklin, J. and Ohlrogge, J.B. (1994) D6 hexadecenoic acid is synthesized by the activity of a soluble D6 palmitoyl-acyl carrier protein desaturase in *Thunbergia alata* endosperm. *Journal of Biological Chemistry*, **269**(44), 27519–27526.

Cahoon, E.B., Hitz, W.D. and Ripp, K.G. (2001) Plant genes for fatty acid modifying enzymes (*FAD2*) associated with conjugated double bond formation and transgenic plants having altered lipid profiles. World Patent Application.

Cahoon, E.B., Lindqvist, Y., Schneider, G. and Shanklin, J. (1997) Redesign of soluble fatty acid desaturases from plants for altered substrate specificity and double bond position. *Proceedings of the National Academy of Sciences USA*, **94**, 4872–4877.

Cahoon, E.B., Ripp, K.G., Hall, S.E. and Kinney, A.J. (2001) Formation of conjugated D8, D10-double bonds by D12-oleic-acid desaturase-related enzymes. *Journal of Biological Chemistry*, **276**(4), 2637–2643.

Cahoon, E.B., Ripp, K.G., Hall, S.E. and McGonigle, B. (2002) Transgenic production of epoxy fatty acids by expression of a cytochrome P450 enzyme from *Euphorbia lagascae* seed. *Plant Physiology*, **128**(2), 615–624.

Cahoon, E.B., Shah, S., Shanklin, J. and Browse, J.A. (1998) A determinant of substrate specificity predicted from the acyl-acyl carrier protein desaturase of developing cat's claw seed. *Plant Physiology*, **117**(2), 593–598.

Cahoon, E.B., Shanklin, J. and Ohlrogge, J.B. (1992) Expression of a coriander desaturase results in petroselinic acid production in transgenic tobacco. *Proceedings of the National Academy of Sciences USA*, **89**, 11184–11188.

Callan, E.J. and Kennedy, C.W. (1995) Tolerance of *Stokes aster* to selected herbicides. *Industrial Crops and Products*, **4**(4), 285–290.

Carlson, K.D., Chaudhry, A. and Bagby, M.O. (1990) Analysis of oil and meal from *Lesquerella fendleri* seed. *Journal of the American Oil Chemists' Society*, **67**(7), 438–442.

Chang, M.K., Conkerton, E.J., Chapital, D.C., Wan, P.J., Vandwa, O.P. and Spiers, J.M. (1996) Chinese melon (*Momordica charantia* L.) seed: composition and potential use. *Journal of the American Oil Chemists' Society*, **73**(2), 263–265.

Chisholm, M.J. and Hopkins, C.Y. (1959). The oxygenated fatty acid of *Tragopogon porrifolius* seed oil. *Chemistry & Industry*, **37**, 1154–1155.

Chisholm, M.J. and Hopkins, C.Y. (1967). Calendic acid in seed oils of the genus Calendula. *Canadian Journal of Biochemistry*, **45**(2), 251–254.

Choi, J.Y., Stuky, J., Hwang, S.Y. and Martin, C.E. (1996) Regulatory elements that control transcription activation and unsaturated fatty acid-mediated repression of the *Saccharomyces cerevisiae* OLE1 gene. *Journal of Biological Chemistry*, **271**, 3581–3589.

Chollet, M.M. (1946) Some constituents of the catalpa tree, *Catalpa bignonioides*. I. Catalposide and catalpic acid. *Bulletin de la Societe de Chimie Biologique*, **28**, 668–671.

Crombie, L. and Jacklin, A.G. (1957) Lipides. VI. Total synthesis of a- and b-eleostearic and punicic (trichosanic) acid. *Journal of the Chemical Society*, (Abstracts), 1632–1646.

Dahlqvist, A., Banas, A. and Stymne, S. (1998) Selective channelling of unusual fatty acids in to triacylglycerols, in *Advances in Plant Lipid Research* (eds J. Sanchez, E. Cerda-Olmedo and E. Martinez-Force), Sevilla, Secretariado de Publicaciones, Universidad de Sevilla, pp. 211–214.

Dailey, H.A. and Strittmatter, P. (1979) Modification and identification of cytochrome $b_5$ carboxyl groups involved in protein-protein interaction with cytochrome $b_5$ reductase. *Journal of Biological Chemistry*, **254**, 5388–5396.

Datta, S.K. (1987) Fatty acid composition in developing seeds of *Trichosanthes cucumarina* L. *Biological Memoirs*, **13**(1), 69–72.

Daulatabad, C., Mahmood, J.D. and Mirajkar, A.M. (1989), Vernolic and cyclopropenoid fatty acids in *Plectranthus mollis*, syn. *Plectranthus incanus* Link, seed oil: a rich source of oil. *Journal of Chemical Technology and Biotechnology*, **45**(2), 143–146.

Daulatabad, C., Mahmood, J.D., Desai, V.A. and Hosamani, K.M. (1991) New source of oil with novel fatty acids for industrial utilization. *Industrial & Enginering Chemistry Research*, **30**(12), 2596–2598.

Daulatabad, C.D., Mulla, G.M. and Mirajkar, A.M. (1995) Vernolic and cyclopropenoic fatty acids in *Piper nigrum* seed oil. *Fett Wissenschaft Technologie-Fat Science Technology*, **97**(12), 453–454.

de Vries, E. (1957) Boleko (Isano) oil. *Oleagineux*, **12**, 427–431.

Dyer, J.M., Chapital, D.C., Kuan, J.C., Mullen, R.T., Turner, C., McKeon, T.A. and Pepperman, A.B. (2002). Molecular analysis of a bifunctional fatty acid conjugase/desaturase from tung. Implications for the evolution of plant fatty acid diversity. *Plant Physiology*, **130**, 2027–2038.

Dyer, J.M., Chapital, D.C., Mullen, R.T., Kuan, J.-C.W. and Pepperman, A.B., Jr. (2003) *Identification and Functional Analysis of Enzymes Required for the Synthesis of Tung Oil, a High Value Industrial Drying Oil*, 226th American Chemical Society National Meeting, New York, NY, American Chemical Society.

Earle, F.R. (1970) Epoxy oils from plant seeds. *Journal of the American Oil Chemists' Society*, **47**(12), 510–513.

Earle, F.R., Barclay, A.S. and Wolff, I.A. (1966) Compositional variation in seed oils of the *Crepis genus*. *Lipids*, **1**(5), 325–327.

Eccleston, V., Cranmer, A., Voelker, T. and Ohlrogge, J. (1996) Medium-chain fatty acid biosynthesis and utilization in Brassica napus plants expressing lauroyl-acyl carrier protein thioesterase. *Planta*, **198**, 46–53.

Eccleston, V.S. and Ohlrogge, J.B. (1998) Expression of lauroyl-acyl carrier protein thioesterase in *Brassica napus* seeds induces pathways for both fatty acid oxidation and biosynthesis and implies a set point for triacylglycerol accumulation. *Plant Cell*, **10**, 613–622.

Facciotti, M.T., Bertain, P.B. and Yuan, L. (1999) Improved stearate phenotype in transgenic canola expressing a modified acyl acyl-carrier protein thioesterase. *Nature Biotechnology*, **6**, 593–597.

Fairbairn, D.J., Bowra, S. and Murphy, D.J. (1998–1999). *Expression of Unusual Fatty Acids in Transgenic Rapeseed Causes Induction of Glyoxylate Cycle Genes*, John Innes Centre Annual Report.

Frentzen, M. (1998) Acyltransferases from basic science to modified seed oils. *Fett-Lipid*, **100**, 161–166.

Fritsche, K., Hornung, E., Peitzsch, N., Renz, A. and Feussner, I. (1999) Isolation and characterization of a calendic acid producing (8,11)-linoleoyl desaturase. *FEBS Letters*, **462**(3), 249–253.

Fukuchi-Mizutani, M., Savin, K., Cornish, E. *et al.* (1995). Senescence induced expression of a homologue of D9 desaturase in rose petals. *Plant Molecular Biology*, **29**, 627–635.

Gargano, S., di Lallo, G., Kobayashi, G.S. and Maresca, B. (1995), A temperature-sensitive strain of *Histoplasma capsulatum* has an altered D9-fatty acid desaturase gene. *Lipids*, **30**(10), 899–906.

Gaydou, E.M., Miralles, J. and Rasoazanakolona, V. (1987) Analysis of conjugated octadecatrienoic acids in *Momordica balsamina* seed oil by GLC and $^{13}$C NMR spectroscopy. *Journal of the American Oil Chemists' Society*, **64**(7), 997–1000.

Gonzalez, C.I. and Martin, C.E. (1996), Fatty acid-responsive control of mRNA stability. Unsaturated fatty acid-induced degradation of the Saccharomyces ole1 transcript. *Journal of Biological Chemistry*, **271**, 25801–25809.

Gowrikumar, G., Mani, V.V.S., Rao, T.C., Kaimal, T.N.B. and Lakshminarayana, G. (1981). *Diplocyclos palmatus* L.: a new seed source of punicic acid. *Lipids*, **16**(7), 558–559.

Green, A., Singh, S., Lenman, M. and Stymme, S. (2002), Sequences of fatty acid epoxygenase genes from Crepis and Vernonia, and uses in modifying fatty acid metabolism. *US Patent Application Publication*, United States, 47.

Gummeson, P.O., Lenman, M., Lee, M., Singh, S. and S. S (2000). Characterisation of acyl-ACP desaturases from *Macadamia integrifolia* Maiden & Betche and *Nerium oleander* L. *Plant Science*, **154**(1), 53–60.

Gunn, C.R. and White, G.A. (1974) *Stokesia laevis*: taxonomy and economic value. *Economic Botany*, **28**(2), 130–135.

Gunstone, F.D. (1954) Fatty acids. II. The nature of the oxygenated acid present in *Vernonia anthelmintica* seed oil. *Journal of the Chemical Society*, Abstracts, 1611–1616.

Gupta, S.C., G.S.S. and Aggarwal, J.S. (1953). Chemical examination of the seeds of *Mallotus philippinensis*. III. Constitution of kamlolenic acid isolated from the oil. *Journal of Scientific & Industrial Research*, **12B**, 240–242.

Gurr, M.I., Blades, J. and Appleby, R. (1972), Studies on seed-oil triglycerides: the composition of *Crambe abyssinica* triglycerides during seed maturation. *European Journal of Biochemistry*, **29**, 362–368.

Haas, J.A. and Fox, B.G. (1999) Role of hydrophobic partitioning in substrate selectivity and turnover of the *Ricinus communis* stearoyl acyl carrier protein D9 desaturase. *Biochemistry*, **38**(39), 12833–12840.

Hackett, C.S. and Strittmatter, P. (1984) Covalent cross-linking of the active sites of vesicle-bound cytochrome $b_5$ and NADH cytochrome $b_5$ reductase. *Journal of Biological Chemistry*, **259**, 3275–3282.

Hayes, D.G., Kleiman, R. and Phillips, B.S. (1995) The triglyceride composition, structure, and presence of estolides in the oils of Lesquerella and related species. *Journal of the American Oil Chemists' Society*, **72**(5), 559–569.

Hildebrand, D.F. and Yu, K. (2003) Genetics of crop improvement: primary metabolism – acyl lipids, in *Encyclopedia of Applied Plant Sciences* (eds B. Thomas, D. Murphy and B. Murray), Oxford, UK, Elsevier Science Ltd.

Hildebrand, D.F., Fukushige, H., Afitlhile, M. and Wang, C. (1998) Lipoxygenases in plant development and senescence, in *Eicosanoids and Related Compounds in Plants and Animals* (eds A.F. Rowley, H. Kuhn and T. Schewe) London, Portland Press, pp. 151–181.

Hildebrand, D.F., Rao, S. and Hatanaka, T. (2001) Redirecting lipid metabolism in plants, in *Lipid Biotechnology* (eds T.M. Kuo and H.W. Gardner) New York, Marcel Dekker, Inc.

Hilditch, T.P. and Williams, P.N. (1964) *The Chemical Constitution of Natural Fats*. London, Chapman & Hall.

Hitz, W.D. (1998) Fatty acid modifying enzymes from developing seeds of *Vernonia galamensis*. United States Patent #5,846,784 to E.I. du Pont de Nemours and Company.

Hornung, E., Pernstich, C. and Feussner, I. (2002) Formation of conjugated D11 D13 linoleic acid (1,4)-acyl-lipid-desaturase in pomegranate seeds. *European Journal of Biochemistry*, **269**, 4852–4859.

Hosamani, K.M. (1994) *Anogiessus latifolia* seed oil: co-occurrence of 12-hydroxyoctadec-*cis*-9-enoic acid, 7-(2-octylcyclopropen-1-yl)heptanoic acid and 8-(2-octylcyclopropen-1-yl)

octanoic acid. *Journal of the Oil Technologists' Association of India (Mumbai, India)*, **26**(3), 69–72.
Hosamani, K.M. and Pattanashettar, R.S. (2000) Industrial utilization of *Celsia coromandeliana* seed oil: a moderate source of isoricinoleic acid. *Industrial & Engineering Chemistry Research*, **39**(12), 5017–5019.
Hosamani, K.M. and Pattanashettar, R.S. (2003) Occurrence of unusual fatty acids in *Ficus benghalensis* seed oil. *Industrial Crops and Products*, **18**(2), 139–143.
Hosamani, K.M. and Sattigeri, R.M. (2003) Analysis of *Cassia marginata* and *Cassia corymbosa* seed oils: an approach for the industrial utilization. *Industrial Crops and Products*, **17**(1), 57–60.
Husain, S., Ahmad, M.U. and Osman, S.M. (1980) New hydroxy fatty acid from seed oil of *Baliospermum axillare*. *Phytochemistry*, **19**(1), 75–77.
Ichihara, K. (1984) *sn*-Glycerol-3-phosphate acyltransferase in a particulate fraction from maturing safflower seeds. *Archives of Biochemistry and Biophysics*, **232**, 685–698.
Itoh, R., Toda, K., Takahashi, H., Takano, H. and Kuroiwa, T. (1998) D9 fatty acid desaturase gene containing a carboxyl-terminal cytochrome $b_5$ domain from the red alga *Cyanidioschyzon merolae*. *Current Genetics*, **33**(3), 165–170.
Iwabuchi, M., Kohno-Murase, J. and Imamura, J. (2003) D12-oleate desaturase-related enzymes associated with formation of conjugated trans-D 11, cis-D 13 double bonds. *Journal of Biological Chemistry*, **287**(7), 4603–4610.
Jackson, F.M., Michaelson, L., Fraser, T.C.M., Stobart, A.K. and Griffiths, G. (1998) Biosynthesis of triacylglycerol in the filamentous fungus *Mucor circinelloides*. *Microbiology*, **144**, 2639–2645.
Jacobson, B.S., Jaworski, J.G. and Stumpf, P.K. (1974) Fat metabolism in higher plants: LXII. Stearyl-acyl carrier protein desaturase from spinach chloroplasts. *Plant Physiology*, **54**, 484–486.
Jain, P. and Iqbal, S.A. (1989) Composition of the plants *Acacia cancinna*, *Acacia senegal*, *Acacia leucopholea*, *Acacia modesta* and the physico-chemical studies of their seed oils. *Oriental Journal of Chemistry*, **5**(2), 200–203.
Jaworski, J. and Cahoon, E.B. (2003) Industrial oils from transgenic plants. *Current Opinion in Plant Biology*, **6**, 178–184.
Joh, Y.-G., Kim, S.-J. and Christie, W.W. (1995) The structure of the triacylglycerols, containing punicic acid, in the seed oil of *Trichosanthes kirilowii*. *Journal of the American Oil Chemists' Society*, **72**(9), 1037–1042.
Kajiwara, S. (2002) Molecular cloning and characterization of the D9 fatty acid desaturase gene and its promoter region from *Saccharomyces kluyveri*. *FEMS Yeast Research*, **2**(3), 333–9.
Kappelmeier, C.P.A. (1936) Oiticica oil and its fundamental difference from tung oil. *Paint and Varnish Production Manager*, **14**, 7–8,10.
Kapseu, C. and Tchiegang, C. (1995) Chemical properties of *Ricinodendron heudelotii* (Bail.) seed oil. *Journal of Food Lipids*, **2**(2), 87–98.
Kinney, A.J. (2001) Perspectives on the production of industrial oils in genetically engineered oilseeds, in *Lipid Biotechnology* (eds T.M. Kuo and H.W. Gardner), New York, Marcel Dekker, Inc., pp. 85–93
Kinney, A.J., Hiz, W.D., Knowlton, S. and Cahoon, E.B. (1998) Re-engineering oilseed crops to produce industrially useful fatty acids, in *Advances in Plant Lipid Research* (eds J. Sanchez, E. Cerda-Olmedo and E. Martinez-Force), Sevilla, Secretariado de Publicaciones, Universidad de Sevilla, pp. 623–628.
Kishore, G.M. and Somerville, C.R. (1993) Genetic engineering of commercially useful biosynthetic pathways in transgenic plants. *Current Opinion in Biotechnology*, **4**, 152–158.
Kleiman, R., Plattner, R.D. and Spencer, G.F. (1977) *Alchornea cordifolia* seed oil: a rich source of a new $C_{20}$ epoxide, (+)*cis*-14,15-epoxy-*cis*-11-eicosenoic acid. *Lipids*, **12**(7), 610–612.
Kleiman, R., Smith, C.R., Jr., Yates, S.G. and Jones, Q. (1965) Search for new industrial oils. XII. Fifty-eight Euphorbiaceae oils, including one rich in vernolic acid. *Journal of the American Oil Chemists' Society*, **43**(3), 169–172.
Kleiman, R., Spencer, G.F., Earle, F.R., Nieschlag, H.J. and Barclay, A.S. (1972) Tetra-acid triglyceride containing a new hydroxyeicosadienoyl moiety in *Lesquerella auriculata* seed oil. *Lipids*, **7**(10), 660–665.

Knutzon, D.S., Lardizabal, K.D., Nelsen, J.S., Bleibaum, J.L., Davies, H.M. and Metz, J.G. (1995) Cloning of a coconut endosperm cDNA encoding a 1-acyl-*sn*-glycerol-3-phosphate acyltransferase that accepts medium-chain-length substrates. *Plant Physiology*, **109**(3), 999–1006.
Knutzon, D.S., Thompson, G.A., Radke, S.E., Johnson, W.B., Knauf, V.C. and K.J.C. (1992) Modification of Brassica seed oil by antisense expression of a stearoyl-acyl carrier protein desaturase gene. *Proceedings of the National Academy of Sciences of the United States of America*, **89**(7), 2624–2628.
Kunst, L., Browse, J. and Somerville, C. (1989) A mutant of Arabidopsis deficient in desaturation of palmitic acid in leaf lipids. *Plant Physiology*, **90**, 943–947.
Lacey, D.J. and Hills, M.J. (1996) Heterogeneity of the endoplasmic reticulum with respect to lipid synthesis in developing seeds of *Brassica napus* L. *Planta*, **199**, 545–551.
Lakshminarayana, G., Rao, K.S., Kittur, M.H. and Mahajanshetty, C.S. (1988) Occurrence of punicic acid in *Trichosanthes bracteata* and *Trichosanthes nervifolia* seed oils. *Journal of the American Oil Chemists' Society*, **65**(3), 347–348.
Lee, M., Lenman, M., Banas, A., Bafor, M., Singh, S., Schweizer, M., Nilsson, R., Liljenberg, C., Dahlqvist, A., Gummeson, P.O., Sjodahl, S., Green, A. and Stymne, S. (1998). Identification of non-heme diiron proteins that catalyze triple bond and epoxy group formation. *Science*, **280**, 915–918.
Lenman, M., Lee, M., Banas, A. *et al.* (1998). In *Advances in Plant Lipid Research* (eds J. Sanchez, E. Cerda-Olmedo and E. Martinez-Force), Sevilla, Secretariado de Publicaciones, Universidad de Sevilla, pp. 657–660.
Lester, R.L., Wells, G.B., Oxford, G. and Dickson, R.C. (1993) Mutant strains of *Saccharomyces cerevisiae* lacking sphingolipids synthesize novel inositol glycerophospholipids that mimic sphingolipid structures. *Journal of Biological Chemistry*, **268**(2), 845–856.
Lindqvist, Y., Huang, W., Schneider, G. and Shanklin, J. (1996) Crystal structure of D9 stearoyl-acyl carrier protein desaturase from castor seed and its relationship to other diiron proteins. *EMBO Journal*, **15**(16), 4081–4092.
Liu, L., Hammond, E.G. and Nikolau, B.J. (1997) *In vivo* studies of the biosynthesis of a-eleostearic acid in the seed of *Momordica charantia L. Plant Physiology*, **113**(4), 1343–1349.
Machado, A. (1938) Oiticica oil. *Revista da Sociedade Brasileira de Quimica-Journal of the Brazilian Society of Chemical Science*, **7**, 73–81.
Matthaus, B., Aitzetmuller, K. and Friedrich, H. (2003) The new database Seed Oil Fatty Acids (SOFA). *GRASAS Y. ACEITES*, **54**, 188–193.
McKinney, S. and Jamieson, G.S. (1936) Composition of oiticica oil. *Oil and Soap-Chicago*, **13**(1), 10–11.
Meesters, P.A.E.P. and Eggink, G. (1996) Isolation and characterization of a D9 fatty acid desaturase gene from the oleaginous yeast *Cryptococcus curvatus* CBS 570. *Yeast*, **12**(8), 723–730.
Meier zu Beerentrup, H. and Roebbelen, G. (1987) Calendula and Coriandrum – new potential oilcrops for industrial uses. *Fett Wissenschaft Technologie-Fat Science Technology*, **89**(6), 227–230.
Mekhedov, S., Cahoon, E.B. and Ohlrogge, J. (2001) An unusual seed-specific 3-ketoacyl-ACP synthase associated with the biosynthesis of petroselinic acid in coriander. *Plant Molecular Biology*, **47**, 507–518.
Mielke, T. (2003) The world outlook for major oilseeds (www.oilworld.biz). *Inform*, **14**(12), 712–713.
Mikolajczak, K.L. and Smith, C.R., Jr. (1967) Optically active trihydroxy acids of Chamaepeuce seed oils. *Lipids*, **2**(3), 261–265.
Mikolajczak, K.L., Freidinger, R.M., Smith, C.R., Jr. and Wolff, I.A. (1968) Oxygenated fatty acids of oil from sunflower seeds after prolonged storage. *Lipids*, **3**(6), 489–494.
Millar, A., Smith, M.A., Kunst, L. (2000) All fatty acids are not equal: discrimination in plant membrane lipids. *Trends in Plant Science*, **5**, 95–101.
Millar, A., Wrischer, M. and Kunst, L. (1998) Accumulation of very-long-chain fatty acids in membrane glycerolipids is associated with dramatic alterations in plant morphology. *Plant Cell*, **11**, 1889–1902.

Minshull, J. and Stemmer, W.P.C. (1999) Protein evolution by molecular breeding. *Current Opinion in Chemical Biology*, **3**(3), 284–290.

Mitchell, A.G. and Martin, C.E. (1995) A novel cytochrome $b_5$–like domain is linked to the carboxyl terminus of the *Saccharomyces cerevisiae* D9 fatty acid desaturase. *Journal of Biological Chemistry*, **270**, 29766–29772.

Moire, L., Rezzonico, E., Goepfert, S. and Poirier, Y. (2004) Impact of unusual fatty acid synthesis on futile cycling through {beta}-oxidation and on gene expression in transgenic plants. *Plant Physiology*, **134**(1), 432–442.

Moon, H., Smith, M.A. and Kunst, L. (2001) A condensing enzyme from the seeds of *Lesquerella fendleri* that specifically elongates hydroxy fatty acids. *Plant Physiology*, **127**(4), 1635–1643.

Moreau, R.A. and Stumpf, P.K. (1981) Recent studies of the enzymic synthesis of ricinoleic acid by developing castor beans. *Plant Physiology*, **67**(4), 672–676.

Murase, J. and Imamura, O. (2002) Conjugated trienoic fatty acid-synthesizing enzyme from *Catalpa ovata*. *Jpn. Kokai Tokkyo Koho*, Japan, 17.

Murphy, D.J. (2002) Biotechnology and the improvement of oil crops – genes, dreams and realities. *Phytochemistry Reviews*, **1**, 66–77.

Murphy, D.J., Hernandez-Pinzon, I. and Patel, K. (2001) Roles of lipid bodies and lipid-body proteins in seeds and other tissues. *Journal of Plant Physiology*, **158**, 471–478.

Muuse, B.G., Cuperus, F.P. and Derksen, J.T.P. (1994) Extraction and characterization of *Dimorphotheca pluvialis* seed oil. *Journal of the American Oil Chemists' Society*, **71**(3), 313–317.

Napier, J.A., Michaelson, L.V. and Sayanova, O. (2003) The role of cytochrome $b_5$ fusion desaturases in the synthesis of polyunsaturated fatty acids. *Prostaglandins Leukotrienes and Essential Fatty Acids*, **68**(2), 135–143.

Napier, J.A., Sayanova, O., Stobart, A.K. and Shewry, P.R. (1997) A new class of cytochrome $b_5$ fusion proteins. *Biochemical Journal*, **1**(328), 717–718.

Nozzolillo, C., Rahal, H. and Liljenberg, C. (1986) Lipid reserves of seeds of *Impatiens capensis* and *I. pallida* (Balsaminaceae): developmental aspects. *American Journal of Botany*, **73**(1), 96–102.

Ohlrogge, J.B. (1994) Design of new plant products: engineering of fatty acid metabolism. *Plant Physiology*, **104**, 821–826.

Ohlrogge, J.B. and Browse, J. (1995) Lipid biosynthesis. *Plant Cell*, 7, 957–970.

Padley, F.B., Gunstone, F.D. and Harwood, J.L. (1994) Occurrence and characteristics of oil and fats, in *The Lipid Handbook* (eds F.D. Gunstone, J.L. Harwood and F.B. Padley), London, Chapman & Hall, pp. 49–170.

Paschke, R.F., Trolberg, W. and Wheeler, D.H. (1953) *Cis-trans* isomers of the eleostearate isomers. *Journal of the American Oil Chemists' Society*, **30**, 97–99.

Pascual-Villalobos, M.J., Robbelen, G., Correal, E. and Ehbrect-von Witzke, S. (1992) Performance test of *Euphorbia lagascae* Spreng., an oilseed species rich in vernolic acid, in southeast Spain. *Industrial Crops and Products*, **1**(2–4), 185–190.

Phillips, B.E., Smith, C.R., Jr. and Hagemann, J.W. (1969) Glyceride structure of *Erlangea tomentosa* seed oil, a new source of vernolic acid. *Lipids*, **4**(6), 473–477.

Powell, R.G., Smith, C.R., Jr. and Wolff, I.A. (1967) *cis*-5, *cis*-9,*cis*-12-Octadecatrienoic and some unusual oxygenated acids in *Xeranthemum annuum* seed oil. *Lipids*, **2**(2), 172–177.

Prasad, M.R. and Joshi, V.C. (1979) Regulation of rat hepatic stearoyl coenzyme A desaturase. The roles of insulin and carbohydrate. *Journal of Biological Chemistry*, **254**, 997–999.

Pugh, E.L. and Kates, M. (1977) Direct desaturation of eicosatrienoyl lecithin to arachidonoyl lecithin by rat liver microsomes. *Journal of Biological Chemistry*, **252**, 68–73.

Qiu, X., Reed, D.W., Hong, H.P., MacKenzie, S.L. and Covello, P.S. (2001) Identification and analysis of a gene from *Calendula officinalis* encoding a fatty acid conjugase. *Plant Physiology*, **125**, 847–855.

Que, L. and Dong, Y. (1996) Modeling the oxygen activation chemistry of methane monooxygenase and ribonucleotide reductase. *Accounts of Chemical Research*, **29**, 190–196.

Radunz, A., He, P. and Schmid, G.H. (1998) Analysis of the seed lipids of *Aleurites montana*. *Zeitschrift fur Naturforschung, C: Biosciences*, **53**(5–6), 305–310.

Reed, D., W., Hammerlindl, J.K., Palmer, C.E., Taylor, D.C., Keller, W. and Covello, P.S. (1997) Hydroxy fatty acid biosynthesis and genetic transformation in the genus Lesquerella

(Brassicaceae), in *Physiology, Biochemistry and Molecular Biology of Plant Lipids* (eds J.P. Williams, M.U. Khan and N.W. Lem), Dordrecht, the Netherlands, Kluwer, pp. 63–65.
Rose, W.G. and Jamieson, G.S. (1943) *Licania arborea* (Cacahuananche) seed oil. *Oil and Soap-Chicago*, **20**, 227–231.
Ruuska, S.A., Girke, T., Benning, C. and Ohlrogge, J.B. (2002) Contrapuntal networks of gene expression during Arabidopsis seed filling. *Plant Cell*, **14**, 1191–1206.
Sakamoto, T., Wada, H., Nishida, I., Ohmori, M. and Murata, N. (1994) D9-acyl lipid desaturases of cyanobacteria: molecular cloning and substrate specificities in terms of fatty acids, *sn*-position, and polar head groups. *Journal of Biological Chemistry*, **269**, 25576–25580.
Sakuradani, E., Kobayashi, M. and Shimizu, S. (1999) D9-fatty acid desaturase from arachidonic acid-producing fungus: unique gene sequence and its heterologous expression in a fungus, Aspergillus. *European Journal of Biochemistry*, **260**, 208–216.
Sarkar, S. and Chakrabarty, M.M. (1956) Component fatty acids of *Impatiens balsamina* seed fat. *Science and Culture*, **21**, 616–618.
Sayanova, O., Beaudoin, F., Libisch, B., Castel, A., Shewry, P.R. and Napier, J.A. (2001) Mutagenesis and heterologous expression in yeast of a plant D6-fatty acid desaturase. *Journal of Experimental Botany*, **52**(360), 1581–1585.
Schneiter, R. and Kohlwein, S.D. (1997) Organelle structure, function, and inheritance in yeast: a role for fatty acid synthesis. *Cell*, **88**, 431–434.
Schultz, D.J. and Ohlrogge, J.B. (2000) Biosynthesis of triacylglycerol in *Thunbergia alata*: additional evidence for involvement of phosphatidylcholine in unusual monoenoic oil production. *Plant Physiology and Biochemistry*, **38**, 169–175.
Schultz, D.J. and Ohlrogge, J.B. (2001) Metabolic engineering of fatty acid biosynthesis, in *Lipid Biotechnology* (eds T.M. Kuo and H.W. Gardner) New York, Marcel Dekker, Inc.
Schultz, D.J., Cahoon, E.B., Shanklin, J., Craig, R., Cox-Foster, D.L., Mumma, R.O. and Medford, J.I. (1996) Expression of a D9 14:0-acyl carrier protein fatty acid desaturase gene is necessary for the production of w5 anacardic acids found in pest-resistant geranium (*Pelargonium xhortorum*). *Proceedings of the National Academy of Sciences USA*, **93**(16), 8771–8775.
Seither, C., Avdiushko, S. and Hildebrand, D.F. (1996) *Isolation of cytochrome P450 genes from Vernonia galamensis*. The 12th International Symposium on Plant lipids, Toronto, Canada, University of Toronto.
Seither, C. Avdiushko, S. and Hildebrand, D.F. (1997) Isolation of Cytochrome P450 genes from *Vernonia galamensis*, in *Physiology, Biochemistry and Molecular Biology of Plant Lipids* (eds J.P. Williams, M.U. Khan and N.W. Lem), Dordrecht, The Netherlands, Kluwer Academic Publishers, pp. 389–391.
Shanklin, J. and Cahoon, E.B. (1998) Desaturation and related modifications of fatty acids. *Annual Review of Plant Physiology and Plant Molecular Biology*, **49**, 611–641.
Shanklin, J. and Somerville, C. (1991) Stearoyl-acyl-carrier-protein desaturase from higher plants is structurally unrelated to the animal and fungal homologs. *Proceedings of the National Academy of Sciences USA*, **88**, 2510–2514.
Shanklin, J., Whittle, E. and Fox, B.G. (1994) Eight histidine residues are catalytically essential in a membrane-associated iron enzyme, stearoyl-CoA desaturase, and are conserved in alkane hydroxylase and xylene monooxygenase. *Biochemistry*, **33**, 12787–12794.
Shu, L., Nesheim, J.C., Kauffmann, K., Munck, E., Lipscomb, J.D. and Que, J.L. (1997) An Fe2 IVO2 diamond core structure for the key intermediate Q of methane monooxygenase. *Science*, **275**, 515–518.
Siddiqi, S.F., Ahmad, F., Siddiqi, M.S. and Osman, S.M. (1980) *Wrightia coccinea* seed oil: a rich source of isoricinoleic acid. *Chemistry & Industry*, (3), 115–116.
Siddiqi, S.F., Ahmad, F., Siddiqi, M.S., Osman, S.M. and Fenwick, G.R. (1984) *Vernonia volkameriaefolia* seed oil: a rich source of epoxy acid. *Journal of the American Oil Chemists' Society*, **61**(4), 798–800.
Singh, C. and Kaul, B.L. (1993) *Vernonia galamensis* (Cass.) a potent new plant source of industrial oil. *Research and Industry*, **38**(1), 1–2.
Sklar, L.A., Hudson, B.S. and Simoni, R.D. (1976), Conjugated polyene fatty acids as fluorescent membrane probes: model system studies. *Journal of Supramolecular Structure*, **4**(4), 449–465.

Smith, C.R., Jr., Koch, K.F. and Wolff, I.A. (1959) Evidence for a new oxygenated fatty acid in the seed oil of *Chrysanthemum coronarium*. *Chemistry & Industry*, **8**, 259–260.

Smith, C.R., Jr., Wilson, T.L., Melvin, E.H. and Wolff, I.A. (1960) Dimorphecolic acid-a unique hydroxydienoid fatty-acid. *Journal of the American Chemical Society*, **82**, 1417–1421.

Smith, C.R., Jr., Wilson, T.L., Bates, R.B. and Shcholfield, C.R. (1962) Densipolic acid: a unique hydroxydienoic acid from *Lesquerella densipila* seed oil. *Journal of Organic Chemistry*, **27**, 3112–3117.

Smith, C.R., Jr., Wilson, T.L., Miwa, T.K., Zobel, H., Lohmar, R.L. and Wolff, I.A. (1961) Lesquerolic acid. A new hydroxy acid from Lesquerella seed oil. *Journal of Organic Chemistry*, **26**, 2903–2905.

Spatz, L. and Strittmatter, P. (1971) A form of cytochrome $b_5$ that contains an additional hydrophobic sequence of 40 amino acid residues. *Proceedings of the National Academy of Sciences USA*, **68**, 1042–1046.

Spencer, G.F., Earle, F.R., Wolff, I.A. and Tallent, W.H. (1973) Oxygenation of unsaturated fatty acids in seeds during storage. *Chemistry and Physics of Lipids*, **10**(2), 191–202.

Sperling, P., Ternes, P., Zank, T.K. and Heinz, E. (2003) The evolution of desaturases. *Prostaglandins Leukotrienes and Essential Fatty Acids*, **68**(2), 73–95.

Spitzer, V., Aitzetmueller, K. and Vosmann, K. (1996) The seed oil of *Bernardia pulchella* (Euphorbiaceae) – a rich source of vernolic acid. *Journal of the American Oil Chemists' Society*, **73**(12), 1733–1735.

Spitzer, V., Marx, F., Maia, J.G.S. and Pfeilsticker, K. (1991) Occurrence of conjugated fatty acids in the seed oil of *Couepia longipendula* (Chrysobalanaceae). *Journal of the American Oil Chemists' Society*, **68**(6), 440–442.

Spitzer, V., Marx, F., Maia, J.G.S. and Pfeilsticker, K. (1992) Occurrence a-eleostearic acid in the seed oil of *Parinari montana* (Chrysobalanaceae). *Fett Wissenschaft Technologie-Fat Science Technology*, **94**(2), 58–60.

Sreenivas, A. and Sastry, P.S. (1994) Synthesis of trilaurin by developing pisa seeds (*Actinodaphne hookeri*). *Archives of Biochemistry and Biophysics*, **311**, 229–234.

Stahl, U., Banas, A. and Stymne, S. (1995) Plant microsomal phospholipid acyl hydrolases have selectivities for uncommon fatty acids. *Plant Physiology*, **107**, 953–962.

Stobart, K., Mancha, M., Lenman, M., Dahlqvist, A. and S. S (1997) Triacylglycerols are synthesised and utilized by transacylation reactions in microsomal preparations of developing safflower (*Carthamus tinctorius* L.) seeds. *Planta*, **203**, 58–66.

Strittmatter, P., Spatz, L., Corcoran, D., Rogers, M.J., Setlow, B. and Redline, R. (1974) Purification and properties of rat liver microsomal stearoyl coenzyme A desaturase. *Proceedings of the National Academy of Sciences USA*, **71**, 4565–4569.

Stukey, J.E., McDonough, V.M. and Martin, C.E. (1989) Isolation and characterization of OLE1, a gene affecting fatty acid desaturation from *Saccharomyces cerevisiae*. *Journal of Biological Chemistry*, **264**, 16537–16544.

Stukey, J.E., McDonough, V.M. and Martin, C.E. (1990) The OLE1 gene of *Saccharomyces cerevisiae* encodes the 19 fatty acid desaturase and can be functionally replaced by the rat stearoyl-CoA desaturase gene. *Journal of Biological Chemistry*, **265**, 20144–20149.

Stymne, S. and Stobart, A.K. (1986) Biosynthesis of g-linolenic acid in cotyledons and microsomal preparations of the developing seeds of common borage (*Borago officinalis*). *Biochemical Journal*, **240**(2), 385–393.

Stymne, S., Bafor, M., Banas, A., Dahlqvist, A., Gummeson, P.O., Lee, M., Lenman, M., Stahl, U. and Wiberg, E. (1998) Metabolism of unusual fatty acids in transgenic and non-transgenic oil seeds, in *Advances in Plant Lipid Research* (eds J. Sanchez, E. Cerda-Olmedo and E. Martinez-Force), Sevilla, Secretariado de Publicaciones, Universidad de Sevilla, pp. 204–210.

Stymne, S., Green, A., Singh, S. and Lenman, M. (1998) Genes for fatty acid D12-epoxygenase of fatty acid epoxide-containing plants and the development of useful producers of fatty acid epoxides. *World Patent*, Australia.

Suh, M.C., Schultz, D.J. and Ohlrogge, T.B. (2002) What limits production of unusual monoenoic fatty acids in transgenic plants? *Planta*, **215**, 584–595.

Sun, C., Cao, Y.Z. and Huang, A.H.C. (1988) Acyl coenzyme A preference of the glycerol phosphate pathway in the microsomes from the maturing seeds of palm, maize, and rapeseed. *Plant Physiology*, **88**, 56–60.

Takagi, T. (1965) Steric configuration of parinaric acid. *Yukagaku*, **14**(7), 370–371.

Tallent, W.H., Harris, J., Wolff, I.A. and Lundin, R.E. (1966) (*R*)-13-Hydroxy-*cis*-9,*trans*-11-octadecadienoic acid, the principal fatty acid from *Coriaria napalensis* seed oil. *Tetrahedron Letters*, **36**, 4329–4334.

Taylor, D.C., Weber, N., Barton, D.L., Underhill, E.W., Hogge, L.R., Weselake, R.J. and Pomeroy, M.K. (1991) Triacylglycerol bioassembly in microspore-derived embryos of *Brassica napus* L. cv Reston. *Plant Physiology*, **97**(1), 65–79.

Thiede, M.A., Ozols, J. and Strittmatter, P. (1986) Construction and sequence of cDNA for rat liver stearoyl coenzyme A desaturase. *Journal of Biological Chemistry*, **261**, 13230–13235.

Thomaeus, S., Carlsson, A.S. and Stymne, S. (2001) Distribution of fatty acids in polar and neutral lipids during seed development in *Arabidopsis thaliana* genetically engineered to produce acetylenic, epoxy and hydroxyl fatty acids. *Plant Science*, **161**(5), 997–1003.

Thompson, G.A. and Li, C. (1997) Altered fatty acid composition of membrane lipids in seeds and seedling tissues of high-saturate canolas, in *Physiology, Biochemistry and Molecular Biology of Plant Lipids* (eds J.P. Williams, M.U. Khan and N.W. Lem), Dordrecht, Netherlands, Kluwer, pp. 313–315.

Thompson, G.A., Scherer, D.E., Foxall-Van Aken, S., Kenny, J.W., Young, H.L., Shintani, D.K., Kridl, J.C. and Knauf, V.C. (1991). Primary structures of the precursor and mature forms of stearoyl-acyl carrier protein desaturase from safflower embryos and requirement of ferredoxin for enzyme activity. *Proceedings of the National Academy of Sciences USA*, **88**, 2578–2582.

Toyama, Y. and Tsuchiya, T. (1935) A new stereoisomer of eleostearic acid in pomegranate seed oil. *Kogyo Kagaku Zasshi*, **38(Suppl. binding)**, 182–185.

Tsujimoto, M. (1936) The naming of the highly unsaturated acid of the kernel fat of Akarittom, *Parinarium laurinum*. *Kogyo Kagaku Zasshi*, **39(Suppl. binding)**, 116–117.

Uzzan, A. (1961) Natural fatty acids of uncommon structure. *J. Inform. Acides Gras Derives, Paris*, 47–54.

van de Loo, F.J., Broun, P., Turner, S. and Somerville, C. (1995) An oleate 12-hydroxylase from *Ricinus communis* L. is a fatty acyl desaturase homolog. *Proceedings of the National Academy of Sciences USA*, **92**, 6743–6747.

van de Loo, F.J., Fox, B.G. and Somerville, C. (1993) Unusual fatty acids, in *Lipid Metabolism in Plants* (ed. T.S. Moore), Boca Raton, CRC Press, pp. 91–126.

van Kuijk, F.J.G.M., Sevanian, A., Handelman, G.J. and Dratz, E.A. (1987) A new role for phospholipase $A_2$: protection of membranes from lipid peroxidation damage. *Trends in Biochemical Science*, **12**, 31–34.

van Loon, J. and Steger, A. (1931) Couepic acid, a new isomer of the eleostearic acids. *Recueil des Travaux Chimiques des Pays-Bas et de la Belgique-Royal Netherlands Chemical Society*, **50**, 936–942.

Vioque, E. and Maza, M.P. (1971) Seed oils of *Vernonia anthelmintica* and *Euphorbia lagascae*. *Grasas y Aceites (Sevilla, Spain)*, **22**(1), 25–30.

Voelker, T. and Kinney, A. (2001) Variations in the biosynthesis of seed-storage lipids. Annual *Review of Plant Physiology and Plant Molecular Biology*, **52**, 335–361.

Voelker, T.A. and Davies, H.M. (1994) Alteration of the specificity and regulation of fatty acid synthesis of *Escherichia coli* by expression of a plant medium-chain acyl-acyl carrier protein thioesterase. *Journal of Bacteriology*, **176**, 7320–7327.

Vogel, G. and Browse, J. (1996) Choline phospho transferase and diacylglycerol acyl transferase:substrate specificities at a key branch point in seed lipid metabolism. *Plant Physiology*, **110**, 923–931.

Wada, K., Onda, M. and Matsubara, H. (1989) Amino acid sequences of ferredoxin isoproteins from radish roots. *Journal of Biochemistry*, **105**, 619–625.

Wallar, B.J. and Lipscomb, J.D. (1996) Dioxygen activation by enzymes containing binuclear nonheme iron clusters. *Chemical Reviews*, **96**, 2625–2657.

Wang, H., *et al.* (1998). Fatty-acid composition and application for seed oil of *Punica granatum* (ed. L. Zhongguo Youzhi). **23**, 54–56.

Weismann, G. (1976) The seed oils of Aleurites species. *Seifen Oele Fette Wachse*, **102**(3), 77–78.
Weselake, R.J. (2001) Biochemistry and biotechnology of triacylglycerol accumulation in plants, in *Lipid Biotechnology* (eds T.M. Kuo and H.W. Gardner), New York, Marcel Dekker, Inc., pp. 27–56.
Whittle, E. and Shanklin, J. (2001) Engineering D9-16:0-acyl carrier protein (ACP) desaturase specificity based on combinatorial saturation mutagenesis and logical redesign of the castor D9–18:0-ACP desaturase. *Journal of Biological Chemistry*, **276**(24), 21500–21505.
Wiberg, E. and Bafor, M. (1995) Medium chain length fatty acids in lipids of developing oil palm kernel endosperm. *Phytochemistry*, **39**, 1325–1327.
Wiberg, E., Banas, A. and Stymne, S. (1997) Fatty acid distribution and lipid metabolism in developing seeds of laurate-producing rape (*Brassica napus* L.). *Planta*, **203**, 341–348.
Wiberg, E., Edwards, P., Byrne, J., Stymne, S. and Dehesh, K. (2000) The distribution of caprylate, caprate and laurate in lipids from developing and mature seeds of transgenic *Brassica napus* L. *Planta*, **212**, 33–40.
Wiberg, E., Tillberg, E. and S.S. (1994) Substrates of diacylglycerol acyltransferase in microsomes from developing oil seeds. *Phytochemistry*, **36**(3), 573–577.
Wongwathanarat, P., Michaelson, L.V., Carter, A.T., Lazarus, C.M., Griffiths, G., Stobart, A.K., Archer, D.B. and MacKenzie, D.A. (1999). Two fatty acid D9-desaturase genes, ole1 and ole2, from *Mortierella alpina* complement the yeast ole1 mutation. *Microbiology*, **145**, 2939–2946.
Zou, J., Katavic, V., Giblin, E.M., Barton, D.L., MacKenzie, S.L., Keller, W.A., Hu, X. and Taylor, D.C. (1997). Modification of seed oil content and acyl composition in the Brassicaceae by expression of a yeast *sn*-2 acyltransferase gene. *Plant Cell*, **9** 909–923.

# 4 Non-food lipids

Sevim Z. Erhan and Atanu Adhvaryu

## 4.1 Introduction

### *4.1.1 Structure and composition of lipids*

Lipids are a diverse group of biological substances made up primarily, or exclusively, of nonpolar groups. They can be readily classified on the basis of their chemical structure and solubility (Burton and Guerra,1972). As a result of their nonpolar character, lipids typically dissolve more readily in nonpolar solvents, such as acetone, ether, chloroform and benzene, than in water. There are three major classes of lipids:

1. Simple lipid – in which a fatty acid (FA) is esterified with glycerol or fatty alcohol (e.g. mono-, di- or triacylglycerols).
2. Compound lipid – simple lipids with an additional group esterified to glycerol (e.g. phospholipids, sphingolipids and glycolipids).
3. Derived lipid – lipid derived from simple or compound lipid (e.g. fatty alcohols).

To emphasize on non-food application of lipids, we shall limit our discussion to the glycerol-containing lipids.

#### *4.1.1.1 Simple lipids*

Simple lipids (also described as neutral lipids) are commonly found as storage fats/oils and they bear no charged groups. Generally, they are completely nonpolar with no affinity for water. Almost all neutral lipids are a combination of FAs with the alcohol – glycerol. FAs are primarily long-chain, unbranched aliphatic acids, with the carbon atoms attached to hydrogens and other groups and the chain terminating with a carboxylic acid. The carboxyl (-COOH) group at one end gives the molecule its acidic properties. Most naturally occurring FAs contain an even number of carbon atoms in their backbone chains. Although a few with odd numbers are found in all organisms, these make up only a minor fraction of the total. The carbon chains that constitute most of the available FAs found in nature vary in length from as few as 4–24 or more carbons. Most FAs found in neutral lipids have even-numbered chains with 14–22 carbons; those with either 16 or 18 carbons occur most frequently (Table 4.1). The polar -COOH group is enough to make the shortest FA chains water-soluble. As chain length increases, the FA becomes progressively less water-soluble and takes on oily or fatty characteristics.

**Table 4.1** Saturated and unsaturated fatty acids.

| Saturated fatty acids | | Unsaturated fatty acids | |
|---|---|---|---|
| Butyric acid (C4) | $CH_3(CH_2)_2CO_2H$ | Crotonic acid | $CH_3CH{=}CHCO_2H$ |
| Caproic acid (C6) | $CH_3(CH_2)_4CO_2H$ | Palmitoleic acid ($C16{:}1^{\Delta 9}$)[b] | $CH_3(CH_2)_5CH{=}CH(CH_2)_7CO_2H$ |
| Caprylic acid (C8) | $CH_3(CH_2)_6CO_2H$ | Oleic acid ($C18{:}1^{\Delta 9}$)[b] | $CH_3(CH_2)_7CH{=}CH(CH_2)_7CO_2H$ |
| Capric acid (C10) | $CH_3(CH_2)_8CO_2H$ | Linoleic acid ($C18{:}2^{\Delta 9,12}$)[b] | $CH_3(CH_2)_3(CH_2CH{=}CH)_2(CH_2)_7CO_2H$ |
| Lauric acid (C12)[a] | $CH_3(CH_2)_{10}CO_2H$ | Linolenic acid ($C18{:}3^{\Delta 9,12,15}$)[b] | $CH_3(CH_2CH{=}CH)_3(CH_2)_7CO_2H$ |
| Myristic acid (C14)[a] | $CH_3(CH_2)_{12}CO_2H$ | Arachidonic acid | $CH_3(CH_2)_3(CH_2CH{=}CH)_4(CH_2)_3CO_2H$ |
| Palmitic acid (C16)[a] | $CH_3(CH_2)_{14}CO_2H$ | Erucic acid | $CH_3(CH_2)_{11}CH{=}CH(CH_2)_7CO_2H$ |
| Stearic acid (C18)[a] | $CH_3(CH_2)_{16}CO_2H$ | Nervonic acid | $CH_3(CH_2)_7CH{=}CH(CH_2)_{13}CO_2H$ |
| Arachidic acid (C20) | $CH_3(CH_2)_{18}CO_2H$ | | |
| Behenic acid (C20) | $CH_3(CH_2)_{20}CO_2H$ | | |
| Lignoceric acid (C24) | $CH_3(CH_2)_{22}CO_2H$ | | |

[a]Common saturated fatty acids. [b]Common unsaturated fatty acids.

If the carbon of a FA chain binds to a maximum possible number of hydrogen atoms, the FA is saturated. If the number of hydrogen atoms bound by the carbons is less than the possible maximum, the FA is unsaturated. At points where hydrogen atoms are missing from adjacent carbon atoms, the carbons share a double instead of a single bond. If double bonds occur at multiple sites (up to a maximum of about six), the FA is polyunsaturated (Perkins and Visek, 1983). Unsaturated FAs have lower melting points than saturated FAs and are more abundant in living organisms.

The carbon chain of a fully saturated FA is more or less straight, without major bends. An unsaturated FA may take one or two forms at a double bond (Figure 4.1). In the *cis* form, both the hydrogen atoms of the double bond are positioned on the same side of the C–C bond, and the chain bends at an angle of about 30°, producing a kink. In the *trans* form (hydrogen atoms on opposite sides of the double bond), the chain is twice bent so that is continues in the same direction without a pronounced kink, after the double bond. The kink of the *cis* form affects the packing of unsaturated FA chains, making them more disordered and, consequently, more fluid at normal temperatures. Unsaturated FAs with the *trans* form are thermodynamically more stable than the *cis* form and, therefore, melt at a higher temperature. During a chemical reaction, such as oxidation, the *cis* configuration at the double bond frequently changes to *trans*, resulting in a chemical modification in the chain structure.

Stearic acid, $COOH\ (CH_2)_{16}CH_3$

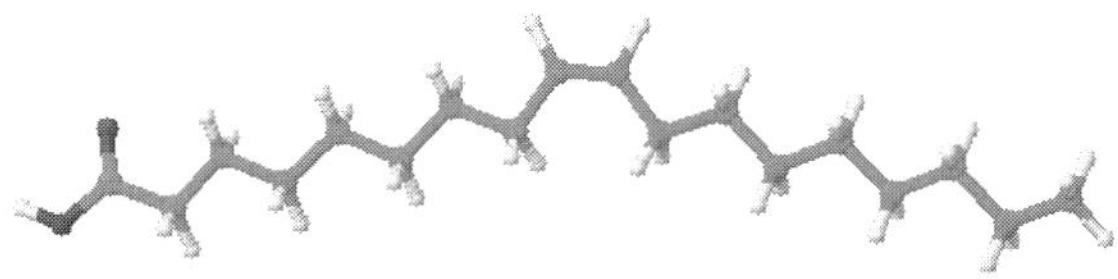

Oleic acid, $COOH\ (CH_2)_7CH{=}CH(CH_2)_7CH_3$

**Figure 4.1** Saturated and unsaturated fatty acids.

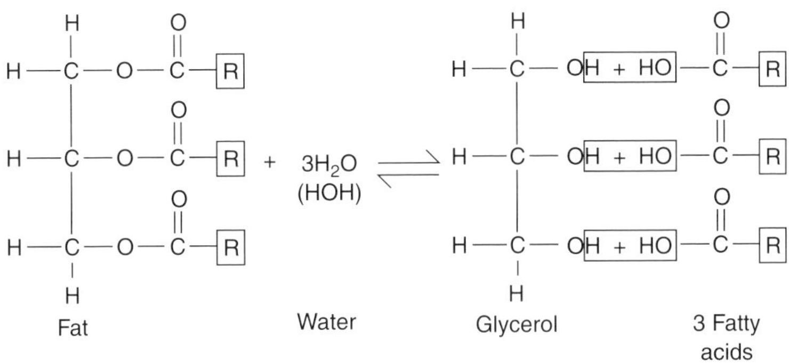


**Figure 4.2** Triacylglycerol from glycerol and fatty acids.

### *4.1.1.2 Triacylglycerols*

The different FAs discussed above can exist in a free state or as esters – most abundantly with a commonly found alcohol in lipid structure, namely glycerol (propane-1,2,3-triol). If three FAs bind to the three –hydroxyl groups of the alcohol, the resulting compound is known as triacylglycerol (formerly known as triglycerides) (Figure 4.2). They are the fully acylated derivative of glycerol. Similarly, when one and two of the hydroxyl groups are esterified with alcohol, it results in monoacylglycerols (2 isomeric forms: α- and β-monoacylglycerols) and diacylglycerols (2 isomeric forms: αβ and αα′), respectively. The FA constituents of a triacylglycerol molecule may be all identical. This is relatively rare and such compounds are called simple or monoacid triacylglycerols e.g. triolein in olive oil and tripalmitin in palm oil. More normally, two or three of the FAs are different and such compounds are termed complex or mixed triacylglycerols.

Free FAs are produced by hydrolysis of oils and are catalyzed by acids, enzymes or metals. Fully refined good quality oils (refined, bleached and deodorized products of high quality) have free FA contents of less than 0.1% (normal range 0.01–0.05%, in most cases related to oleic acid content). The quality of crude oils largely depends on the content of free FAs and, in very general terms, the quality deteriorates as the acid content rises. Good crude oils have free FA ranges of less than 5% (mostly in the range of 0.5–3.0%). Crude oils with contents higher than 10% free FA are regarded as spoiled, as they are difficult to refine.

The oils constitute mostly of triacylglycerols (98%), diacylglycerols (0.5%), free FAs (0.1%), sterols (0.3%) and tocopherols (0.1%). Tocopherols are known to have certain antioxidant properties. They are a series of benzopyranols with one, two or three methyl groups attached to the phenolic ring along with a $C_{16}$ side chain on the pyran ring. Antioxidant activity as a result of tocopherol content is high in corn oil, soybean oil, walnut oil and cottonseed oil (Rossell and Pritchard, 1991; Gunstone *et al.*, 1994).

***Oxidative stability.*** It is a well-known fact that when vegetable oils are exposed to an oxidizing environment, they undergo oxidative degradation. Oxidation is the single most important reaction leading to polymerization and degradation of oils resulting in increased acidity, corrosion, viscosity and volatility when used in industrial base oils. Vegetable oil triacylglycerol is a complex association of different FA molecules attached to a single triacylglycerol structure (Figure 4.3). The double bonds present in the triacylglycerol molecule from oleic, linoleic and linolenic acid moieties, act as active sites for various oxidation reactions.

Saturated FAs have relatively high oxidation stabilities (Brodnitz, 1968), which decrease with increasing unsaturation in the molecule. The rate of oxidation depends on the degree of unsaturation of a fatty acyl chain. In general, the rate of oxidation of linoleic ($C_{18:2}$) is 10 times greater than oleic ($C_{18:1}$), while linolenic ($C_{18:3}$)

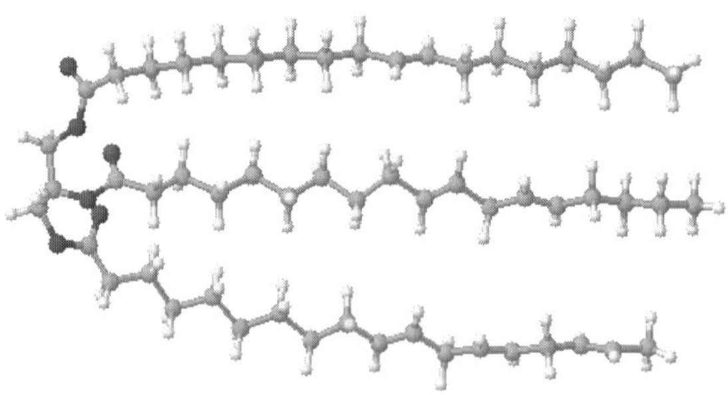


**Figure 4.3** Typical structure of a triacylglycerol molecule.

is twice as great as the linoleic ($C_{18:2}$) fatty acyl chain. Oxidation usually takes place through a radical initiated chain mechanism involving: initiation, propagation, branching, chain-stopping inhibition and peroxide decomposition. The free radicals generated during the initiation stage react with $O_2$ to form peroxy free radicals and hydroperoxides (Privett and Blank, 1962). During this period, $O_2$ is consumed in a zero-order process (Labuza and Bergquist, 1983), leading to intermediates that are not too well-characterized, prior to the formation of peroxides (Privett and Blank, 1962). The latter undergo further reactions to form alcohols, ketones, aldehydes, carboxylic acids (Shahidi, 1997), leading to rancidity and toxicity (Moll, 1976), thereby, accelerating the oil degradation process (Harman, 1982; Steinberg *et al.*, 1989). The extent of oxidation and formation of oxidation products are further complicated by the amount of unsaturation, structural differences in the various triacylglycerol molecules and presence of antioxidants. The increased oxidative instability of the polyunsaturated fatty acyl chains is mainly due to *bis* allylic methylenes present between the double bonds in linoleic and linolenic acids. The cumulative effect of various structural parameters in the triacylglycerol molecule makes oxidation a highly complex process (Adhvaryu *et al.*, 2000). Elimination of *bis* allylic methylenes by hydrogenation will increase the oxidative stability of an oil, but this will also adversely affect the low temperature properties like pour point [ASTM D-97, the lowest temperature at which movement of the oil is observed on tilting the test jar (a clear cylindrical flat bottom glass jar) is recorded as the pour point of the sample] and the viscosity index of the oil.

***Low temperature fluidity.*** The fluidity of a material is mainly determined by the efficiency of molecular packing, intermolecular interactions and molecular weight. Vegetable oils, when subjected to a low temperature environment, undergo solidification through crystallization, which poses a major drawback for use in industrial applications. The FA chains of triacylglycerol molecules have a bend 'tuning fork' conformation and undergo molecular stacking during the cooling process. The relatively poor low temperature flow properties of vegetable oils arise from the appearance of waxy crystals that rapidly agglomerate resulting in the solidification of oil. Wax crystallization at low temperature is controlled by steric and geometrical constrains in these molecules (Adhvaryu *et al.*, 2003). In vegetable oil triacylglycerols, the presence of double bonds in the fatty acyl chain influences the low temperature behavior, as shown in Table 4.2. The decreased melting temperatures of these compounds are a result of disorganization of the crystalline lattice by the presence of double bonds. This demonstrates the contradiction in simultaneously requiring both low temperature properties and the best possible oxidative stability in a given triacylglycerol molecule. Due to the presence of various triacylglycerol polymorphisms ($\alpha$, $\beta'$, $\beta$) (Chapman, 1962; Hoerr and Paulicka, 1968; Hagemann and Rothfus, 1983) in crystal packing and the energy involved in low temperature crystallization, seed oils do not crystallize over a narrow temperature range (Hagemann and Rothfus, 1988). Instead, it is a slow continuous process when

**Table 4.2** Low temperature and oxidation rate of simple triacylglycerols.

| | Iodine value | Melt temperature (°C) | Oxidation | Relative rate |
|---|---|---|---|---|
| Tristearin $C_{18:0}$ | 0 | 74 | Low | 1 |
| Triolein $C_{18:1}$ | 86 | 5 | Moderate | 10 |
| Trilinolein $C_{18:2}$ | 173 | −11 | High | 100 |
| Trilinolenin $C_{18:3}$ | 261 | −24 | Very high | 200 |

the microcrystalline structures initially formed become macrocrystalline and rapidly change to a solid-like consistency. Deposition of waxy materials from oil results in a rapid increase in viscosity leading to poor pumpability, lubrication and rheological behavior.

***Iodine value and melting point.*** The FAs in triacylglycerol molecules can be individually saturated or unsaturated, resulting in a straight chain or bent chain configuration, respectively. The straight chain FAs are more pliable and, therefore, harden at lower temperatures. With increasing numbers of double bonds in the chain, the ability to gain a closely packed conformation is prevented and, therefore, the oil remains liquid. To determine the amount of double bonds in a given sample of oil, iodine is introduced; this binds to the unsaturated carbons at the C=C sites. The amount of iodine in grams absorbed per 100 ml of oil is called the iodine value (IV). The higher the IV, the more unsaturated (the greater the number of double bonds) the oil and, therefore, higher is the potential for the oil to polymerize through oxidation. Any attempt to increase the saturation content of the oil (through hydrogenation) will increase the melting temperature of the oil (Table 4.3) (Lide, 1996). It has been observed that an IV of less than 25 is required if the neat oil is to be used for long-term applications in unmodified diesel engines. Triacylglycerols in the range of IV 50–100 may result in decreased engine life, and, in particular, will reduce the life of fuel pump and injector.

Most vegetable oils are unsuitable for industrial applications due to their high saturated or polyunsaturated FA content. The oxidation stability of polyunsaturated FAs can be improved significantly by converting them to saturated fats. However, this will also adversely affect the performance of the oil at low temperatures. Optimally, vegetable oils with high oxidation stability and low pour points can be produced by converting all the FAs into monounsaturated FAs. Through genetic manipulation in plant breeding, FA profiles of vegetable oils can be altered to obtain high oleic varieties. The oxidation stability of high oleic oils is three to six times greater than conventional vegetable oils. Oils thus produced will provide both high thermo-oxidation stability and reasonable low temperature flow properties.

***Industrial applications.*** Non-food uses of vegetable oils have grown very little during the past 40 years. Although some markets have expanded and new ones have been developed, other markets have been lost to more competitive petroleum

**Table 4.3** Melting point and iodine values of some common oils.

| Oil | Approx. melting point (°C) | Iodine value |
|---|---|---|
| Coconut oil | 25 | 10 |
| Palm kernel oil | 24 | 37 |
| Mutton tallow | 42 | 40 |
| Beef tallow | — | 50 |
| Palm oil | 35 | 54 |
| Olive oil | −6 | 81 |
| Castor oil | −18 | 85 |
| Peanut oil | 3 | 93 |
| Rapeseed oil | −10 | 98 |
| Cotton seed oil | −1 | 105 |
| Sunflower oil | −17 | 125 |
| Soybean oil | −16 | 130 |
| Tung oil | −2.5 | 168 |
| Linseed oil | −24 | 178 |
| Sardine oil | — | 185 |

products. Vegetable oils are currently being used in various industrial applications such as emulsifiers, lubricants, plasticizers, surfactants, plastics, solvents and resins. Research and development approaches take advantage of the natural properties of these oils: unsaturation, amphiphilic character, high molecular weight, high viscosity index and being eco-friendly among others. These oils, derived from renewable resources, also do not contribute to volatile organic chemicals (VOC) that are regarded as undesirable pollutants detrimental both to the environment and to human health. Some of the major industrial applications of plant oils are:

*Lubricants.* This is a well-established and highly competitive market growing at an average rate of less than 1% per year (Padavich and Honary, 1995). More than 70% of total lubricant volume is used as motor oils for automotive engines and approximately 10% as hydraulic fluids. Vegetable oil basestocks and other lipid derivatives have shown significant increases in use as biodegradable lubricants over the last decade, although they still contribute less than 2% of all basestocks used in the market. Other areas of usage are relatively small: cutting oils, two-stroke engine oils, chainsaw bar oils, wire-rope oils, bicycle chain oils, railroad oils, pump oils, outboard engine oils, drilling oils and other niche markets. These oils have several advantages for industrial lubricant applications, such as non-toxicity (Battersby *et al.*, 2000), low volatility, having an ester linkage to deliver inherent lubricity, good solubilizing power for additive molecules and excellent viscosity/temperature characteristics.

*Printing inks.* In the printing industry, the processes in current use are lithography (43%), gravure (23%), flexography (21%), letterpress (4%), screen printing (3%) and electronic desktop publishing (6%) (The Rauch Guide to the US Ink Industry, 1998). Alkali-refined seed oils such as soybean, cottonseed, canola, safflower and sunflower oils have been used in ink formulations. Biodegradability and VOC characteristics of seed-oil-based fluids are superior to petroleum-based products (Erhan and Bagby, 1995).

*Paints and coatings.* Various seed oils and lipid derivatives are currently used in the production of alkyd resins, which are, in turn, a primary constituent of so-called oil-based or solvent-borne paints. These oils represent a large, stable supply of natural, renewable raw material and also meet the paint industry's major environmental challenge to eliminate volatile organic compounds.

*Surfactants.* Today, the $10 billion (annual worldwide market) surfactant industry is in search of environmentally acceptable surfactants (Longhini, 1996). Surfactants based on lipids and their derivatives can potentially compete technically and economically with many petroleum- and oleochemical-based commercial surfactants used now. Emulsion, polymerization, textile specialties and oil field chemicals are some important areas of seed-oil-based surfactants used in industry.

*Solvents.* Methyl esters derived from various seed oils offer good potential as industrial solvents for use in cold cleaning, carrier solvents, resin removal and cleanup. They are low-cost, biodegradable, low-toxicity, high-flash point alternatives to the traditional chlorinated, hydrocarbon and fluorocarbon solvents being regulated out of markets as ozone-depleting chemicals, hazardous air pollutants and/or source of volatile organic compounds.

## 4.2 Industrial applications

### *4.2.1 Industrial commodity seed oils*

#### *4.2.1.1 Soybean oil*

Soybean is the second highest value cash crop in the United States. The farm value of soybean production in the crop year 2000 was $13 billion. Soy oil is typically 18–19% of the weight of soybean. Once soybean oil is extracted from the bean, it can be used in its raw or refined form in a variety of industrial products. About two-thirds of this oil is economically extracted by crushing and pressing operations. Soy oil (extracted via solvent extraction) is also used for a number of industrial applications. Bioengineered (high oleic and/or low linoleic) soybean may provide highly desirable improvements for fuels and other industrial products. Currently, different industries are using soybean to manufacture a variety of industrial products. A comprehensive list is presented in Table 4.4 below.

Current research priorities are focused on developing improved soy-oil-based lubricants with improved oxidation stability and cold-weather pour properties. Developments in the area of total loss lubricants, hydraulic fluids and crankcase lubricants are showing promising results. Products, such as biodegradable greases and two-cycle engine oils based on soybean oil, are currently available. These are very environmentally friendly when used in outboard motors, lawnmowers and other small engines. Soybean oil used in crankcase must exhibit properties such as high lubricity, viscosity index, flash point and low evaporation loss.

**Table 4.4** Industrial uses of soybean oil.

| | |
|---|---|
| Biodiesel fuel | Dust control agents |
| Printing inks | Electrical insulation |
| Paints | Epoxies |
| Soaps | Linoleum backing |
| Shampoos | Metal casting/working |
| Detergents | Pesticides |
| Anti-corrosion agents | Plasticizers |
| Anti-static agents | Protective coatings |
| Caulking compounds | Putty |
| Disinfectants | Vinyl plastics |

Soybean oil has been effectively modified for use as an active diluent for alkyd coatings. Development of rapid drying techniques utilizing UV-curing has effectively reduced the rate of unwanted color development in soybean-oil-based alkyd paints. Soybean oil phosphate-ester-polyol in water- and/or solvent-borne inks could replace gel ink in inexpensive ballpoint pens. Replacing a portion of the more expensive acrylate oligomer resins with up to 20% soybean oil would also decrease the amount of VOCs released during curing. Soybean alkyds may become the next binder in the pigment printing technique employed in textile printing. Soy alkyds produce bright, crisp prints on fabric with depth of shade values similar to acrylic binder. With a drying speed between the popular linseed and tung oils, conjugated soybean oil can replace the more expensive, imported drying oils and be used in printing inks and surface coatings. Conjugated oil has a low odor, and also appears to produce a harder ink film (Nichols *et al.*, 1951).

Several derivatives of soybean oil are currently manufactured and sold on the open market. Of these, the use of methylsoyate as a direct fuel substitute in diesel engines has been thoroughly researched and practically applied (Borgelt *et al.*, 1994). Unlike traditional fuels, soybean oil does not evaporate; it is cleaner and acts as an anti-caking agent.

Plastic composites are a combination of filler, polymer, reinforcement and performance additives. When exposed to a catalytic initiator or a source of heat, these composites cure into a permanent nonreversible shape. The composite polymer systems that currently utilize soybean oil are unsaturated polyesters and polyurethanes (Williams and Wool, 2000). These composite polymers find utility in a variety of applications in the transportation, construction and agricultural equipment markets. Polyurethanes are composed of a multifunctional alcohol, known as a polyol and a di- or tri-isocyanate plus various additives. A polyol product based on soybean oil, which is useful as a substitute for petroleum-based polyols has been developed. Polyols have been commercialized in polyurethane formulations that are used to produce elastomers, flexible molded foams, structural and rigid polyurethane foams (Dweib *et al.*, 2004).

*4.2.1.2 Canola oil*

Canola (normally *Brassica napus* or less commonly *Brassica rapa*) is a cool-season, annual oilseed crop. It is a member of the mustard family with the seed containing 40% oil and a residual animal feed meal containing 37–38% crude protein. Canola oil is more commonly known as 'rape oil', and is a semi-drying oil that is used as a lubricant, fuel, soap and synthetic rubber base, and as an illuminant to give color pages in magazines their slick look. It is typically referred to in industry as penetrating oil and has a higher level of linolenic FA than soybean oil. High α-tocopherol content (19 mg/100 gm) and trace amounts of erucic FA contribute to the oil stability as compared to soybean oil (Takagi and Ando, 1991; Leonard, 1994). With the rapid development of high oleic variety oils, various technical properties of the oil can be significantly improved meeting industrial specifications. Table 4.5 shows a comparison between conventional and high oleic canola oil.

Canola-based motor oils have rapidly evolved into a competitive product as a potential substitute for mineral-oil-based products. In terms of pricing, they are highly competitive with synthetic motor oils. They are also the most 'environmentally friendly' of the motor oils available and have excellent properties of non-toxicity and biodegradability. In terms of functionality, they have exceeded expectations by surpassing both conventional and synthetic oils in the tests conducted. There has been significant reduction in tailpipe gas emissions of nitrous oxides, carbon monoxide, non-methane hydrocarbons, and hydrocarbons, therefore providing an easy and effective way to reduce air pollution (Reece and Peterson, 1985).

*4.2.1.3 Sunflower oil*

Sunflower, *Helianthus annuus*, is the only oilseed native to the northern Great Plains of North America. In the former Soviet Union, it was developed as an oilseed crop and today about 25% of the world sunflower oil is produced there, mainly in the Ukraine. Sunflower may be chosen as an alternative oilseed crop to canola since it is not attacked by most canola insect pests, except for cutworm. Traditional sunflower oil consists of 68%, linoleic acid ($C_{18:2}$) and about 20% oleic acid ($C_{18:1}$). Two

**Table 4.5** Physical properties of natural and high oleic canola oil.

| Fatty acid distribution/technical properties | High oleic canola oil | Conventional canola oil |
|---|---|---|
| Palmitic acid ($C_{16:0}$) | 3.5 | 3.5 |
| Stearic acid ($C_{18:0}$) | 1.5 | 1.5 |
| Oleic acid ($C_{18:1}$) | 84.0 | 61.5 |
| Linoleic acid ($C_{18:2}$) | 3.5 | 19.5 |
| Linolenic acid ($C_{18:3}$) | 4.0 | 10.5 |
| Melting temperature (°C) | −5.5 | −9.5 |
| PDSC* (minutes) at 130 °C | 71 | 25 |

*Pressurized differential scanning calorimetry.

varieties of the oil are currently available, namely the original high linoleic acid type and a newer high oleic type. The high oleic variety, which contains as much as 60–70% oleic acid, has much higher oxidation stability and is suitable for industrial applications. Both canola and sunflowers are popular alternative oilseeds in the US Sunflower oil is considered to be a premium oil (35–45% oil) due to its light color, mild flavor, low level of saturated fats, and ability to withstand high cooking temperatures. Although the sunflower has the potential for many industrial uses, in the United States it is mostly used for food or feed purposes. Sunflower hulls have a limited market for specialty purposes such as poultry litter, fireplace logs, and other high fiber products. Sunflower and soybean oil have been researched as potential diesel fuel substitutes, while the meal is used as a livestock feed and the glycerol as an industrial chemical (Haumann, 1994).

For sunflower, it is possible to produce enough oil to replace diesel fuel use without producing excessive meal for on-farm livestock. Fuel applications largely depend on the physical and chemical characteristics of the oil. General, polyunsaturated oils break down faster than saturated oils, but increased saturation leads to solidification at near-room temperatures. Ignition quality (cetane number) and heat of combustion for vegetable oils are generally lower than diesel fuels. The most serious problem related to using these oils as diesel fuels is their viscosity. Viscosity is critically dependent on temperature, and the viscosity of seed oils is more affected by temperature than the viscosity of conventional diesel fuels. Higher pour points of seed oils also pose problem in low temperature applications. These oils can be used straight or blended with diesel fuel. Short-term testing of oilseed fuels indicated that these fuels were roughly equivalent to diesel fuel. Longer-term tests have had problems of deposit build-up in the combustion chamber and injector nozzle due to the poor thermal stability of vegetable oils and piston ring sticking and engine failure due to decreased fuel atomization (Gupta, 1998; Hardin, 1998; Kleingartner and Warner, 2001).

#### *4.2.1.4 Safflower oil*

Different varieties of safflower, *Carthamus tinctorius*, may have levels of seed oil ranging from 38–44% total seed weight. The actual oil content of the de-hulled safflower seed is relatively constant at 59–64%. There are two types of safflower varieties; one that produces oil which is high in monounsaturated FAs (oleic acid), and the other with high concentrations of polyunsaturated FAs (linoleic acid) (Christie *et al.*, 1991). The high linoleic safflower oil contains nearly 75% linoleic acid, which is considerably higher than corn, soybean, cottonseed, peanut or olive oils. The high oleic safflower varieties, which may contain up to 80% oleic acid are comparable to olive oil, and are stable when heated. As an industrial oil, it is considered as a drying or semi-drying oil, with properties intermediate between soybean and linseed oils, and is extensively used in manufacturing paints and other surface coatings. The oil is light in color and will not yellow with aging; hence it is used in white and light-colored paints. This oil can also be used as a diesel fuel substitute, but like

**Table 4.6** Typical fatty acid composition (%) of industrial linseed oil.

| | | |
|---|---|---|
| $C_{16:0}$ | 6.0 | Palmitic acid |
| $C_{18:0}$ | 2.5 | Stearic acid |
| $C_{20:0}$ | 0.5 | Arachidic acid |
| $C_{18:1}$ | 19.0 | Oleic acid |
| $C_{18:2}$ | 24.1 | Linoleic acid |
| $C_{18:3}$ | 47.4 | Linolenic acid |

most vegetable oils, is currently too expensive for this application. Other industrial uses include non-yellowing drying paints, alkyd resins in enamels, caulks and putties. High oleic safflower oil is rapidly gaining recognition as one of nature's most valuable vegetable oil with extensive industrial applications.

#### *4.2.1.5 Linseed oil*

Seed of the linseed (or flax) plant, *Linum usitatissimum*, normally produces an oil (linseed oil) that is rich in α-linolenic acid, a polyunsaturated FA that makes the oil highly susceptible to oxidation. The oil is a high quality drying oil as a result of its ability to form a durable film upon exposure to air. This is the result of a high content of unsaturated FAs, oleic, linoleic and α-linolenic, a combination not found in any other vegetable oil (Table 4.6). Due to these unique properties, traditionally linseed has been grown for its oil (as opposed to flax, which is the same species grown as a fiber crop). Linseed oil is used in the manufacture of coatings, paints, lacquers, varnishes, and stains because of its drying and hardening characteristics when exposed to air and sunlight, as well as in soap, putty, printing ink and linoleum (Chan *et al.*, 1991).

#### *4.2.1.6 Tung oil*

Tung oil is obtained from the seeds of a tropical tree, the tung tree, *Aleurites fordii* of the spurge family, and from seeds of some related species. It is known also as China wood oil and nut oil. The poisonous seeds found in the heart of tung fruit (which is the size of a small apple) contain more than 50% tung oil, which is readily obtained when the seeds are heated, ground, and pressed. The fruit contains from 14–20% oil, the kernel 53–60% and the nut has an oil content of 30–40%. The components are: eleostearic (80%), oleic (8%), palmitic (about 4%), linoleic (4%), linolenic (3%) and stearic acids (about 1%) (Gunstone *et al.*, 1994). The oil is amber-colored and contains a high proportion of eleostearic acid. The physical and chemical properties of tung oil make it useful as a protective coating, solvent, and/or drying agent in various paints, varnishes, lacquers, resins, fiberboard, concrete sealers, electronic circuit boards, and printing inks. Its superior drying properties allow it to be sold at a price premium compared to other vegetable drying oils such as linseed oil. Tung oil products are used to coat containers for food, beverages and medicines. They are also used for insulating wires and other metallic surfaces such as radios, radar, telephone and other communications hardware. Various new applications for tung oil and its byproducts are also being developed for use in products such as cosmetics, insecticides, and lubricants.

### *4.2.2 New industrial oilseed crops*

Most alternative oilseeds have high oil contents, typically 40% or more of the seed weight, compared to about 20% in soybean seeds. Certain specialty oilseeds, often with industrial uses, have been domesticated from wild plants and are gaining a commercial market. Examples include meadowfoam, jojoba, vernonia, lesquerella and crambe.

#### *4.2.2.1 Meadowfoam oil*

Meadowfoam, *Limnanthes alba*, seeds (nutlets) contain 20–30% oil. This oil contains three previously unknown long-chain FAs and is most similar to high erucic acid rapeseed oil. It is unusually high in long-chain FAs (over 90% $C_{20}$–$C_{22}$ FAs) with very high levels of mono-unsaturation and very low levels of poly-unsaturation. Typical major FA composition is as follows: $C_{20:1}$ is 63%, $C_{22:1}$ is 16%, and $C_{22:2}$ are 17% (Kamel and Kakuda, 1994). These characteristics make meadowfoam oil very stable, even when heated or exposed to air. Therefore, crude meadowfoam oil is the most oxidatively stable vegetable oil known. Meadowfoam oil has the added benefit of enhancing the properties of other oils when mixed with them. Less expensive oils can be mixed with meadowfoam oil without the loss of the qualities of either oil and it can increase the stability of the oils to which it is added. Meadowfoam oil and its derivatives can be chemically transformed into a liquid wax ester that is a substitute for sperm whale oil and jojoba oil. The oil can be converted to a light-colored premium grade solid wax, a sulfur compound valuable to the rubber industry, or used as a lubricant, detergent or plasticizer. Derivatives of meadowfoam oil, such as estoloides and silicone esters have potential as coatings and adhesives.

#### *4.2.2.2 Lesquerella oil*

The oil in the seeds of Lesquerella, *Lesquerella fendleri*, contains a $C_{20}$ hydroxy FA, lesquerolic acid, which is similar to ricinoleic acid, the $C_{18}$ hydroxy FA found in castor oil. These hydroxy FAs are important raw materials used for the manufacture of resins, waxes, nylons, plastics, corrosion inhibitors, coatings, lubricating greases and cosmetics (Smith *et al.*, 1961, 1962). Lesquerolic acid can be chemically modified to obtain dodecanedioic and hydroxydodecanoic acids. Lesquerolic acid produces 12-tridecenoic acid that can be used in the production of the monomer for nylon-13. The saturated hydroxyl FAs, via hydrogenation process, could be useful in the production of greases by converting them into their lithium soaps. The triacylglycerols from Lesquerella are suitable for many uses. For example, they could substitute for dehydrated and oxidized castor oil. Many of the properties may be enhanced over those of castor oil because of the increased chain length of this new crop oil.

*4.2.2.3 Cuphea oil*

The seeds of most members of the genus Cuphea contain 30–36% oil. The seed oil contains high amounts of short- and medium-chain saturated FAs, $C_8$, $C_{10}$, $C_{12}$ and $C_{14}$, which are used in the production of solvents, detergents and emulsifiers (Miller *et al.*, 1964; Wolf *et al.*, 1983). Many different species of Cuphea are available with a very diverse FA pattern. *Cuphea painteri,* for instance, is very rich in caprylic ($C_{8:0}$) acid (73%) while *Cuphea carthagenensis* has lauric acid ($C_{12:0}$) as its major FA (81%). *Cuphea koehneana* is probably the best example of monoacid seed oil, with more than 95% of its acyl groups as capric acid ($C_{10:0}$). At present, the only commercially available sources of these FAs are coconut and palm kernel oils. Cuphea oil offers medium-chain length glycerides based on capric ($C_{10:0}$) or lauric ($C_{12:0}$) acid that could be used in temperate countries as a replacement for coconut and palm kernel oils that are imported from tropical regions. Other uses of the seed oil include cosmetics and in motor oils.

*4.2.2.4 Crambe oil*

Crambe, *Crambe maratima*, is a cool-season oilseed that grows best in semi-arid regions with warm days, cool nights and low humidity. The crop is grown for its oil, which contains high amounts of erucic acid (*cis*-13-docosenoic), a $C_{22}$ FA. Erucic acid is used to make intermediate chemicals, emollients and surfactants that are used as inputs in the manufacture of such items as plastic bags, cosmetics, personal care products and laundry detergents. Erucic acid is converted to erucamide for use as a slip agent in the manufacture of polyethylene sheets, so that the sheets will extrude well and not stick together after extrusion (Ackman, 1983).

*4.2.2.5 Jojoba wax*

Perhaps the most commercially advanced of the new crops is jojoba, *Simmondsia chinensis*. The oil is obtained from the fruit, which is a nut (about the size of peanut), about 50% of which is oil. Jojoba 'oil' is chemically a liquid wax rather than the familiar triacylglycerol produced in well-known oil seeds like soybean or canola. The oil is made up of a wax ester of 40–42 carbon atoms and is formed from the esterification of a very-long-chain ($C_{20}$ or $C_{22}$) fatty alcohol to a FA of similar chain length. The principal market for jojoba products continues to be for the oil pressed from jojoba seed for use within the cosmetic industry. An estimated 2000 tonnes per annum is consumed by the cosmetic industry in the form of whole jojoba oil, hydrogenated jojoba oil, jojoba esters, hydrolyzed jojoba oil, ethoxylated jojoba oil and other value-added jojoba oil derivatives. Jojoba oil in lubricant applications provides a market for approximately 100 tonnes annually. In general, the price of jojoba is too high for this market compared to other available lubricant oils. The molecular structure of this oil is very tough; therefore, it is stable even at high temperatures and pressures, unlike many other lubricants (Bhatia *et al.*, 1990).

### *4.2.3 Use of tallow and yellow grease for industrial applications*

Tallow is inedible grease derived from animal fat renderings. Products such as soap, cosmetics, lubricants and plastics, are made from tallow. It has found use as a bio-fuel, where a bio-diesel formulation was developed by combining 65% of tallow methyl esters with 35% of fuel ethanol. Mixing 20% of this blend with 80% No. 2 diesel resulted in a fuel with a viscosity identical to diesel alone, with good engine performance. Tallow is gradually being replaced by yellow grease (waste vegetable oils). Yellow grease comes from used cooking oil from fast-food restaurants. It contains a mix of unsaturated and saturated FAs (in a ratio of about 2.8 : 1); the dietary energy value is greater than saturated fats, such as tallow, and comparable to that of poly-unsaturated fats, such as soybean oil. Therefore, yellow grease is used primarily in livestock feed and also as a component in bio-diesel fuel. Bio-diesel is usually sold as a blend of 30% recycled and refined yellow grease and 70% petroleum-based products. Other application areas of yellow grease are manufacture of soap, makeup, clothing, rubber and detergents.

## 4.3 Structural modifications

### *4.3.1 Interesterification*

Natural fats do not have an even distribution of FAs on the glyceride molecules. The concentration and relative distribution of certain acid in the triacylglycerol molecule varies with species, environment and location of the plant. Physical characteristics of lipids are greatly affected not only by the nature of constituent FAs (i.e. chain length and unsaturation) but also by their distribution in the triacylglycerol molecules. In some cases, the unique FA distribution patterns of some natural lipids limit their industrial applications. Interesterification is one of the processes that can improve the consistency of such fats and improve their usefulness. The process involves ester interchange within a single triacylglycerol molecule (intraesterification) as well as ester exchange among different molecules. Catalysts used in the high temperature interesterification process are KOH and NaOH while low temperature catalyst is sodium methoxide, $NaOCH_3$.

Considering two (n) FAs (A and B) available for esterification with glycerol, eight possible triacylglycerol species can result ($n^3$), as shown below.

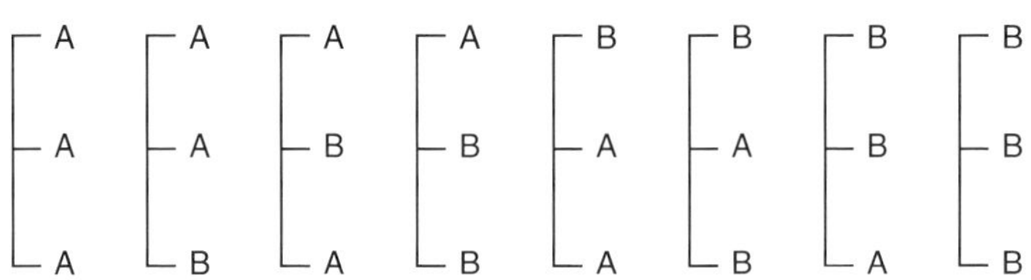

**Table 4.7** Melting point of the oil in °C.

| Fat | Before esterification | After esterification |
|---|---|---|
| Soybean oil | −7.0 | 5.5 |
| Cottonseed oil | 10.5 | 34.0 |
| Lard | 43.0 | 43.0 |
| Coconut oil | 26.0 | 28.2 |
| 25% Tristearin + 75% soybean oil | 60.0 | 32.2 |

Quantitative proportions of the different species depend only on the amount of each acid in the original fat, and can be predicted. This process involves rearranging the FAs in such a way that their distribution among the triacylglycerol molecules of the fat becomes random (random interesterification) or conforms to some special pattern (directed interesterification) (Howley and Holman, 1956). There is significant improvement in various physical properties of the oil through interesterification, such as melting point (Table 4.7) and it also has wide use in industrial applications, such as the production of bio-diesel.

### 4.3.2 *Fractionation*

After lipid extraction, the next stage involves the fractionation of the complex mixture into the various classes of lipids. Lipids from natural sources contain various components and precise analysis of each one needs some preliminary purification. Fractionation procedures must be adapted to the situation and the nature of the components to be separated. The most commonly used and efficient methods are described below:

#### 4.3.2.1 *Solvent fractionation*

This is the simplest and, perhaps, the most efficient for separating a group of lipids of interest when the others are without value. The process is largely based on the differential solubility of lipids in organic solvents. It is however, important to operate the whole procedure at a constant temperature to obtain reproducible results.

#### 4.3.2.2 *Column chromatography*

Chromatography of lipids using a glass column filled with a suitable stationary phase is a common and useful method for fractionation of lipid classes either on an analytical or a semi-preparative scale. This method generates fractions, which need to be quantified later. Two types of column (or liquid) chromatography generally used are adsorption and ion-exchange chromatography. In the former, lipid mixtures are separated on the basis of their polarity, while in the latter separation is based on the ionic groups present in some lipid molecules.

*4.3.2.3 Thin-layer chromatography*
This is perhaps the most efficient and versatile technique for separation of complex lipid mixtures as well as for individual lipids. The method is rapid and allows the characterization of the separated fractions by elution and the fractions can be studied by other procedures.

### *4.3.3 Hydrogenation*

Hydrogenation is the addition of hydrogen to double bonds in the FA chains, and, therefore, increases the degree of saturation in the molecule. As a result of hydrogenation process, liquid oils are converted into semi-solids with an improvement in their oxidation stability (Coenen and Linssen, 1970). Hydrogenation involves mixing the oil (refined, bleached, low in soap and dry) with a nickel catalyst (most commonly used), heating it to between 140–225°C and then exposing it to hydrogen at high pressure (60 psig) with stirring. The reaction mixture is heterogeneous in nature where the catalyst and reactants exist in different physical states and the reaction takes place on the catalyst surface. The reaction goes through an agitation process whereby $H_2$ gas is uniformly dissolved in the oil and the catalyst evenly dispersed. The agitation also allows effective heat dissipation during the process. The process rate is determined by: nature of the substrate, pressure, temperature, catalyst concentration and agitation rate. During hydrogenation, double bond migration can occur, and when it does, there are 2 *trans-* bonds formed for every *cis-* bond. The results of hydrogenation are three-fold, namely, the saturation of double bonds, *cis/trans* isomerization and double bond migration. The stages in linoleate hydrogenation to stearate are shown below.

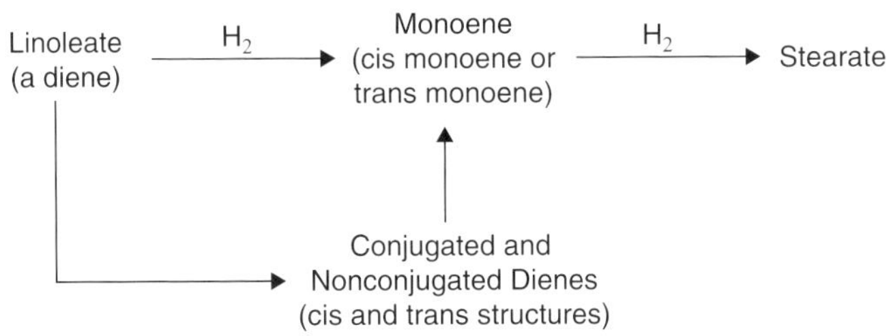


## 4.4 Concluding remarks

Plant lipids have enormous potential to substitute for conventional mineral oils in a wide variety of industrial applications. Major benefits expected to be derived from such uses are environmental, such as access to renewable sources of raw materials that are non-toxic but still have comparable physiochemical properties with petroleum-based products. This drive toward the use of renewable resources was given major boost in the United States of America with the signing, in 1999, of the executive order by President Bill Clinton that encouraged Federal agencies to triple their use of bio-based products and bio-energy by 2010. Also, non-food uses of lipids are expected to gain predominance in areas where stiff regulations require expensive clean-up and disposals. For example, the use of non-toxic lubricating oils and diesel fuels is now required in many non-urban areas, especially parks and forests, in Europe. This favors the use of plant oils in such applications, even if they are still more expensive than petroleum-derived products.

The widespread use of plant-derived lipids will ultimately depend on how well they perform in a wide range of temperature, pressure, metal surface, in presence of other functional fluids and existing technology. Lipids draw much of their important functional properties from their molecular structures namely: the presence of mono- and poly-unsaturation, polar groups, long hydrocarbon chain structure and easy functionalization to other important chemical derivatives. Plant lipids and their numerous chemically modified derivatives now have significantly improved thermo-oxidative, low temperature stability and lubrication properties, therefore, increasing their use in a variety of industrial applications, such as lubricants, functional fluids, specialty chemicals, polymers, composites, inks, etc. Development of new types of lipids from genetically modified plants is a lengthy process and such lipids are currently not extensively available as all-purpose inexpensive materials capable of delivering all performance qualities for non-food uses. Nevertheless, it is evident that over the last decade there has been a significant progress in the research and development of lipid technology for innovative non-food uses (Murphy, 2002, 2003).

## References

Ackman, R.G. (1983) Chemical composition of rapeseed oil, II. Fatty acids, in *High and Low Erucic Acid Rapeseed Oils. Production, Usage, Chemistry, and Toxicological Evaluation* (eds J.K.G. Kramer, F.D. Sauer and W.J. Pigden), Academic Press, Toronto, Canada, p. 86.

Adhvaryu, A., Erhan, S.Z., Liu, Z.S. and Perez, J.M. (2000) Oxidation kinetic studies of oils derived from unmodified and genetically modified vegetables using pressurized differential scanning calorimetry and nuclear magnetic resonance spectroscopy. *Thermochimica Acta*, **364**(1–2), 87–97.

Adhvaryu, A., Erhan, S.Z. and Perez, J.M. (2003) Wax appearance temperatures of vegetable oils determined by differential scanning calorimetry: effect of triacylglycerol structure and its modification. *Thermochimica Acta*, **395**(1–2), 191–200.

Battersby, N.S., Pack, S.E. and Watkinson, R.J. (2000) The biodegradability and microbial toxicity testing of lubricants – some recommendations. *Chemosphere*, **41**(7), 1011–1027.

Bhatia, V.K., Chaudhary, A., Sivashankaran, G.A., Bisht, R.P.S. and Kashyap, M. (1990) Modification of jojoba oil for lubricant formulations. *Journal of the American Oil Chemists' Society*, **67**, 1–7.

Borgelt, S.C., Kolb, T.S. and Schumacher, L.G. (1994) Biodiesel: world status, in *Proceedings of an Alternative Energy Conference* (ed. B.E. Dale), ASAE, St. Joseph, MI, pp. 67–76.

Brodnitz, M.H. (1968) Auto-oxidation of saturated fatty acids. *Journal of Agricultural and Food Chemistry*, **16**, 994–999.

Burton, R.M. and Guerra, F.C., eds (1972) *Fundamentals of Lipid Chemistry*, Bi-Science Publication Division, Webster Groves, MO, pp. 1–10.

Chan, J.K., Bruce, V.M. and McDonald, B.E. Chan (1991) Dietary alpha-linolenic acid is as effective as oleic acid and linoleic acid in lowering blood cholesterol in normolipidemic men. *The American Journal of Clinical Nutrition*, **53**, 1230–1234.

Chapman, D. (1962) The polymorphism of glycerides. *Chemical Reviews*, **62**, 433–456.

Christie, W.W., Nikolova-Damyanova, B., Laakso, P. and Herslof, B. (1991) Stereospecific analysis of triacyl-*sn*-glycerols via resolution of diasetreomeric diacylglycerol derivatives by high performance liquid chromatography on silica. *Journal of the American Oil Chemists' Society*, **68**, 695–701.

Coenen, J.W.E. and Linssen, B.G. (1970) *Physical and Chemical Aspects of Adsorbents and Catalysts* (ed. B.G. Linssen), Academic, New York.

Dweib, M.A., Hu, B., O'Donnell, A., Shenton, H.W. and Wool, R.P. (2004) All natural composite sandwich beams for structural applications. *Composite Structures*, **63**, 147–157.

Erhan, S.Z. and Bagby, M.O. (1995) Biodegradation of news ink with modified sturm test, in *TAGA Proceedings*, Vol. 1, Book Crafters, Inc., Chelsea, MI, pp. 184–203.

Gunstone, F.D., Harwood, J.L. and Padley, F.B., eds (1994) *The Lipid Handbook*, 2nd edn, Chapman and Hall, London.

Gupta, M.K. (1998) NuSun – the future generation of oils. *INFORM*, **9**, 1150–1154.

Hagemann, J.W. and Rothfus, J.A. (1983) Computer modelling of theoretical structures of monoacid triglyceride α-forms in various subcell arrangements. *Journal of the American Oil Chemists' Society*, **60**, 1308–1314.

Hagemann, J.W. and Rothfus, J.A. (1988) Effect of chain length, conformation and α-form packing arrangement on theoretical monoacid triglyceride β′ -forms. *Journal of the American Oil Chemists' Society*, **65**, 4, 638–646.

Hardin, B. (1998) Mid-oleic sunflower hybrids debut. *Agricultural Research*, **46**(6), 14–15.

Harman, D. (1982) *Free Radical in Biology* Vol. 5 (ed. W.A. Pryor), Academic Press, N.Y, p. 255.

Haumann, B.F. (1994) Modified oil may be key to sunflower's future. *INFORM*, **5**, 110–111.

Hoerr, C.W. and Paulicka, F.R. (1968) The role of X-ray diffraction in studies of the crystallography of monoacid saturated triglycerides. *Journal of the American Oil Chemists' Society*, **45**, 793–797.

Howley, H.K. and Holman, G.W. (1956) Directed interesterification as a new processing tool for lard. *Journal of the American Oil Chemists' Society*, **33**, 29–35.

Kamel, B.S. and Kakuda, Y., eds. (1994) *Technological Advances in Improved and Alternate Sources of Lipids*. Blackie, London.

Kleingartner, L. and Warner, K. (2001) A look at a new sunflower oil. *Cereal Foods World*, **46**, 399–404.

Labuza, T.P. and Bergquist, S. (1983) Kinetics and oxidation of potato chips under constant temperature and sine wave temperature conditions. *Journal of Food Science*, **48**, 712–715.

Leonard, C. (1994) Sources and commercial applications of high-erucic vegetable oils. *Lipid Technology*, **6**, 79–83.

Lide, D.R., ed. (1996) *Handbook of Chemistry and Physics*, 76th Edn, CRC Press, Boco Raton, USA.

Longhini, D. (1996) Surfactants on cleanup detail in paper mills. *Chemical and Engineering News*, **6**, 80–81.

Miller, R.W., Earle, F.R., Wolff, I.A. and Jones, Q. (1964) Search for new industrial oil. IX. Cuphea, a versatile source of fatty acid. *Journal of the American Oil Chemists' Society*, **41**, 279–285.

Moll, C., Biermann, U. and Grosch, W. (1979) Occurrence and formation of bitter-testing trihydroxy fatty acids in soybean. *Journal of Agricultural and Food Chemistry*, **27**, 239–243.

Murphy, D.J. (2002) Novel oils from plants – genes, dreams and realities. *Phytochemistry Reviews*, **1**, 67–77.

Murphy, D.J. (2003) Agricultural biotechnology and oil crops: current uncertainties and future prospects. *Applied Biotechnology, Food Science and Policy*, **1**, 25–38

Nichols, P.L., Herb, S.F. and Reimenschneider, R.W. (1951). Isomers of conjugated fatty acids. I. Alkali-isomerised ricinoleic acid. *Journal of the American Chemical Society*, **73**, 247–252.

Padavich, R.A. and Honary, L. (1995) A market research and analysis report on vegetable-based industrial lubricants. *Society of Automotive Engineers, Technical Paper 952077*, 13–14.

Perkins, E.G. and Visek, W.J., eds. (1983) *Dietary Fats and Health*, American Oil Chemists' Society, Champaign, IL.
Privett, O.S. and Blank, M.L. (1962) The initial stages of autoxidation. *Journal of the American Chemical Society*, **39**, 465–469.
Reece, D.L. and Peterson, C.L. (1985) A report on the Idahoon-road vehicle test with RME and neat rapeseed oil as an alternative to diesel fuel. ASAE Paper No. 93-5018, ASAE, St. Joseph, MI.
Rossell, J.B. and Pritchard, J.L.R., eds. (1991) *Analysis of Oil Seeds, Fats and Fatty Foods*, Elsevier, London.
Shahidi, F., ed. (1997) *Natural Antioxidants Chemistry, Health Effects and Application*, AOCS Press, Champaign, IL.
Smith, C.R., Jr., Wilson, T.L., Bates, R.B. and Scholfield, C.R. (1962) Densipolic acid: a unique hydroxyl dienoid acid from lesquerella densilila seed oil. *The Journal of Organic Chemistry*, **27**, 3112–3117.
Smith, C.R., Jr., Wilson, T.L., Miwa, T.K., Zobel, H.F., Lohmar, R.L. and Wolff, I.A. (1961) Lesquerolic acid: a new hydroxyl acid from lesquerella seed oil. *The Journal of Organic Chemistry*, **26**, 2903–2905.
Steinberg, D., Parthasarathy, S., Carew, T.E., Khoo, J.C. and Witztum, J.L. (1989) Beyond cholesterol. Modifications of low-density lipoprotein that increase its atherogenicity. *New England Journal of Medicine*, **320**, 915–924.
Takagi, T. and Ando, Y. (1991) Stereospecific analysis of triacylglycerols by chiral high-performance liquid chromatography. *Lipids*, **26**, 542–547.
*The Rauch Guide to the US Ink Industry* (1998) Impact Marketing Consultants, Inc., Manchester Center, VT, pp. 1.
Williams, G.I. and Wool, R.P. (2000) Composites from natural fibers and soy oil resins. *Applied Composite Materials*, **7**, 421–432.
Wolf, R.B., Graham, S.A. and Kleiman, R. (1983) Fatty acid composition of cuphea seed oils. *Journal of the American Oil Chemists' Society*, **60**, 103–104.

# 5 Membrane lipids

Peter Dörmann

## 5.1 Introduction

Biological membranes of living cells consist of a bilayer of polar lipids, predominantly glycerolipids, into which hydrophobic proteins are embedded. In plant membranes, two structurally distinct groups of glycerolipids are found: phosphoglycerolipids ('phospholipids'), which contain phosphate in their head group, and glycoglycerolipids ('glycolipids') harbouring a carbohydrate moiety in the head group. Phospholipids are also abundant in membranes of animals and yeast. The major phospholipid classes were identified in lipid extracts from eggs and animal tissues more than 100 years ago (for a historical overview see Chapter 1 and www.cyberlipid.org). However, glycoglycerolipids are found only as major membrane lipids in plants and some bacteria, and these non-phosphorous lipids were discovered in lipid extracts from wheat in the 1950s.

The function of glycolipids in higher plants in photosynthesis and during phosphate deprivation was discovered only very recently. In addition to phospholipids and glycolipids, non-glycerolipids (i.e. sphingolipids and sterol lipids), which are described in other chapters of this book (especially Chapter 11), are found in plants. It should be noted that sphingolipids and sterol lipids as well as their glycosylated derivatives (cerebrosides, sterol glycosides) are highly abundant in extraplastidic membranes, e.g. tonoplast and plasma membranes. Furthermore, membranes with specialised functions, such as thylakoids of chloroplasts and the inner mitochondrial membrane, contain lipid-like molecules, which are in part associated with protein complexes (e.g., chlorophyll, carotenoids, plastoquinone, phylloquinone and ubiquinone). Different glycerolipid-derived metabolites involved in cell signalling are present in plant membranes in low amounts. These molecules originate from glycerolipid breakdown, e.g. by action of phospholipases. The presence of distinct acyl groups esterified to the *sn*-1 and *sn*-2 positions of the glycerol backbone adds to the complexity of membrane lipids, and gives rise to numerous molecular species for each glycerolipid class. Many of the fatty acids found in plant lipids contain two or three double bonds. These polyunsaturated fatty acids are particularly enriched in photosynthetic membranes, and they have been shown to play critical roles in different physiological processes of plant cells.

Membrane glycerolipid biosynthesis and function has been the focus of several excellent reviews in the past (e.g. Joyard *et al.*, 1993; Kinney, 1993). However, many biosynthetic pathways were previously only described on a biochemical level because the isolssation of the corresponding genes was hampered by the low abundance and the hydrophobicity of enzymes involved in lipid metabolism. The use of *Arabidopsis thaliana* as a model organism has greatly advanced our understanding of lipid metabolism and function in plants down to a molecular level (e.g. Joyard *et al.*, 1998; Vijayan *et al.*, 1998). The isolation of lipid-deficient Arabidopsis *mutants* combined with the capability of isolating genes via map-based cloning, as well as the sequencing of the entire Arabidopsis genome (Arabidopsis Genome Initiative, 2000) has laid the groundwork for the identification of many lipid biosynthesis genes. Based on the sequence information available, a classification system was recently established covering the entire set of lipid-related genes in Arabidopsis (Beisson *et al.*, 2003; www.plantbiology.msu.edu/lipids/genesurvey/index.htm). This chapter focuses on the characterisation of phospholipid and glycolipid synthesis, degradation and function, taking advantage of the wealth of information particularly derived from research on Arabidopsis, which became available in the recent past.

## 5.2 Structures and localisation of glycerolipids

### *5.2.1 Phosphatidic acid*

Phosphatidic acid (PA; 1,2-diacyl-*sn*-glycerol-3-phosphate), which was first isolated from cabbage leaf by Chibnall and Channon (1927) represents a central intermediate in phospholipid metabolism (Figure 5.1). However, there is still debate about whether or not this glycerolipid indeed represents an authentic constituent of plant membranes or is merely an artefact of lipid extraction from plant tissues. It has been shown that PA accumulation can originate from phospholipase D (PLD) activity during lipid extraction. On the other hand, PA has repeatedly been identified as an authentic lipid of extraplastidic membranes, e.g. the plasma membrane, where it accounts for about 5% of glycerolipids (see Table 5.1 and Chapter 9).

### *5.2.2 Galactolipids*

Galactolipids were first discovered by Carter *et al.* (1956) in wheat flour. The structures of the two galactolipids were identified as 1,2-diacyl-3-O-(β-D-galactopyranosyl)-*sn*-glycerol (MGDG) and 1,2-diacyl-3-O-(β-D-galactopyranosyl-1→6-α-D-galactopyranosyl)-*sn*-glycerol (DGDG; Figure 5.1). Oligogalactolipids containing three or more galactose moieties in the head group (TriGDG and TetraGDG) were also detected in plants, although in only minor amounts (Benson *et al.*, 1958; Fujino and Miyazawa, 1979; Kojima *et al.*, 1990). Polyunsaturated

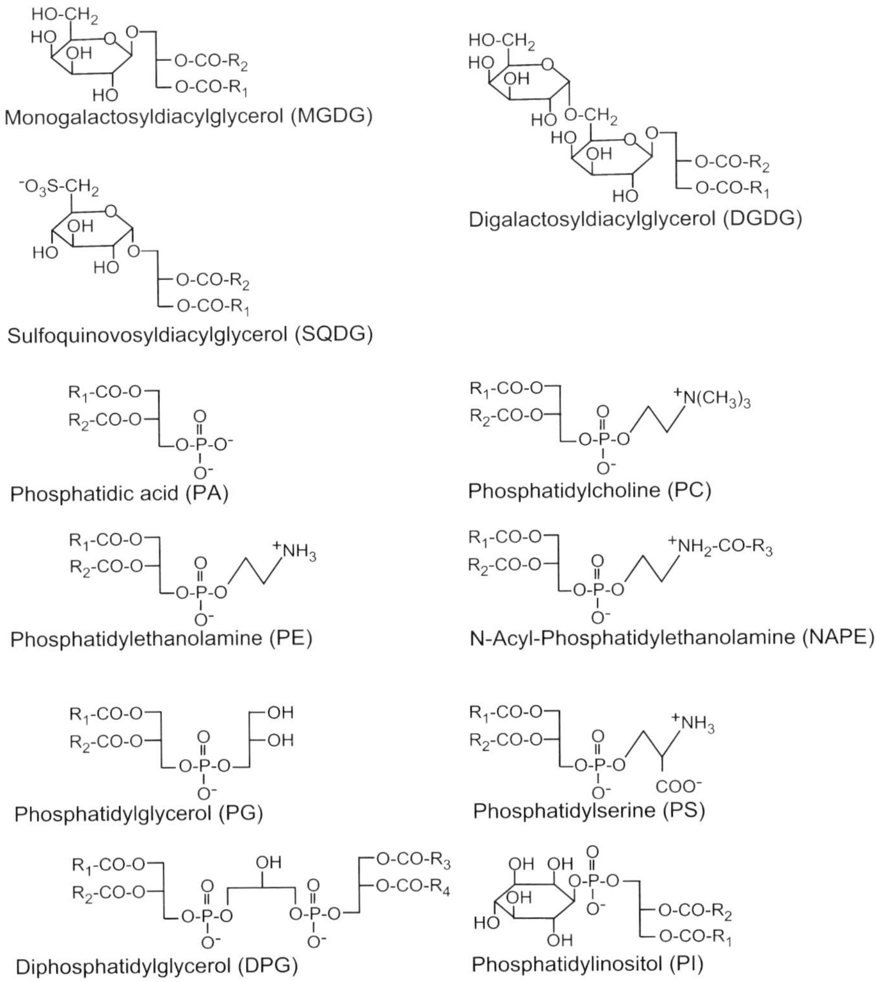


**Figure 5.1** Structures of membrane glycerolipids. In higher plants, the non-phosphorous glycolipids MGDG, DGDG and SQDG are the predominant lipid classes in chloroplasts. Phospholipids (PC, PE and PG) are abundant in extraplastidic membranes. Additional phospholipids (PA, PS, NAPE, DPG and PI) are minor components of plant membranes, and are restricted to specific organelles or accumulate under certain environmental conditions.

fatty acids such as α-linolenic acid (18:3) and hexadecatrienoic acid (16:3) are abundant in galactolipids. Hexadecatrienoic acid is only found in a certain group of plants ('16:3 plants'; see Section 5.3.1) and is restricted to the *sn*-2 position of MGDG. Galactolipids are the predominant lipids in photosynthetic membranes of plants and cyanobacteria, where they account for about 85% of total glycerolipids

**Table 5.1** Glycerolipid composition of different cellular membranes.

| | Chloroplast (spinach leaf; Block *et al.*, 1983) | | | Mitochondrion (sycamore cells; Bligny and Douce, 1980) | | Endomembranes (barley roots; Brown and DuPont, 1989) | | Plasma membrane (oat roots; Andersson *et al.*, 2003) | |
|---|---|---|---|---|---|---|---|---|---|
| | Thylakoids | Outer envelope | Inner envelope | Outer membrane | Inner membrane | ER | Tonoplast[1] | +Pi | −Pi |
| MGDG | 52 | 17 | 49 | 0 | 0 | tr[2] | 1 | 0 | 0 |
| DGDG | 26 | 29 | 30 | 0 | 0 | tr | 5 | 8 | 70 |
| SQDG | 6.5 | 6 | 5 | 0 | 0 | —[3] | — | — | — |
| PC | 4.5 | 32 | 6 | 68 | 29 | 58 | 53 | 38 | 15 |
| PE | 0 | 0 | 0 | 24 | 50 | 21 | 23 | 16 | 5 |
| PS | — | — | — | tr | tr | 1 | 3 | 5[5] | 1[5] |
| PI | 1.5[4] | 5[4] | 1[4] | 5 | 2 | 10[6] | 9[6] | | — |
| PG | 9.5 | 10 | 8 | 2 | 1 | 6 | 6 | 25 | 4 |
| DPG | 0 | 0 | 0 | — | 17 | 0 | tr | — | — |
| PA | — | — | — | — | — | 1 | 2 | 5 | 2 |

Data are presented in mol% of total glycerolipids. [1]contains Golgi membranes; [2]traces; [3]not measured; [4]tentatively identified; [5]PS and PI were not separated; [6]includes lyso-PE

(Block *et al.*, 1983). The inner envelope membrane of chloroplasts shows a lipid composition similar to thylakoids. The outer envelope, however, contains relatively more DGDG as compared to MGDG (Table 5.1). In roots and seeds, the amounts of galactolipids are reduced as compared to phospholipids. This reflects the fact that galactolipids are abundant in thylakoid membranes, which are absent from non-photosynthetic tissues. Galactolipids were previously believed to be restricted to plastids and excluded from all other cellular membranes. However, recent evidence clearly demonstrates that their exclusive localisation to plastids is not conserved during phosphate deprivation, because under these conditions, DGDG, but not MGDG, accumulates in extraplastidic membranes, e.g. in the plasma membrane (Härtel *et al.*, 2000; Andersson *et al.*, 2003; Table 5.1).

### *5.2.3 Sulfolipid*

Sulfolipid was first isolated and its structure determined as 1,2-diacylglycerol-3-(6-sulfo-α-D-quinovopyranosyl)-*sn*-glycerol (SQDG) by Benson *et al.* (1959, 1963) (Figure 5.1). Plant sulfolipid is rich in saturated fatty acids, particularly palmitic acid (Siebertz *et al.*, 1979). SQDG is restricted to plastid membranes where it accounts for about 5% of glycerolipids (Table 5.1). Within plastids, sulfolipid is found in thylakoids and in the two envelope membranes of chloroplasts but there are no reports of extraplastidic occurrence of SQDG in higher plants.

### *5.2.4 Phosphatidylglycerol and diphosphatidylglycerol*

The structure of phosphatidylglycerol (PG) was first described as 1,2-diacyl-*sn*-glycerol-3-(phospho-1′-*sn*-glycerol) by Benson and Maruo (1958) (Figure 5.1). PG is the only phospholipid in thylakoid membranes, but is also present in extraplastidic membranes including endoplasmic reticulum (ER), plasma membrane and mitochondria (Table 5.1). In most plants species, the plastidic form of PG contains a *trans* fatty acid (16:1$\delta$3 *trans*), which is absent from all other lipids.

A related lipid, diphosphatidylglycerol (DPG), was isolated from bovine heart muscle and therefore designated 'cardiolipin' (Pangborn, 1942). Later, it was found in all tissues and in other organisms, including plants. DPG is restricted to the inner membrane of mitochondria (Table 5.1; Bligny and Douce, 1980). Its structure was determined as 1,3-bis(*sn*-3′-phosphatidyl)-*sn*-glycerol by MacFarlane *et al.* (1958) (Figure 5.1).

### *5.2.5 Phosphatidylcholine, phosphatidylethanolamine, phosphatidylserine and N-acyl-phosphatidylethanolamine*

Phosphatidylcholine (PC: 1,2-diacyl-*sn*-glycerol-3-phosphocholine) was first isolated from egg yolk and from brain tissue and designated 'lecithin' (Gobley, 1874). Phosphatidylethanolamine (PE: 1,2-diacyl-*sn*-glycerol-3-phosphoethanolamine) and phosphatidylserine (PS: 1,2-diacyl-*sn*-glycerol-3-(phospho-L-serine)), which both carry a primary amino group, were originally isolated as a mixed lipid preparation ('cephalin') from brain tissue. It was Folch (1942) who separated three phospholipids, PE, PS and PI (phosphatidylinositol – see Section 5.2.6), from the cephalin fraction. The structures of PC, PE and PS are related because they carry an aminoalcohol in their head group, i.e. choline, ethanolamine or L-serine, respectively (Figure 5.1). Furthermore, it has been shown that in many organisms, these lipids can be inter-converted by various enzymatic reactions, i.e. base exchange (PS and PE), decarboxylation (PS and PE) and methylation (PE and PC) (see Section 5.3.6). PC and PE represent the most abundant phospholipid classes in many organisms, including plants. Together, these two lipids usually account for more than 50% of glycerolipids in extraplastidic membranes, including ER, plasma membrane, mitochondrion, tonoplast and Golgi (Table 5.1). However, PE is not found in chloroplasts, and PC is also absent from thylakoid membranes, but accumulates in the outer envelope membrane of chloroplasts. PS represents a minor lipid in plants, accounting for less than 1% of glycerolipids. Its distribution resembles that of PE, i.e. it is absent from chloroplasts, but is found in many extraplastidic membranes (Table 5.1).

Plants, as well as other organisms, contain a derivative of PE carrying an acyl group covalently bound to the ethanolamine head group (NAPE, *N*-acyl phosphatidylethanolamine) (Bomstein, 1965; Figure 5.1). However, this lipid is of very low abundance and usually not considered as a structural component of plant membranes. NAPE and its hydrolysis product, *N*-acyl-ethanolamine, are believed to be involved in lipid signalling (see Section 5.4.1). Holmbäck *et al.* (2001) identified 16:0, 18:1 and 18:2 as the major fatty acids attached to the amino group of NAPE.

### 5.2.6 *Phosphatidylinositol*

Phosphatidylinositol was discovered in soybean by Klenk and Sakai (1939), and it is also part of the cephalin fraction isolated from brain tissue (Folch, 1942; see Section 5.2.5). Its structure was determined as 1,2-diacyl-*sn*-glycerol-3-(phospho-1-D-myo-inositol) by Pizer and Ballou (1959; Figure 5.1). PI was detected in low amounts in almost all membranes of plant cells (Table 5.1). However, while its presence in extraplastidic membranes (particularly in the ER and plasma membrane) is generally accepted, the identification of PI in thylakoids of chloroplasts has not been confirmed. Although PI represents a minor glycerolipid of plant membranes, it plays a major role in signal transduction. PI is converted into phosphatidylinositol-bisphosphate ($PIP_2$) by two sequential phosphorylation reactions. $PIP_2$ is cleaved by PI-specific phospholipase C (PLC) into myo-inositol-trisphosphate and diacylglycerol, which serve as second messengers triggering different responses in the cell (see Section 5.4.1 and Chapter 9).

## 5.3 Biosynthesis of membrane glycerolipids

### 5.3.1 *Biosynthesis of phosphatidic acid*

Glycerolipids of plant cells contain two fatty acids, which are esterified to the *sn*-1 and *sn*-2 (stereochemical numbering) positions of glycerol, and a polar head group, which is bound to the *sn*-3 position. The fatty acid distribution in glycerolipids is not random, but different acyl groups are specifically found in the *sn*-1 or *sn*-2 positions of each lipid class (Heinz, 1977). The major sites in plants that are known to contribute to the synthesis of glycerolipids are chloroplasts and the ER. Due to different substrate specificities of the enzymes involved in acylation of the glycerol backbone ('acyltransferases'), two distinct structures of glycerolipids exist in plants. These are the 'prokaryotic' form (C18 and minor amounts of C16 fatty acids at *sn*-1, exclusively C16 at *sn*-2), which is derived from plastids (normally chloroplasts), and the 'eukaryotic' form (C18 at both *sn*-positions, minor amounts of C16 at *sn*-1) originating from the ER (Roughan and Slack, 1982; Browse *et al.*, 1986). Further desaturation of acyl groups at the chloroplast or ER membranes adds to the complexity of fatty acid composition in each lipid class. Eukaryotic lipid precursors derived from the ER are also used for lipid synthesis in the chloroplasts, i.e. galactolipids and sulfolipid. The lipid molecule that is transported back from the ER to chloroplast membranes has not been clearly identified, but it is generally accepted that it is derived from PC (i.e. phosphatidylcholine, diacylglycerol or lysophosphatidylcholine; Mongrand *et al.*, 2000). Plants that are capable of synthesising chloroplast lipids from both prokaryotic and eukaryotic lipids are designated '16:3 plants', because they contain 16:3 in the *sn*-2 position of MGDG (e.g. spinach, Arabidopsis, rape; Heinz and Roughan, 1983). Other plant species, however, have almost entirely lost the capability to synthesise galactolipids via the prokaryotic (plastid) pathway, and, therefore, the eukaryotic (ER) pathway is predominantly used for galactolipids synthesis. As a consequence, these plants are devoid of 16:3, and are thus called '18:3 plants' (e.g. pea). Depending on the plant species, galactolipids or sulfolipids can be synthesised by either or both of the two pathways, but the plastidic form of PG is always synthesised from prokaryotic lipid precursors. The exact molecular mechanism responsible for the differences between 16:3 and 18:3 plants is unknown, but the loss of the prokaryotic pathway seems to have occurred independently more than once during evolution (Mongrand *et al.*, 1998). Furthermore, the contribution of the two pathways can vary in different tissues and under different growth conditions, even in a single species.

The glycerol backbone of glycerolipids originates from dihydroxyacetone-phosphate, which in turn is derived from glycolysis. In many plants, glycerol-3-phosphate is directly synthesised from dihydroxyacetone-phosphate by action of glycerol-3-phosphate dehydrogenase (GPDH; dihydroxyacetone-phosphate reductase). Different isoforms of this enzyme are present in Arabidopsis, where they were localised to the plastid and mitochondrion (Wei *et al.*, 2001; Shen *et al.*, 2003). Furthermore, the Arabidopsis genome contains several genes predicted to encode ER/cytosolic forms of GPDH (Beisson *et al.*, 2003). In the loss of function mutant *gly1* that is deficient in GPDH activity, the flux through the plastid glycerolipid synthesis pathway is affected, resulting in a reduction of 16:3 content (Miquel *et al.*, 1998). An alternative pathway, i.e. the sequential conversion of dihydroxyacetone-phosphate to glyceraldehyde-3-phosphate by triose phosphate isomerase, dephosphorylation and reduction yields glycerol, which can be phosphorylated by glycerol kinase to produce glycerol-3-phosphate. The glycerol kinase pathway was suggested to be operating in some plant tissues, e.g. germinating seeds, where large amounts of glycerol are released during triacylglycerol breakdown (Huang, 1975). In Arabidopsis, only one gene is present with sequence similarity to glycerol kinases from human and *E. coli*. Interestingly, an Arabidopsis mutant (*nho1*, for non-host resistance) was isolated recently. This carries a mutation in the glycerol kinase gene and is affected in its resistance against certain *Pseudomonas syringae* pathovars (Kang *et al.*, 2003). The fact that germination and growth of *nho1* plants are similar to the wild type suggests that the glycerol kinase pathway is largely dispensable for membrane lipid synthesis in Arabidopsis.

The major site of fatty acid *de novo* synthesis in plants is the plastid (Ohlrogge *et al.*, 1979). Therefore, the predominant fraction of fatty acids in glycerolipids originates from the plastid, although it is not entirely clear whether the mitochondrion might contribute to a small extent to fatty acid synthesis (Wada *et al.*, 1997). The major products of plastidic fatty acid synthesis are 16:0 and 18:1. Fatty acyl chains are attached to the *sn*-1 and *sn*-2 positions of glycerol-3-phosphate by subsequent action of glycerol-3-phosphate acyltransferase (GPAT) and lysophosphatidic acid acyltransferase (LPAAT), respectively, finally resulting in the production of phosphatidic acid (Figures 5.2 and 5.3). Acyltransferases of the chloroplast directly employ acyl-ACP esters as substrates. The plastidic GPAT, which shows highest activity with oleoyl-ACP, was first isolated from squash (Bertrams and Heinz, 1981; Ishizaki *et al.*, 1988). LPAAT is almost exclusively active with palmitoyl-ACP, which is directed to the *sn*-2 position of lysophosphatidic acid (Frentzen, 1993). Therefore, the major form of phosphatidic acid synthesised in the plastid is 1-oleoyl-2-palmitoyl-*sn*-glycerol-3-phosphate, the substrate of prokaryotic lipid synthesis (Figure 5.2). A block in plastidic GPAT, as observed in the *act1* mutant of Arabidopsis, inhibits the synthesis of prokaryotic lipid precursors (Kunst *et al.*, 1988). In this plant, flux through the eukaryotic pathway is increased, thus complementing for the loss of the prokaryotic pathway. Therefore, membrane lipids in *act1* are virtually devoid of prokaryotic lipids and lack the MGDG-specific fatty acid 16:3.

Acyl-CoA esters serve as acyl donors for acylation of glycerol-3-phosphate at the ER (Figure 5.3). The acyl-chains incorporated into ER lipids originate from the chloroplast where they are released from ACP esters by action of acyl-ACP thioesterases and are subsequently re-esterified to Coenzyme A by acyl-CoA synthetases after export from the chloroplast (see Section 5.4.3). GPAT activity of the ER shows highest specificity for 16- and 18-carbon acyl groups, whereas LPAAT is almost exclusively active with C18 unsaturated acyl-CoA esters (Frentzen, 1993). Therefore, ER-derived phosphatidic acid, the precursor of eukaryotic lipid synthesis, contains mostly 18-carbon fatty acids in the two *sn* positions, and small amounts of 16:0 at *sn*-1.

### 5.3.2 *Synthesis of glycerolipids from diacylglycerol or CDP-diacylglycerol*

Glycerolipids in plants are synthesised from phosphatidic acid following two different pathways: the 'diacylglycerol pathway' or the 'CDP-diacylglycerol pathway'. In *Escherichia coli*, all phospholipids (PS, PG and DPG) depend on the synthesis via the CDP-diacylglycerol pathway (Raetz *et al.*, 1990). PE in *E. coli* is derived from PS by decarboxylation. Plants have preserved the CDP-diacylglycerol pathway but, in addition, use diacylglycerol and activated head groups for the synthesis of the other major glycerolipids (galactolipids, SQDG, PC, PE).

In the diacylglycerol pathway, phosphatidic acid is hydrolysed by phosphatidic acid phosphatase (PAP). Three genes encoding PAP have been identified in Arabidopsis, which are predicted to be localised to the chloroplast, ER and mitochondrion (Pierrugues *et al.*, 2001; Beisson *et al.*, 2003; Figures 5.2, 5.3 and 5.4). Subsequently, an activated head group is directly coupled with diacylglycerol resulting in the synthesis of MGDG, SQDG and of the phospholipids PE and PC.

In the second pathway, phosphatidic acid and CTP are converted to CDP-diacylglycerol by catalysis of CDP-diacylglycerol synthase (CDS, phosphatidic acid cytidylyltransferase). This activity has been localised to various membranes,

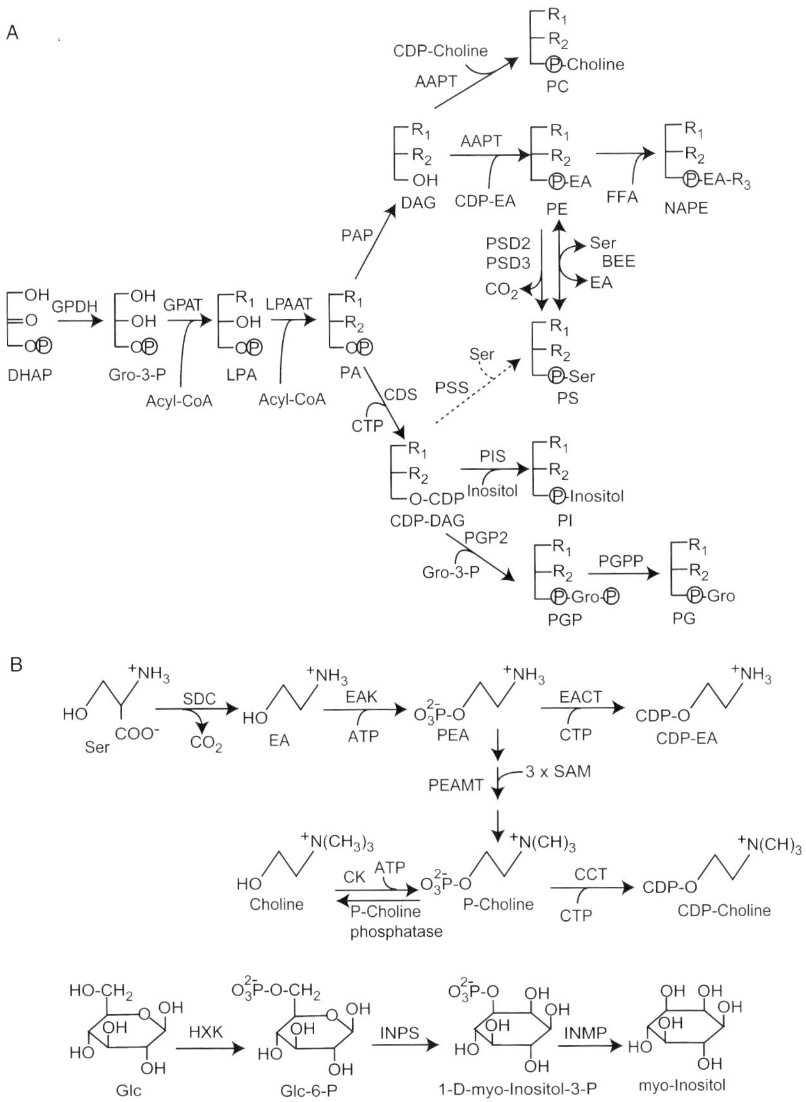


**Figure 5.2** Phospholipid and glycolipid synthesis in plastids. A. Glycerolipid synthesis in plastid envelope membranes. Plastid lipid synthesis depends on two precursors – diacylglycerol, which gives rise to galactolipid and sulfolipid biosynthesis, and CDP-diacylglycerol, which is the precursor of PG. B. Synthesis of the glycolipid head group precursors, UDP-galactose and UDP-sulfoquinovose. The conversion of UDP-glucose to UDP-galactose is catalysed by UDP-glucose 4-epimerase in the cytosol. UDP-sulfoquinovose is synthesised from UDP-glucose and sulfite by action of UDP-sulfoquinovose synthase (SQD1), which is localised to the stroma of plastids. CDP-DAG, CDP-diacylglycerol; CDS, CDP-diacylglycerol synthase; DAG, diacylglycerol; DGDG, digalactosyldiacylglycerol; DGD1, DGD2, DGDG synthase 1 or 2; DHAP, dihydroxyacetone-phosphate; Gal, D-galactose; Glc, D-glucose; GGGT, galactolipid:galactolipid galactosyltransferase; GPAT, glycerol-3-phosphate acyltransferase; GPDH, glycerol-3-phosphate dehydrogenase; Gro, glycerol; Gro-3-P, glycerol-3-phosphate; LPA, lysophosphatidic acid; LPAAT, lysophosphatidic acid acyltransferase; MGD1, MGD2, MGD3, MGDG synthase 1, 2 or 3: MGDG, monogalactosyldiacylglycerol; P, phosphate; PA, phosphatidic acid; PAP, phosphatidic acid phosphatase; PG, phosphatidylglycerol; PGP, phosphatidylglycerol-3-phosphate; PGP1, phosphatidylglycerol-3-phosphate synthase 1; PGT, processive galactosyltransferase; SQD1, UDP-sulfoquinovose synthase; SQD2, sulfoquinovosyldiacylglycerol synthase; UDP-Gal, UDP-D-galactose; UDP-Glc, UDP-D-glucose; UDP-SQ, UDP-sulfoquinovose; UGE, UDP-D-glucose 4-epimerase.

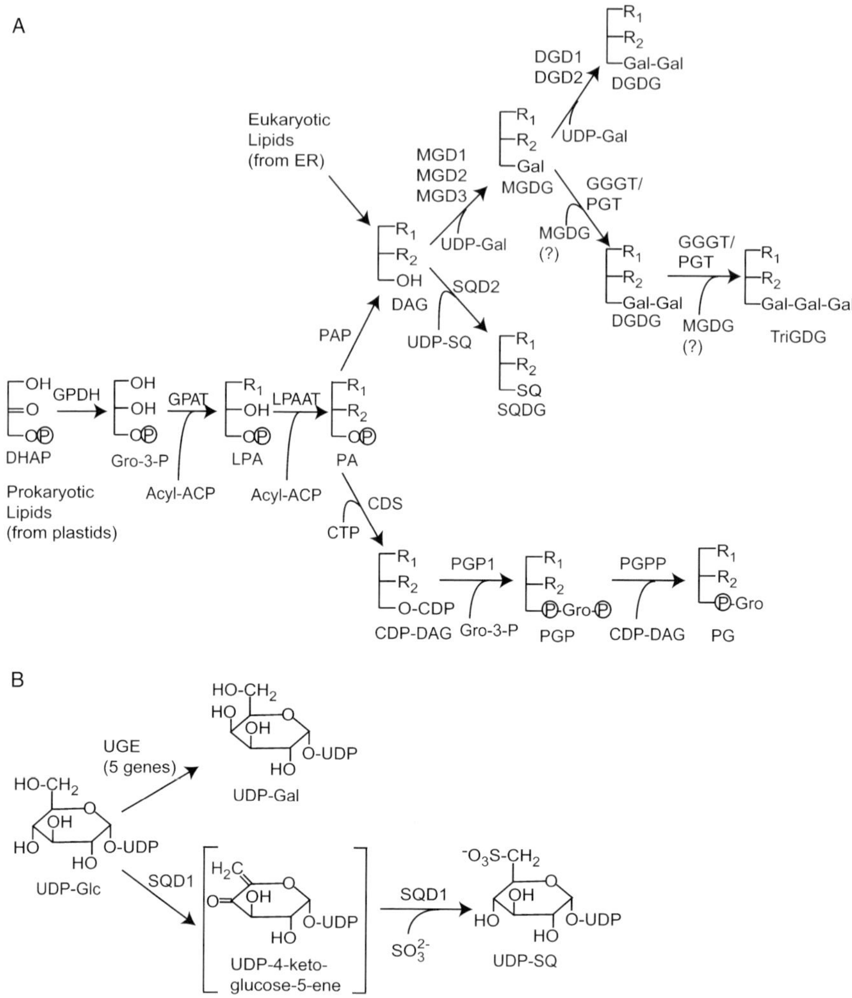


**Figure 5.3** Phospholipid synthesis at the endoplasmic reticulum (ER). A. Phospholipid synthesis in ER membranes. As in plastids, two substrates, diacylglycerol and CDP-diacylglycerol, serve as precursors for glycerolipid synthesis in ER membranes. Phosphatidylserine is synthesised from phosphatidylethanolamine by the base-exchange enzyme. In some species, e.g. wheat, phosphatidylserine can be produced from CDP-diacylglycerol by phosphatidylserine synthase. However, this enzyme seems to be absent from *Arabidopsis* (dashed arrow). B. Synthesis of the phospholipid head group precursors CDP-ethanolamine,

i.e. plastids, mitochondria and microsomes, and different genes are known in Arabidopsis, which encode proteins predicted to be localised to chloroplasts or the ER (Kopka *et al.*, 1997; Beisson *et al.*, 2003). In plants, CDP-diacylglycerol is used as precursor for the synthesis of PI, PS, PG and DPG.

In addition to plastids and the ER, mitochondria were shown to be capable of glycerolipid synthesis, because they harbour all enzyme activities necessary for phosphatidic acid production (Frentzen, 1993). In addition, the Arabidopsis genome contains sequences predicted to encode a mitochondrial form of CDP-diacylglycerol synthase (Beisson *et al.*, 2003). Thus, mitochondria can probably synthesise PG and DPG from CDP-diacylglycerol. Furthermore, mitochondria contain a PS decarboxylase activity capable of PE production from PS. In the Arabidopsis genome, sequences predicted to encode mitochondrial forms of PA phosphatase and aminoalcohol-phosphotransferase were found (Beisson *et al.*, 2003), but it is not clear whether or not mitochondria indeed contain a full set of enzymes required for PE and PC synthesis from diacylglycerol. It seems that the predominant amounts of the two major mitochondrial phospholipids, PE and PC are imported from the ER (Figures 5.4 and 5.5). Furthermore, the origin of fatty acyl groups of mitochondrial glycerolipids is not well understood. Mitochondria contain all of the enzymes required for fatty acid synthesis *de novo*. However, the fatty acyl chain elongation in mitochondria, which is linked to ACP as in plastids (Chuman and Brody, 1989) is assumed to result in the production of C8 to C14 acyl groups. C8 acyl groups are used for lipoic acid production, which is a cofactor of many enzymes including pyruvate dehydrogenase (Wada *et al.*, 1997). Because mitochondrial fatty acid synthesis predominantly results in the production of medium chain fatty acids, which are not incorporated into membranes lipids, a large fraction

---

**Figure 5.3** (*continued*)
CDP-choline and myo-inositol. The synthesis of phospholipid head group precursors is presumed to be localised to the cytosol. CDP-ethanolamine is derived from ethanolamine, which is produced from L-serine by decarboxylation. CDP-choline is synthesised from phosphocholine by cytidinylation. In Arabidopsis, phosphoethanolamine can be converted to phosphocholine by three methylation reactions. AAPT, aminoalcoholphosphotransferase; BEE, base-exchange enzyme; CDP-DAG, CDP-diacylglycerol; CDP-EA, CDP-ethanolamine; CDS, CDP-diacylglycerol synthase; CK, choline kinase; DAG, diacylglycerol; EA, ethanolamine; EAK, ethanolamine kinase; FFA, free fatty acid; Glc, D-glucose; GPAT, glycerol-3-phosphate acyltransferase; GPDH, glycerol-3-phosphate dehydrogenase; Gro-3-P, glycerol-3-phosphate; LPA, lysophosphatidic acid; LPAAT, lysophosphatidic acid acyltransferase; NAPE, *N*-acyl-phosphatidylethanolamine; P, phosphate; PA, phosphatidic acid; PAP, phosphatidic acid phosphatase; CCT, phosphocholine cytidylyltransferase; PEA, phosphoethanolamine; EACT, phosphoethanolamine cytidylyltransferase; PEAMT, phosphoethanolamine methyltransferase; PG, phosphatidylglycerol; PGP, phosphatidylglycerol-3-phosphate; PGP2, phosphatidylglycerol-3-phosphate synthase 2; PI, phosphatidylinositol; PIS, phosphatidylinositol synthase; PS, phosphatidylserine; PSD2, PSD3, phosphatidylserine decarboxylase 2, 3; PSS, phosphatidylserine synthase; SAM, S-adenosylmethionine; SDC, serine decarboxylase; Ser, L-serine.

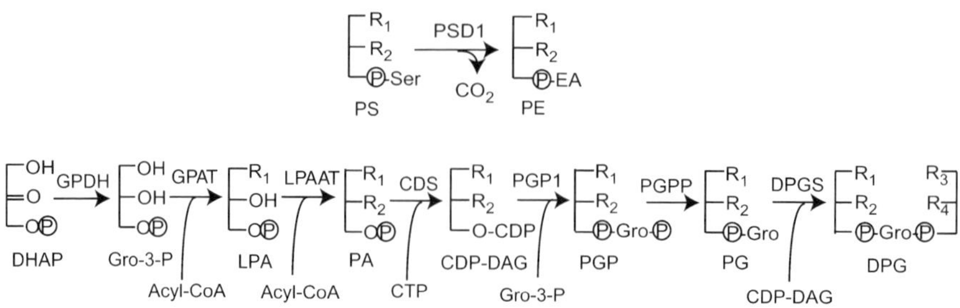


**Figure 5.4** Phospholipid synthesis in the mitochondrion. The synthesis of the minor glycerol-containing lipids, phosphatidylglycerol and diphosphatidylglycerol has been shown to occur on mitochondrial membranes. However, the more abundant phospholipids (PE and PC) are probably imported from the ER. CDP-DAG, CDP-diacylglycerol; CDS, CDP-diacylglycerol synthase; DAG, diacylglycerol; DPG, diphosphatidylglycerol (cardiolipin); DPGS, diphosphatidylglycerol synthase; GPAT, glycerol-3-phosphate acyltransferase; GPDH, glycerol-3-phosphate dehydrogenase; Gro-3-P, glycerol-3-phosphate; LPA, lysophosphatidic acid; LPAAT, lysophosphatidic acid acyltransferase; P, phosphate; PA, phosphatidic acid; PG, phosphatidylglycerol; PGP, phosphatidylglycerol-3-phosphate; PGP1, phosphatidylglycerol-3-phosphate synthase 1; PS, phosphatidylserine; PSD1, phosphatidylserine decarboxylase 1.

of acyl groups in mitochondrial glycerolipids might be derived from the chloroplast and/or ER.

### *5.3.3 Biosynthesis of galactolipids*

The galactolipids MGDG and DGDG are synthesised from UDP-galactose and diacylglycerol in the envelope membranes of chloroplasts (Figure 5.2; for a review see Dörmann and Benning, 2002). UDP-galactose, the precursor of galactolipid head groups, is derived from UDP-glucose via an epimerase reaction localised to the cytosol (UDP-glucose 4-epimerase, UGE; Königs and Heinz, 1974). The Arabidopsis genome has five *UGE* genes (Reiter and Vanzin, 2001). The contribution of the individual UGE isoforms for UDP-galactose synthesis is not known. Reduction of UGE activity by antisense expression (*UGE1*; Dörmann *et al.*, 1999) or by mutation (*uge4*, synonymous: *rhd1*, root hair deficient; Seifert *et al.*, 2002) does not affect galactolipid content. UDP-galactose is believed to be cytosolic and to diffuse to the outer envelope and into the inter-envelope membrane space of chloroplasts, where it is incorporated into galactolipids by MGDG synthases and DGDG synthases.

The MGDG synthases transfer a galactose moiety from UDP-α-D-galactose onto diacylglycerol. During this reaction, the anomeric configuration of the glycosidic carbon atom is inverted because in MGDG, galactose is bound to diacylglycerol in a β-glycosidic linkage. The MGDG synthase, MGD1, was isolated from cucumber and spinach and localised to the inner chloroplast envelope (Shimojima *et al.*, 1997;

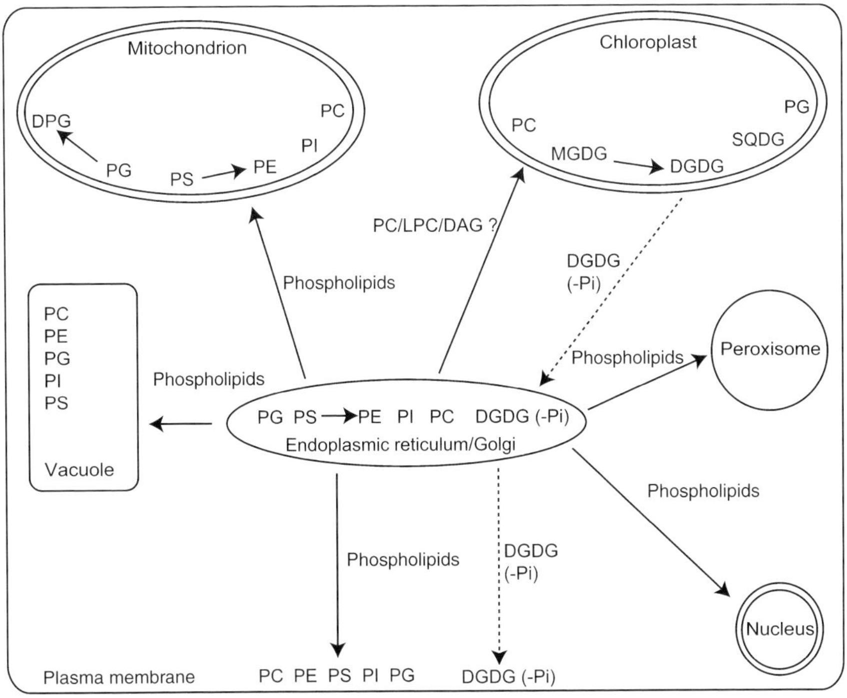


**Figure 5.5** Phospholipid and glycolipid trafficking in plant cells. Plastids and the endoplasmic reticulum (ER) are the major sites of membrane glycerolipid assembly in higher plants. Phospholipids generated at the ER are presumably transported to the Golgi apparatus and to the plasma membrane in the form of vesicles. The ER was also suggested to be the site of synthesis for phospholipids determined for transport to the vacuole, the peroxisome, the nucleus and, at least to some extent, to the mitochondrion. Furthermore, a PC-derived lipid (PC; LPC, lysophosphatidic acid or DAG, diacylglycerol) is transported from the ER back to the plastid were it serves as precursor for eukaryotic lipid synthesis. Plastids are capable of assembling their membrane lipids (galactolipids, sulfolipid, PG) from eukaryotic or prokaryotic precursors. Under phosphate deficient conditions, DGDG is synthesised in plastid envelope membranes in large amounts, and a major fraction of it is transported to extraplastidic membranes (dashed arrows).

Miège *et al.*, 1999). In Arabidopsis, three genes encoding MGDG synthases are known (Jorasch *et al.*, 2000; Awai *et al.*, 2001). Analysis of the *mgd1* mutant of Arabidopsis demonstrated that this enzyme is responsible for the synthesis of the predominant fraction of MGDG (Jarvis *et al.*, 2000). MGD2 and MGD3, which localise to the outer site of chloroplast envelopes, are believed to synthesise MGDG as a precursor for DGDG synthesis (Awai *et al.*, 2001).

Two DGDG synthases are known in Arabidopsis, DGD1 and DGD2 (Dörmann *et al.*, 1999; Kelly *et al.*, 2002). Analysis of the *dgd1* mutant of Arabidopsis demonstrated that DGD1 is the enzyme responsible for the synthesis of the predominant fraction of chloroplast DGDG (Dörmann *et al.*, 1995). The two DGDG synthases represent UDP-galactose dependent galactosyltransferases and localise to the outer site of chloroplast envelopes (Froehlich *et al.*, 2001; Kelly *et al.*, 2003). During galactosylation by DGD1 and DGD2, the anomeric configuration of the galactose moiety is retained. Therefore, the second galactose molecule in DGDG is linked in α-configuration.

For a long time, it was assumed that yet another enzymatic activity, designated galactolipid: galactolipid galactosyltransferase (GGGT) is the sole enzyme responsible for DGDG synthesis in plants (van Besouw and Wintermans, 1978; Heemskerk *et al.*, 1990). GGGT was proposed to employ another MGDG molecule, but not UDP-galactose as galactose donor, and was found to be processive, i.e. to synthesise oligogalactolipids such as TriGDG and TetraGDG in *in vitro* reactions (Dorne *et al.*, 1982; Cline and Keegstra, 1983; Figure 5.2). These oligogalactolipids are absent from most plant tissues, but are found in low amounts in some plant species (Fujino and Miyazawa, 1979; Kojima *et al.*, 1990). However, recent studies by Xu *et al.* (2003) and Kelly *et al.* (2003) clearly demonstrated that in Arabidopsis, the GGGT activity (or 'processive galactosyltransferase', PGT) is independent from DGD1 and DGD2. A double mutant devoid of DGD1 and DGD2 was shown to contain only trace amounts of DGDG in leaves. However, *in vitro* assays with radioactive UDP-galactose or radioactive MGDG done with chloroplasts isolated from this plant resulted in the synthesis of DGDG and oligogalactolipids, and thus demonstrated that a third DGDG synthase activity (GGGT or PGT) must exist. This enzyme does not contribute to DGDG synthesis *in vivo*, but is strongly stimulated during membrane isolation from chloroplasts (Kelly *et al.*, 2003). The isolation of the *tgd1* mutant of Arabidopsis, which is deficient in a permease-like protein normally found in chloroplast membranes, has shed some light on the regulation of the GGGT/PGT activity and lipid trafficking (Xu *et al.*, 2003). In this plant, the GGGT/PGT enzyme is permanently active, resulting in the accumulation of TriGDG and TetraGDG *in vivo*. In contrast to DGD1 and DGD2, this activity is capable of synthesising DGDG and oligogalactolipids with all glycosidic bonds in β-configuration (Xu *et al.*, 2003). However, the nature of the galactose donor for this reaction (MGDG or UDP-galactose) and the gene encoding this activity remain unknown, and the function of this additional galactolipid synthesising activity is enigmatic.

### *5.3.4 Biosynthesis of sulfolipid*

The sulfolipid, SQDG, is synthesised by transfer of the sulfur containing sugar sulfoquinovose from UDP-sulfoquinovose onto diacylglycerol by SQDG synthase (SQD2) in the envelope membranes of chloroplasts (Heinz *et al.*, 1989; Figure 5.2). A mutation in the *SQD2* gene results in complete loss of SQDG synthesis (Yu *et al.*, 2002). The head group of SQDG is synthesised from UDP-glucose and inorganic sulfite ($SO_3^{2-}$) by UDP-sulfoquinovose synthase (SQD1; Essigmann *et al.*, 1998; Sanda *et al.*, 2001). In most plants, including Arabidopsis, ADP-glucose is the predominant form of activated glucose in the stroma of chloroplasts because this is the glucose donor for starch synthesis. However, SQD1 depends on UDP-glucose for UDP-sulfoquinovose production. Because SQD1 was localised to the stroma of plastids, low amounts of UDP-glucose might be present in this compartment. A carbon–sulfur bond is formed in the unique reaction catalysed by SQD1, after reversible oxidation of the C4 carbon and addition of inorganic sulfite to the UDP-4-ketoglucose-5-ene intermediate (Mulichak *et al.*, 1999; Figure 5.2). For this reason SQD1, which shows sequence similarities to sugar nucleotide epimerases and dehydrogenases, harbours a redox active $NAD^+$ cofactor (Mulichak *et al.*, 1999).

### 5.3.5 *Biosynthesis of PG and DPG*

The first step in PG synthesis is the transfer of diacylglycerol from CDP-diacylglycerol onto glycerol-3-phosphate by PG-3-phosphate synthase. PG is present in microsomal membranes, chloroplasts and mitochondria, and its synthesis has also been localised to these compartments (Block *et al.*, 1983; Griebau and Frentzen, 1994). In Arabidopsis, two PG-3-phosphate synthase genes are known (PGP1, PGP2; Müller *et al.*, 2001). PGP1 is crucial for PGP activity in chloroplasts and mitochondria. This enzyme contains an *N*-terminal dual targeting sequence for import into these two different organelles (Müller *et al.*, 2001; Xu *et al.*, 2002; Babiychuk *et al.*, 2003). A mutation in *pgp1* causes a severe reduction in PG synthesis in chloroplasts. As a consequence, *pgp1* null alleles are incapable of photoautotrophic growth (Xu *et al.*, 2002; Hagio *et al.*, 2002; Babiychuk *et al.*, 2003). The PGP2 protein presumably is localised to microsomal membranes (Müller *et al.*, 2001). PG is derived from dephosphorylation of PG-3 phosphate, a reaction catalysed by PG-3-phosphate phosphatase (PGPP). The *PGPP* gene has not been identified in higher plants.

DPG (cardiolipin) is synthesised from PG and another molecule of CDP-diacylglycerol in the inner mitochondrial membrane (Frentzen and Griebau, 1994). The gene encoding DPG synthase was recently identified in Arabidopsis, and its sequence was shown to be similar to PGP1 and PGP2 (Nowicki *et al.*, 2003).

### 5.3.6 *Biosynthesis of PS, PC, PE and NAPE*

The assembly of PS, PC and PE in plants is complex because different pathways contribute to their biosynthesis and several inter-converting reactions are known for these lipids. PC and PE are predominantly synthesised from diacylglycerol and CDP-choline/CDP-ethanolamine following the diacylglycerol pathway (diacylglycerol aminoalcoholphosphotransferase; AAPT; Figures 5.3 and 5.4). Two AAPT genes are present in the Arabidopsis genome, predicted to be localised to the ER and mitochondrion (Beisson *et al.*, 2003). The two enzymes, AAPT1 and AAPT2, are active with both substrates, CDP-ethanolamine and CDP-choline, but AAPT2 is more specific for CDP-ethanolamine (Goode and Dewey, 1999). The precursors for PC and PE synthesis are produced in the cytosol of the plant cell (Figure 5.3). CDP-ethanolamine is derived from L-serine by a set of reactions including serine decarboxylase (SDC), ethanolamine kinase (EAK) and CTP: phosphoethanolamine cytidylyltransferase (EACT; CDP-ethanolamine synthase). Arabidopsis contains one gene encoding SDC (Rontein *et al.*, 2001). This enzyme is responsible for the synthesis of the predominant fraction of ethanolamine. The other two enzymes, EAK and EACT, are represented by single copy genes in Arabidopsis. Whereas EAK is presumed to be localised to the cytosol, the EACT activity was found in microsomes and in the mitochondria of castor bean (Tang and Moore, 1997). Similar to PE, PC can be synthesised from choline following a pathway starting with choline kinase (CK). Phosphocholine is the substrate of CTP:phosphocholine cytidylyltransferase (CDP-choline synthase, CCT), which produces CDP-choline, the precursor of PC synthesis. This step is generally believed to be rate limiting for PC production in plants. Four cDNAs from rape encoding CCT were isolated by complementation of the yeast *cho1* mutant (Nishida *et al.*, 1996) and two genes, *CCT1* and *CCT2*, were isolated from Arabidopsis (Inatsugi *et al.*, 2002).

In contrast to animals and yeast, where PC can also be synthesised by sequential methylation of PE, this pathway seems to be absent from Arabidopsis. However, it is not clear whether or not in other plant species the PE methylation pathways contributes to PC synthesis. A phosphoethanolamine-specific methylase (PEAMT or NMT1) was isolated from Arabidopsis, which produces phosphocholine from phosphoethanolamine using three molecules of *S*-adenosylmethionine (Bolognese and McGraw, 2000; Nuccio *et al.*, 2000; Charron *et al.*, 2002). This pathway seems to be a major route for phosphocholine, and (after dephosphorylation) for choline production in plants.

The minor phospholipid NAPE is derived from PE after acylation by NAPE synthase. The enzyme from cottonseed was characterised and shown to utilise free fatty acids for this acylation reaction (Chapman and Moore, 1993). The acylation activity was found in the microsomal fraction, but the protein has not been isolated. Therefore, the identity of the NAPE synthase gene of higher plants remains unknown.

PS represents a minor membrane lipid in plants, and its synthesis is believed to be localised to the ER (Figure 5.3). PS can be synthesised from L-serine via the CDP-diacylglycerol pathway. The corresponding gene, PS synthase (*PSS*), has been isolated from wheat (Delhaize *et al.*, 1999). Overexpression of PSS in tobacco and Arabidopsis results in an increased rate of PS synthesis accompanied with altered growth and the occurrence of necrotic lesions on leaves. Interestingly, PSS activity in Arabidopsis wild type is very low, and the Arabidopsis genome contains no gene with sequence similarity to the wheat *PSS* gene (Delhaize *et al.*, 1999; Beisson *et al.*,

2003). Therefore, another enzymatic activity must be involved in PS synthesis in Arabidopsis. In higher plants, a phospholipid base-exchange enzyme (BEE) has been described which is capable of replacing the head group of PE with free L-serine in a reversible reaction (Shin and Moore, 1990). Arabidopsis contains one open reading frame with sequence similarities to animal BEEs (Beisson *et al.*, 2003).

It is believed that the predominant fraction of PE in higher plants is synthesised via the diacylglycerol pathway rather than by decarboxylation of PS (Rontein *et al.*, 2001, 2003). Three genes were found in Arabidopsis with sequence similarities to bacterial and animal PS decarboxylases (*PSD1, PSD2, PSD3*). *PSD1* from tomato was recently shown to encode a functional PS decarboxylase localised to mitochondrial membranes (Rontein *et al.*, 2003).

### *5.3.7 Biosynthesis of PI*

The head group of PI, myo-inositol, is derived from glucose by a set of reactions believed to be localised to the cytosol (for a review see Loewus *et al.*, 2000). Glucose is converted into glucose-6-phosphate by hexokinase, and glucose-6-phosphate is the substrate of 1-D-myo-inositol 3-phosphate synthase. This key enzyme of myo-inositol synthesis is responsible for creating the C6 ring of the sugar alcohol. The corresponding cDNA was isolated from different plants including Arabidopsis (Johnson and Sussex, 1995). Dephosphorylation of myo-inositol 3-phosphate by myo-inositol monophosphatase leads to the formation of free myo-inositol. This enzyme (also involved in inositol-trisphosphate breakdown) is encoded by a multigene family in plants (Gillapsy *et al.*, 1995). PI is produced from CDP-diacylglycerol and myo-inositol by action of PI synthase (PIS1) (Collin *et al.*, 1999; Xue *et al.*, 2000). Interestingly, a head group exchange reaction between PI and free myo-inositol was observed in plant extracts, which was recently shown to be associated with the very same enzyme, PIS1 (Justin *et al.*, 2002). PI synthesis in plants is localised to the ER, from where the lipid is believed to be transported to other membranes (Figure 5.5).

## 5.4 Membrane lipid turnover

Glycerolipids of plants are subject to constant turnover by lipid hydrolysing enzymes, phospholipases and galactolipases. The mechanisms underlying the regulation of membrane lipid synthesis, degradation and homeostasis are only poorly understood. Based on the substrate specificity and the position of the cleavage site within the lipid molecule, different lipase activities can be distinguished. Phospholipases which cleave acyl groups or head groups from phospholipids are the focus of intense studies, because they are critical for the generation of lipid-derived signal molecules (see Chapters 9 and 10). Different phospholipase classes were characterised in higher plants, which cleave acyl groups (phospholipases A1, A2, B;

lysophospholipases and acyl hydrolases) or the head group (phospholipases C, D) of phospholipids. Glycolipase and glycosidase activities, which are involved in galactolipid and sulfolipid breakdown were described in plant extracts on a biochemical level. A large number of lipase-like sequences were identified in the Arabidopsis genome whose function still is enigmatic. The present chapter summarises the lipid hydrolysing activities, following a classification system, which was recently presented by Wang (2001) and Beisson *et al.* (2003).

### *5.4.1 Hydrolysis of phospholipid head groups: phospholipases C and D*

#### *5.4.1.1 Phospholipase C*

Phospholipase C hydrolyses the ester bond between the glycerol and the phosphate group of phospholipids, thus releasing diacylglycerol and the respective phosphorylated head group (Figure 5.6). Three classes of PLCs are known in plants – the non-specific PLCs mostly active with PC, the GPI-specific PLCs and the PI-specific PLCs involved in phosphoinositide signalling (see Wang, 2001 and Chapter 9). The non-specific PLCs are implicated in breakdown of phospholipids such as PC, PE and PG. PC-specific PLC activity was detected in several plant species (e.g. Kates, 1955), and different genes predicted to encode non-specific PLCs are present in the Arabidopsis genome. However, this lipase class has not been analysed in plants at the molecular level. GPI-specific PLCs hydrolyse the lipid moiety of glycosylated PI (GPI) anchored proteins. GPI anchored proteins are localised to the outer membrane of plant cells where they act as enzymes or receptors. In Arabidopsis, several genes with sequence similarities to GPI-specific PLCs are present, but the respective activity has only been characterised in crude plant extracts (Butikofer and Brodbeck, 1993). The class of PI-specific PLCs has been characterised in greatest detail. These enzymes cleave phosphoinositides, particularly $PIP_2$, into inositol-triphosphate and diacylglycerol, two molecules, that serve as second messengers in the plant cell. Whereas $PIP_2$ diffuses to intracellular organelles where it stimulates calcium release, diacylglycerol remains in the membrane where it activates protein kinase C (for reviews see Mueller-Roeber and Pical, 2002; Meijer and Munnik, 2003). The first higher plant cDNA encoding a PI-PLC was isolated by Hirayama *et al.* (1995), and several additional *PI-PLC* genes are present in Arabidopsis. Because these lipases are only poorly active on other phospholipids, PI-PLCs are believed to be exclusively involved in $PIP_2$ signalling, rather than in bulk membrane lipid turnover (see Chapter 9).

#### *5.4.1.2 Phospholipase D*

Phospholipase D cleaves the ester bond between the phosphate and the alcohol of the head group, thereby producing phosphatidic acid and choline, ethanolamine, L-serine, glycerol or myo-inositol, respectively. The PLD class in Arabidopsis encompasses five different gene families ($\alpha$, $\beta$, $\gamma$, $\delta$ and $\zeta$), each with different

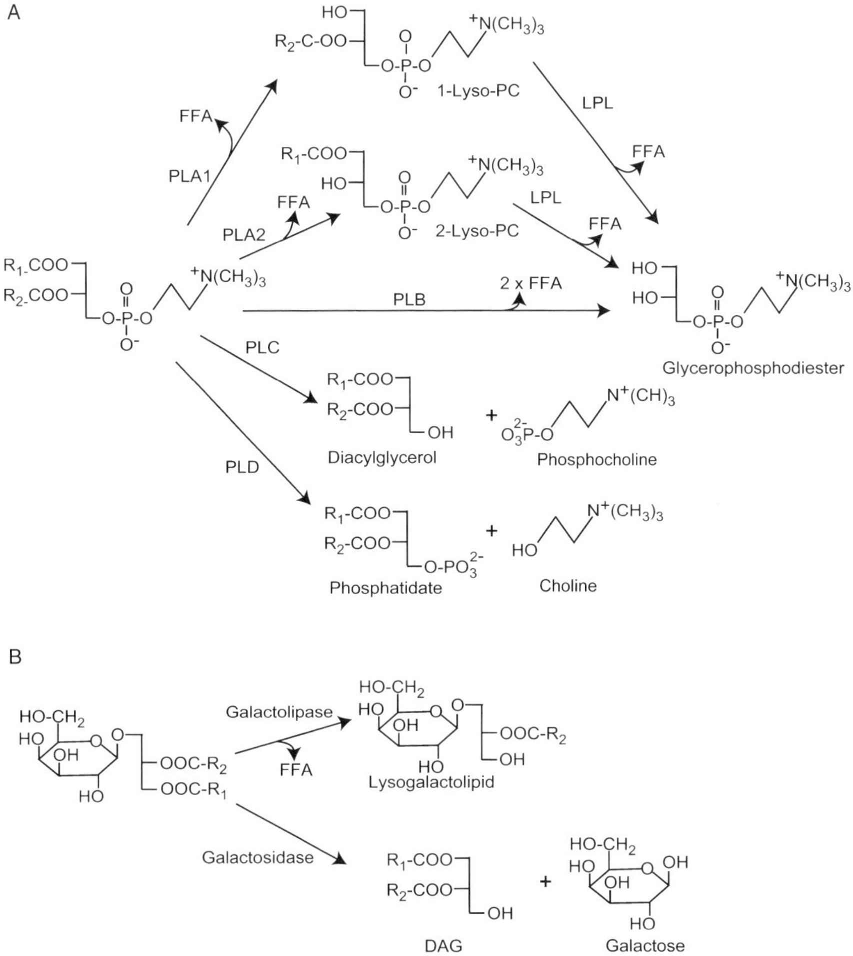


**Figure 5.6** Turnover of phospholipids and galactolipids. A. Phospholipases. Hydrolysis of the fatty acyl groups at the *sn*-1 or *sn*-2 positions of PC by phospholipase A1 or A2 results in the production of 1-lyso-PC or 2-lyso-PC, respectively. The remaining acyl group can be cleaved by lysophospholipase. Phospholipase B non-specifically cleaves acyl groups from the *sn*-1 or *sn*-2 position, finally producing a glycerophosphodiester. Phospholipase C and D cleave the phosphoester group between diacylglycerol and the phosphate or between the phosphate and the head group, respectively. B. Galactolipases and galactosidases. Whereas galactolipases cleave acyl groups from MGDG or DGDG, galactosidases hydrolyse the head group, thereby releasing diacylglycerol and galactose from MGDG. DAG, diacylglycerol; FFA, free fatty acid; LPL, lysophospholipase; PLA, PLB, PLC, PLD, phospholipase A, B, C or D.

enzyme characteristics including requirements for $Ca^{2+}$, $PIP_2$ and free fatty acids (Qin and Wang, 2002). The enzymes are mostly active with PC, PE and PG, but not with phosphoinositides. Stimulation of PLD activity has been observed under different stress conditions, e.g. senescence, freezing, drought, wounding, pathogen attack and nutrient deficiency (Wang, 2001; Meijer and Munnik, 2003). However, it is not clear to which extent PLD activity is involved in membrane lipid degradation or in lipid signalling. Phosphatidic acid, one of its reaction products, has been identified as a critical component in stress responses (Wang *et al.*, 2002; Meijer and Munnik, 2003). Furthermore, PLD activity is involved in the generation of the signal molecule *N*-acylethanolamine from *N*-acyl-PE, which was shown to accumulate in plants under stress. *N*-acylethanolamine plays a role during seed germination and plant defence responses (Chapman, 2000). The mechanism of action of NAPE or *N*-acylethanolamines is unknown and the genes involved in NAPE hydrolysis still await their identification.

### 5.4.2 *Hydrolysis of acyl groups from membrane lipids*

In plant extracts, several hydrolase activities acting on the *sn*-1 or *sn*-2 positions of phospholipids and glycolipids have been described. The isolation of the first genes encoding specific lipases and the recent completion of the sequencing of the Arabidopsis genome have greatly advanced our understanding of the role of these enzymes in cellular signalling and lipid turnover. Phospholipases A1 and A2 cleave acyl groups from the *sn*-1 or *sn*-2 positions of phospholipids, respectively, thereby, releasing free fatty acids and lysophospholipid (Figure 5.6). Lysophospholipids are the substrates for lysophospholipases. Orthologous genes encoding enzymes belonging to these lipase classes were identified by sequence comparison with lipases from animals and yeast. Furthermore, plants contain additional gene families whose function is less well understood, i.e. patatin-like acyl hydrolases and DAD1-like acyl hydrolases (Beisson *et al.*, 2003).

#### 5.4.2.1 *Phospholipase A1*

Phospholipase A1 cleaves the acyl groups from the *sn*-1 position of phospholipids, thereby, releasing free fatty acids and *sn*-2-lysophospholipid. An Arabidopsis mutant (*sgr2* for shoot gravitropism deficient), which carries a mutation in a gene orthologous to PLA1 from animals was isolated (Kato *et al.*, 2002). Only one PLA1-like gene locus is known in Arabidopsis, but the PLA1 activity of the corresponding protein has not been confirmed.

#### 5.4.2.2 *Phospholipase A2*

Phospholipase A2 cleaves acyl groups from the *sn*-2 position of phospholipids and lysophospholipids. In the Arabidopsis genome, several gene loci are present which show strong sequence similarity to PLA2 from animals and yeast. A subfamily of low molecular weight PLA2 enzymes is localised outside the cell and shows sequence similarity to PLA2 enzymes from animals, e.g. snake venom PLA2. The first plant PLA2 was isolated from the elm tree (Ståhl *et al.*, 1998, 1999). Elm seed triacylglycerols are rich in medium chain fatty acids, but these fatty acids are excluded from all membrane lipids. Plant PLA2 activity was suggested to be involved in the removal of unusual fatty acids from membrane lipids in different plant species, some of them accumulating unusual fatty acids (castor, elm) and some not (e.g. rapeseed) (e.g. Ståhl *et al.*, 1998). However, the phospholipase isolated from elm is related to secretory PLA2 enzymes from animals. For this reason, the elm enzyme might be localised to the extracellular matrix, and therefore is supposedly unrelated to the microsomal PLA2 activity involved in lipid acylation/deacylation. Recent evidence for the function of secretory PLA2 in plants came from a transgenic study in Arabidopsis (Lee *et al.*, 2003). These authors demonstrated that secretory PLA2 activity is critical for cell elongation and gravitropism.

The Arabidopsis genome contains two genes with sequence similarity to mammalian cytosolic PLA2 (Beisson *et al.*, 2003). However, it is not known whether these enzymes are involved in lipid breakdown, turnover or signalling. Hydrolysis of PE by PLA2 results in the production of lysophosphatidylethanolamine (LPE), which was found to retard senescence in plants (e.g. Farag and Palta, 1993). Thus, at least some PLA2 isoforms seem to be directly involved in cell signalling.

#### *5.4.2.3 Lysophospholipase*

The Arabidopsis genome contains nine genes with sequence similarities to yeast lysophospholipases (Figure 5.6; Beisson *et al.*, 2003). These enzymes cleave the second acyl group of lysophospholipids resulting in the release of free fatty acid and a glycerophosphodiester (Figure 5.6). Due to the presence of other lipases (i.e. LPA1 and LPA2, phospholipase B which cleaves acyl groups from the *sn*-1 and *sn*-2 positions), it is difficult to clearly identify LPL activity in crude plant extracts. As a consequence, biochemical characterisation of LPL was hampered by the difficulty in obtaining pure enzyme preparations.

#### *5.4.2.4 Patatin-like acyl hydrolases with phospholipase and glycolipase activities*

Patatin is an abundant storage glycoprotein in vacuoles of potato tuber cells (Höfgen and Willmitzer, 1990). *In vitro* assays of recombinant patatin protein demonstrated that it harbours strong acyl hydrolase activity, and its preferred substrates are phospholipids (Andrews *et al.*, 1988). Additional patatin-like acyl hydrolase cDNAs were isolated from cucumber and tobacco, and the respective enzymes displayed PLA2 activity (May *et al.*, 1998; Dhondt *et al.*, 2000). Thus, a subset of patatin-like genes seems to encode acyl hydrolases with phospholipase A2 activity, and hence can be referred to as an additional PLA2 family (see above; Wang, 2001).

Other patatin-like enzymes have different substrate specificities for phospholipids or galactolipids (Huang, 1993). Interestingly, patatin-like PLA2 enzymes were suggested to be involved in α-linolenic acid hydrolysis for jasmonic acid production (see below). A patatin-like acyl hydrolase was isolated from cowpea:this was specifically active with galactolipids but not with phospholipids (Matos *et al.*, 2001). The corresponding mRNA was strongly induced during drought stress. It is tempting to speculate that the genes encoding galactolipase and sulfolipase (see Section 5.4.3) belong to the family of patatin-like acyl hydrolases.

#### *5.4.2.5 DAD1-like acylhydrolases*

In 2001, Ishiguro *et al.* (2001) identified a gene in Arabidopsis designated DAD1 ('defective in anther dehiscence'), which was shown to encode an acyl hydrolase with activity towards PC, and to some extent towards MGDG. The corresponding mutant *dad1* is deficient in pollen maturation and anther dehiscence, and can be rescued by application of free α-linolenic acid and jasmonic acid. DAD1 harbours a putative chloroplast targeting sequence, and therefore might be involved in the release of α-linolenic acid from phospholipids or glycolipids in the plastid. Lipid hydrolysis is considered the first step of the octadecanoid signalling pathway. Triunsaturated fatty acids are the precursors for the lipoxygenase reaction, which results in the synthesis of jasmonic acid-related molecules (for recent reviews see Weber, 2002; Farmer *et al.*, 2003). Jasmonic acid is a member of the group of oxylipins that were shown to be involved in different plant stress responses, e.g. wounding, plant-pathogen interactions. A large fraction of 12-oxophytodienoic acid (OPDA), a functionally active intermediate of jasmonic acid synthesis, was found to be esterified to MGDG (Stelmach *et al.*, 2001). Therefore, in addition to free fatty acids, lipid-bound fatty acids might also serve as precursors for oxylipin synthesis. Alternatively, an exchange reaction of free OPDA and MGDG-bound OPDA might exist in chloroplast membranes. The details of the synthesis and function of this important lipid class are presented in Chapter 10. Interestingly, *DAD1* belongs to a gene family, which has 12 members in Arabidopsis (Ishiguro *et al.*, 2001). The functions of the respective enzymes are unknown, but they might be involved in phospholipid and galactolipid acyl hydrolysis (Ishiguro *et al.*, 2001).

#### *5.4.2.6 SAG101-like acyl hydrolases and PDAT-like acyltransferases*

An acyl hydrolase-like gene (SAG101, senescence-associated gene) was isolated from Arabidopsis and shown to encode a protein capable of cleaving fatty acyl groups from triacylglycerol (He and Gan, 2002). The gene is strongly induced and might thus be critical for lipid turnover during senescence. Interestingly, Arabidopsis contains at least 21 SAG101-like genes whose functions are unknown.

The gene encoding phospholipid:diacylglycerol acyltransferase (PDAT), which catalyses the acyl transfer from phospholipids onto DAG-yielding triacylglycerol, was isolated from yeast and plants by Dahlqvist *et al.* (2000). Several additional PDAT-like genes are present in Arabidopsis, and it is possible that they are involved in triacylglycerol synthesis as well as membrane lipid acylation/deacylation (Beisson *et al.*, 2003).

### *5.4.3 Glycolipases*

Physiological studies on head group turnover of glycolipids are scarce, but it is generally accepted that the head groups from plant glycolipids are eventually hydrolysed during lipid breakdown (Sastry and Kates, 1964; Helmsing, 1967). Furthermore, it is not entirely clear whether glycosidases involved in glycolipid breakdown directly act on glycolipids (i.e. MGDG, DGDG or sulfolipid) or whether they hydrolyse the sugar moiety from glycosylglycerol, the deacylated product of galactolipase and sulfolipase reactions. α-Galactosidase activity is required for the cleavage of the second galactose moiety of DGDG, and a β-galactosidase is involved in hydrolysing the galactose moiety from diacylglycerol in MGDG (Sastry and Kates, 1964). Consequently, these two galactosidase activities were detected in plant extracts, e.g. in runner bean leaves (Sastry and Kates, 1964). Similarly, an α-glycosidase activity is required for sulfolipid breakdown, but as for the galactosidase activities, the identity of the corresponding genes is not known. Almost 400 gene loci with sequence similarities to glycosylhydrolases/glycosidases are present in the Arabidopsis genome (afmb.cnrs-mrs.fr/CAZY/; Henrissat and Bairoch, 1996). This high number of glycosylhydrolases represents a major obstacle for bioinformatics-based approaches of identifying hydrolases acting on glycolipids.

Galactolipid acyl hydrolase activities have been characterized in crude plant extracts of runner bean (Sastry and Kates, 1964; Helmsing, 1967). These enzymatic activities hydrolyse acyl groups from the *sn*-1 and *sn*-2 position of MGDG and DGDG (Figure 5.6). High galactolipase activity has been associated with chilling-sensitivity because due to their detergent-like properties, high amounts of free fatty acids are known to destabilise biological membranes (e.g. Gemel and Kaniuga, 1987). However, recent results by Terasaki and Itabshi (2003) suggest that a brown alga from the Pacific Ocean (*Cladosiphon*) containing elevated galactosidase activity can tolerate extremely high amounts of non-esterified fatty acids in its cells.

A sulfolipid acyl hydrolase activity was detected in *Scenedesmus*, but the identity of the corresponding enzyme is unknown (Yagi and Benson, 1962).

### *5.4.4 Fatty acyl turnover and acyl-CoA synthetases*

Fatty acids are hydrolysed from membrane lipids by action of phospholipases or glycolipases. Different stress factors were shown to increase the rate of phospholipase activity, thus releasing additional amounts of free fatty acids (e.g. Conconi *et al.*, 1996). Although the free fatty acid content in non-stressed leaves is very low, a high rate of ‘retailoring’, i.e. acylation/deacylation of fatty acyl groups on glycerolipids was detected. By labelling experiments with $H_2^{18}O$, Pollard and Ohlrogge (1999) estimated that the rate of acyl hydrolysis and re-acylation is in the range of fatty acid *de novo* synthesis in leaves. Using $H_2^{18}O$ labelling, Pollard and Ohlrogge (1999) estimated that the rate of acyl hydrolysis and re-acylation is in the range of fatty acid *de novo* synthesis in leaves. Using radioactive feeding experiments with pumpkin leaves, the turnover of radioactivity in galactolipids was found to be much slower as compared to PC (Roughan, 1970). It was concluded that fatty acids in PC are acylated/deacylated at a much higher rate and thus might serve as precursors for galactolipid synthesis in the plastids. Instead of being used for acylation reactions, free fatty acids released from membrane lipids can be subjected to degradation via β-oxidation. By feeding Arabidopsis leaves with carbon dioxide labelled with different carbon isotopes, Bao *et al.* (2000) determined the rate of fatty acid degradation at around 4% per day. Therefore, in only 25 days an amount equal to the entire fatty acid content in a leaf cell is degraded. The reason for the high turnover rates (acylation/deacylation and degradation) in plants is unknown, but it is possible that this process is important to channel specific fatty acids into different lipid classes and to remove oxidised fatty acyl groups from membrane lipids. Free fatty acids have to be quickly removed from plant cells, e.g. by degradation or re-incorporation into other lipids, because due to their detergent-like characteristics, their accumulation is detrimental to membranes. As an initial step, free fatty acids are activated by conversion into their respective Coenzyme A esters. In this reaction (catalysed by acyl-CoA synthetases), an acyl-adenylate intermediate is formed from fatty acids and ATP, which is subsequently converted into acyl-CoA thioesters. Nine long-chain acyl-CoA synthetases (LACS) were found in the Arabidopsis genome (Shockey *et al.*, 2002), which are involved in activation of free fatty acids in the chloroplast (Schnurr *et al.*, 2002) or in the peroxisome (Fulda *et al.*, 2002; Hayashi *et al.*, 2002). Whereas the chloroplast LACS enzymes might be responsible for activation of fatty acids derived from plastidic *de novo* synthesis or from chloroplast lipid turnover (Schnurr *et al.*, 2002), the peroxisomal enzymes are most likely crucial for the activation of free fatty acids derived from triacylglycerol hydrolysis. Acyl-CoAs are subsequently degraded by peroxisomal β-oxidation (Fulda *et al.*, 2002; Hayashi *et al.*, 2002).

## 5.5 Physiological roles of membrane lipids

### *5.5.1 Growth at high and low temperatures*

Under normal physiological conditions, the lipid phase of biological membranes is in a so-called 'liquid crystalline state'. At low temperatures, however, the acyl groups can undergo a change of state whereby they become organised into straight chains resulting in a more rigid state designated 'gel phase'. The transition of membranes into the gel phase under low temperatures is believed to contribute to cell damage and can even result in cell death. Different factors can influence the phase transition temperature of the membrane, and thus its fluidity/rigidity. The presence of *cis*-double bonds in fatty acids has a strong impact on their biophysical characteristics because the double bond introduces a bent configuration into the acyl chain. In contrast, *trans* fatty acids have more rigid chains and resemble saturated fatty acids in their physical properties. As a rule of thumb, high amounts of saturated fatty acids (which have straight acyl chains) favour membrane rigidity, whereas an increasing number of *cis* double bonds leads to an increase in fluidity. Following this hypothesis, it is expected that the presence of saturated fatty acids is particularly important at high temperatures, but that unsaturated fatty acids are critical for survival under low temperatures ('chilling tolerance'). However, the impact of lipid unsaturation on lipid fluidity in plants seems to be more complex. The study of Arabidopsis *mutants* and of transgenic plants with alterations in fatty acid desaturation has greatly advanced our understanding of the role of fatty acid unsaturation for thermal tolerance (for a recent review see Wallis and Browse, 2002).

#### *5.5.1.1 Unsaturated fatty acids*

The triunsaturated fatty acids 16:3 and 18:3 are the most abundant fatty acid classes in plants of the '16:3 group', including Arabidopsis and tobacco. For a long time, it has been speculated that these fatty acids are crucial for photosynthesis and plant growth – in particular, at low temperatures. In accordance with this assumption, the amounts of 16:3 and 18:3 increase during growth at low temperatures (Graham and Patterson, 1982). Interestingly, expression of one desaturase gene (*FAD8*) is induced at low temperature in Arabidopsis (Gibson *et al.*, 1994). Therefore, this gene might be involved in the synthesis of polyunsaturated fatty acids during chilling stress. However, an Arabidopsis triple mutant (*fad3 fad7 fad8*) devoid of 16:3 and 18:3 shows no severe phenotype at normal growth conditions, and at low temperatures, photosynthesis in these plants is only slightly affected (McConn and Browse, 1996). Furthermore, overexpression of the Arabidopsis FAD7 desaturase in tobacco seedlings resulted in an increase in 16:3 and 18:3, but in only minor protection against chilling-induced damage (Kodama *et al.*, 1994). Whereas a reduction in the amounts of triunsaturated fatty acids had only minor effects on photosynthesis and growth at normal and low temperatures, it strongly improved plant performance at elevated temperatures. Transgenic tobacco plants reduced in desaturase activity by co-suppression of the *FAD7* gene, as well as the Arabidopsis *fad7 fad8* double mutant, were both reduced in the amounts of 16:3 and 18:3, and as a consequence, showed strongly enhanced growth and higher photosynthetic activity at high temperature (Murakami *et al.*, 2000). Therefore, a decrease in lipid desaturation seems to have a stabilising effect on membranes at elevated temperatures.

The *fad2 fad6* double mutant of Arabidopsis is deficient in the synthesis of fatty acids with two double bonds, and as a consequence, is devoid of di- and triunsaturated fatty acids (McConn and Browse, 1998). In contrast to the *fad3 fad7 fad8* plant, the additional reduction in diunsaturated fatty acids in *fad2 fad6* results in the loss of the capability of photoautotrophic growth even at normal growth conditions.

Reduction in the degree of desaturation as in the Arabidopsis mutant *fad2* results in reduced growth rates at low temperatures, and eventually to plant death after cold treatment for a prolonged time (Miquel *et al.*, 1993). However, the low temperature response of *fad2* seems to be different from the classic chilling response, and thus other factors might contribute to the observed phenotype.

Two additional *mutants* (*fad5* and *fad6*) with a reduced degree of fatty acid unsaturation show a decrease in growth at low temperatures, but improved plant performance at elevated temperatures (Hugly *et al.*, 1989, 1992; Kunst *et al.*, 1989).

The *fab2* mutant of Arabidopsis, which carries a mutation in one of the stearoyl-ACP desaturase genes, shows an accumulation of 18:0 in its membrane lipids accompanied by a reduction of 18:2 and 18:3 (Lightner *et al.*, 1994). As a consequence, the plant shows a dramatic reduction in growth, which is caused by the arrest of cell expansion. Interestingly, this strong phenotype is ameliorated under high temperatures, where wild-type and *fab2* plants look much more similar. The interpretation of this effect is difficult, but possibly the accumulation of saturated fatty acids in *fab2* results in increased membrane rigidity at normal temperatures, but this effect might be compensated for by higher acyl chain mobility at elevated temperatures.

### *5.5.2 The role of unsaturated molecular species of PG in chilling sensitivity*

The plastidic pool of PG in plants contains large amounts of palmitic acid and 16:1δ3 *trans* at the *sn*-2 position. Because the *trans* double bond (in contrast to the *cis* double bond) does not introduce a bent configuration into the acyl chain, the structure of this fatty acid is similar to saturated fatty acids. Therefore, the fatty acyl group attached to the *sn*-1 position of plastidic PG is considered to be crucial for the determination of the melting point of PG, and as a consequence of the chilling sensitivity of the respective plant species. In many chilling sensitive plants (e.g. squash, tobacco), saturated fatty acids, such as 16:0, are attached to the *sn*-1 position of PG, whereas in chilling tolerant plants (Arabidopsis, spinach) polyunsaturated fatty acids with 18 carbon atoms dominate (Murata *et al.*, 1982; Murata, 1983). Therefore, the substrate specificity of the plastidic GPAT is considered to be in causal relationship with the nature of the acyl group attached to the *sn*-1 position of PG, and thus with the degree of chilling tolerance. In accordance with this hypothesis, genetic engineering of transgenic tobacco plants by overexpression of GPAT from the chilling sensitive plant squash or GPAT from the chilling tolerant plant Arabidopsis, resulted in a reduced or improved tolerance to chilling stress, respectively (Murata *et al.*, 1992; Moon *et al.*, 1995). Furthermore, overexpression of the *E. coli* GPAT in plastids of transgenic Arabidopsis resulted in an accumulation of saturated fatty acids (mostly 16:0) at the *sn*-1 position, thus strongly reducing chilling tolerance (Wolter *et al.*,

1992). An Arabidopsis mutant affected in β-ketoacyl-ACP synthase II expression (*fab1*) contains increased amounts of 16:0 in prokaryotic lipids. As a consequence, the plastidic PG pool contains large amounts of saturated fatty acid (16:0) at the *sn*-1 position of PG. Still, chilling sensitivity of *fab1* plants was very similar to wild-type, and only after prolonged incubation at low temperature, a difference in growth and photosynthesis was observed (Wu and Browse, 1995). Therefore, the role of high melting point molecular species of PG might be just one of many parameters involved in chilling tolerance.

#### *5.5.2.1 Increase of PC synthesis during cold treatment*

PC and PE are the major phospholipids in extraplastidic membranes, and their synthesis largely depends on the ER diacylglycerol pathway. Thus, the relative abundance of the head group precursors CDP-ethanolamine and CDP-choline is crucial to control PE and PC synthesis (Figure 5.3). The amount of PC increases during growth at low temperature, and this increase was suggested to be critical for the development of freezing tolerance (Yoshida, 1984; Uemura *et al.*, 1995). In plants, expression of two genes involved in PC biosynthesis is induced during cold acclimation. The expression of phosphoethanolamine methyltransferase (PEAMT), which is critical for choline production is induced in wheat and spinach during cold and salt treatment (Nuccio *et al.*, 2000; Charron *et al.*, 2002). Increased production of choline is the prerequisite for the synthesis of the osmoprotectant betaine as well as for PC synthesis. Antisense inhibition of PEAMT in transgenic Arabidopsis plants resulted in reduced choline synthesis and in increased stress sensitivity (Mou *et al.*, 2002). Furthermore, CDP-choline synthase (CCT) activity was strongly increased during cold treatment in Arabidopsis, which could be traced down to an enhanced expression of the *CCT2* gene (Figure 5.3; Inatsugi *et al.*, 2002). These studies suggest that the regulation of the PC-to-PE ratio in extraplastidic membranes might be causally related to adaptation to growth at low temperatures.

### *5.5.3 The role of thylakoid lipids in photosynthesis*

The pigment–protein complexes of photosynthesis are embedded in a matrix of glycerolipids in the thylakoids of chloroplasts. The two galactolipids, MGDG and DGDG, and the phospholipid PG are predominant lipids in thylakoids, and, therefore, were suggested to be crucial in supporting photosynthetic light reactions. This hypothesis was confirmed by the findings that galactolipids and PG occur as integral components of the pigment–protein complexes. Four lipid molecules were found in the crystal structure of photosystem I from cyanobacteria (Jordan *et al.*, 2001). These lipids (one MGDG and three molecules of PG) form hydrogen bonds with the core proteins PsaA and PsaB at the stromal side of thylakoid membranes (Klukas *et al.*, 1999; Jordan *et al.*, 2001). The crystal structures of PSII from cyanobacteria or from plants did not reveal the presence of lipid molecules (Rhee *et al.*, 1997; Zouni *et al.*,

2001). Two lipid molecules (PG and DGDG) were detected in the crystal structure of the major light harvesting complex LHCII from the dinoflagellate *Amphidinium carterae* (Hofmann *et al.*, 1996). These two lipids were also found in LHCII preparations isolated from pea chloroplasts (but not in the crystal structure of LHCII from pea; Kühlbrandt *et al.*, 1994), and were crucial for the stability of LHCII complexes *in vitro* (Nußberger *et al.*, 1993; Reinsberg *et al.*, 2000).

The study of lipid deficient Arabidopsis *mutants* laid the ground for our better understanding of the role of glycerolipids in photosynthesis. Inhibition of MGDG synthase activity in the *mgd1* mutant of Arabidopsis resulted in a reduction of MGDG by about 50% (Jarvis *et al.*, 2000). As a consequence, chloroplast ultrastructure and chlorophyll content were affected in the leaves. However, the *mgd1* mutant still contains substantial amounts of MGDG, which might be derived from residual MGD1 activity or from the other two Arabidopsis MGDG synthases MGD2 and MGD3.

Arabidopsis *mutants* were obtained for the two DGDG synthases, DGD1 and DGD2. The amounts of DGDG in *dgd1* are reduced down to 10% of wild-type levels (Dörmann *et al.*, 1995). This plant shows a severe reduction in growth and alterations in photosynthesis (Härtel *et al.*, 1997; Reifarth *et al.*, 1997). In *dgd1*, chloroplast ultrastructure is changed, and the plants contain less chlorophyll. An Arabidopsis mutant carrying a T-DNA insertion in the *DGD2* gene contains normal amounts of DGDG, and growth and photosynthesis are indistinguishable from wild-type (Kelly *et al.*, 2003). However, in the double mutant, *dgd1 dgd2*, the amount of DGDG was reduced below detection limit, indicating that DGD2 is responsible for the synthesis of the residual amount of DGDG in *dgd1*. Double mutant plants are even further reduced in growth and photosynthesis as compared to *dgd1* single *mutants*, suggesting that the residual amount of DGDG in *dgd1* is still important for normal plant development (Kelly *et al.*, 2003).

In Arabidopsis, mutations of the *PGP1* gene encoding the plastidic/mitochondrial form of PG-3-phosphate synthase result in a severe reduction of plastidic PG content (Xu *et al.*, 2002; Hagio *et al.*, 2003; Babiychuk *et al.*, 2003). Null alleles of *pgp1* are pale green and unable to survive on soil. These plants show a strong reduction in photosynthetic pigments and an altered chloroplast structure, emphasising the importance of the anionic phospholipid PG for photosynthesis. In all photosynthetic organisms, chloroplast PG contains high amounts of the *trans* fatty acid, 16:1$\delta$3 *trans*. Therefore, this fatty acyl group was suggested to be particularly important for the association of PG with protein–pigment complexes. However, isolation of the *fad4* mutant of Arabidopsis devoid of 16:1$\delta$3 *trans* clearly demonstrated that this fatty acid is dispensable for photosynthetic light reactions (Browse *et al.*, 1985). Only subtle alterations in the stability of oligomeric PSI and LHCII complexes were found in *fad4* (McCourt *et al.*, 1985).

In addition to PG and the two galactolipids, thylakoid membranes contain low amounts of SQDG, an anionic sulfolipid. However, in contrast to the other three thylakoid lipids, SQDG is not essential for photosynthesis. Inactivation of the *SQD2* gene (SQDG synthase) in Arabidopsis by insertional mutagenesis demonstrated that total loss of SQDG results in only minor changes of photosynthesis, but strongly affects growth under phosphate limitation (Yu *et al.*, 2002; see Section 5.5.4). Therefore, SQDG seems to play an important role as surrogate lipid for anionic phospholipids during phosphate deprivation, rather than being critical for photosynthesis.

### 5.5.4 *Growth during phosphate deprivation*

In all living organisms, large amounts of organically bound phosphate are found in nucleic acids (DNA, RNA), proteins, nucleotides and sugar phosphates, but also in phospholipids. In contrast to animals, higher plants and some bacteria contain large amounts of glycoglycerolipids and change their membrane lipid composition according to phosphate availability. In *Pseudomonas*, anionic phospholipids are replaced by anionic glycolipids during growth under phosphate limitation (Minnikin

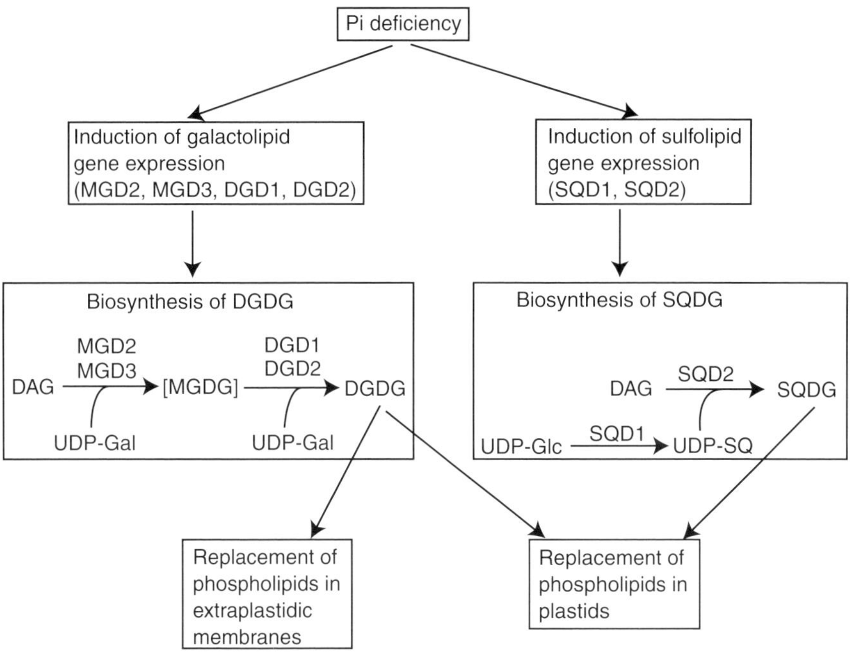


**Figure 5.7** The role of DGDG and SQDG during phosphate deprivation. Phosphate deficiency results in induction of gene expression for galactolipid and sulfolipid synthesis. The additional amounts of DGDG and SQDG synthesised during phosphate deprivation replace phospholipids in plastid membranes. Furthermore, DGDG is transported to extraplastidic membranes (e.g. plasma membrane). The reduction in phospholipid content in cellular membranes allows the incorporation of phosphate into biological molecules essential for plant survival (e.g. DNA and RNA).

*et al.*, 1974). Similarly, the amounts of DGDG and SQDG strongly increase under low phosphate conditions in plants (Essigmann *et al.*, 1998; Härtel *et al.*, 2000). The two glycolipids replace phospholipids in the membranes, and the phosphate released is channelled into other important pathways (e.g. nucleic acid synthesis). In Arabidopsis, the increase in DGDG and SQDG is mediated by induction of gene expression, because expression of two MGDG synthase genes (*MGD2*, *MGD3*), two DGDG synthase genes (*DGD1*, *DGD2*), UDP-sulfoquinovose synthase (*SQD1*) and SQDG synthase (*SQD2*) is induced during phosphate deprivation (Essigmann *et al.*, 1998; Awai *et al.*, 2001; Kelly *et al.*, 2002, 2003; Yu *et al.*, 2002; Figure 5.7). SQDG, which increases during phosphate deprivation in chloroplasts is believed to replace plastidic PG, whereas DGDG is thought to serve as a surrogate lipid for PC. Interestingly, DGDG does not only increase in the chloroplast, but also in extraplastidic membranes during phosphate deprivation (Härtel *et al.*, 2000). In oat roots, DGDG is of low abundance in plasma membranes under normal growth conditions, but increases to up to 70% of plasma membrane glycerolipids under phosphate deprivation (Table 5.1; Andersson *et al.*, 2003). The DGDG synthase DGD1 is considered to be responsible for DGDG synthesis for the chloroplast under normal and phosphate deprived conditions, whereas DGD2 synthesises DGDG for the extraplastidic membranes during phosphaste deprivation (Kelly *et al.*, 2003). Because the two proteins, DGD1 and DGD2, were localised to the outer side of chloroplast membranes, DGDG synthesised under phosphate deprivation has to be transported to the ER/plasma membrane via an unknown mechanism (Figure 5.5). The findings that different bacteria and plants change their membrane lipid composition in response to phosphate limitation can be explained by convergent evolution of gene functions.

## 5.6 Summary and future perspectives

In addition to phospholipids present in all organisms, plants contain large amounts of glycoglycerolipids in their membranes. Membrane lipids are subject to constant turnover by phospholipases and galactolipases. Furthermore, lipid-derived metabolites play a major role in cell signalling. The identification of enzymes and genes of glycerolipid biosynthesis in plants has resulted in impressive progress in the last few years, and many unexpected aspects of lipid biosynthesis and function have been unravelled. The characterisation of lipid-deficient Arabidopsis *mutants* has been essential for our understanding of the function of different lipid classes and fatty acyl groups in many key physiological processes, including photosynthesis, adaptation to high and low temperatures and phosphate deprivation. These studies emphasise the importance of membrane lipids for plant cells beyond their structural role of providing ‘building blocks’ for the formation of biological membranes. However, many important aspects of lipid biochemistry and function still remain unclear. In particular, lipid trafficking between membranes and organelles as well as lipid turnover are only poorly understood, and will be the focus of future research (see Chapter 7 for more discussion of this topic).

## Acknowledgements

I would like to thank John B. Ohlrogge and Frédéric Beisson (Michigan State University, East Lansing, MI, USA) for critically reading this manuscript.

## References

Arabidopsis Genome Initiative (2000) Analysis of the genome sequence of the flowering plant, *Arabidopsis thaliana*. *Nature*, **408**, 796–815.

Andersson, M.X., Stridh, M.H., Larsson, K.E., Liljenberg, C. and Sandelius, A.S. (2003) Phosphate-deficient oat replaces a major portion of the plasma membrane phospholipids with the galactolipid digalactosyldiacylglycerol. *FEBS Letters*, **537**, 128–132.

Andrews, D.L., Beames, B., Summers, M.D. and Park, D. (1988) Characterization of the lipid acyl hydrolase activity of the major potato (*Solanum tuberosum*) tuber protein, patatin, by cloning and abundant expression in a baculovirus vector. *Biochemical Journal*, **252**, 199–206.

Awai, K., Maréchal, E., Block, M.A. *et al.* (2001) Two types of MGDG synthase genes, found widely in both 16:3 and 18:3 plants, differentially mediate galactolipid synthesis in photosynthetic and nonphotosynthetic tissues in *Arabidopsis thaliana*. *Proceedings of the National Academy of Sciences USA*, **98**, 10960–10965.

Babiychuk, E., Müller, F., Eubel, H., Braun, H.-P., Frentzen, M. and Kushnir, S. (2003) *Arabidopsis* phosphatidylglycerophosphate synthase 1 is essential for chloroplast differentiation, but is dispensable for mitochondrial function. *Plant Journal*, **33**, 899–909.

Bao, X., Focke, M., Pollard, M. and Ohlrogge, J. (2000) Understanding *in vivo* carbon precursor supply for fatty acid synthesis in leaf tissue. *Plant Journal*, **22**, 39–50.

Beisson, F., Koo, A.J.K., Ruuska, S. *et al.* (2003) Arabidopsis genes involved in acyl lipid metabolism. A 2003 census of the candidates, a study of the distribution of expressed sequence tags in organs, and a web-based database. *Plant Physiology*, **132**, 681–697.

Benson, A.A. (1963) The plant sulfolipid. *Advances in Lipid Research*, **1**, 387–394.

Benson, A.A. and Maruo, B. (1958) Plant phospholipids. I. Identification of the phosphatidyl glycerols. *Biochimica et Biophysica Acta*, **27**, 189–195.

Benson, A.A., Danie, H. and Wiser, R. (1959) A sulfolipid in plants. *Proceedings of the National Academy of Sciences USA*, **45**, 1582–1587.

Benson, A.A., Wiser, R., Ferrari, R.A. and Miller, J.A. (1958) Photosynthesis of galactolipids. *Journal of the American Chemical Society*, **80**, 4740.

Bertrams, M. and Heinz, H. (1981) Positional specificity and fatty acid selectivity of purified *sn*-glycerol 3-phosphate acyltransferases from chloroplasts. *Plant Physiology*, **68**, 653–657.

Bligny, R. and Douce, R. (1980) A precise localization of cardiolipin in plant cells. *Biochimica et Biophysica Acta*, **617**, 254–263.

Block, M.A., Dorne, A.-J., Joyard, J. and Douce, R. (1983) Preparation and characterization of membrane fractions enriched in outer and inner envelope membranes from spinach chloroplasts. II. Biochemical characterization. *Journal of Biological Chemistry*, **258**, 13281–13286.

Bolognese, C.P. and Patricia McGraw, P. (2000) The isolation and characterization in yeast of a gene for Arabidopsis *S*-adenosylmethionine:phosphoethanolamine *N*-methyltransferase1. *Plant Physiology*, **124**, 1800–1813.

Bomstein, R.A. (1965) A new class of phosphatides isolated from soft wheat flour. *Biochemical and Biophysical Research Communications*, **21**, 49–54.

Brown, D.J. and DuPont, F.M. (1989) Lipid composition of plasma membranes and endomembranes prepared from roots of barley (*Hordeum vulgare* L.). *Plant Physiology*, **90**, 955–961.

Browse, J., McCourt, P. and Somerville, C.R. (1985) A mutant of *Arabidopsis thaliana* lacking a chloroplast-specific lipid. *Science*, **227**, 763–765.

Browse, J., Warwick, N., Somerville, C.R. and Slack, C.R. (1986) Fluxes through the prokaryotic and eukaryotic pathways of lipid synthesis in the 16:3 plant *Arabidopsis thaliana. Biochemical Journal*, 235, 25–31.

Butikofer, P. and Brodbeck, U. (1993) Partial purification and characterization of a (glycosyl) inositol phospholipid-specific phospholipase C from peanut. *Journal of Biological Chemistry*, **268**, 17794–17802.

Carter, H.E., McCluer, R.H. and Slifer, E.D. (1956) Lipids of wheat flour. I. Characterization of galactosylglycerol components. *Journal of the American Chemical Society*, **78**, 3735–3738.

Chapman, K.D. (2000) Emerging physiological roles for *N*-acylphosphatidylethanolamine metabolism in plants: signal transduction and membrane protection. *Chemistry and Physics of Lipids*, **108**, 221–230.

Chapman, K.D. and Moore, T.S. (1993) Catalytic properties of a newly discovered acyltransferase that synthesizes *N*-acylphosphatidylethanolamine in cottonseed (*Gossypium hirsutum* L.) microsomes. *Plant Physiology*, **102**, 761–769.

Charron, J.-B.F., Breton, G., Danyluk, J., Muzac, I., Ibrahim, R.K. and Sarhan, F. (2002) Molecular and biochemical characterization of a cold-regulated phosphoethanolamine *N*-methyltransferase from wheat. *Plant Physiology*, **129**, 363–373.

Chibnall, A.C. and Channon, H.J. (1927) The ether-soluble substances of cabbage leaf cytoplasm. II. Calcium salts of glyceridephosphoric acids. *Biochemical Journal*, **21**, 233–246.

Cline, K. and Keegstra, K. (1983) Galactosyltransferases involved in galactolipid biosynthesis are located in the outer membrane of pea chloroplast envelopes. *Plant Physiology*, **71**, 366–372.

Chuman, L. and Brody, S. (1989) Acyl carrier protein is present in the mitochondria of plants and eucaryotic micro-organisms. *European Journal of Biochemistry*, **184**, 643–649.

Collin, S., Justin, A.-M., Cantrel, C., Arondel, V. and Kader, J.-C. (1999) Identification of AtPIS, a phosphatidylinositol synthase from Arabidopsis. *European Journal of Biochemistry*, **262**, 652–658.

Conconi, A., Miquel, M., Browse, J.A. and Ryan, C.A. (1996) Intracellular levels of free linolenic and linoleic acids increase in tomato leaves in response to wounding. *Plant Physiology*, **111**, 797–803.

Dahlqvist, A., Ståhl, U., Lenman, M. *et al* (2000) Phospholipid:diacylglycerol acyltransferase: an enzyme that catalyzes the acyl-CoA-independent formation of triacylglycerol in yeast and plants. *Proceedings of the National Academy of Sciences USA*, **97**, 6487–6492.

Delhaize, E., Hebb, D.M., Richards, K.D., Lin, J.-M., Ryan, P.R. and Gardner, R.C. (1999) Cloning and expression of a wheat (*Triticum aestivum* L.) phosphatidylserine synthase cDNA. Overexpression in plants alters the composition of phospholipids. *Journal of Biological Chemistry*, **274**, 7082–7088.

Dhondt, S., Geoffroy, P., Stelmach, B.A., Legrand, M. and Heitz, T. (2000) Soluble phospholipase A2 activity is induced before oxylipin accumulation in tobacco mosaic virus-infected tobacco leaves and is contributed by patatin-like enzymes. *Plant Journal*, **23**, 431–440.

Dörmann P. and Benning C. (2002) Galactolipids rule in seed plants. *Trends in Plant Science*, **7**, 112–118.

Dörmann, P., Balbo, I. and Benning, C. (1999) Arabidopsis galactolipid biosynthesis and lipid trafficking mediated by DGD1. *Science*, **284**, 2181–2184.

Dörmann, P., Hoffmann-Benning, S., Balbo, I. and Benning, C. (1995) Isolation and characterization of an Arabidopsis mutant deficient in the thylakoid lipid digalactosyl diacylglycerol. *Plant Cell*, **7**, 1801–1810.

Dorne, A.-J., Block, M.A., Joyard, J. and Douce, R. (1982) The galactolipid:galactolipid galactosyltransferase is located on the outer membrane of the chloroplast envelope. *FEBS Letters*, **145**, 30–34.

Essigmann, B., Gueler, S., Narang, R.A., Linke, D. and Benning, C. (1998) Phosphate availability affects the thylakoid lipid composition and the expression of SQD1, a gene required for sulfolipid biosynthesis in *Arabidopsis thaliana. Proceedings of the National Academy of Sciences USA*, **95**, 1950–1955.

Farag, K.M. and Palta, J.P. (1993) Use of lysophosphatidylethanolamine, a natural lipid, to retard tomato leaf and fruit senescence. *Physiologia Plantarum*, **87**, 515–521.
Farmer, E.E., Alméras, E. and Krishnamurtzy, V. (2003) Jasmonates and related oxylipins in plant responses to pathogenesis and herbivory. *Current Opinion in Plant Biology*, **6**, 372–378.
Folch, J. (1942) Brain cephalin, a mixture of phosphatides. Separation from it of phosphatidyl serine, phosphatidyl ethanolamine and a fraction containing an inositol phosphatide. *Journal of Biological Chemistry*, **146**, 35–44.
Frentzen, M. (1993) Acyltransferases and triacylglycerols, in *Lipid Metabolism in Plants* (ed. T.S. Moore, Jr.) CRC Press, Boca Raton, FL, pp. 195–230.
Frentzen, M. and Griebau, R. (1994) Biosynthesis of cardiolipin in plant mitochondria. *Plant Physiology*, **106**, 1527–1532.
Froehlich, J., Benning, C. and Dörmann, P. (2001) The digalactosyldiacylglycerol synthase DGD1 is inserted into the outer envelope membrane of chloroplasts in a manner independent of the General Import Pathway and does not depend on direct interaction with MGDG synthase for DGDG biosynthesis. *Journal of Biological Chemistry*, **276**, 31806–31812.
Fujino, Y. and Miyazawa, T. (1979) Chemical structures of mono-, di-, tri- and tetraglycosyl gycerides in rice bran. *Biochimica et Biophysica Acta*, **572**, 442–451.
Fulda, M., Shockey, J., Werber, M., Wolter, F.P. and Heinz, E. (2002) Two long-chain acyl-CoA synthetases from *Arabidopsis thaliana* involved in peroxisomal fatty acid β-oxidation. *Plant Journal*, **32**, 93–103.
Gemel, J. and Kaniuga, Z. (1987) Comparison of galactolipase activity and free fatty acid levels in chloroplasts of chill-sensitive and chill-resistant plants. *European Journal of Biochemistry*, **166**, 229–233.
Gibson, Y., Arondel, V., Iba, K. and Somerville, C. (1994) Cloning of a temperature-regulated gene encoding a chloroplast omega-3 desaturase from *Arabidopsis thaliana*. *Plant Physiology*, **106**, 1615–1621.
Gillapsy, G.E., Keddie, J.S., Oda, K. and Gruissem, W. (1995) Plant inositol monophosphatase is a Lithium-sensitive enzyme encoded by a multigene family. *Plant Cell*, **7**, 2175–2185.
Gobley, M. (1874) Recherches chimiques sur le cerveau. *J. Pharm. Chim.*, **20**, 98–99.
Goode, J.H. and Dewey, R.E. (1999) Characterization of aminoalcoholphospho-transferases from *Arabidopsis thaliana* and soybean. *Plant Physiology and Biochemistry*, **37**, 445–457.
Graham, D. and Patterson, B.D. (1982) Responses of plants to low, non-freezing temperatures: proteins, metabolism, and acclimation. *Annual Review of Plant Physiology*, **33**, 347–372.
Griebau, R. and Frentzen, M. (1994) Biosynthesis of phosphatidylglycerol in isolated mitochondria of etiolated mung bean (*Vigna radiata* L.) seedlings. *Plant Physiology*, **105**, 1269–1274.
Hagio, M., Sakurai, I., Sato, S., Kato, T., Tabata, S. and Wada, H. (2002) Phosphatidylglycerol is essential for the development of thylakoid membranes in *Arabidopsis thaliana*. *Plant & Cell Physiology*, **43**, 1456–1464.
Hayashi, H., De Bellis, L., Hayashi, Y., Nito, K., Kato, A., Hayashi, M., Hara-Nishimura, I. and Nishimura, M. (2002) Molecular characterization of an Arabidopsis acyl-Coenzyme A synthetase localized on glyoxysomal membranes. *Plant Physiology*, **130**, 2019–2026.
Härtel, H., Dörmann, P. and Benning, C. (2000) DGD1-independent biosynthesis of extraplastidic galactolipids following phosphate deprivation in *Arabidopsis*. *Proceedings of the National Academy of Sciences USA*, **97**, 10649–10654.
Härtel, H., Lokstein, H., Dörmann, P., Grimm, B. and Benning, C. (1997) Changes in the composition of the photosynthetic apparatus in the galactolipid deficient *dgd1* mutant of *Arabidopsis thaliana*. *Plant Physiology*, **115**, 1175–1184.
He, Y. and Gan, S. (2002) A gene encoding an acyl hydrolase is involved in leaf senescence in Arabidopsis. *Plant Cell*, **14**, 805–815.
Heemskerk, J.W.M., Storz, T., Schmidt, R.R. and Heinz, E. (1990) Biosynthesis of digalactosyldiacylglycerol in plastids from 16:3 and 18:3 plants. *Plant Physiology*, **93**, 1286–1294.
Heinz, E. (1977) Enzymatic reactions in galactolipid biosynthesis, in *Lipids and Lipid Polymers in Higher Plants* (eds M. Tevini and H.K. Lichtenthaler), Springer-Verlag, Berlin/New York, pp. 102–120.

Heinz, E. and Roughan, P.G. (1983) Similarities and differences in lipid metabolism of chloroplasts isolated from 18:3 and 16:3 plants. *Plant Physiology*, **72**, 273–279.
Heinz, E., Schmidt, H., Hoch, M., Jung, K.H., Binder, H. and Schmidt, R.R. (1989) Synthesis of different nucleoside 5′-diphosphosulfoquinovoses and their use for studies on sulfolipid biosynthesis in chloroplasts. *European Journal of Biochemistry*, **184**, 445–454.
Helmsing, P.J. (1967) Hydrolysis of galactolipids by enzymes in spinach leaves. *Biochimica et Biophysica Acta*, **144**, 470–472.
Henrissat, B. and Bairoch, A. (1996) Updating the sequence-based classification of glycosyl hydrolases. *Biochemical Journal*, **316**, 695–696.
Hirayama, T., Ohto, C., Mizoguchi, T. and Shinozaki, K. (1995) A gene encoding a phosphatidylinositol-specific phospholipase C is induced by dehydration and salt stress in *Arabidopsis thaliana*. *Proceedings of the National Academy of Sciences USA*, **92**, 3903–3907.
Höfgen, R. and Willmitzer, L. (1990) Biochemical and genetic analysis of different patatin isoforms expressed in various organs of potato (*Solanum tuberosum*). *Plant Science*, **66**, 221–230.
Hofmann, E., Wrench, P.M., Sharples, F.P., Hiller, R.G., Wilte, W. and Diederichs, K. (1996) Structural basis of light harvesting by carotenoids: peridinin-chlorophyll-protein from *Aphidinium carterae*. *Science*, **272**, 1788–1744.
Holmbäck J., Karlsson, A.Å. and Arnoldsson, K.C. (2001) Characterization of *N*-acylphosphatidylethanolamine in oats. *Lipids*, **36**, 153–165.
Huang, A.H.C. (1975) Enzymes of glycerol metabolism in the storage tissues of fatty seedlings. *Plant Physiology*, **55**, 555–558.
Huang, A.H.C (1993) Lipases, in *Lipid Metabolism in Plants* (ed. T.S. Moore Jr.) CRC Press, Boca Raton, FL, pp. 473–503.
Hugly, S. and Somerville, C. (1992) A role for membrane lipid polyunsaturation in chloroplast biogenesis at low temperature. *Plant Physiology*, **99**, 197–202.
Hugly, S., Kunst, L., Browse, J. and Somerville, C. (1989) Enhanced thermal tolerance of photosynthesis and altered chloroplast ultrastructure in a mutant of *Arabidopsis* deficient in lipid desaturation. *Plant Physiology*, **90**, 1134–1142.
Inatsugi, R., Nakamura, M. and Nishida, I. (2002) Phosphatidylcholine biosynthesis at low temperature: differential expression of CTP: phosphorylcholine cytidylyltransferase isogenes in *Arabidopsis thaliana*. *Plant & Cell Physiology*, **43**, 1342–1350.
Ishiguro, S., Kawai-Oda, A., Ueda, J., Nishida, I. and Okada, K. (2001) The *DEFECTIVE IN ANTHER DEHISCENCE1* gene encodes a novel phospholipase A1 catalyzing the initial step of pollen maturation, anther dehiscence, and flower opening in Arabidopsis. *Plant Cell*, **13**, 2191–2209.
Ishizaki, O., Nishida, I., Agata, K., Eguchi, G. and Murata, N. (1988) Cloning and nucleotide sequence of cDNA for the plastid glycerol-3-phosphate acyltransferase from squash. *FEBS Letters*, **238**, 424–430.
Jarvis, P., Dörmann, P., Peto, C.A., Lutes, J., Benning, C. and Chory, J. (2000) Galactolipid deficiency and abnormal chloroplast development in the *Arabidopsis* MGD synthase 1 mutant. *Proceedings of the National Academy of Sciences USA*, **97**, 8175–8179.
Johnson, M.D. and Sussex, I.M. (1995) 1 L-myo-lnositol 1-phosphate synthase from Arabidopsis thaliana. *Plant Physiology*, **107**, 613–619.
Jorasch, P., Warnecke, D.C., Lindner, B., Zähringer, U. and Heinz, E. (2000) Novel processive and nonprocessive glycosyltransferases from *Staphylococcus aureus* and *Arabidopsis thaliana* synthesize glycoglycerolipids, glycophospholipids, glycosphingolipids and glycosylsterols. *European Journal of Biochemistry*, **267**, 3770–3783.
Jordan, P., Fromme, P., Witt, H.T., Klukas, O., Saenger, W. and Krauss, N. (2001) Three-dimensional structure of cyanobacterial photosystem I at 2.5 Å resolution. *Nature*, **411**, 909–917.
Joyard, J., Block, M.A., Malherbe, A., Maréchal, E. and Douce, R. (1993) Origin and synthesis of galactolipid and sulfolipid head groups, in *Lipid Metabolism in Plants* (ed. T.S. Moore Jr.) CRC Press, Boca Raton, FL, pp. 231–258.
Joyard, J., Teyssier, E., Miège, C. *et al.* (1998) The biochemical machinery of plastid envelope membranes. *Plant Physiology*, **118**, 715–723.

Justin, A.-M., Kader, J.-C. and Collin, S. (2002) Phosphatidylinositol synthesis and exchange of the inositol head are catalysed by the single phosphatidylinositol synthase from *Arabidopsis*. *European Journal of Biochemistry*, **269**, 2347–2352.

Kang, L., Li, J., Zhao, T. *et al.* (2003) Interplay of the *Arabidopsis* nonhost resistance gene *NHO1* with bacterial virulence. *Proceedings of the National Academy of Sciences USA*, **100**, 3519–3524.

Kates, M. (1955) Hydrolysis of lecithin by plant plastid enzymes. *Canadian Journal of Biochemistry and Physiology*, **33**, 575–589.

Kato, T., Morita, M.T., Fukaki, H. *et al.* (2002) SGR2, a phospholipase-like protein, and ZIG/SGR4, a SNARE, are involved in the shoot gravitropism of Arabidopsis. *Plant Cell*, **14**, 33–46.

Kelly, A.A. and Dörmann, P. (2002). *DGD2*, an *Arabidopsis* gene encoding a UDP-galactose dependent digalactosyldiacylglycerol synthase is expressed during growth under phosphate limiting conditions. *Journal of Biological Chemistry*, **277**, 1166–1173.

Kelly, A.A., Froehlich, J.E. and Dörmann, P. (2003) Disruption of the two digalactosyldiacylglycerol synthase genes *DGD1* and *DGD2* in Arabidopsis reveals the existence of an additional enzyme of galactolipid synthesis. *Plant Cell*, **15**, 2694–2706.

Kinney, A.J. (1993) Phospholipid head groups, in *Lipid Metabolism in Plants* (ed. T.S. Moore Jr.), CRC Press, Boca Raton, FL, pp. 259–284.

Klenk, E. and Sakai, R. (1939) Inositmonophosphorsäure, ein Spaltprodukt der Sojabohnenphosphatide (14. Mitteilung über Phosphatide). *Hoppe-Seyler's Zeitschrift fur Physiologische Chemie*, **258**, 33–38.

Klukas, W.-D. Schubert, P.J., Krauß, N., Frommei, P., Witti, H.T. and Saenger, W. (1999) Localization of two phylloquinones, $Q_K$ and $Q'_K$, in an improved electron density map of photosystem I at 4-Å resolution. *Journal of Biological Chemistry*, **274**, 7361–7367.

Kodama, H., Hamada, T., Horiguchi, C., Nishimura, M. and Iba, K. (1994) Genetic enhancement of cold tolerance by expression of a gene for chloroplast omega-3 fatty acid desaturase in transgenic tobacco. *Plant Physiology*, **105**, 601–605.

Kojima, M., Seki, K., Ohnishi, M., Ito, S. and Fujino, Y. (1990) Structure of novel glyceroglycolipids in Adzuki bean (*Vigna angularis*) seeds. *Biochemistry and Cell Biology*, **68**, 59–64.

Könings, B. and Heinz, E. (1974) Investigation of some enzymatic activities contributing to the biosynthesis of galactolipid precursors in *Vicia faba*. *Planta*, **118**, 159–169.

Kopka, J., Ludewig, M. and Müller-Röber, B. (1997) Complementary DNAs encoding eukaryotic-type cytidine-5′-diphosphate-diacylglycerol synthases of two plant species. *Plant Physiology*, **113**, 997–1002.

Kühlbrandt, W., Wang, D.N. and Fujiyoshi, Y. (1994) Atomic model of plant light harvesting complex by electron crystallography. *Nature*, **367**, 614–621.

Kunst, L., Browse, J. and Somerville, C. (1988) Altered regulation of lipid biosynthesis in a mutant of *Arabidopsis* deficient in chloroplast glycerol-3-phosphate acyltransferase activity. *Proceedings of the National Academy of Sciences USA*, **85**, 4143–4147.

Kunst, L., Browse, J. and Somerville, C. (1989) Enhanced thermal tolerance in a mutant of Arabidopsis deficient in palmitic acid unsaturation. *Plant Physiology*, **91**, 401–408.

Lee, H.Y., Bahn, S.C., Kang, Y.-M. *et al.* (2003) Secretory low molecular weight phospholipase A2 plays important roles in cell elongation and shoot gravitropism in Arabidopsis. *Plant Cell*, **15**, 1990–2002.

Lightner, J., James D.W., Jr., Dooner, H.K. and Browse, J. (1994) Altered body morphology is caused by increased stearate levels in a mutant of *Arabidopsis*. *Plant Physiology*, **106**, 1443–1451.

Loewus, F.A. and Murthy, P.P.N. (2000) *myo*-Inositol metabolism in plants. *Plant Science*, **150**, 1–19.

Matos, A.R., d'Arcy-Lameta, A., França, M. *et al.* (2001) A novel patatin-like gene stimulated by drought stress encodes a galactolipid acyl hydrolase. *FEBS Letters*, **491**, 188–192.

May, S., Preisig-Müller, R., Höne, M., Grau, P. and Kindl, H. (1998) A phospholipase A2 is transiently synthesized during seed germination and localized to lipid bodies. *Biochimica et Biophysica Acta*, **1393**, 267–276.

McConn, M. and Browse, J. (1996) The critical requirement for linolenic acid is pollen development, not photosynthesis, in an Arabidopsis mutant. *Plant Cell*, **8**, 403–416.
McConn, M. and Browse, J. (1998) Photosynthetic membranes are required for photosynthetic competence in a mutant of *Arabidopsis*. *Plant Journal*, **15**, 521–530.
McCourt, P., Browse, J., Watson, J., Arntzen, C.J. and Somerville, C.R. (1985) Analysis of photosynthetic antenna function in a mutant of *Arabidopsis thaliana* (L.) lacking trans-hexadecenoic acid. *Plant Physiology*, **78**, 853–858.
MacFarlane, M.G. (1958) Structure of cardiolipin. *Nature*, **182**, 946.
Meijer, H.J.G. and Munnik, T. (2003) Phospholipid-based signaling in plants. *Annual Review of Plant Biology*, **54**, 265–306.
Miège, C., Maréchal, E., Shimojima, M., Awai, K., Block, M.A., Ohta, H., Takamiya, K.-i., Douce, R. and Joyard, J. (1999) Biochemical and topological properties of type A MGDG synthase, a spinach chloroplast envelope enzyme catalyzing the synthesis of both prokaryotic and eukaryotic MGDG. *European Journal of Biochemistry*, **265**, 990–1001.
Minnikin, D.E., Abdolrahimyadeh, H. and Baddiley, J. (1974) Replacement of acidic phospholipids by acidic glycolipids in *Pseudomonas diminuta*. *Nature*, **249**, 268–269.
Miquel, M., Cassagne, C. and Browse, J. (1998) A new class of Arabidopsis mutants with reduced hexadecatrienoic acid fatty acid levels. *Plant Physiology*, **117**, 923–930.
Miquel, M., James, D., Dooner, H. and Browse, J. (1993) *Arabidopsis* requires polyunsaturated lipids for low-temperature survival. *Proceedings of the National Academy of Sciences USA*, **90**, 6208–6212.
Mongrand, S., Bessoule, J.-J., Cabantous, F. and Cassagne, C. (1998) The C16:3/C18:3 fatty acid balance in photosynthetic tissues from 468 plant species. *Phytochemistry*, **49**, 1049–1064.
Mongrand, S., Cassagne, C. and Bessoule, J.-J. (2000) Import of lyso-phosphatidylcholine into chloroplasts likely at the origin of eukaryotic plastidial lipids. *Plant Physiology*, **122**, 845–852.
Moon, B.Y., Higahis, S.-I., Gombos, Z. and Murata, N. (1995) Unsaturation of the membrane lipids of chloroplasts stabilizes the photosynthetic machinery against low-temperature photoinhibition in transgenic tobacco plants. *Proceedings of the National Academy of Sciences USA*, **92**, 6219–6223.
Mou, Z., Wang, X., Fu, Z. *et al.* (2002) Silencing of phosphoethanolamine *N*-methyltransferase results in temperature-sensitive male sterility and salt hypersensitivity in Arabidopsis. *Plant Cell*, **14**, 2031–2043.
Müller, F. and Frentzen, M. (2001) Phosphatidylglycerophosphate synthases from *Arabidopsis thaliana*. *FEBS Letters*, **509**, 298–302.
Mueller-Roeber, B. and Pical, C. (2002) Inositol phospholipid metabolism in Arabidopsis. Characterized and putative isoforms of inositol phospholipid kinase and phosphoinositide-specific phospholipase C. *Plant Physiology*, **130**, 22–46.
Mulichak, A.M., Theisen, M.J., Essigmann, B., Benning, C. and Garavito, R.M. (1999) Crystal structure of SQD1, an enzyme involved in the biosynthesis of the plant sulfolipid headgroup donor UDP-sulfoquinovose. *Proceedings of the National Academy of Sciences USA*, **96**, 13097–13102.
Murakami, Y., Tsuyama, M., Kobayashi, Y., Kodama, H. and Iba, K. (2000) Trienoic fatty acids and plant tolerance of high temperature. *Science* **287**, 476–479.
Murata, N. (1983) Molecular species composition of phosphatidylglycerols from chilling-sensitive and chilling-resistant plants. *Plant & Cell Physiology*, **24**, 81–86.
Murata, N., Ishizaki-Nishizawa, O., Higashi, S., Hayashi, H., Tasaka, Y. and Nishida, I. (1992) Genetically engineered alteration in the chilling sensitivity of plants. *Nature*, **356**, 710–713.
Murata, N., Sato, N., Takahashi, N. and Hamazaki, Y. (1982) Compositions and positional distributions of fatty acids in phospholipids from leaves of chilling-sensitive and chilling-resistant plants. *Plant & Cell Physiology*, **23**, 1071–1079.
Nishida, I., Swinhoe, R., Slabas, A.R. and Murata, N. (1996) Cloning of *Brassica napus* CTP:phosphocholine cytidylyltransferase cDNAs by complementation in a yeast *cct* mutant. *Plant Molecular Biology*, **31**, 205–211.

Nowicki, M., Müller, F. and Frentzen, M. (2003) Cardiolipin synthase of *Arabidopsis thaliana*. 1st European symposium on plant lipids, Aachen, Germany.

Nuccio, M.L., Ziemak, M.J., Henry, S.A., Weretilnyk, E.A. and Hanson, A.D. (2000) cDNA Cloning of phosphoethanolamine *N*-methyltransferase from spinach by complementation in *Schizosaccharomyces pombe* and characterization of the recombinant enzyme. *Journal of Biological Chemistry*, **275**, 14095–14101.

Nußberger, S., Dörr, K., Wang, N. and Kühlbrandt, W. (1993) Lipid-protein interactions in crystals of plant light-harvesting complex. *Journal of Molecular Biology*, **234**, 347–356.

Ohlrogge, J.B., Kuhn, D.N. and Stumpf, P.K. (1979) Subcellular localization of acyl carrier protein in leaf protoplasts of *Spinacia oleracea*. *Proceedings of the National Academy of Sciences USA*, **76**, 1194–1198.

Pangborn, M.C. (1942) Isolation and purification of a serologically active phospholipid from beef heart. *Journal of Biological Chemistry*, **143**, 247–256.

Pierrugues, O., Brutesco, C., Oshiro, J. *et al.* (2001) Lipid phosphate phosphatases in *Arabidopsis*. Regulation of the *AtLPP1* gene in response to stress. *Journal of Biological Chemistry*, **276**, 20300–20308.

Pizer, F.L. and Ballou, C.E. (1959) Studies on myo-inositol phosphates of natural origin. *Journal of the American Chemical Society*, **81**, 915–921.

Pollard, M. and Ohlrogge, J. (1999) Testing models of fatty acid transfer and lipid synthesis in spinach leaf using *in vivo* oxygen-18 labeling. *Plant Physiology*, **121**, 1217–1226.

Qin, C. and Wang, X. (2002) The Arabidopsis phospholipase D family. Characterization of a Calcium-independent and phosphatidylcholine-selective PLDζ1 with distinct regulatory domains. *Plant Physiology*, **128**, 1057–1068.

Raetz, C.R.H. and Dowhan, W. (1990) Biosynthesis and function of phospholipids in *Escherichia coli*. *Journal of Biological Chemistry*, **265**, 1235–1238.

Reifarth, F., Christen, G., Seeliger, A.G., Dörmann, P., Benning, C. and Renger, G. (1997) Modification of the water oxidizing complex in leaves of the *dgd1* mutant of *Arabidopsis thaliana* deficient in the galactolipid digalactosyldiacylglycerol. *Biochemistry*, **36**, 11769–11776.

Reinsberg, D., Booth P.J., Jegerschöld, C., Khoo, B.J. and Paulsen, H. (2000) Folding, assembly and stability of the major light-harvesting complex of higher plants, LHCII, in the presence of native lipids. *Biochemistry*, **39**, 14305–14313.

Reiter, W.-D. and Vanzin, G.F. (2001) Molecular genetics of nucleotide sugar interconversion pathways in plants. *Plant Molecular Biology*, **47**, 95–113.

Rhee K.-H., Morris, E.P., Barber, J. and Kühlbrandt, W. (1997) Two-dimensional structure of plant photosystem II at 8 Å resolution. *Nature*, **389**, 522–526.

Rontein, D., Nishida, I., Tashiro, G. *et al.* (2001) Plants synthesize ethanolamine by direct decarboxylation of serine using a pyridoxal phosphate enzyme. *Journal of Biological Chemistry*, **276**, 35523–35529.

Rontein, D., Wu, W.-I., Voelker, D.R. and Hanson, A.D. (2003) Mitochondrial phosphatidylserine decarboxylase from higher plants. Functional complementation in yeast, localization in plants, and overexpression in Arabidopsis. *Plant Physiology*, **132**, 1678–1687.

Roughan, P.G. (1970) Turnover of the glycerolipids of pumpkin leaves. The importance of phosphatidylcholine. *Biochemical Journal*, **117**, 1–8.

Roughan, P.G. and Slack, C.R. (1982) Cellular organization of glycerolipid metabolism. *Annual Review of Plant Physiology*, **33**, 97–132.

Sanda, S., Leustek, T., Theisen, M.J., Garavito, R.M. and Benning, C. (2001) Recombinant *Arabidopsis* SQD1 converts UDP-glucose and sulfite to the sulfolipid head group precursor UDP-sulfoquinovose *in vitro*. *Journal of Biological Chemistry*, **276**, 3941–3946.

Sastry, P.S. and Kates, M. (1964) Hydrolysis of monogalactosyl and digalactosyl diglycerides by specific enzymes in runner-bean leaves. *Biochemistry*, **3**, 1280–1287.

Schnurr, J.A., Shockey, J.M., de Boer, G.-J. and Browse, J.B. (2002) Fatty acid export from the chloroplast. Molecular characterization of a major plastidial acyl-Coenzyme A synthetase from Arabidopsis. *Plant Physiology*, **129**, 1700–1709.

Seifert, G.J., Barber, C., Wells, B., Dolan, L. and Roberts, K. (2002) Galactose biosynthesis in *Arabidopsis*: genetic evidence for substrate channeling from UDP-D-galactose into cell wall polymers. *Current Biology*, **12**, 1840–1845.

Shen, W., Wei, Y., Dauk, M., Zheng, Z. and Zou, J. (2003) Identification of a mitochondrial glycerol-3-phosphate dehydrogenase from *Arabidopsis thaliana*: evidence for a mitochondrial glycerol-3-phosphate shuttle in plants. *FEBS Letters*, **536**, 92–96.

Shimojima, M., Ohta, H., Iwamatsu, A., Masuda, T., Shioi, Y. and Takamiya, K.-I. (1997) Cloning of the gene for monogalactosyldiacylglycerol synthase and its evolutionary origin. *Proceedings of the National Academy of Sciences USA*, **94**, 333–337.

Shin, S. and Moore, T.S., Jr. (1990) Phosphatidylethanolmaine synthesis by castor bean endosperm: a base exchange reaction. *Plant Physiology*, **93**, 148–153.

Shockey, J.M., Fulda, M.S. and Browse, J.A. (2002) Arabidopsis contains nine long-chain acyl-Coenzyme a synthetase genes that participate in fatty acid and glycerolipid metabolism. *Plant Physiology*, **129**, 1710–1722.

Siebertz, H.P., Heinz, E., Linscheid, M., Joyard, J. and Douce, R. (1979) Characterization of lipids from chloroplast envelopes. *European Journal of Biochemistry*, **101**, 429–438.

Ståhl, U., Ek, B. and Stymne, S. (1998) Purification and characterization of a low-molecular-weight phospholipase A2 from developing seeds of elm. *Plant Physiology*, **117**, 197–205.

Ståhl, U., Lee, M., Sjödahl, S. *et al.* (1999) Plant low-molecular-weight phospholipase A2s (PLA2s) are structurally related to the animal secretory PLA2s and are present as a family of isoforms in rice (*Oryza sativa*). *Plant Molecular Biology*, **41**, 481–490.

Stelmach, B.A., Müller, A., Hennig, P., Gebhardt, S., Schubert-Zsilavecz, M. and Weiler, E.W. (2001) A novel class of oxylipins, *sn*1-*O*-(12-oxophytodienoyl)-*sn*2-O-(hexadecatrienoyl)-monogalactosyl diglyceride, from *Arabidopsis thaliana*. *Journal of Biological Chemistry*, **276**, 12832–12838.

Tang, F. and Moore, T.S., Jr. (1997) Enzymes of the primary phosphatidylethanolamine biosynthetic pathway in postgermination castor bean-endosperm. *Plant Physiology*, **115**, 1589–1597.

Terasaki, M. and Itabashi, Y. (2003) Glycerolipid acyl hydrolase activity in the brown alga *Cladosiphon okamuranus* TOKIDA. *Bioscience, Biotechnology, and Biochemistry*, **67**, 1986–1989.

Uemura, M., Joseph, R.A. and Steponkus, P.L. (1995) Cold acclimation of *Arabidopsis thaliana*. *Plant Physiology*, **109**, 15–30.

van Besouw, A. and Wintermans, J.F.G.M. (1978) Galactolipid formation in chloroplast envelopes. I. Evidence for two mechanisms in galactosylation. *Biochimica et Biophysica Acta*, **529**, 44–53.

Vijayan, P., Routaboul, J.-M. and Browse, J. (1998) A genetic approach to investigating membrane lipid structure and function, in *Lipids in Photosynthesis: Structure, Function and Genetics* (eds P.A. Siegenthaler and N. Murata), Kluwer Academic Press (Dordrecht, The Netherlands), pp. 263–285.

Wada, H., Shintani, D. and Ohlrogge, J. (1997) Why do mitochondria synthesize fatty acids? Evidence for involvement in lipoic acid production. *Proceedings of the National Academy of Sciences USA*, **94** 1591–1596.

Wallis, J.G. and Browse, J. (2002) Mutants of *Arabidopsis* reveal many roles for membrane lipids. *Progress in Lipid Research*, **41**, 254–278.

Wang, X. (2001) Plant phospholipases. *Annual Review of Plant Physiology and Plant Molecular Biology*, **52**, 211–231.

Wang, X., Wang, C., Sang, Y., Qin, C. and Welti, R. (2002) Networking of phospholipases in plant signal transduction. *Physiologia Plantarum*, **115**, 331–335.

Weber, H. (2002) Fatty acid-derived signals in plants. *Trends in Plant Science*, **7**, 217–224.

Wei, Y., Periappuram, C., Datla, R., Selvaraj, G. and Zou, J. (2001) Molecular and biochemical characterizations of a plastidic glycerol-3-phosphate dehydrogenase from *Arabidopsis*. *Plant Physiology and Biochemistry*, **39**, 841–848.

Wolter, F.P., Schmidt, R. and Heinz, E. (1992) Chilling sensitivity of *Arabidopsis thaliana* with genetically engineered membrane lipids. *EMBO Journal*, **11**, 4685–4692.

Wu., J. and Browse, J. (1995) Elevated levels of high-melting-point phosphatidylglycerols do not induce chilling sensitivity in an Arabidopsis mutant. *Plant Cell*, **7**, 17–27.

Xu, C., Fan, J., Riekhof, W., Froehlich, J.E. and Benning, C. (2003) A permease-like protein involved in ER to thylakoid lipid transfer in *Arabidopsis*. *EMBO Journal*, **22**, 2370–2379.

Xu, C., Härtel, H., Wada, H. *et al.* (2002) The *pgp1* mutant locus of Arabidopsis encodes a phosphatidylglycerol-phosphate synthase with impaired activity. *Plant Physiology*, **129**, 594–604.
Xue, H.-W., Hosaka, K., Plesch, G. and Mueller-Roeber, B. (2000) Cloning of *Arabidopsis thaliana* phosphatidylinositol synthase and functional expression in the yeast *pis* mutant. *Plant Molecular Biology*, **42**, 757–764.
Yagi, T. and Benson, A.A. (1962) Plant sulfolipid. V. Lysosulfolipid formation. *Biochimica et Biophysica Acta*, **57**, 601–603.
Yoshida, S. (1984) Chemical and biophysical changes in the plasma membrane during cold acclimation of mulberry bark cells (*Morus bombycis* Koidz. cv Goroji). *Plant Physiology*, **76**, 257–265.
Yu, B., Xu, C. and Benning, C. (2002) *Arabidopsis* disrupted in *SQD2* encoding sulfolipid synthase is impaired in phosphate-limited growth. *Proceedings of the National Academy of Sciences USA*, **99**, 5732–5737.
Zouni, A., Witt, H.-T., Ker, J. *et al.* (2001) Crystal structure of photosystem II from *Synechococcus elongatus* at 3.5 Å resolution. *Nature*, **409**, 739–743.

# 6 Storage lipids

Randall J. Weselake

## 6.1 Introduction

Plant storage lipids represent a highly reduced form of carbon that can serve as a source of energy during germination and early seedling development. The storage lipids of oleaginous crops also serve as a source of edible oils for human consumption, for both adequate nutrition and nutraceutic value. Diets having a low amount of fat but containing a high ratio of polyunsaturated:saturated fatty acids (FAs) have been shown to reduce the incidence of cardiovascular disease, cancer and diabetes (Broun *et al.*, 1999). Global production of oils and fats amounts to about 120 megatonnes. About 20 megatonnes of this is derived from vegetable oils, with soybean, oil palm, oilseed rape and sunflower accounting for about 65% of this production (Gunstone, 2001). Most of this lipid is used for human consumption and about 15% of it is used by the chemical industry for soaps and detergents. As petrochemical reserves begin to diminish and we become more environmentally conscious, there will be an increasing demand for plant oils and their derivatives as substitutes for petrochemicals in industrial applications, such as biolubricants, biofuels, nylon precursors and detergent feedstocks. Although major oilseed crops are composed mainly of palmitic acid (16:0), stearic acid (18:0), oleic acid (18:1$cis\Delta^{9}$), linoleic acid (18:2$cis\Delta^{9,12}$) and α-linolenic acid (α-18:3) (18:3$cis\Delta^{9,12,15}$), a number of unusual and novel FAs with limited natural occurrence can be found in exotic plant species, which might serve as a source of genes for transformation of major oilseed crops to produce triacylglycerols (TAGs) containing industrially useful FAs (Jaworski and Cahoon, 2003).

TAGs and the liquid wax of the jojoba (*Simmondsia chinensis*) (Voelker and Kinney, 2001) are the major storage lipids found in plants. Structures for these lipids are depicted in Figure 6.1. TAGs consist of three FA chains esterified via the hydroxyl groups of a glycerol backbone (Figure 6.1A). The carbon atoms can be distinguished from one another because glycerol lacks rotational symmetry (Gurr, 1980). Stereospecific numbering can be used to label the three positions on the glycerol derivative as *sn*-1, *sn*-2 and *sn*-3 from top to bottom, with the secondary hydroxyl positioned to the left of the central carbon (IUPAC-IUP, 1967; Brockerhoff, 1971). Based on the analyses of numerous seed oils, there is a tendency for saturated FAs to occupy the *sn*-1 position, whereas unsaturated FAs tend to occupy the *sn*-2 position (Gurr, 1980; Norton and Harris, 1983; Rossell *et al.*, 1985). More variation in FA composition occurs at the *sn*-3 position. In contrast, the liquid wax of the jojoba

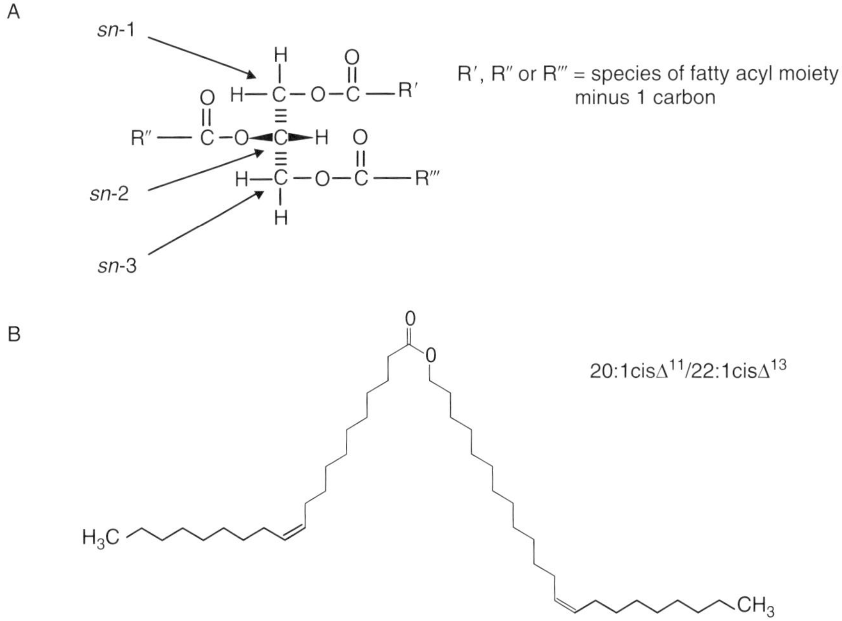


**Figure 6.1** General structure of triacylglycerol (A) and a molecular species of a liquid wax ester (B).

seed is composed of esters of very long-chain monounsaturated FAs and alcohols (Figure 6.1B).

This chapter mainly deals with TAG biosynthesis in higher plants and the intimate relationship of this process with membrane metabolism. Although TAG biosynthesis has been studied mainly in the endoplasmic reticulum (ER), the chapter also features some discussion on the enzymology of TAG accumulation in plastids. Information based on TAG biosynthesis in microorganisms and mammals will be discussed where the findings have been relevant to advances in understanding storage lipid metabolism in higher plants. The properties of the enzymes involved in this process are discussed with a focus on substrate specificity/selectivity and regulation. In some cases, genetic engineering has had a considerable impact on TAG composition through inter-species gene transfer. Plant tissue and cell culture have been particularly useful in advancing our knowledge of the biochemistry and molecular biology of TAG accumulation (Weselake and Taylor, 1999; Weselake, 2000, 2002). Culture systems have the advantage of providing a continuous supply of fresh oil-forming material as opposed to maturing seeds, which have to be harvested at specific stages of development to obtain membrane fractions active in oil synthesis. TAGs appear to be produced in specialized regions of the ER leading to the formation of microscopic oil droplets or lipid bodies. The review moves on to examine research in this area along with discussion of the characteristics, biogenesis and composition of oil bodies. Since TAG serves as an energy store, the chapter also presents information on TAG mobilization, and an overview of β-oxidation of FAs and conversion of lipid to carbohydrate, highlighting some of the recent advances in these areas. Thereafter, the chapter deals with storage lipids in developing pollen grains followed by a discussion of environmental effects on TAG metabolism. The final three sections of the chapter deal with TAG metabolism during membrane turnover, liquid wax as a storage lipid and possible transport of lipids in phloem.

## 6.2 Pathways leading to triacylglycerols

A generalized scheme for pathways leading to TAG in developing oleaginous seeds is depicted in Figure 6.2. Both acyl-CoA-dependent and acyl-CoA-independent

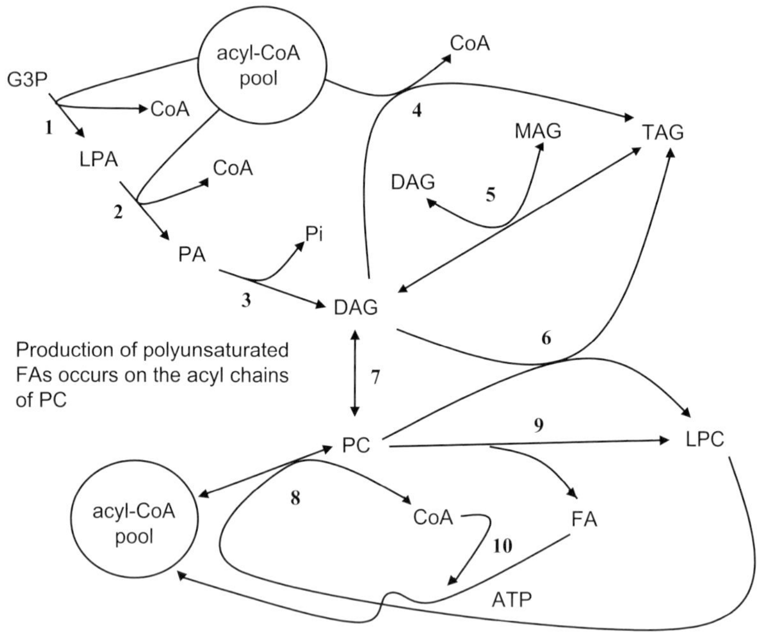


**Figure 6.2** Generalized scheme depicting pathways leading to triacylglycerol (TAG), and the relationship between TAG synthesis and membrane metabolism. Enzymes: (1) *sn*-glycerol-3-phosphate acyltransferase, (2) lysophosphatidic acid acyltransferase, (3) phosphatidate phosphatase, (4) diacylglycerol acyltransferase, (5) diacylglycerol transacylase, (6) phospholipid:diacylglycerol acyltransferase, (7) CDP-choline:1,2-diacylglycerol cholinephosphotransferase, (8) lysophosphatidylcholine acyltransferase, (9) phospholipase $A_2$, (10) acyl-CoA synthetase (EC 6.2.1.3). Additional abbreviations: DAG, *sn*-1,2-diacylglycerol; G3P, *sn*-glycerol-3-phosphate; LPA, lysophosphatidic acid; LPC, lysophosphatidylcholine; MAG, monoacylglycerol; PA, phosphatidate; PC, phosphatidylcholine. Adapted from the following references: Stymne and Stobart (1987), Mancha and Stymne (1997), Stobart *et al.* (1997), Banás *et al.* (2000), Dahlqvist *et al.* (2000), Lin and McKeon (2000).

reactions that involve microsomal enzymes are known to lead to the formation of TAG (Stymne and Stobart, 1987; Weselake, 2002), although the extent to which these reactions occur in different oilseeds and at different stages of seed development may be variable. *sn*-Glycerol-3-phosphate (G3P) is generally regarded as the starting point of the G3P or Kennedy (1961) pathway leading to TAG. G3P can be produced from dihydroxyacetone phosphate (DHAP) through the action of L-G3P:$NAD^+$ oxidoreductase (EC 1.1.18) (Finlayson and Dennis, 1980). Lysophosphatidic acid (LPA) is produced from G3P and acyl-CoA through the action of *sn*-G3P acyltransferase (GPAT, EC 2.3.1.15). A second acyl-CoA-dependent acylation catalyzed by lysophosphatidic acid acyltransferase (LPAAT, EC 2.3.1.51) leads to the formation of phosphatidate (PA). The dephosphorylation of PA catalyzed by PA phosphatase (EC 3.1.3.4) then leads to production of *sn*-1,2-diacylglycerol (DAG). TAG is then produced through a third acyl-CoA-dependent acylation catalyzed by diacylglycerol acyltransferase (DGAT, EC 2.3.1.20). In a reaction catalyzed by cytidine diphosphate (CDP)-choline:1,2-diacylglycerol cholinephosphotransferase (CPT, EC 2.7.8.2), DAG can also accept a phosphocholine from CDP-choline to generate phosphatidylcholine (PC) and cytidine monophosphate. The CPT-catalyzed reaction has been shown to be a reversible process that could potentially regenerate a remodeled DAG for use by DGAT in the mainstream of TAG biosynthesis. Formation of PC containing polyunsaturated FAs occurs via the action of FAD2 and FAD3 desaturases, which use PC as substrate (Ohlrogge and Browse, 1995). The DAG component of PC can also undergo acyl exchange at the *sn*-2 position catalyzed through the forward and reverse reactions of lysophosphatidylcholine acyltransferase (LPCAT, EC 2.3.1.23). The action of phospholipase $A_2$ ($PLA_2$, EC 3.1.1.4), also known as phosphatide 2-acylhydrolase (Ståhl *et al.*, 1995; Wang, 2001), could potentially lead to lysophosphatidylcholine (LPC) for utilization in the forward reaction of LPCAT through removal of the middle FA from PC.

More recently, it has been demonstrated that DAG and PC can also serve as FA donors in non-acyl-CoA-dependent synthesis of TAG. Diacylglycerol transacylase (DGTA) catalyzes the transfer of an FA between two DAG molecules to generate TAG and monoacylglycerol (MAG) (Stobart *et al.*, 1997). The process might be reversible, providing a mechanism to remodel TAG (Mancha and Stymne, 1997). Phospholipid:diacylglycerol acyltransferase (PDAT, EC 2.3.1.158) catalyzes the formation of TAG and LPC from PC and DAG (Dahlqvist *et al.*, 2000; Oelkers *et al.*, 2000). Thus, PDAT action could also potentially provide a substrate for the forward reaction of LPCAT. In another recent study, a minor route for TAG formation was identified in yeast (*Saccharomyces cerevisiae*) that involved the acylation of DAG with free FA (Sorger and Daum, 2002). Thus far, this activity has not been reported in oilseeds. The potential deacylation–reacylation reactions affecting PC, acyl-exchange at the *sn*-2 position of PC and production of polyunsaturated FAs at the level of PC represent processes that influence the molecular species of *sn*-1,2-DAG that can return to the mainstream of the G3P pathway via the reverse reaction of CPT.

In a recent study with developing peanut (*Arachis hypogaea*) cotyledons, a pathway for DAG production involving soluble enzymes was suggested; this involved the sequential action of LPA phosphatase and acyl-CoA:monoacylglycerol acyltransferase (MGAT, EC 2.3.1.22) (Tumaney *et al.*, 2001; Shekar *et al.*, 2002). LPA phosphatase catalyzes the dephosphorylation of LPA to produce MAG, which is the substrate for MGAT. The existence of this pathway in other plant systems, however, has not been reported. Shekar *et al.* (2002) have suggested that the DAG generated through the action of these soluble enzymes could potentially be utilized by either microsomal DGAT or a cytosolic DGAT such as the 10 S TAG biosynthetic multienzyme complex from oleaginous yeast (*Rhodotorula glutinis*) (Gangar *et al.*, 2001). The complex utilized both free FAs and acyl-CoAs. MAG for DAG production could conceivably be obtained through the direct acylation of glycerol as recently described by Lee *et al.* (2001) in studies with pig heart microsomes. Acyl-CoA:glycerol acyltransferase activity, however, has not been reported in a plant system.

## 6.3 Properties and regulation of enzymes involved in triacylglycerol biosynthesis and associated phospholipid metabolism

### *6.3.1 sn-Glycerol-3-phosphate acyltransferase*

Microsomal GPATs from oil-forming tissues have been shown to utilize a range of acyl-CoA species with many oilseed species having enhanced specificity and selectivity for 16:0-CoA (Ichihara, 1984; Griffiths *et al.*, 1985; Bernerth and Frentzen, 1990; Manaf and Harwood, 2000; Weselake, 2002), which in many cases accounts for the tendency of the *sn*-1 position of TAG to be occupied by saturated FAs (Gurr, 1980; Norton and Harris, 1983; Rossell *et al.*, 1985). Specificity determinations are based on assays with one acyl-CoA species at a time whereas selectivity refers to the preference of an acyltransferase for a particular acyl-CoA species within a mixture of two or more different species of acyl-CoA. Selectivity studies are more reflective of the *in vivo* situation where cellular acyltransferase action occurs in an environment containing a mixture of endogenous acyl-CoA species. It is often difficult to assay microsomal GPAT directly because of the considerably higher activity of LPAAT, which rapidly consumes the LPA generated by the GPAT-catalyzed reaction. Subsequently, some investigators have resorted to determining GPAT activity based on the analysis of the total FAs at the *sn*-1 positions of LPA and PA (Bernerth and Frentzen, 1990; Stobart and Stymne, 1990). The presence of 5 mM EDTA in GPAT reaction mixtures appears to increase the enzyme activity (Eccleston and Harwood, 1995; Manaf and Harwood, 2000). Higher plant microsomal GPATs have been solubilized and partially purified (Frentzen, 1990; Eccleston and Harwood, 1995; Manaf and Harwood, 2000). The microsomal enzyme, however, has been purified 1308-fold from the membrane fraction of *Mortierella ramanniana* var. angulispora (Mishra and Kamisaka, 2001). The thiol-reagent-sensitive enzyme displayed a molecular mass of 45 kDa based on SDS-PAGE and was strictly dependent on anionic phospholipids for activity.

The first genes, designated *GAT1* and *GAT2*, encoding microsomal GPATs (Gat1p and Gat2p, respectively) were cloned from yeast (*S. cerevisiae*) using a molecular genetic approach (Zheng and Zou, 2001). Identification of the candidate genes involved detecting the presence of deduced amino acid sequence segments with similarities to conserved domains of known acyltransferases. In both cases, disruption of the genes resulted in reduced GPAT activity. *GAT1* was functionally expressed in *Escherichia coli* (*plsB*) strain BB26-36, which is known to have low GPAT activity. Expression of *GAT2* in *E. coli*, however, was deleterious to the host cells. The genes were also expressed, using yeast expression vector pYES2, in the yeast mutant *gat1*Δ, which showed low residual GPAT activity. Both Gat1p and Gat2p utilized G3P and DHAP as acyl acceptors. Gat1p was capable of effectively utilizing 16:0-, 16:1-, 18:0- and 18:1-CoA, whereas Gat2p exhibited a distinct preference for acyl-CoAs consisting of 16-carbon FAs.

Recently, Zheng *et al.* (2003) used a series of BLAST searches based on a partial sequence encompassing a region conserved between yeast GPATs and other FA acyltransferases to identify a gene family in *Arabidopsis thaliana*, designated *AtGPAT*. The deduced amino acid sequences of AtGPAT1–AtGPAT7, all contained the four previously defined acyltransferase domains. The *AtGPAT* cDNAs were cloned by reverse transcriptase-mediated PCR. Northern blotting studies indicated that *AtGPAT1* transcript accumulated at high levels in developing siliques and flower buds, whereas *AtGPAT6* mainly accumulated in flower buds. *AtGPAT3* was expressed at low levels in various tissues and *AtGPAT4* was abundantly expressed in all tissues examined (roots, seedlings, leaves, flower buds and siliques). The remaining transcripts were not detected in any tissue. The *AtGPAT* cDNAs were expressed in yeast mutant *gat1*Δ. AtGPAT1 and AtGPATs 4–7 exhibited GPAT activity, with isoforms 2 and 3 showing no activity.

AtGPAT1, which exhibited the highest specific activity, specifically used G3P as an FA acceptor. The recombinant enzyme showed the highest preferences for 16:1- and 18:1-CoA compared to 16:0-, 18:0- and 20:1-CoA. Incorporation of 5 mM EDTA into the reaction mixture, however, shifted the relative preference markedly to 18:0-CoA clearly indicating that the assay conditions used can influence acyl-CoA specificity. AtGPAT1 had two membrane spanning segments (138–158 and 334–354) based on sequence analysis with transmembrane prediction programs. The recombinant enzyme could be incorporated into the membranes of dog pancreatic microsomes and imported into isolated pea (*Pisum sativum*) mitochondria following proteolytic cleavage of a leader sequence.

The first cDNAs that were cloned, encoding plant GPATs, were for the soluble plastidial form of the enzyme in the stroma, which utilized acyl–acyl carrier protein (ACP) as the FA donor (Ishizaki *et al.*, 1988; Murata and Tasaka, 1997). Plastidial GPAT has been purified to electrophoretic homogeneity (Douady and

Dubacq, 1987; Nishida *et al.*, 1987). The natural substrate of plastidial GPAT is acyl-ACP, although the enzyme also accepts acyl-CoAs (Frentzen, 1993). Purified plastidial GPAT was selective for 18:1 moieties over 16:0 moieties even when high ratios of 16:0-CoA/18:1-CoA prevailed in the reaction mixture, and the selectivity for 18:1 moieties was increased with acyl-ACP forms (Douady and Dubacq, 1987). A nuclear gene encodes plastidial GPAT and the preprotein formed in the cytosol is delivered to the plastid via a cleavable *N*-terminal transit peptide (Frentzen and Wolter, 1998). Jain *et al.* (2000) have increased the seed oil content of *A. thaliana* through cytosolic expression of cDNA encoding plastidial GPAT from safflower (*Carthamus tinctorius*) and a gene encoding *E. coli* GPAT.

Advanced structural and kinetic analyses have been conducted with recombinant plastidial GPAT from squash. One of the domains of the crystallized enzyme was shown to consist of an extended cleft, which was bordered with hydrophobic amino acid residues having a cluster of positive charges at one end flanked by a $H(X)_4D$ motif (Turnball *et al.*, 2001). The investigators suggested that the hydrophobic and positively charged residues represent interaction sites for acyl-CoA and the phosphate group of G3P, respectively. The $H(X)_4D$ motif was conserved among many glycerolipid acyltransferases. In kinetic studies of recombinant plastidial GPAT, the reaction was shown to proceed via a compulsory-ordered ternary complex with the binding of acyl-ACP occurring before G3P (Hayman *et al.*, 2002). Prior to this, the same research group generated a mutant squash GPAT with enhanced selectivity for 18:1-ACP (Hayman *et al.*, 2000). The enzyme from wild-type squash used 18:1- and 16:0-ACP at similar rates.

### *6.3.2 Lysophosphatidic acid acyltransferase*

Microsomal LPAAT activity has been characterized in a number of plant sources (Weselake, 2002). In general, low activities are observed with acyl-CoAs consisting of saturated FAs. The microsomal enzyme from *Brassica napus* does not effectively utilize erucoyl (*cis*$\Delta^{13}$22:1)-CoA or lauroyl (12:0)-CoA during *in vitro* assays (Sun *et al.*, 1988; Bernerth and Frentzen, 1990; Taylor *et al.*, 1991; Frentzen, 1993). The low specificity for 22:1 is in agreement with the absence of this FA at the *sn*-2 position of TAG in the seed of high-erucic acid *B. napus* (Norton and Harris, 1983). It is interesting to note that selected high-erucic acid breeding lines of *Brassica oleracea* have been shown to contain over 23% (wt) of 22:1 at the *sn*-2 position of TAG (Taylor *et al.*, 1994). The molecular species of the acyl acceptor, LPA, also influences LPAAT activity. Microsomal LPAAT from maize (*Zea mays*) and *B. napus* have also been shown to exhibit a higher preference for LPA containing 18:1 acyl chains compared to 12:0 acyl chains (Oo and Huang, 1989).

In contrast to *B. napus*, microsomal LPAAT from palm (*Butia capitata* Becc. and *Syagrus cocoides* Martius) is capable of effectively utilizing 12:0-CoA (Sun *et al.*, 1988; Oo and Huang, 1989), and microsomal LPAAT from maturing seeds of *Limnanthes alba* can utilize 22:1-CoA. Recently, Brown *et al.* (2002a) determined that microsomal LPAATs from several plant species displayed little selectivity between 18:1- and 18:2-CoA donor substrates, whereas 16:0 and 18:0 were generally discriminated against.

Knutzon *et al.* (1995) partially purified microsomal LPAAT from coconut (*Cocos nucifera*) endosperm. The enzyme preparation was enriched in a 29-kDa polypeptide, which was microsequenced to develop a probe for the cloning of an LPAAT cDNA. The cDNA was functionally expressed in *E. coli.* cDNA clones encoding LPAAT have been obtained from a number of plant species (Brown *et al.*, 1994, 1995; Hanke *et al.*, 1995; Knutzon *et al.*, 1995; Lassner *et al.*, 1995; Frenzten and Wolter, 1998). Six of the cDNAs have been shown to encode microsomal forms of the enzyme. One of the cDNAs, *pRAT1*, encoded a microsomal LPAAT with the same substrate specificity as LPAAT activity in *B. napus* microsomes (Frentzen and Wolter, 1998). The deduced amino acid sequences of the plant LPAAT cDNA clones have been compared to cDNAs encoding the enzyme from other organisms. Sequence alignments have indicated that two regions, denoted box 1 (NHXXXXD) and box 2 (FP/VEGTR), were conserved relative to their sequence and spacing (Frentzen and Wolter, 1998). Also conserved, was the spacing between the boxes. Brown *et al.* (2002b) have raised antibodies against recombinant truncations of two membrane-bound LPAATs, LAT1 and LAT2, from *Limnanthes douglasii*. The antibodies were used to monitor LAT1 and LAT2 polypeptides during seed development. The highest activity of LAT2 coincided with the maximal level of erucic acid biosynthesis in the seeds.

The inability of microsomal LPAAT from *B. napus* to accept 12:0- and 22:1-CoA has had important implications in the genetic engineering of this crop to produce oils enriched in 12:0 and 22:1, respectively (Sun *et al.*, 1988; see Weselake, 2002). *Brassica napus* enriched in 12:0 moieties could potentially provide for a source of structured TAGs for the production of cocoa-butter-like oils (Del Vecchio, 1996). Although the seed specific introduction of a medium-chain thioesterase (EC 3.1.2.14) from the California bay plant (*Umbellularia californica*) into *B. napus* resulted in oil containing 12:0, high levels could not be obtained because of the inability of the resident LPAAT to utilize 12:0-CoA (Voelker *et al.*, 1996). The LPAAT of coconut, however, was known to be specific for medium-chain FAs (Davies *et al.*, 1995). Subsequently, *B. napus* expressing the medium-chain thioesterase was crossed with *B. napus* expressing LPAAT from coconut endosperm (Knutzon *et al.*, 1999). Developing seed from *B. napus* expressing both foreign enzymes, incorporated 12:0 at the *sn*-2 position of the glycerol backbone to produce trilaurin and overall higher levels of 12:0 in the seed oil compared to plants that only expressed the medium-chain thioesterase. A similar approach has been used to increase the 22:1 content of high-erucic acid *B. napus*. High-erucic acid oils represent a valuable feedstock in the syntheses of slip-promoting/antiblocking agents, detergents, plasticizers and surface coatings, with trierucin serving as a high-temperature lubricant (Princen and Rothfus, 1984). The LPAAT from developing meadowfoam (*L. alba*) has been shown to catalyze the incorporation of 22:1 into the *sn*-2 position of LPA

(Bafor *et al.*, 1990; Cao *et al.*, 1990; Laurent and Huang, 1992). Transformation of *B. napus* with cDNAs encoding meadowfoam LPAAT has been shown to result in transgenic seeds producing some trierucin (Lassner *et al.*, 1995; Brough *et al.*, 1996). In studies using polyclonal antibodies raised against LPAAT2, Wilmer *et al.* (2000) demonstrated that there was no correlation between the level of 22:1 at the *sn*-2 position of the TAG of *B. napus* seed expressing *L. douglassii* LPAAT2 and the level of LPAAT2 protein on a Western blot. Although transformed seeds produced oil with 22:1 at the *sn*-2 position, other metabolic factors affected the FA distribution.

The mutated yeast *sn*-2 acyltransferase gene (*SLC1-1*) described by Nagiec *et al.* (1993) has been shown to accept 22:1-CoA (Zou *et al.*, 1997). Engineering of *B. napus* L. cv Hero with the *SLC1-1* gene resulted in transformants with increased seed oil content and average seed weight, and increased 22:1 content in the seed oil. The proportions of 22:1 in the transgenic seeds ranged from about 49 to 56% (wt) compared to about 45% (wt) for non-transformed controls. There were small increases in the proportions of 22:1 and total very long-chain FAs at the *sn*-2 position of TAG from transgenic seed. The increased seed oil and 22:1 content of these transgenic plants were confirmed in field studies with $T_4$ and $T_5$ generations of *B. napus* (Taylor *et al.*, 2001). The increases in seed oil content might be attributable to differences in the regulation of meadowfoam LPAAT compared to the resident enzyme (Zou *et al.*, 1997).

Acyl-ACP:LPAAT, which is associated with the envelope of plastids, shows an increased preference for 16:0-ACP while discriminating against 18:1-ACP (Frentzen *et al.*, 1983; Frentzen, 1993). The selectivity of plastidial LPAAT is consistent with the selectivity of the prokaryotic pathway that incorporates 16:0 as the predominant species of FA at the *sn*-2 position of plastidial glycerolipids (Ohlrogge and Browse, 1995). Bourgis *et al.* (1999) cloned a cDNA encoding plastidial LPAAT by functional complementation of the *E. coli* mutant *p1sC* with a developing embryo cDNA library of *B. napus*. LPAAT activity from bacteria expressing the plant cDNA showed enhanced specificity for 16:0-ACP, which was consistent with the specificity of the enzyme activity in plastids. Chloroplast import studies indicated that a 39-kDa precursor-polypeptide was post-translationally processed to produce a 32-kDa mature protein that localized to chloroplast membranes. The deduced amino acid sequences of plastidial LPAAT shared a high level of identity with microsomal LPAATs from seeds. Investigations of plastidial LPAAT using site-directed mutagenesis have shown that alanine substitutions of residues found in two conserved motifs, which included a possible catalytic histidine, resulted in loss of enzyme activity (Maisonneuve *et al.*, 2000). Sequences in the *C*-terminal section were also essential for function.

### *6.3.3 Phosphatidic acid phosphatase*

Plant microsomal PA phosphatase associated with glycerolipid synthesis has been characterized in developing safflower seed and peanuts, microspore-derived (MD) cultures of *B. napus* and mesocarp of avocado (*Persea americana*) (Kocsis and Weselake, 1996; Weselake, 2002). Safflower PA phosphatase probably has little effect on the FA composition of TAG because the enzyme has been shown to accept a wide range of molecular species of PA (Stymne *et al.*, 1983; Ichihara, 1991). Increasing concentrations of oleate have been shown to induce the translocation of safflower PA phosphatase activity from the soluble fraction into microsomes (Ichihara *et al.*, 1990). Microsomal PA phosphatase activity from both developing peanut and safflower seeds reaches a maximum during the active phase of oil accumulation (Sukumar and Sastry, 1987; Ichihara *et al.*, 1990). In safflower, the amount of microsomal PA phosphatase diminished considerably as the seed oil content reached a plateau. The microsomal PA phosphatase from developing safflower seed exhibits an absolute requirement for $Mg^{2+}$ (Griffiths *et al.*, 1985; Ichihara *et al.*, 1990). In developing seeds and MD cultures of *B. napus*, only 10–20% of the total PA phosphatase activity has been shown to be associated with microsomes (10 000–100 000 × g particulate fraction) (Kocsis *et al.*, 1996). In contrast, the soluble fraction from developing seeds and MD embryos of *B. napus* contained 36 and 61%, respectively, of the total PA phosphatase activity recovered after subcellular fractionation of tissue. The microsomal and soluble PA phosphatases exhibited pH optimums of 6–7 and 5, respectively, suggesting that these were different enzymes. Furukawa-Stoffer *et al.* (1998) demonstrated that PA phosphatase activity in microsomes from MD embryos and MD cell suspension cultures of *B. napus* could be partially inhibited following preincubation with 0.5 mM *N*-ethylmaleimide (NEM). Also, the activity of the microsomal enzyme from MD embryos of *B. napus* was partially dependent on $Mg^{2+}$. The PA phosphatase activity in microsomal fractions from MD cultures appeared to have characteristics associated with two forms of the enzyme identified in mammalian systems – one form involved in glycerolipid synthesis and the other in signal transduction. In mammalian systems, PA phosphatase-1 is involved in glycerolipid synthesis through FA-mediated translocation from the cytosol to the ER, and is sensitive to NEM and dependent on $Mg^{2+}$ for activity (Brindley, 1984; Kocsis and Weselake, 1996; Coleman and Lee, 2004). In contrast, mammalian PA phosphatase-2 isoform, which appears to play a role in signal transduction, is associated with the plasma membrane, is NEM-insensitive and is not dependent on $Mg^{2+}$ for activity (Jamal *et al.*, 1991; Waggoner *et al.*, 1995, 1996). It is possible that microsomal fractions from MD cultures of *B. napus* may contain two forms of PA phosphatase or the microsomal enzyme is only partially inhibited by NEM.

Effective solubilization of microsomal PA phosphatase from MD embryos of *B. napus* was achieved using Tween 20 at a concentration of 0.4% (w/v) at a detergent : protein ratio of 1:1 (w/w) (Kocsis *et al.*, 1996). The solubilized enzyme was quite unstable, losing 60% of its activity after storage at 4°C for one day (Furukawa-Stoffer *et al.*, 1998). The solubilized PA phosphatase could catalyze the hydrolysis of a number of different molecular species of PA and various phosphate-containing compounds including G3P (Kocsis *et al.*, 1996). Although the microsomal PA phosphatase of *B. napus* appears to be a promiscuous enzyme, its organization in the membrane of the ER may limit the enzyme to only accepting PA generated via the LPAAT-catalyzed reaction. Pearce and Slabas (1998) purified avocado (*P. americana* L. cv Hass) mesocarp PA phosphatase about 7000-fold over the activity in the microsome in a procedure that involved three chromatography steps. The enzyme was solubilized from the 30 000–100 000 × g particulate fraction using 0.5% (w/v) 3-[(3-cholamidopropyl)dimethylammonio]-1-propanesulfonate (CHAPS) at a microsomal protein concentration of 5 mg/mL. The purified enzyme had a molecular mass of 49 kDa based on SDS-PAGE and analysis by gel filtration chromatography under non-denaturing conditions indicated that the enzyme was a monomer. Gel filtration chromatography of solubilized microsomal PA phosphatase from MD embryos of *B. napus* displayed a similar minimum apparent molecular mass of about 40 kDa (Kocsis *et al.*, 1996). Avocado PA phosphatase was investigated using a surface dilution kinetic model using substrate in Triton X-100 mixed micelles (Pearce and Slabas, 1998). The enzyme could also utilize *sn*-1-LPA, *sn*-2-LPA and ceramide-1-phosphate, with *sn*-1-LPA acting as a better substrate than PA. Avocado PA phosphatase was, however, both NEM-insensitive and $Mg^{2+}$-independent, which suggested that the enzyme might be a type-2 PA phosphatase. Although the microsomal PA phosphatase of the avocado mesocarp does not conform to all the characteristics of a type-1 PA phosphatase, it is probably involved in glycerolipid synthesis given the high TAG content of this tissue.

PA and diacylglycerol-pyrophosphate (DAGPP) may serve as important signaling roles in the stress response in plants (see Chapter 10). For example, in studies with suspension-cultured tomato (*Lycopersicon esculentum*), PA and DAGPP have been shown to transiently accumulate in response to the application of elicitors suggesting that these phospholipids play a role in plant signal transduction (van der Luit *et al.*, 2000). While examining global responses of genes to radiation stress in *A. thaliana*, Pierrugues *et al.* (2001) identified a gene called *AtLPP1* whose deduced protein structure was similar to lipid phosphatase enzymes from yeast and mammalian sources. When expressed in yeast membranes, the gene product displayed both PA phosphatase and DAGPP phosphatase activity. Expression of *AtLPP1* in *A. thaliana* could be induced by radiation or elicitor treatments suggesting that encoded lipid phosphatase enzyme may attenuate the signal transduction functions of PA and/or DAGPP in response to stress. In another study, Marcel *et al.* (2000) cloned two cDNAs encoding putative PA phosphatases from cowpea (*Vigna unguiculata*). The predicted protein structures for VuPAP-α and VuPAP-β were similar to the transmembrane topology of mammalian members of the PA phosphatase-2 family. VuPAP-β was expressed during rapid desiccation whereas VuPAP-α expression was stimulated with rewatering of drought-treated plants. *In vitro* studies demonstrated that VuPAP-α was imported into and processed in chloroplasts. In contrast to the advances made on the purification and cloning of type-2 PA phosphatases, type-1 PA phosphatases have not been purified or cloned (Kocsis and Weselake, 1996; Weselake, 2002; Coleman and Lee, 2004).

The biochemical properties of PA phosphatase from the chloroplasts of spinach (*Spinacia oleracea*) leaves have been studied (Block *et al.*, 1983; Malherbe *et al.*, 1992). The enzyme is associated with the inner envelope of the chloroplast where it serves in the regulation of glycerolipid biosynthesis. PA can serve as a precursor in the biosynthesis of phosphatidylglycerol and DAG, and the latter can lead to production of galactolipids and sulfolipids (Joyard *et al.*, 1994). In contrast to plant microsomal PA phosphatases, which have enzyme activity pH optimums near neutrality (Ichihara *et al.*, 1989; Kocsis *et al.*, 1996; Pearce and Slabas, 1998), plastidial PA phosphatase exhibits maximum activity at pH 9.0 (Block *et al.*, 1983). *sn*-1,2-DAG has been shown to be a potent inhibitor of plastidial PA phosphatase whereas *sn*-1,3-DAG shows low inhibition (Malherbe *et al.*, 1992). The enzyme has been solubilized with CHAPS and partially purified using hydroxyapatite chromatography (Malherbe *et al.*, 1995).

### *6.3.4 Diacylglycerol acyltransferase*

DGAT has been investigated extensively in microsomal fractions from numerous species of developing seeds (Weselake, 2002) and MD cultures of *B. napus* (Weselake and Taylor, 1999; Weselake, 2000, 2002), with some studies of the enzyme in microsomal membranes from germinating seeds (Wilson and Kwanyuen, 1986; Settlage *et al.*, 1995; Lu *et al.*, 2003). The enzyme activity is effectively assayed using [$^{14}$C]acyl-CoA as the donor substrate. Assays that use radiolabeled DAG are subject to possible interference by DGTA and PDAT activity. If labeled DAGs are used, however, it is better to have the label on the glycerol backbone rather than the FA chains because of possible acyl-exchange reactions (Vogel and Browse, 1995). In many cases, endogenous DAG has also been shown to interfere with the assay of DGAT (Cao and Huang, 1986; Weselake *et al.*, 1991; Little *et al.*, 1994; Byers *et al.*, 1999; Valencia-Turcotte and Rodríguez-Sotres, 2001) making it difficult to assess the effect of defined molecular species of *sn*-1,2-DAG. However, Valencia-Turcotte and Rodríguez-Sotres (2001) have demonstrated that repeated washings of freeze-dried microsomes of developing maize embryos with benzene can be used to remove endogenous DAGs. Exogenous DAG + PC or Tween 20 was then introduced in benzene to the washed membranes followed by removal of solvent with nitrogen flow. With the addition of buffer and radiolabeled acyl-CoA to this fraction, DGAT activity, which was dependent solely on exogenous DAG, could then be assayed effectively.

In early studies with maturing safflower, it was demonstrated that particulate DGAT displayed a broad substrate specificity for various molecular species of acyl-CoA and *sn*-1,2-DAG (Ichihara and Noda, 1982; Ichihara *et al.*, 1988). It was also suggested that the FA composition of the *sn*-3 position of TAG in safflower might be dependent on the nature of the endogenous acyl-CoA pool. Microsomal DGATs from both low- and high-erucic acid cultivars of *B. napus* are able to utilize 12:0-, 18:1- or 22:1-CoA as substrates, with a similar pattern of acyl-CoA preferences for each cultivar, suggesting that the plant breeding process used to produce low-erucic acid cultivars did not affect the genes encoding DGAT (Cao and Huang, 1987). The reduced 22:1 content of low-erucic acid oil has been linked to a single amino acid substitution in the FA elongation-1 polypeptide, encoded by the *FAE1* gene, which is involved in the elongation of 18:1-CoA to 22:1-CoA (Katavic *et al*., 2002). In an earlier study, however, Bernerth and Frentzen (1990) demonstrated that microsomal DGAT from a variety of *B. napus* (Lenora × CHR 1775/82) containing 54% erucic acid in the seed oil showed enhanced specificity for 22:1-CoA while a variety (Krokos) containing 35% erucic acid in the seed oil displayed a much lower preference for 22:1-CoA relative to 18:1-CoA. Studies with microsomes from MD embryos of low- and high-erucic acid cultivars of *B. napus* have shown that the DGAT from both sources displayed an enhanced preference for 22:1-CoA over 18:1-CoA at a thioester concentration of 14 μM (Weselake *et al*., 1991). Also, microsomes from high-erucic acid *B. napus* have been shown to effectively utilize *sn*-1,2-dierucin indicating that the DAG specificity of DGAT would not cause problems in attempts to genetically engineer *B. napus* producing trierucin (Taylor *et al*., 1992). Little *et al.* (1994) have demonstrated that solubilized microsomal DGAT has a considerably enhanced specificity for 18:1- and 16:0-CoA over 18:0-CoA at thioester concentrations in excess of 2 μM.

Microsomal DGAT has been solubilized from germinating soybean (Kwanyuen and Wilson, 1986), MD cultures of *B. napus* (Little *et al*., 1994) and developing maize embryos (Valencia-Turcotte and Rodríguez-Sotres, 2001). The *B. napus* enzyme was solubilized in the presence of 1% (w/v) octanoyl-*N*-methylglucamide and 2 M NaCl (Little *et al*., 1994). Combinations of detergent and high salt concentration have also been successfully used in the solubilization of other acyltransferases of the G3P pathway (Frentzen, 1990; Davies *et al*., 1995). Valencia-Turcotte and Rodríguez-Sotres (2001) have solubilized DGAT from delipidated lipid bodies of developing maize (*Z. mays*) embryos using a solution of methanol/acetic acid/water (1:1:1, v/v/v). The solution was clarified by centrifugation and subjected to chromatography on a column of porous glass beads equilibrated in the same solution. DGAT activity was assayed in the column fractions following evaporation of the solvent and introduction of exogenous DAG, PC and [$^{14}$C]oleoyl-CoA. The enzyme eluted with a molecular mass of about 100 kDa but protein was not detected suggesting that the purified enzyme had a very high specific activity.

Photoaffinity labeling has also been used in attempts to identify a DGAT polypeptide. The photoreactive substrate analogs 12-[(4-azidosalicyl)amino]dodecanoyl-CoA (ASD-CoA) and 12-azidooleoyl-CoA appeared to competitively displace 18:1-CoA in assays with membrane bound DGAT from MD embryos of *B. napus* L. cv Reston (Weselake *et al*., 1995). Photoaffinity labeling studies with [$^{125}$I]ASD-CoA and particulate and solubilized membrane fractions followed by SDS-PAGE and autoradiography indicated the enrichment of a 39 kDa radiolabeled polypeptide in the solubilized fraction.

In 1998, Cases *et al.* reported the cloning of a cDNA for a mouse DGAT, which swiftly led to the cloning of cDNAs encoding DGAT1 from *A. thaliana* (Hobbs *et al.*, 1999; Routaboul *et al.*, 1999; Zou *et al.*, 1999; Bouvier-Navé *et al.*, 2000). The predicted molecular mass of the 520-amino acid protein was 59 kDa and the sequence shared about 38% identity with the predicted amino sequence of mouse DGAT (Hobbs *et al.*, 1999). cDNAs encoding BnDGAT1 and BnDGAT2 from *B. napus* were cloned from MD cell suspension cultures of the cultivar Jet Neuf (Nykiforuk *et al.*, 1999a,b, 2002). The cell suspension cultures expressed both mRNAs and the cDNAs were functionally expressed in *Pichia pastoris*. *BnDGAT1*and *BnDGAT2* encoded proteins with predicted molecular masses of 56.9 and 39.5 kDa, respectively. BnDGAT1 was 96% homologous to BnDGAT2 across 341 overlapping amino acids, sharing 85% identity with *A. thaliana* DGAT and 38% identity with mouse DGAT. Interestingly, the molecular mass of BnDGAT2 was similar in size to the polypeptide identified by photoaffinity-labeling of membranes from MD embryos of *B. napus* (Weselake *et al.*, 1995). The BnDGAT2 of MD cell suspension cultures, however, lacks the hydrophilic *N*-terminal segment present in BnDGAT1. cDNAs encoding type-1 DGATs have been cloned from other sources including tobacco (*Nicotiana tabacum*) (Bouvier-Navé *et al.*, 2000), olive (*Olea europaea*) (Giannoulia *et al.*, 2000) and castor bean (*Ricinus communis*) (He *et al.*, 2003).

The hydrophilic *N*-terminal segment of BnDGAT1 (from *B. napus*) (residues 1–116) has been expressed in *E. coli* as a polypeptide with a poly-histidine tag to facilitate purification by immobilized nickel ion chromatography (Weselake *et al.*, 2000a). Binding studies indicated that recombinant *N*-terminal fragment interacted with radiolabeled acyl-CoA (Weselake *et al.*, 2000a), which was consistent with the presence of an acyl-CoA binding signature (Nykiforuk *et al.*, 2002). Analysis of the *N*-terminal segment of the predicted amino acid sequence of DGAT1 from *A. thaliana*, using a program that predicts transmembrane helices by comparison with a database of membrane bound proteins, indicated that the hydrophilic *N*-terminus was on the cytosolic side of the ER (Hobbs *et al.*, 1999). Assuming the situation is similar for BnDGAT1, the *N*-terminus would presumably be free to interact with cytosolic acyl-CoA. This acyl-CoA binding site in the *N*-terminal region, however, is probably not part of the catalytic site that binds acyl-CoA because BnDGAT2 can be functionally expressed in yeast. A recent study with the recombinant *N*-terminal fragment of BnDGAT1 has indicated that the polypeptide segment also self-associates into a tetramer (Weselake, 2003), which is consistent with ability of human DGAT1 to self-associate into a homotetramer via the *N*-terminal regions of the polypeptide (Cheng *et al.*, 2001). Lardizabal *et al.* (2001) used a combination of protein purification, partial peptide sequencing and PCR techniques to clone a second DGAT gene family from the oleaginous fungus *Mortierella ramanniana*. Two cDNAs, which corresponded to related proteins (54% homology), were identified; these had molecular masses of 36 and 36.5 kDa, respectively, with members in many other species including *A. thaliana* and *B. napus*. The proteins shared no homology to the DGAT1 family. The DGAT2 family identified by Lardizabal *et al.* (2001), however, should not be confused with the BnDGAT2 from *B. napus*, which is a truncated form of BnDGAT1 (Nykiforuk *et al.*, 1999b). In an earlier work, Kamisaka *et al.* (1997) also used *M. ramanniana* in the purification of DGAT. The investigators used various chromatography methods to purify a DGAT with molecular mass of 53 kDa based on SDS-PAGE. The enzyme could utilize either *sn*-1,2-DAG or *sn*-2-monooleoylglycerol as a substrate.

In developing seeds of *B. napus*, DGAT activity reaches a maximum during the active phase of oil accumulation and then declines markedly as the seed lipid level reaches a plateau indicating that the appearance of enzyme activity is coordinated with storage lipid production (Tzen *et al.*, 1993; Weselake *et al.*, 1993). The specific activity of DGAT has also been shown to increase steadily during development of MD embryos of *B. napus* L. cv Topas reaching a maximum level at mid-cotyledonary stage (Weselake *et al.*, 1993). Numerous studies point to the DGAT-catalyzed reaction as being a critical, and perhaps limiting, step in the oil formation process. Harwood and colleagues have demonstrated that DAG is a major intermediate of the G3P pathway that accumulates during the active phase of oil formation in developing seeds of *B. napus* (Perry and Harwood, 1993a,b; Perry *et al.*, 1999). The investigators suggested that DGAT activity exerts significant flux control during periods of rapid TAG accumulation. Recent metabolic studies with olive tissue cultures have shown that DGAT may exert significant flux control only when the rate of lipid synthesis is high (Ramli *et al.*, 2002a). An *A. thaliana* mutant (AS11) with reduced DGAT activity has been shown to have a decreased TAG/DAG ratio compared to the wild-type (Katavic *et al.*, 1995). A few years later, this effect was reported to be due to a mutant allele of the *DGAT1* gene (Zou *et al.*, 1999). Seed-specific overexpression of *DGAT1* in *A. thaliana* has been shown to lead to both increased seed oil content and increased seed size (Jako *et al.*, 2001). In addition, studies with soybean have indicated that genotypic differences in DGAT activity contribute to expression of genetic variation in oil content among various types of germplasm (Settlage *et al.*, 1998). The possible limiting effect of DGAT activity in oil formation has also been inferred from measurements of the specific activity of the enzymes of the G3P pathway. In both developing safflower seed and *B. napus* seed, DGAT showed the lowest specific activity of the four enzymes in the G3P pathway leading to TAG (Ichihara *et al.*, 1988; Perry *et al.*, 1999).

Other studies have not supported the concept that DGAT has a major effect on the flow of carbon into seed oil. In an investigation with developing embryos of *Cuphea lanceolata, Ulmus carpinifolia* and *Ulmus parvifolia*, Bao and Ohlrogge (1999) have shown that the amount of FA available can limit TAG accumulation. PA phosphatase has also been identified as having a limiting role in TAG biosynthesis (Stymne and Stobart, 1987; Wilmer *et al.*, 1998). In enzyme assays with [U-$^{14}$C]G3P and 18:1-CoA added to particulate fractions from MD embryos of *B. napus* L. cv Reston, Wilmer *et al.* (1998) reported substantial production of PA, with DAG being the next most abundant lipid class. In this case, the data suggested that PA phosphatase activity was limiting in TAG production, possibly sharing some control with DGAT. Francki *et al.* (2002) have assessed the maximum activities of acetyl-CoA carboxylase (EC 6.4.1.2) and DGAT during seed development in two species of *Lupinus* seeds differing substantially in seed oil content. *Lupinus mutabilis* contained 16% oil in the mature seed and had twice as much acetyl-CoA carboxylase activity as *Lupinus angustifolius*, which displayed 8% oil content in the mature seed. In contrast, the maximum DGAT activity was the same for each species during seed development. The investigators concluded that DGAT is not involved in regulating TAG biosynthesis in *Lupinus* spp.

In some plant systems, changes in DGAT activity are paralleled by changes in the activity of other enzymes of TAG biosynthesis. LPAAT and GPAT have been shown to display similar activity profiles to DGAT during seed development in *B. napus* (Tzen *et al.*, 1993). Also, the activities of microsomal DGAT and PA phosphatase have been shown to reach a maximum during the active oil accumulation phase in developing peanut seeds (Sukumar and Sastry, 1987). A similar result was reported for microsomal PA phosphatase in maturing safflower seeds (Ichihara *et al.*, 1990). In studies with developing seeds of *B. napus*, Slabas and colleagues have shown that the components of the FA synthase complex are largely induced before TAG accumulation and the activities of these enzymes remain high as the rate of lipid accumulation decreases (Slabas *et al.*, 1986, 1987; MacKintosh *et al.*, 1989). The results suggested that the process of TAG assembly was more tightly controlled compared to FA synthesis. Recently, Ramli *et al.* (2002b) have described the use of Top-Down (Metabolic) Control Analysis to examine, quantitatively, lipid biosynthesis in tissue cultures of olive and oil palm. Results indicated that control of lipid biosynthesis was distributed between both FA biosynthesis and glycerolipid assembly. The investigators indicated that the results of flux control analysis on oil-forming plant systems may provide valuable information for genetic engineering of oilseed metabolism.

A number of electrolytes, small molecular weight organic compounds and proteins have been shown to enhance the activity of DGAT in particulate fractions of MD cultures of *B. napus* (Little *et al.*, 1994; Byers *et al.*, 1999; Weselake *et al.*, 2000b). The largest activation in microsomal DGAT activity has been observed with 25 mM $MgSO_4$ or $MgCl_2$ (Byers *et al.*, 1999). Hobbs and Hills (2000) have demonstrated that recombinant AtDGAT1 expressed in microsomes of an insect culture system is also stimulated by $Mg^{2+}$. Although increasing concentrations of $MgCl_2$ activated *B. napus* DGAT, the stimulatory effects of ATP and CoA (e.g. at 1 mM) were diminished (Byers *et al.*, 1999). A recent study with mice has suggested that two isozymes of DGAT have different dependencies on $MgCl_2$ concentration for activity (Cases *et al.*, 2001). For example, mouse DGAT1 exhibited relatively high activity at 5 mM $MgCl_2$ while DGAT2 required about 20 mM $MgCl_2$ for full manifestation of activity. The strong stimulatory effect of ATP and/or CoA observed at low $MgCl_2$ concentration in the investigation of Byers *et al.* (1999) with MD cell suspension cultures might have been attributable to specific activation of DGAT2 activity. PA has been shown to enhance microsomal DGAT activity from MD cell suspension cultures by about 2-fold at a concentration of 500 μM (Byers *et al.*,

1999). Perhaps addition of PA to the reaction mixture fueled the production of DAG via PA phosphatase, thereby, providing a DAG substrate that was more easily utilized by DGAT. Attempts to demonstrate that microsomal DGAT from MD cell suspension cultures of *B. napus* was down-regulated through phosphorylation catalyzed by cytosolic DGAT kinase, as in mammalian systems (Rodriguez *et al.*, 1992), revealed the presence of a low molecular weight organic factor that stimulated the enzyme (Byers *et al.*, 1999).

Microsomal DGAT activity from MD cultures of *B. napus* is enhanced by bovine serum albumin (BSA) and human acylation stimulating protein (ASP) (Little *et al.*, 1994; Weselake *et al.*, 2000b). A concentration of 3–4 mg BSA/mL stimulated microsomal DGAT activity from MD embryos of *B. napus* L. cv Reston by 4 to 5-fold. BSA might stimulate acyltransferase activity by preventing the formation of acyl-CoA micelles that could inhibit enzyme activity (Stobart and Stymne, 1990) and/or bound acyl-CoAs may be more effective substrates for acyltransferases (Hershenson and Ernst-Fonberg, 1983). ASP is a low molecular mass (~9 kDa), basic protein in the bloodstream, which is a cleavage product of the complement C3 (Baldo *et al.*, 1995). Yasruel *et al.* (1991) demonstrated that ASP could stimulate DGAT activity in human adipose tissue by interacting directly with membrane containing DGAT. In addition, ASP has been shown to enhance TAG biosynthesis in adipocytes though a protein kinase C signaling pathway (Baldo *et al.*, 1995). ASP has been shown to stimulate microsomal DGAT activity from MD cell suspension cultures of *B. napus* L. cv Jet Neuf by 50% at a concentration of 11 μM and over 100% at 44 μM (Weselake *et al.*, 2000b). The degree of activation of *B. napus* DGAT activity by ASP remained unchanged regardless of the concentration of thioester, suggesting that ASP did not improve substrate delivery to the active site of the enzyme. Amino acid sequences homologous to human ASP are either absent in plants or have not been identified. It has been hypothesized that the expression of BSA or ASP during seed development in *B. napus* might lead to alterations in seed oil production (Weselake *et al.*, 2000b). Recently, *B. napus* L. cv Westar was transformed with cDNA encoding either BSA or ASP using a seed specific promoter (Weselake *et al.*, 2003). Two of the ASP transformants displayed enhanced seed size and increased lipid content on a per seed basis. The only BSA transformant that produced the encoding transcript in the developing seed, exhibited a marked increase in the proportions of 18:2 and α-18:3 in the total acyl lipids of the mature seed. In light of this observation, it is interesting to note that Bafor *et al.* (1990) have shown that BSA has a concentration-dependent influence on the acyl selectivity of microsomal LPAAT from developing safflower seeds.

TAG has also been shown to accumulate in plastids (Martin *et al.*, 1984; Xue *et al.*, 1997). An early study by Martin and Wilson (1984) indicated that acyl-CoA-dependent DGAT activity could be assayed in spinach leaf plastids. Sakaki *et al.* (1990a,b) used [1-$^{14}$C]acetate feeding coupled with analyses of lipid classes to demonstrate that plastidial TAG was formed from monogalactosyldiacylglycerol

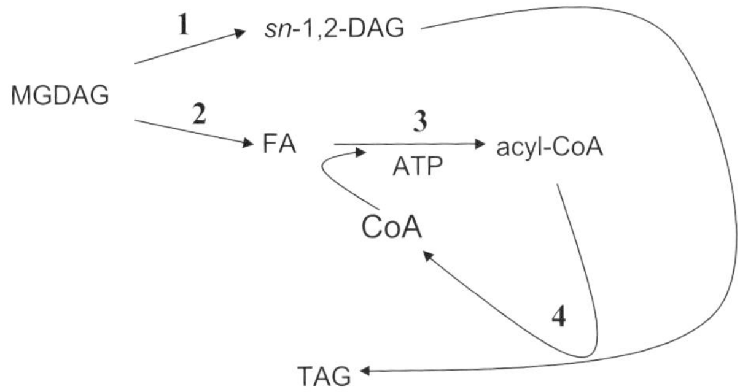


**Figure 6.3** Proposed pathway for the synthesis of triacylglycerol (TAG) from monogalactosyldiacylglycerol (MGDAG) in spinach chloroplasts. Enzymes: (1) galactolipid:galactolipid galactosyltransferase (EC 2.4.1.184), (2) galactolipase (EC 3.1.1.26), (3) acyl-CoA synthetase, (4) diacylglycerol acyltransferase. Additional abbreviations: FA, fatty acid; *sn*-1,2-DAG, *sn*-1,2-diacylglycerol. Adapted from Sakaki *et al.* (1990a,b).

(MGDAG), based on the steps outlined in Figure 6.3. Northern blotting experiments have revealed DGAT transcripts in the leaves of *A. thaliana* and *B. napus* (Hobbs *et al.*, 1999; Zou *et al.*, 1999), which may encode a plastidial form of the enzyme. Kaup *et al.* (2002) investigated TAG production during leaf senescence in *A. thaliana*. Both *AtDGAT1* transcript and AtDGAT1 protein increased with TAG levels during leaf senescence. AtDGAT1 was present in the membranes of the chloroplast. Western blotting indicated that the molecular mass of plastidial AtDGAT1 was 29 kDa, considerably smaller than the 51 kDa molecular mass deduced from the cDNA encoding the enzyme. The investigators ascribed this molecular mass change to proteolysis. Assuming AtDGAT1 is synthesized outside the chloroplast, the mechanism whereby enzyme is imported into the plastid is in need of investigation. As indicated in Section 6.3.1, pea mitochondria have been shown to import and process microsomal AtGPAT1 (Zheng *et al.*, 2003).

### *6.3.5 Enzymes catalyzing acyl-CoA-independent synthesis of triacylglycerol*

Within the last 8 years non-acyl-CoA-dependent transacylations (see Figure 6.2) have been identified in developing seeds from a number of oleaginous plants. Lehner and Kuksis (1993) first reported a DGTA activity in rat liver and intestinal microsomes, which transfers an FA chain between two *sn*-1,2-DAG molecules to form *sn*-2-MAG and TAG. The same investigators purified and characterized the DGTA protein from rat intestine. Rat DGTA was specific toward the 1(3)-positions of *rac* 1,2-DAGs but did not utilize 1,3-DAGs. Stobart *et al.*(1997) presented evidence for both the forward and reverse reactions of DGTA in studies with microsomes from developing safflower seeds. The specific activity of DGTA was about 25% of the DGAT activity. Fraser *et al.* (2000) later reported that DGTA might in fact represent the major TAG-generating enzyme in developing safflower seed. In the presence of $Mg^{2+}$, acyl-CoA had little effect on the conversion of radiolabeled PA into TAG in safflower microsomes, thus implicating the action of another TAG-forming pathway. DGTA action has also been implicated in the remodeling of TAGs in microsomal preparations of developing castor bean endosperm (Mancha and Stymne, 1997). TAG in oil bodies, however, was not rearranged. Although DGTA has been purified from rat intestine, the cDNA encoding the enzyme has not been cloned from any source.

Recently, an acyltransferase catalyzing the transfer of FAs from PC to *sn*-1,2-DAG to form TAG was identified in yeast (*S. cerevisiae*) (Dahlqvist *et al.*, 2000; Oelkers *et al.*, 2000) and developing oil seeds (Dahlqvist *et al.*, 2000). The gene encoding the enzyme had a significant similarity to lecithin:cholesterol acyltransferase (EC 2.3.1.43), which catalyzed the transfer of an FA from lecithin to cholesterol. The activity of the plant enzyme, known as PDAT, was further characterized in microsomes from developing seeds of sunflower (*Helianthus annuus*), castor bean and *Crepis palaestina* (Banaś *et al.*, 2000; Dahlqvist *et al.*, 2000). PDAT played a major role in removing ricinoleic acid (HO-18:1) and vernolic acid (*cis*-12-epoxyoctadeca-*cis*-9-enoic acid) from phospholipids in developing castor bean and *C. palaestina* seeds, respectively. The specificity of PDAT for the acyl group in phospholipid varied between species. The specificity of yeast PDAT was shown to depend on the nature of the head-group of the acyl donor, the acyl species transferred and the molecular species of DAG (Dahlqvist *et al.*, 2000). Overexpression of yeast PDAT resulted in a seven-fold increase in microsomal PDAT activity compared to the control strain generated with the empty expression vector. TAG content was two-fold higher in PDAT-overexpressing yeast in early logarithmic phase cells suggesting the potential of using the PDAT gene to increase seed oil content. Banaś *et al.* (2000) identified an ortholog to the yeast PDAT gene in *A. thaliana* (GenBank accession no. T04806). Shoots and roots of *A. thaliana* showed substantial microsomal PDAT activity with HO-18:1-PC and 18:1-PC.

### *6.3.6 CDP-choline:-1,2-diacylglycerol cholinephosphotransferase*

Labeling studies with microsomes from various developing oil seeds have shown that CPT catalyzes the reversible movement of *sn*-1,2-DAG between PC and the mainstream of TAG synthesis via the G3P pathway (Slack *et al.*, 1983, 1985; Stobart and Stymne, 1985). The reverse reaction of CPT would also provide DAGs for transacylation reactions (Stobart *et al.*, 1997). Slack *et al.* (1985) have reported that maximum CPT activity correlates with the greatest rates of PC and TAG production in developing safflower seeds. Vogel and Browse (1996) have shown that the microsomal enzyme from developing seeds of *B. napus* L. cv Indore exhibits little or no preference for various molecular species of DAG. G3P has been shown to enhance the conversion of [$^3$H]PC to [$^3$H]*sn*-1,2-DAG in studies with microsomes from MD embryos of *B. napus* suggesting that CPT might be allosterically regulated by G3P (Taylor *et al*., 1991).

cDNAs, encoding proteins with CPT activity, have been cloned from yeast (cDNAs *EPT1* and *CPT1*) (Hjelmstad and Bell, 1990, 1991; McMaster and Bell, 1997). Dewey *et al*. (1994) cloned a cDNA (*AAPT1*) encoding soybean CPT by complementing a CPT-deficient mutant of yeast. The deduced amino acid sequences of both yeast and soybean CPT exhibited seven potential membrane-spanning segments (Hjelmstad and Bell, 1990, 1991; Dewey *et al*., 1994; McMaster and Bell, 1997). Soybean *AAPT* encodes a deduced amino acid sequence that is 32–33% identical to the sequences predicted for yeast *EPT1* and *CPT1* (Dewey *et al*., 1994; McMaster and Bell, 1997).

### *6.3.7 Lysophosphatidylcholine acyltransferase*

LPCAT, which catalyzes the acyl-CoA-dependent acylation of LPC (Lands and Hart, 1965; Moreau and Stumpf, 1982), also appears to catalyze the reverse process under some *in vitro* conditions (Stymne and Glad, 1981; Stymne *et al*., 1983; Stymne and Stobart, 1984a,b, 1987). The forward reaction, catalyzed by LPCAT, might acylate LPC generated via the action of $PLA_2$ or PDAT. The combination of the forward and reverse reactions of LPCAT results in acyl-exchange with the acyl-CoA pool leading to entry of polyunsaturated FAs into the acyl-CoA pool and modified PC (Stymne and Stobart, 1987). New opportunities are thus created for incorporating polyunsaturated FAs into TAG. Also, incorporation of 18:1 moieties from the acyl-CoA pool into PC creates new opportunities for desaturation at the level of PC (Griffiths *et al*., 1985; Stymne and Stobart, 1987). Ichihara *et al*. (1995) investigated the specificity and selectivity of LPCAT in the direction of PC formation using a particulate fraction from developing safflower seeds. The relative order of preference of LPCAT for acyl-CoAs was 18:2 > 18:1 > 18:0 = 16:0. Acyl-CoAs consisting of 12:0 and 22:1 moieties were not utilized. LPCs containing 16:0, 18:0, 18:1 and 18:2 moieties were all effective as acceptor substrates. Recently, similar results on LPCAT specificity were reported for the microsomal enzyme from MD cell suspension cultures of *B. napus* L. cv Jet Neuf (Furukawa-Stoffer *et al*., 2003). Bernerth and Frentzen (1990) had previously reported that 22:1-CoA was a poor substrate for microsomal LPCAT from developing seeds of high-erucic acid *B. napus*. The inability of the *B. napus* LPCAT to utilize 12:0-CoA or 22:1-CoA in the direction of PC formation has important implications in attempts to genetically engineer the FA composition at the *sn*-2 position of TAG from this crop because the reverse reaction catalyzed by CPT could potentially move the DAG skeleton from PC, modified by reacylation of LPC, into the *sn*-1,2-DAG pool that fuels TAG production. Attempts to modify the FA composition of the *sn*-2 position of TAG in *B. napus* seed oil have only focused on alteration of the LPAAT-driven step (Weselake and Taylor, 1999; Weselake, 2002). Thus, *sn*-1,2-DAG stemming from PC that has been modified because of LPCAT action could potentially dilute the *sn*-1,2-DAG generated in the G3P pathway as result of heterologous expression of coconut or meadowfoam LPAAT, which would result in the incorporation of 12:0 or 22:1, respectively, at the *sn*-2 position of the glycerol backbone.

Moreau and Stumpf (1982) have solubilized LPCAT from safflower microsomes using octylglucoside and other detergents. Photoreactive substrate analogs of LPC have been synthesized and shown to be effective in photoinactivation of microsomal LPCAT activity (Rajasekharan and Nachiappan, 1994; Tumaney and Rajasekharan, 1999). Radiolabeled azido-LPCs have been used to label polypeptides, which may be candidates for LPCAT in microsomal membranes from developing soybean cotyledons (Tumaney and Rajasekharan, 1999) and partially purified LPCAT from microsomes of developing sunflower seed (Fraser and Stobart, 2000). In the study by Fraser and Stobart (2000), LPCAT was solubilized in an active form from sunflower microsomes using 7.5 M urea at pH 9.0.

The membranes of plastids are dependent on the import of lipid precursors from the membranes of the ER (Moreau *et al.*, 1998). LPCAT activity in the plastidial envelope might serve to acylate LPC species derived from the ER following deacylation of PC in the ER (Bessoule *et al.*, 1995). Akermoun *et al.* (2000) solubilized about 30% of the total activity of LPCAT from plastidial membranes using a CHAPS : protein ratio of 5. Thus far, neither the microsomal or plastidial forms of LPCAT have been purified to homogeneity, nor has a cDNA encoding the enzyme been cloned.

### *6.3.8 Phospholipases*

Plant phospholipases have a wide range of functions including involvement in cell regulation, lipid catabolism and membrane remodeling (Wang, 2001). Specific phospholipases exhibit specific specificities for removing substituents of phospholipid such as PC. For example phospholipase $A_1$(EC 3.1.1.32) and $PLA_2$ catalyze the cleavage of FAs at the *sn*-1 and *sn*-2 positions of PC, respectively. In contrast, phospholipase C (EC 3.1.4.3) and phospholipase D (EC 3.1.4.4) catalyze the cleavage of phosphocholine and choline head groups, respectively, from PC. Phospholipases could potentially affect the FA composition of TAG by providing free FAs for activation for the acyl-CoA pool, and PA and *sn*-1,2-DAG that could enter TAG biosynthetic pathways. $PLA_2$ action could also provide LPC as substrate for the ER resident LPCAT, or for transfer of LPC to the plastid as described in the previous section. Banaś *et al.* (1992) investigated the ability of microsomal fractions from developing *B. napus* embryos, young wheat (*Triticum aestivum*) roots and ripening avocado mesocarp to remove radiolabeled FAs from specifically synthesized PCs including *sn*-2-[$^{14}$C]oleoyl-PC, *sn*-2-[$^{14}$C]linoleoyl-PC, *sn*-2[$^{14}$C]linoleoyl-hydroperoxide-PC and *sn*-2-[$^{14}$C]ricinoleoyl-PC. In each case, the highest phospholipase activities were found with *sn*-2[$^{14}$C]ricinoleoyl-PC indicating a general preference for PC with oxygenated acyl groups. The results indicated that these species could potentially remove unusual FAs from membranes through phospholipase action. Castor oil is highly enriched in HO-18:1 and the hydroxylation of 18:1 to produce HO-18:1 occurs at the level of PC (van de Loo *et al.*, 1995). $PLA_2$ has been shown to be a key enzyme in driving HO-18:1 into the TAG of castor bean oil (Lin and McKeon, 2000). PC has been shown to be an intermediate in this process in that $PLA_2$ appears to selectively remove HO-18:1 from PC for conversion to HO-18:1-CoA to fuel the acyltransferases of the G3P pathway (Bafor *et al.*, 1991; Ståhl *et al.*, 1995; Lin *et al.*, 1998, 2000, 2002a). Vernolic acid is generated from 18:2 attached to PC in the seed oil of *Euphorbia lagascae* (Bafor *et al.*, 1993). Similar to the hydrolysis of HO-18:1-PC, vernoleoyl moieties could be selectively released from vernoleoyl-PC to produce free vernolic acid (Bafor *et al.*, 1993; Ståhl *et al.*, 1995). Phospholipase has been shown to be the main activity in the removal of this FA from the phospholipid of developing seeds of *E. lagascae* (Banaś *et al.*, 2000). The same study indicated that microsomal phospholipases were also involved in removing crepenynic and capric acids (10:0) from the phospholipids of *Crepis rubra* and elm (*Ulmus glabra*) seeds. In contrast, microsomal PDAT activity had the major role in removing vernoleoyl moieties from phospholipids of developing seeds of *C. palaestina*.

Ståhl *et al.* (1998) purified a soluble $Ca^{2+}$-dependent 13.9 kDa $PLA_2$ from developing elm seeds. The enzyme was related to the animal secretory $PLA_2$s. Shortly thereafter, the full length sequences of two distinct but homologous rice (*Oryza sativa*) cDNAs encoding $PLA_2$ were reported along with the purification of a soluble $PLA_2$ from green rice shoots (Ståhl *et al.*, 1999). cDNAs for other $PLA_2$s resembling animal secretory $PLA_2$s have also been cloned from carnation during flower development (Kim *et al.*, 1999) and *A. thaliana* (Bahn *et al.*, 2003). The recombinant *A. thaliana* enzyme displayed maximal $PLA_2$ activity at pH 8.0 and was $Ca^{2+}$-dependent, but showed head-group specificity to phosphatidylethanolamine rather than PC. Jung and Kim (2000) have purified a 48 kDa membrane-associated $PLA_2$ from the leaves of broad bean (*Vicia faba*). The enzyme was also $Ca^{2+}$-dependent and was suggested to have a role in releasing $\alpha$-18:3, the precursor of jasmonic acid, via a signal transduction pathway.

### *6.3.9 Soluble lysophosphatidic acid phosphatase and monoacylglycerol acyltransferase in developing peanut*

Tumaney *et al.* (2001) have purified a soluble MGAT of 43 kDa by about 6600-fold from developing peanut cotyledons. The acyl-CoA-dependent MGAT was specific for MAG, but displayed lower $K_m$ values for *sn*-1-16:0-MAG and *sn*-1-18:1-MAG than with equivalent *sn*-2-MAGs. In another study, Shekar *et al.* (2002) purified a soluble LPA phosphatase of 39 kDa by about 5000-fold from the same source. The LPA phosphatase was specific to LPA and did not catalyze the dephosphorylation of other phosphate-containing compounds tested. The two studies suggested the presence of an alternative pathway for forming *sn*-1,2-DAG as a substrate for TAG production in developing peanut cotyledons. In mammals, the MAG pathway for

DAG biosynthesis is predominant in intestinal enterocytes (Coleman and Lee, 2004). Recently, genes encoding acyl-CoA-dependent MGAT1 and MGAT2 have been identified in mice (Yen *et al.*, 2002, 2003; Cao *et al.*, 2003). Previously, Hiroyama and Takenawa (1998) had purified and characterized an LPA-specific phosphatase from an animal system.

## 6.4 Complex metabolic processes can affect the fatty acid composition of triacylglycerol

From the above discussion of various enzymes of TAG biosynthesis, it is apparent that the FA composition of TAG may provide clues on the specificity and selectivity of the enzymes of the Kennedy pathway, but this is not always the case because of the involvement of other reactions in the modeling of TAG and aspects of phospholipid metabolism that may affect substrate pools available to enzymes of the G3P pathway.

It becomes difficult to generalize about mechanisms of TAG formation because of species differences in metabolism. Although *Tropaeolum majus* and *L. douglasii* both contain high levels of 22:1 in their seed oil, *T. majus* appears to use a different mechanism to acylate the *sn*-2 position of TAG with 22:1 (Löhden and Frentzen, 1992). The microsomal LPAAT of *T. majus* was inactive with 22:1-CoA suggesting that acyl-exchange might represent a route for introducing 22:1 into the seed oil of this species. Recently, Ichihara and Suda (2003) reported that acyl-exchange at the *sn*-2 position of PC may represent the main mechanism for α-18:3 accounting for 65% of the total FAs in perilla (*Perilla frutescens*) seed oil. The FA composition of TAG is generally more diverse than the FA composition of membranes (Ohlrogge and Browse, 1995). Unusual FAs, including 12:0, petroselinic acid (18:1*cis*$\Delta^6$) and HO-18:1 and 22:1 do not accumulate in membrane, but instead these FAs appear transiently in PC before being ultimately incorporated into TAG (Taylor *et al.*, 1991; Cahoon and Ohlrogge, 1994; Ohlrogge and Browse, 1995; Schultz and Ohlrogge, 2000). Further insights into the operation and limitations of TAG biosynthetic machinery have come from investigations with transgenic *B. napus* engineered to produce high levels of medium-chain FAs in their seed oil (Wiberg *et al.*, 1997, 2000; Larson *et al.*, 2002). The exclusion of 12:0 from the membranes of transgenic *B. napus* has been shown to be less effective than in plant systems that normally accumulate 12:0 in their seed oil (Wiberg *et al.*, 1997). Larson *et al.* (2002) found that the acyl-CoA pool of *B. napus* engineered to produce medium-chain FAs was considerably more enriched in medium-chain FAs than the resulting TAG, suggesting limitations in acyltransferase action.

Two hypotheses have emerged regarding *sn*-1,2-DAG as a precursor in TAG biosynthesis. The first suggests one pool of DAG exists for the biosynthesis of both membrane and TAG (Stymne, 1993; Wiberg *et al.*, 1997), whereas the second suggests that the DAG pools for membrane and TAG may be spatially separated in the ER (Vogel and Browse, 1996). The first hypothesis receives support from the identification of DGATs with differential selectivity to molecular species of DAG (Wiberg *et al.*, 1994, 1997) and the ability of certain phospholipases to remove unusual FAs from membranes (Ståhl *et al.*, 1995). In support of the second hypothesis, Vogel and Browse (1996) have reported that microsomal CPTs from various developing oil seeds showed little or no specificity for a range of molecular species of *sn*-1,2-DAG. Also, some of the DGATs examined in these sources showed modest specificities. The investigators suggested that the exclusion of unusual FAs from membranes at the level of DAG was not based on the specificity of CPT and DGAT, but may involve different pools of DAG that are spatially separated. Lacey and Hills (1996) used sucrose density-gradient centrifugation to identify a low-density membrane fraction of the ER with more enhanced TAG-biosynthetic capacity than the bulk ER suggesting that there may be regions of the ER that specialize in the formation and secretion of TAG. The possible spatial separation of TAG biosynthesis is discussed further in the context of lipid body biogenesis in the next section.

The selectivity of acyltransferases may also be modulated by acyl-CoA binding proteins (ACBPs). The presence of BSA, which is known to bind acyl-CoAs, has been shown to alter the selectivity of microsomal LPAAT from developing safflower seed when present at sufficiently high concentration in the reaction mixture (Bafor *et al.*, 1990). There was a shift in preference for 18:2-CoA over 18:1-CoA when the BSA concentration was raised from 0.25 to 0.5 mg/mL. In contrast, Brown *et al.* (2002a) found that the presence of 1 mg BSA/mL did not alter the selectivity of LPAAT in *B. napus* microsomes in studies with equimolar mixtures of 18:1/18:2 and 18:1/16:0 acyl-CoAs. Micromolar concentrations of recombinant ACBP from *B. napus* have been shown to stimulate microsomal GPAT activity when assayed with [$^{14}$C]18:1-CoA, but amounts in excess of a 1 : 1 ratio of ACBP:18:1-CoA led to enzyme inhibition (Brown *et al.*, 1998). A similar effect of recombinant ACBP concentration has been reported for microsomal LPAAT activity from developing sunflower seeds (Brown *et al.*, 2002a). A recent study with mitochondrial GPAT from rats has shown that the enzyme has a preference for 16:0-CoA over 18:1-CoA when albumin or recombinant ACBP is present in the assay (Kannan *et al.*, 2003). The possible role of ACBPs in the modulation of acyltransferase selectivity in TAG-forming plant systems is in need of further investigation.

## 6.5 Structure, composition and biogenesis of lipid bodies

In lipid-storing seeds, TAGs are packaged in discrete lipid bodies (also known as oil bodies or oleosomes) of 0.2–2.5 μm in the cytoplasm, which are encased by a half-unit membrane and specialized proteins (Huang, 1992, 1996; Murphy, 1993). Proteins embedded in lipid bodies include oleosins (Huang, 1992; Murphy, 2001), caleosins (Frandsen *et al.*, 2001) and the more recently characterized stero-leosin (Lin *et al.*, 2002b). The three types of protein share the common feature of having a long internal hydrophobic segment that serves to anchor them to lipid bodies. In recent years, a number of excellent reviews have appeared on the topic of lipid bodies and their associated proteins (e.g. Huang, 1996; Napier *et al.*, 1996; Zweytick *et al.*, 2000; Frandsen *et al.*, 2001; Murphy, 2001).

The most thoroughly characterized of the lipid proteins are oleosins, which are alkaline proteins ranging in size from about 14 to 26 kDa (Huang, 1992; Murphy, 2001). Further aspects of oleosin structure/function, and that of other lipid-associated proteins, are also discussed in Chapter 7. Oleosins are relatively abundant in oil seeds. For example, oleosins make up about 8% of the total seed protein in *Brassica* (Huang, 1996). The smallest lipid bodies (0.6 μm in diameter) are also found in *Brassica* spp., thus providing a very large surface area per unit TAG to facilitate effective lipase (EC 3.1.1.3) action during germination. Treatment of MD embryos of *B. napus* with abscisic acid (ABA), a plant hormone associated with seed maturation and desiccation (Finkelstein and Somerville, 1990), has been shown to enhance the production of oleosins and encoding transcripts (Taylor *et al.*, 1990; Holbrook *et al.*, 1991; van Rooijen *et al.*, 1992). Although oleosins are normally found in seeds that undergo desiccation through the maturation process (Murphy, 2001), there has been some recent evidence for their transient expression in root tips (Murphy *et al.*, 2001). Oleosins consist of three domains which include an *N*-terminal hydrophilic region of 30–60 amino acid residues, a central hydrophobic segment of about 70–80 highly conserved amino acid residues and a *C*-terminal hydrophilic α-helix of about 25–30 residues (Huang, 1992; Zweytick *et al.*, 2000; Murphy, 2001). The central hydrophobic domain contains a 'proline knot' sequence (Huang, 1992). Abell *et al.* (1997) substituted three universally conserved proline residues with leucine residues in Arabidopsis oleosin, which resulted in an altered gene product that did not accumulate in lipid bodies. The modified protein, however, was still incorporated into the ER at levels comparable with the native form of the lipid protein. The results indicated that the proline knot motif is not critical in ER integration, but is necessary for targeting to the lipid body. Analyses of the secondary structure of the central hydrophobic domain using biophysical methods have yielded two proposed structures. Recent studies of the central hydrophobic domain of oleosin from safflower and sunflower have reported a structure enriched in α-helix with no β-sheet (Millichip *et al.*, 1996; Lacey *et al.*, 1998; Alexander *et al.*, 2002). A model has been proposed for this scenario, which involves a α-helical hairpin structure with two helices being separated by a turn region (Alexander *et al.*, 2002). In contrast, studies with the central hydrophobic domain of *B. napus* oleosin have led to a model that features an extended parallel β-sheet with small amounts of anti-parallel structure (Li *et al.*, 2002). These investigators further suggested that the *N*- and *C*-terminal domains of oleosins could potentially interact to form dimers, trimers or oligomers, producing a network of oleosin molecules around the surface of the lipid body.

Both oleosins and phospholipids are needed to stabilize lipid bodies (Tzen and Huang, 1992), with oleosins appearing to be particularly important in stabilizing lipid bodies during seed imbibition prior to germination (Leprince *et al.*, 1998). The oleosins keep lipid bodies from coalescing by providing steric hindrance and electronegative repulsion because of their approximate isoelectric point of 9.4 (Tzen *et al.*, 1993). Previous calculations based on the oleosin content of lipid bodies have shown that there is enough oleosin to form a continuous mono-molecular coat around the entire surface of lipid bodies (Huang, 1992; Tzen and Huang, 1992). Ting *et al.* (1996) conducted studies of oleosin genes, oleosin content and lipid body appearance in two maize strains differing greatly in oil content (15% *vs.* 0.5%). Genes encoding oleosins and the defined proportion of three oleosin isoforms were unaltered in both sources, and developing seeds of both strains accumulated TAG and oleosins coordinately. The results indicated that oleosin genes are expressed independently of the genes affecting oil content, and that the size and shape of lipid bodies are governed by the ratio of TAG to oleosin synthesized during seed development. In contrast, Froese *et al.* (2003) have shown that lipid bodies from germinating seedlings of wax bean (*Phaseolus vulgaris*) are only partially coated with oleosin and thus, have exposed regions of naked phospholipid. It was suggested that these lipid bodies were not prone to coalescence because they were interspersed *in situ* among protein bodies and starch granules, which presumably interfered with inter-lipid body contacts.

A recent study has shown that there are commonalities in the interaction of oleosin with lipid bodies and the association of viral proteins with lipid droplets of mammalian tissue culture cells (Hope *et al.*, 2002). When *B. napus* oleosin was expressed in mammalian cells, it interacted with the lipid droplets. Site-directed mutagenesis was used to convert critical proline residues to alanine residues in the hydrophobic lipid body-targeting region (proline knot) of oleosin and an analogous region in the viral core protein of the hepatitis C virus. Both altered proteins failed to interact with lipid droplets. Also, the lipid-body-binding domain of the viral core protein could substitute for the equivalent domain in oleosin. In earlier work, it was also suggested that oleosins exhibited structural and functional similarities with animal serum apolipoproteins (Murphy *et al.*, 1991).

Most models of oleosin interaction with lipid bodies, take into consideration the monolayer of phospholipids, which encases the lipid body. The predominant theory for biogenesis of lipid bodies suggests that they pinch off of the ER and end up being encased by a half-unit membrane (Wanner and Theimer, 1978; Wanner *et al.*, 1981; Huang, 1992). In studies with developing seeds of *B. napus*, Tzen *et al.* (1993) have shown that TAG and oleosin accumulate at the same time suggesting that oleosins accompany phospholipids as they move away from the bulk of the ER as constituents of lipid bodies. A second theory has indicated that lipid bodies arise initially as naked oil droplets on the surface of the ER (Bergfeld *et al.*, 1978;

Stobart *et al.*, 1986; Herman, 1987) and acquire oleosins (Millichip *et al.*, 1996), or oleosins and phospholipids (Murphy *et al.*, 1989; Murphy, 1993) at a later stage. As already discussed, most models that describe the structure of oleosin in relation to the lipid body take the half-unit membrane into consideration. Strong support for the presence of this monolayer of phospholipids has come through electron microscopy and biochemical analyses of lipid bodies from peanuts (Yatsu *et al.*, 1972; Jacks *et al.*, 1990).

As previously discussed, Lacey and Hills (1996) have isolated a low-density membrane fraction with enhanced TAG biosynthetic capacity from developing embryos of *B. napus*. The study has also shown that oleosin may be associated with this low-density membrane fraction, which may represent a specialized region of the ER. Earlier studies have shown that the *in vitro* synthesis of oleosin is stimulated through addition of canine pancreatic microsomes (Hills *et al.*, 1993; Loer and Herman, 1993). Purified canine signal recognition particle (SRP) has been shown to inhibit the *in vitro* synthesis of oleosin in absence of microsomes suggesting that oleosin was inserted into the ER through an SRP-dependent translocation (Thoyts *et al.*, 1995; Napier *et al.*, 1996). Transfer of oleosin to lipid bodies, however, has been shown to be dependent on the presence of the *N*-terminal region of the polypeptide (van Rooijen and Moloney, 1995). In studies with microsomes from developing sunflower seeds, Lacey *et al.* (1999) used $^1$H-NMR spectroscopy to demonstrate that oil droplets were deposited within the lipid bilayer of the ER. In addition, immunochemical analysis of oleosins indicated that the lipid body proteins were directed to specific regions of the ER where the biosynthesis and accumulation of TAG occurred. Further studies with the sunflower system have also shown that oleosin transcript is associated with the rough ER, and that oleosin polypeptide was synthesized on bound ribosomes and co-translationally inserted into microsomes (Beaudoin *et al.*, 1999).

The targeting of seed oleosins has also been investigated using *S. cerevisiae*, which has become a useful model system for examining protein sorting through the endomembrane system (Ting *et al.*, 1997; Beaudoin *et al.*, 2000). Ting *et al.* (1997) have shown that maize oleosin was properly targeted and accumulated in yeast lipid particles. Beaudoin *et al.* (2000) have used *sec mutants* of yeast, which are defective in various aspects of endomembrane trafficking, to gain insight into the targeting pathway for oleosin. It was confirmed that SRP was, in fact, required for oleosin targeting to the ER and deposition on lipid bodies *in vivo*. In further studies of both *in vitro* and *in vivo* oleosin expression in yeast, Beaudoin and Napier (2002) have shown that the information required for targeting and/or membrane insertion was located in the central hydrophobic region of the protein, which was consistent with the earlier work of Abell *et al.* (1997). Both *N*- and *C*-terminal hydrophilic domains, however, were necessary for the correct topology of oleosin in the ER membrane. The results indicated that more than one domain could interact with SRP to target oleosin to the ER. A recent study of the targeting of Arabidopsis oleosin to pancreatic canine microsomes has identified several ER targeting sequences in the central hydrophobic regions and has shown that the insertion of the polypeptide is facilitated by SRP (Abell *et al*., 2002). In contrast, Chen and Tzen (2001) have shown that lipid body proteins are spontaneously targeted to lipid bodies in a co-translational mode without assistance from factors such as SRP. Oleosin produced using a reticulocyte lysate system in the presence of artificial emulsions of TAG and phospholipids resulted in targeting to the artificial oil emulsion. The targeting efficiency was increased using negatively charged phospholipids. Beaudoin and Napier (2000) have also demonstrated that sunflower oleosin can be ectopically expressed in the ER of non-oil storing tissues of *A. thaliana* including leaves, roots and flower petals. Deletion of the first 40 amino acid residues of oleosin had no effect on insertion into the ER. Recently, Wahlroos *et al*. (2003) transiently expressed a green fluorescent protein (GFP)-tagged *Arabidopsis* oleosin in tobacco (*Nicotiana benthamiana*) leaves. Oleosin was tagged via the *C*-terminus because it represented a permissive site for constructing translational gene fusions without compromising lipid body targeting (van Rooijen and Moloney, 1995). Although tobacco leaves are not oil-storing tissues, they did have lipid bodies, which contained TAG of similar compositional spectrum to sunflower oil. Monitoring of oleosin-GFP expression and trafficking by confocal laser scanning microscopy revealed that the non-oil storing leaves could synthesize and transport the fusion protein to the lipid bodies. Thus, early events in lipid body biogenesis were shown to occur outside the embryo and the processes could be monitored in real-time. The lipid bodies coated with oleosin-GFP tended to aggregate, however, suggesting a reduced capacity of tobacco cells to accommodate extensive oil deposition.

Through their association with lipid bodies, oleosins have also served as a vehicle for the production of therapeutic proteins in plants (Cramer *et al*., 1999). In the general strategy, the target therapeutic protein coupled to oleosin as a fusion protein is synthesized in the developing oil seed where the complex is deposited to lipid bodies. A protease cleavage site engineered between the therapeutic protein and oleosin is used to release the therapeutic protein from the lipid body, which is followed by further purification of the target protein to homogeneity. Hirudin, a naturally occurring anti-coagulant in the salivary glands of medicinal leeches (*Hirudo medicinalis*), was used in an earlier demonstration of this biotechnology (Parmenter *et al*., 1995).

More recently, $Ca^{2+}$-binding proteins have been identified at the surface of lipid bodies in seeds of oleaginous crops (Frandsen *et al*., 1996; Chen *et al*., 1999; Næsted *et al*., 2000; Frandsen *et al*., 2001). The proteins were named caleosins, independently, by two research groups (Chen *et al*., 1999; Næsted *et al*., 2000). Unaware of the $Ca^{2+}$-binding properties of the polypeptide, Chen *et al*. (1998) first discovered the lipid body protein in sesame (*Sesamum indicum*) seed as Sop1 along with two other lipid body polypeptides named Sop2 and Sop3, which were all distinct from oleosin. Sop1, Sop2 and Sop3 exhibited molecular masses of 27, 37 and 39 kDa, respectively, based on SDS-PAGE. The appearance of these proteins in lipid bodies during seed formation was similar or slightly delayed to that of oleosins. Sop1 from sesame seed was soon recognized as a $Ca^{2+}$-binding protein (Chen *et al.*, 1999), which was analogous to a $Ca^{2+}$-binding membrane protein whose transcript was induced by ABA in rice seeds (Frandsen *et al.*, 1996). Caleosins contain an *N*-terminal hydrophilic $Ca^{2+}$-binding domain followed by a central hydrophobic domain and then a *C*-terminal hydrophilic phosphorylation domain (Chen *et al.*, 1999; Frandsen *et al.*, 2001). The central hydrophobic domain has a proline knot motif as in oleosin. Caleosins are tenaciously associated with lipid bodies and the protein-lipid interactions can withstand treatment with 1 M NaCl or 9 M urea (Næsted *et al.*, 2000). The caleosins have been shown to make up a multigene family of at least 5 members in Arabidopsis (At*Clo*1-5) (Næsted *et al.*, 2000). Studies with Arabidopsis and *B. napus* indicated that caleosin was strongly expressed during seed maturation and in root tips, and the protein is also present in the ER. Caleosins may have a role in membrane and lipid body $Ca^{2+}$-mediated fusion, which may be modulated by the phosphorylation status of the proteins. For a more detailed discussion about caleosins, see Chapter 7.

Sop2, a minor protein in sesame lipid bodies (Chen *et al.*, 1998), was later shown to have homology to a sterol-binding dehydrogenase/reductase class involved in signal transduction and was named steroleosin (Lin *et al.*, 2002b). Structural predictions have indicated that this protein has a hydrophilic dehydrogenase component that is anchored to the surface of the lipid body via its hydrophobic *N*-terminal segment, which features a 'proline knob' with a slightly different structural organization than the proline knot. The cytosolic part of steroleosin forms a seven-stranded β-sheet surrounded by α-helices and is partitioned into an NADPH-binding subdomain, active site and sterol-binding domain. Arabidopsis has been shown to contain eight steroleosin-homologous genes (Lin *et al.*, 2002b). Steroleosin may be involved in signal transduction events with lipid bodies and plant sterols such as brassinosteroid, and may act as a presignal molecule in association with caleosin, serving as partner receptor.

Lipid bodies are also found within plastids. Chloroplasts contain lipid bodies known as plastoglobuli, which may serve as reservoirs for movement of acyl moieties in and out of the thylakoid compartment (Kessler *et al.*, 1999; Kaup *et al.*, 2002). Chloroplast lipid body proteins referred to as fibrillin or plastid-lipid-associated protein and plastoglobulin 1 may have a similar function to the cytoplasmic oleosins by helping to disperse lipid bodies in the stroma (Pozueta-Romero *et al.*, 1997; Kessler *et al.*, 1999; Rey *et al.*, 2000). Indeed, when fibrillin was over-expressed in transgenic tobacco using a constitutive promoter, the chloroplasts contained an increased number of plastoglobuli, which were organized in clusters (Rey *et al.*, 2000).

Unique lipid bodies are also associated with pollen development. The major portion or internal region of the pollen grain is derived from a haploid genome and contains lipid bodies and oleosins in a situation similar to the seed (Huang, 1996). In the anther, developing microspores receive nutrients from specialized cells known as the tapetum, which is based on a diploid genome (Evans *et al.*, 1992; Huang, 1996). The tapetum contains specialized organelles known as elaioplasts and tapetosomes (Wu *et al.*, 1997, 1999; Piffanelli and Murphy, 1998; Ting *et al.*, 1998; Hernández-Pinzón *et al.*, 1999). The elaioplasts have been shown to contain small lipid bodies of 0.2–0.6 μm mainly consisting of steryl esters and proteins homologous to the described plastid associated proteins. In contrast, the tapetosomes are composed of polyunsaturated TAGs, phospholipids and oleosin-like proteins. Pollen storage lipids are discussed further in Section 6.7.

## 6.6 Mobilization of storage lipids

### *6.6.1 Degradation of triacylglycerols into fatty acids*

FAs are released from TAG via the catalytic action of a true lipase, which catalyzes the hydrolysis of its insoluble substrate at an oil–water interface (Huang, 1987; Huang and Wang, 1992). The FAs produced by lipase action undergo further catabolism in the glyoxysomes, which represent specialized peroxisomes involved in the β-oxidation of FAs and directing acetyl-CoA to carbohydrate to provide soluble carbohydrate to support seedling growth (Canvin and Beevers, 1961; Coopers and Beevers, 1969; Huang, 1987; Somerville *et al.*, 2000). Non-glyoxysomal peroxisomes, which also occur in higher plants, lack glyoxylate cycle activity (Gerhardt, 1992). In the remainder of this chapter, these organelles are generally referred to as peroxisomes. Since lipid bodies and peroxisomes are near each other or in immediate contact *in vivo*, this may facilitate the transfer of FAs from degrading lipid bodies to peroxisomes. Some evidence has been presented to suggest that lipid bodies might represent a source of phospholipids for enlarging cottonseed peroxisomes (Chapman and Trelease, 1991). It was shown that the lipid classes in lipid bodies are similar to those found in peroxisomes.

The effective initiation of TAG degradation in lipid bodies may require some degradation of oleosins (Matsui *et al.*, 1999; Sadeghipour and Bhatla, 2002). Treatment of sunflower lipid bodies with a cytosolic protease from the same source rendered them more susceptible to lipase-catalyzed degradation (Sadeghipour and Bhatla, 2002). Caleosins have been shown to disappear within the first 6 days of germination (Næsted *et al.*, 2000) although later studies showed that it was only the seed and lipid-body specific form that disappeared, while a higher molecular weight isoform located on the ER persisted in the developing seedlings (Hernández-Pinzón *et al.*, 2001). Caleosins may play a role in the docking of peroxisomes to lipid bodies during TAG mobilization (Chen *et al.*, 1999; Næsted *et al.*, 2000). Steroleosins may also be involved in this process (Lin *et al.*, 2002b). In contrast, a study by Beisson *et al.* (2001) suggested that lipid body lipolysis in seedlings might occur without any pre-degradation of the protein coat.

Effective initiation of TAG mobilization, in some species, may also be dependent on converting polyunsaturated FA residues in lipid body phospholipids and TAG into oxygenated FAs through lipoxygenase (LOX) (linoleate:oxygen oxidoreductase, EC 1.13.11.12) action (Feussner *et al.*, 1995; Matsui *et al.*, 1999). LOX products (hydroxy FAs) derived from polyunsaturated FA residues (e.g. 18:2), from the phospholipids monolayer and core TAG, have been shown to accumulate during germination in the lipid bodies of seeds from cucumber (*Cucumis sativus*), tobacco and *B. napus* (Feussner *et al.*, 1995). Mature soybean seeds, however, already contained relatively high levels of LOX products suggesting that, in this case, LOX action on lipid bodies took place during seed development. The concept of LOX-dependent degradation of storage TAG emerged, which utilized a special lipase to preferentially release the hydroxy FA residues from the lipoxygenated TAG into the cytosol for eventual metabolism by β-oxidation (Feussner *et al.*, 1995, 2001; Weichert *et al.*, 2002). See Chapter 11 for further discussion about LOXs and their roles in oxylipin formation and function in plants.

The major lipases involved in degrading the bulk of TAG in lipid bodies have been investigated in a number of plant sources. Lipases associated with microsomes and lipid bodies appear to have a major role in this process, although soluble lipases exhibiting action on insoluble TAGs have been also been studied (Huang, 1987; Hills and Mukherjee, 1990). Lipases exhibiting acidic, neutral and alkaline pH optima have been reported (e.g. Lin and Huang, 1983; Hills and Murphy, 1988; García-Agustín *et al.*, 1992). The involvement of the alkaline glyoxysomal lipase from castor bean (Maeshima and Beevers, 1985) in TAG hydrolysis is questionable due to its localization to this organelle (Huang, 1987). The specificity and selectivity of various plant lipases for different FA moieties have been examined along with some studies of the positional specificity of lipase in removing FA residues from a particular position on the glycerol backbone of TAG. Lin *et al.* (1986) have shown that lipases from a number of seed species are relatively specific for TAGs containing the major FA moieties of the storage TAGs in the same species. In contrast, a recent survey of the FA selectivity of lipases from a wide range of oilseeds with varying FA composition indicated that only seeds with more than 80% (w/w) saturated or unusual FAs may contain lipases, which show selectivity (Hellyer *et al.*, 1999). For example, *Cuphea procumbans*, which is enriched in medium chain FAs, was shown to have over 20-fold selectivity for 10:0. Although the soluble lipase from seedlings of *B. napus* has been shown to catalyze the hydrolysis of TAG into FAs and glycerol, the enzyme initially catalyzes the hydrolysis of FA moieties from the *sn*-1 and *sn*-3 positions to yield 1,2 (2,3)-DAGs (Hills and Mukherjee, 1990). Plant lipases show potential for industrial applications in the production of specialized TAGs and esterfied alcohols (Hills *et al.*, 1990; Hills and Mukherjee, 1990; Wanasundara *et al.*, 2001). In some of the applications, the soluble lipase from *B. napus* seedlings was shown to operate in the presence of organic solvents.

A number of factors have been shown to affect plant lipase activity, both directly and indirectly. Proteinaceous inhibitors of soybean lipase exert their effects through interactions with their lipase substrates (Huang and Wang, 1992). During early seedling growth, castor bean lipid bodies have been shown to acquire a neutral lipase activity, which is stimulated 40-fold by 30 μM free $Ca^{2+}$ (Hills and Beevers, 1987). The $Ca^{2+}$ may be required for the catalytic process. Perhaps the $Ca^{2+}$-dependence of this neutral lipase is in some way related to the function of caleosins. The level of lipase observed during germination may be affected by applications of hormones and changes in environmental conditions. Gibberellic acid and jasmonic acid treatments have been shown to enhance lipase activity in germinating seeds (Ranjan and Lewak, 1992, 1995; Imeson *et al.*, 1993). In an early study with barley (*Hordeum vulgare* L. cv Himalaya) aleurone cells, gibberellic was shown to enhance the transfer of a neutral lipase from protein bodies to lipid bodies via lateral diffusion within the plane of the fused phospholipid monolayer (Fernandez and Staehelin, 1987). Although ABA treatment has been found to lead to decreased lipase activity (Imeson *et al.*, 1993; Ranjan and Lewak, 1995), a recent study with *Arabidopsis* mutants has shown that TAG mobilization is largely ABA-independent (Pritchard *et al.*, 2002). In studies with apple (*Malus domestica*) embryos, the acid lipase of the embryo axis has been shown to be particularly active during cold stratification (dormancy removal), whereas alkaline lipase activity was shown to develop in cotyledons during light-stimulated germination of dormant embryos (Żarska-Maciejewska, 1992).

Microsomal lipase from dark grown germinating seedlings of *B. napus* was solubilized with Triton X-100 and partially purified by gel filtration and ion-exchange chromatography (Weselake *et al.*, 1989). The enzyme exhibited a molecular mass of about 250 kDa in the presence of detergent, which was similar to the size of the soluble lipase from the same *Brassica* species (Hills and Mukherjee, 1990). Lipase from the lipid bodies of scutella from maize seedlings has been purified and characterized (Lin and Huang, 1984) and the biosynthesis of the enzyme has been studied (Wang and Huang, 1987). The enzyme was proposed to be a tetramer because SDS-PAGE indicated a molecular mass of 65 kDa whereas sucrose density gradient centrifugation in the presence of 1% (w/v) deoxycholate indicated a molecular mass 270 kDa. Presumably, the intrasubunit associations were maintained in the presence of deoxycholate. Interestingly, the native molecular mass of maize scutellar lipase was similar to the molecular masses of solubilized microsomal lipase and soluble lipase from *B. napus* seedlings. Advances in the purification, cloning and structural analysis of membrane-associated plant lipases have been limited and this represents an area in need of further research.

Recently, an ethylene-induced cDNA encoding a lipase expressed at the beginning of petal senescence was isolated from carnation flowers (*Dianthus caryophyllus* L. cv Improved White Sim) (Hong *et al.*, 2000). Although the main function of this lipase is probably to deesterify membrane lipids during senescence, the enzyme was capable of catalyzing the hydrolysis of trilinolein.

### *6.6.2 β-Oxidation of fatty acids and conversion of lipid to carbohydrate*

Fatty acid β-oxidation results in the conversion of long-chain acyl-CoA to acetyl-CoA by repeated removal of two-carbon units from the thiol end of acyl-CoA (Somerville *et al.*, 2000). Succinate is generated from acetyl-CoA in the glyoxylate cycle, and is converted to carbohydrate to support seedling growth via gluconeogenesis. This section of the chapter is not intended to provide for a comprehensive discussion of the above processes, but rather to provide a general overview of the topic and to highlight a few recent advances. For example, the oxidation of uncommon FAs containing various functional groups, such as hydroxyls and epoxys, and branched FAs is not discussed. The reader is encouraged to consult excellent reviews on FA degradation by Gerhardt (1992) and Graham and Eastmond (2002).

The process whereby FAs from the breakdown of TAG are imported into peroxisomes is unclear (Graham and Eastmond, 2002), but may involve lipid transfer proteins (Edqvist and Farbos, 2002) and/or an isoform of long-chain acyl-CoA synthetase (LACS; EC 6.2.1.3), which is localized on the membrane of the organelle (Hayashi *et al.*, 2002). The steps and enzymes representing core reactions involved in the oxidation of straight-chain saturated FAs are depicted in Figure 6.4 (Gerhardt, 1992; Graham and Eastmond, 2002). Steps 3 and 4 are catalyzed by enzyme activities belonging to the multifunctional protein (MFP). Initially, FAs are

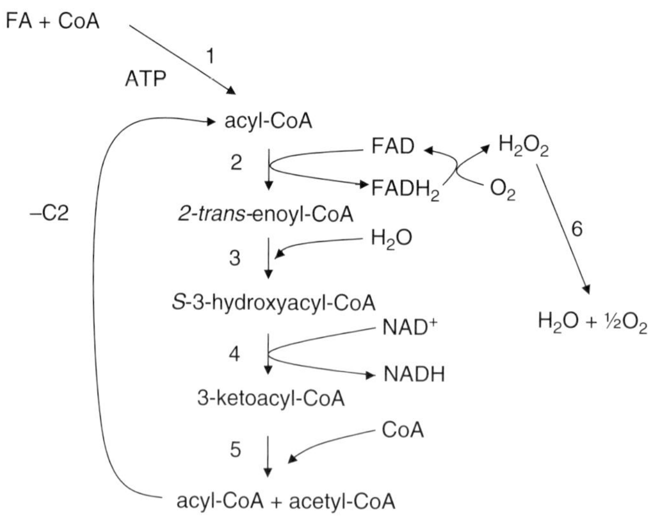


**Figure 6.4** Scheme for fatty acid (FA) activation and the core reactions of β-oxidation of straight-chain saturated FAs. Enzymes: (1) acyl-CoA synthetase, (2) acyl-CoA oxidase, (3) multifunctional protein 2-*trans*-enoyl-CoA hydratase, (4) multifunctional protein *S*-3-hydroxyacyl-CoA dehydrogenase, (5) 3-ketoacyl-CoA thiolase, (6) catalase. Adapted from Gerhardt (1992) and Graham and Eastmond (2002).

activated to acyl-CoAs via the catalytic action of LACS. Acyl-CoA oxidase (ACX; EC 1.3.3.6) catalyzes the conversion of acyl-CoA to 2-*trans*-enoyl-CoA, which represents the first committed step of β-oxidation. FAD is a cofactor in the reaction with molecular oxygen serving as an electron acceptor. The resulting hydrogen peroxide is degraded by a peroxisomal catalase (EC 1.11.1.6). The conversion of 2-*trans*-enoyl-CoA to 3-ketoacyl-CoA is a two-step process that occurs on catalytic sites belonging to MFP. 2-*trans*-Enoyl-CoA hydratase (EC 4.2.1.17) catalyzes the formation of *S*-3-hydroxyacyl-CoA, which in turn is oxidized to 3-ketoacyl-CoA in a $NAD^+$-dependent reaction catalyzed by *S*-3-hydroxyacyl-CoA dehydrogenase (EC 1.1.1.35). Finally, the thiolytic cleavage of 3-ketoacyl-CoA into a two-carbon shortened acyl-CoA and acetyl-CoA is catalyzed by 3-ketoacyl-CoA thiolase (KAT; EC 2.3.1.16).

Additional enzyme activities, combined with reactions of core β-oxidation, are required to oxidize acyl-CoAs consisting of mono- and polyunsaturated FA moieties (Gerhardt, 1992; Graham and Eastmond, 2002). A flow diagram for the β-oxidation of 18:2 $cis\Delta^{9,12}$ is depicted in Figure 6.5. An MFP $\Delta^3,\Delta^2$-enoyl-CoA isomerase (EC 5.3.3.8) is required to accommodate the 3-*cis*-enoyl-CoA (12:2 $cis\Delta^{3,6}$-CoA) generated by β-oxidation of a FA with a *cis* double bond extending from an odd-numbered carbon atom of an unsaturated FA.

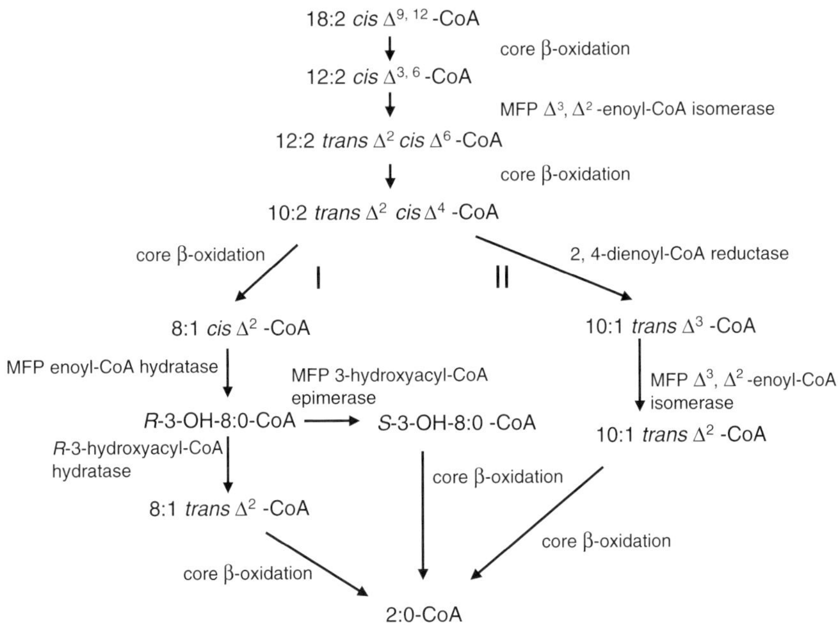


**Figure 6.5** β-Oxidation of linoleic acid. MFP, multifunctional protein. Adapted from Gerhardt (1992) and Graham and Eastmond (2002).

The isomerase catalyzes the conversion of the 3-*cis*-enoyl-CoA to a 2-*trans*-enoyl-CoA (12:2 *trans*$\Delta^2$ *cis*$\Delta^6$-CoA), which is a substrate of the enoyl-CoA hydratase of MFP. The 2-*trans*-enoyl-CoA (10:2 *trans*$\Delta^2$ *cis*$\Delta^4$-CoA) produced through the ACX-catalyzed reaction of core β-oxidation can be further metabolized via two possible pathways (I and II in Figure 6.5) that deal with a FA with a *cis* double bond extending from an even-numbered carbon atom of an unsaturated FA. In pathway I, the 2-*trans*-enoyl-CoA is further metabolized by core β-oxidation to a 2-*cis*-enoyl-CoA (8:1 *cis*$\Delta^2$-CoA) followed by conversion to *R*-3-hydroxyacyl-CoA (instead of *S*-3-hydroxyacyl-CoA) by the catalytic action of the enoyl-CoA hydratase activity of MFP. In turn, *R*-3-hydroxyacyl-CoA is converted to a 2-*trans*-enoyl-CoA (8:1 *trans*$\Delta^2$-CoA) by the catalytic action of *R*-3-hydroxyacyl-CoA hydratase. The 2-*trans*-enoyl-CoA then serves as a substrate for the enoyl-CoA hydratase activity of MFP resulting in production of *S*-3-hydroxyacyl-CoA, which can enter core β-oxidation through the reaction catalyzed by MFP *S*-3-hydroxyacyl-CoA dehydrogenase. In a variation of pathway I, *R*-3-hydroxyacyl-CoA may also be converted to *S*-hydroxyacyl-CoA via the catalytic action of an MFP 3-hydroxyacyl-CoA epimerase. In pathway II, a 2,4-dienoyl reductase (EC 1.3.1.34) is required to catalyze the conversion of the 2-*trans*-enoyl-CoA (10:2 *trans*$\Delta^2$ *cis*$\Delta^4$-CoA) to a 3-*trans*-enoyl-CoA (10:1 *trans*$\Delta^3$-CoA), which can be utilized by the MFP $\Delta^3$, $\Delta^2$enoyl-CoA isomerase to generate a 2-*trans*-enoyl-CoA (10:1 *trans*$\Delta^2$-CoA) for entry into core β-oxidation at the step catalyzed by MFP 2-*trans*-enoyl-CoA hydratase. Pathway II is referred to as the reductase/isomerase pathway.

Kleiter and Gerhardt (1998) have studied the overall β-oxidation pathway in peroxisomes of germinating sunflower seed. The degradation of medium- and short-chain acyl-CoAs was dependent on a system to efficiently remove acetyl-CoA. The study, which utilized radiolabeled FAs, provided evidence that long-chain FAs could be degraded completely to acetyl-CoA by higher plant peroxisomes.

In Arabidopsis, a number of genes encoding various isoforms of LACS have been identified, with two of these (*AtLACS6* and *AtLACS7*) being associated with the peroxisome (Fulda *et al.*, 2002; Hayashi *et al.*, 2002; Shockey *et al.*, 2002). Both of the peroxisomal LACS contained peroxisomal targeting sequences. The subcellular localization of peroxisomal LACS has been studied using fusions of AtLACS6 and AtLACS7 with spectral variants of GFP (Shockey *et al.*, 2002) and immunogold labeling of AtLACS6 (Hayashi *et al.*, 2002). In general, there was a lack of striking substrate specificity among the various LACS suggesting that other factors, such as differences in temporal or cell-type expression, may influence the specific roles of these isozymes (Shockey *et al.*, 2002).

Four Arabidopsis ACX isozymes have been characterized, which have been shown to exhibit substrate specificities encompassing the full range of acyl-CoA chain lengths existing *in vivo* (Eastmond *et al.*, 2000; Graham and Eastmond, 2002). The short-chain ACX4 isoform differed from the three other isoforms in that it had twice as high of a $K_m$ for its preferred acyl-CoA substrates and also formed a tetramer instead of a dimer. The investigation of ACX expression and activity during seedling development revealed that long-chain ACX had a major role in controlling the rate of FA breakdown. Recently, considerable progress into understanding the regulation of β-oxidation and conversion of lipid to carbohydrate has come from biochemical and molecular genetic analyses of seedlings of Arabidopsis with mutations in the various genes encoding enzymes in these pathways. Rylott *et al*. (2003a) investigated the *mutants*, *acx4* and *acx3*, which featured mutations in short-chain ACX3 and medium-chain ACX4, respectively. Short-chain acyl-CoAs accumulated in seedlings of the *acx4* mutant while seedling growth and establishment were unaffected. The same held true for the *acx3* mutant, which accumulated medium-chain acyl-CoAs.

Two MFPs have been cloned from Arabidopsis and enoyl-CoA hydratase activity determined following their over-expression in *E. coli* (Richmond and Bleeker, 1999). Chuong *et al.* (2002) reported that cytosolic microtubule-associated MFP from developing rice seed may have a role in regulating the localization or translocation of mRNAs. Cytosolic MFP was shown to interact with microtubules that are at the periphery of the peroxisome suggesting a role for MFP in facilitating protein import by enriching mRNAs that code for peroxisomal-destined proteins.

Germain *et al*. (2001) investigated the KAT2 isoform, which was encoded by the *KAT* gene on chromosome 2. A homozygous *kat2* mutant was generated in which *KAT2* expression was abolished. In the mutant, seedling growth was dependent upon exogenous sugar, and lipid bodies remained intact in the cotyledons following germination. The peroxisomal compartment increased and long-chain acyl-CoAs accumulated. Since *KAT2* was the only thiolase gene expressed at a substantial level during germination and seedling growth, it was assumed that KAT2 had broad substrate specificity. Through a comparison with glyoxylate cycle *mutants*, the investigators concluded that gluconeogenesis from FAs is not essential to support seedling growth and peroxisomal β-oxidation is necessary for seedling growth and TAG breakdown.

Rylott et al. (2001) have demonstrated the co-ordinate expression of key genes involved in storage lipid mobilization during germination in Arabidopsis. These included genes encoding ACX and KAT of β-oxidation, isocitrate lyase (ICL; EC 4.1.3.1) and malate synthase (MS; EC 2.3.3.9) of the glyoxylate cycle and phosphoenolpyruvate carboxykinase (PEPCK; EC 4.1.1.49) of gluconeogenesis. ICL catalyzes the formation of glyoxylate and succinate from isocitrate, whereas malate synthase catalyzes the formation of malate from glyoxylate and acetyl-CoA. PEPCK catalyzes the conversion of oxaloacetate to phosphoenolpyruvate in gluconeogenesis. A generalized scheme for storage lipid mobilization pathways indicating enzyme activities that are co-ordinated is shown in Figure 6.6. In a more recent study, Rylott *et al.* (2003b) demonstrated that PEPCK was essential for seedling establishment in Arabidopsis. 35S-*PCK1*-antisense Arabidopsis plants were used to characterize the role of PCK1, an isoform of PEPCK, in post-germinative growth. The seedlings of transgenic plants had reduced ability to use both storage lipid and

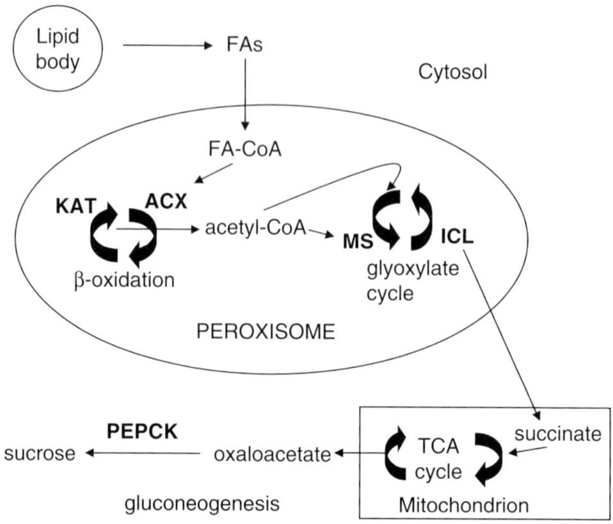


**Figure 6.6** General overview of the conversion of storage lipid into sucrose indicating five key enzymes whose expression is coordinated. Abbreviations used: ACX, acyl-CoA oxidase; FA, fatty acid; ICL, isocitrate lyase; KAT, 3-ketoacyl-CoA thiolase; MS, malate synthase; PEPCK, phosphoenolpyruvate carboxykinase; TCA, tricarboxylic acid. Adapted from Rylott *et al.* (2001).

storage protein through gluconeogenesis to produce soluble carbohydrate. Mutants with disruptions in genes encoding KAT and isocitrate lyase, *kat1* and *icl1*, respectively, have been used to study the effect of ABA on germination and storage lipid mobilization (Pritchard *et al.*, 2002). The study concluded that the ABA block on germination was caused by inhibition of developmental growth related to germination and was independent of TAG mobilization.

### *6.6.3 β-Oxidation during seed maturation*

Eccleston and Ohlrogge (1998) have shown that developing seeds of transgenic *B. napus* producing high levels of 12:0 due to expression of a medium-chain thioesterase, were active in β-oxidation and glyoxylate cycle activity. Levels of ACP and several enzymes of FA synthesis increased to compensate for the lost 12:0 and thus maintained seed oil content. The investigators suggested that metabolic regulation of β-oxidation should be an important consideration in genetic engineering of oilseed crops. In a more recent study with developing embryos of from non-transformed *B. napus*, Chia and Rawsthorne (2000) demonstrated that FA breakdown pathways become more prevalent with embryo maturation. ICL protein was detected in late-stage embryos and PEPCK protein gradually increased during embryo development. Embryo feeding experiments with $[^{14}C]$acetate indicated a shift in carbon usage, at later stages of embryo development, from FA synthesis to FA catabolism.

The accumulation of polyhydroxyalkanoate (PHA) through the action of a microbial PHA synthase on the *R*-isomer of 3-hydroxyacyl-CoA has been shown to be a useful way of assessing the extent of carbon flow through β-oxidation in developing seeds (Mittendorf *et al.*, 1999; Poirier *et al.*, 1999; Poirier, 2002). Co-expression of a PHA synthase gene (*PhaC1*) targeted to the plastid and a plastidial medium-chain thioesterase gene (*FatB3*) in Arabidopsis has been shown to result in enhanced PHA production due to stimulation of the β-oxidation pathway (Poirier *et al.*, 1999). Analysis of the PHA showed that 3-hydroxydecanoic acid was the prevalent monomer, which indicated a large flow of 10:0 toward FA breakdown. In the same study, enhanced PHA accumulation was also observed when PHA synthase was expressed in an Arabidopsis mutant deficient in DGAT activity indicating that unincorporated FAs were channeled into β-oxidation. Although the *S*-isomer of 3-hydroxyacyl-CoA is an intermediate in the β-oxidation of saturated FAs, there are at least three metabolic possibilities for the production of the *R*-isomer to fuel the PHA synthase (Poirier, 2002) (refer to Figure 6.5). One possibility lies with the action of 3-hydroxyacyl-CoA epimerase activity of MFP on *S*-hydroxyacyl-CoA to generate *R*-hydroxyacyl-CoA. Second, the enoyl-CoA hydratase activity of MFP could potentially catalyze the conversion of 2-*cis*-enoyl-CoA into *R*-hydroxyacyl-CoA. In the latter case, the 2-*cis*-enoyl-CoA would be derived from the β-oxidation of unsaturated FAs with a *cis* double bond at an even position. A third possibility is linked to an enoyl-CoA hydratase activity, which can catalyze the conversion of 2-*trans*-enoyl-CoA to *R*-3-hydroxyacyl-CoA. In another recent investigation of β-oxidation during embryo development, Rylott *et al.* (2003a) generated Arabidopsis *acx3acx4* double *mutants* with a complete block in short-chain ACX activity. The developing seeds aborted during the first phase of embryo development suggesting that peroxisomal β-oxidation might have a role in the generation of lipid-based signaling molecules, which are necessary for embryogenesis.

## 6.7 Storage lipids in developing pollen grains

As indicated in Section 6.5, developing pollen grains of the immature anther also accumulate storage lipids. Microspores, produced by meiosis, divide into two identical daughter cells, which in turn lead to the formation of a pollen grain (Reynolds, 1997). Mature pollen grains of *B. napus* consist of about one-third lipid on a dry weight basis, the majority of which is associated with the internal region of the pollen grain (Evans *et al.*, 1987). This reservoir of lipid may facilitate the eventual growth of the pollen tube (Piffanelli *et al.*, 1998). As the pollen grains develop, the cells of the tapetum, which once nurtured the microspores, eventually rupture to discharge lipids and proteins onto the surface of the pollen grain to form the tryphine, which imparts adhesion, insect attractability and pollen recognition characteristics (Pacini *et al.*, 1985; Huang, 1996; Piffanelli *et al.*, 1998; Mayfield and Preuss, 2000).

Piffanelli *et al.* (1997) have shown that the FA composition of TAGs from both the pollen internal region and the tryphine are similar at maturity in *B. napus* L. cv Topas. α-Linolenic acid represented about 65% of the total FAs in TAG from both sources, with 18:1 being a minor constituent. The bulk of storage lipid accumulation in developing pollen grains of *B. napus* has been shown to occur during the late-vacuolate and early-maturation stages of development, when bud lengths range from 3.4 to 4.3 mm (Evans *et al.*, 1992). In contrast, the tapetum was shown to accumulate lipid at an earlier stage of maturation, when bud length was about 2.7 mm. Expression of $\Delta^9$-stearoyl-ACP desaturase (EC 1.14.99.6) and enoyl-ACP reductase (EC 1.3.1.9) has been investigated in both the internal region of the pollen grain and the tryphine (Piffanelli *et al.*, 1997). Although both tissue compartments appeared to involve the same FA biosynthesis genes, the timing of expression was different, with transcript production for these enzymes occurring at somewhat later stage in the internal region of the pollen grain.

Within the last decade, numerous studies have appeared on the expression, targeting, localization and genomics of oleosin-like proteins in pollen (see Huang, 1996 and Piffanelli *et al.*, 1998 for reviews). Kim *et al.* (2002) identified 16 oleosin genes in the Arabidopsis genome, which were divided into three groups based on their sequences and tissue-specific expression. Eight genes were expressed in the tapetum with five genes expressed in maturing seeds, and three genes expressed in developing seeds and microspores. Franco *et al.* (2002) have shown that oleosin glycine-rich protein genes designated *Atgrp-6*, *Atgrp-7* and *Atgrp-8* occur in a cluster in the Arabidopsis genome and are expressed specifically in the tapetum. This may represent a mechanism to co-ordinate protein production of specific oleosin-like proteins at the same time. Partially degraded oleosin-like proteins and steryl esters are deposited onto the outside of the pollen grain following rupture of the tapetum cells (Ross and Murphy, 1996; Wu *et al.*, 1997; Ting *et al.*, 1998). The plastid-associated proteins of the elaioplasts and the tapetosome TAGs, however, appeared to be hydrolyzed during later stages of pollen development (Ting *et al.*, 1998). The role of the plastid-associated proteins appears to be in stabilizing the globules of steryl esters and not in contributing to the pollen protein coat. An alkaline lipase activity appeared to be responsible for catalyzing the hydrolysis of tapetosome TAGs (Ting *et al.*, 1998). Foster *et al.* (2002) have targeted the marker enzyme, GUS, to the pollen coat of transgenic *Brassica carinata* plants, which expressed a fusion of GUS and tapetal oleosin-like protein. Given the potential importance of oleosin-like proteins of the pollen coat in pollen recognition (Mayfield and Preuss, 2000), the study suggested that it may be possible to alter the interaction between pollen and pistil, which could have important ramifications in terms of controlling transgene flow.

TAG biosynthetic enzymes have been investigated to a limited extent in developing pollen grains. Hobbs *et al.* (1999) have shown that developing anthers of *B. napus* express high levels of transcript encoding TAG biosynthetic enzymes, which is consistent with the high TAG content of the pollen grain. Recently, Zheng *et al*., (2003) have shown that a deficiency in AtGPAT1 correlates with impaired ER membrane formation and altered oil-body size in developing pollen grains. Most microspores aborted before reaching maturity indicating that AtGPAT1 has a critical role in proper pollen development and male fertility.

## 6.8 Effect of environmental conditions and carbon source on triacylglycerol accumulation

The process of TAG accumulation and associated events in the ER can be affected by environmental conditions such as light intensity, temperature, mineral deprivation, osmotic agents and alteration of carbon source. In many cases, the application of an altered environmental condition can be revealing about aspects of lipid biosynthesis. Environmental effects leading to increased TAG accumulation have been linked to changes in phospholipid metabolism in the ER. In studies with developing seed of the maize *floury-2* mutant, Shank *et al*. (2001) suggested that the TAG content of some crop species might be affected by the ER stress response pathway during seed development, which included activation of phospholipid biosynthetic enzymes. The *floury-2* mutant is a single point mutation in the α-zein storage protein that stops the signal peptide from being cleaved, causing it to remain attached and anchoring zein to the membrane (Gillikin *et al*., 1997). The mutant gene is only expressed in the developing seed. Unfolded proteins accumulating in the ER are known to trigger a signal transduction pathway involving chaperone induction and changes in membrane phenotype.

Perry *et al*. (1999) measured the endogenous levels of Kennedy pathway intermediates in developing *B. napus* embryos exposed to dark and light conditions. FA production increased under light conditions and resulted in increases in all intermediates except LPA. DAG levels were increased significantly under light conditions pointing to the DGAT-driven step as exerting a substantial level of control in TAG synthesis. Also, an exciting new study by Vigeolas *et al*. (2003) has shown that oxygen supply can be a limiting factor, which affects lipid synthesis more than starch synthesis. Elevated oxygen supply led to two-fold higher rates of lipid synthesis, and DGAT activity was substantially increased in developing seeds from siliques that were exposed to higher oxygen tensions.

Considerable research has been conducted on the effect of temperature on FA desaturation (Mikami and Murata, 2003). Much less, however, is known about the effects of temperature on TAG accumulation. An early study by Canvin (1965) demonstrated that decreasing temperature during seed development in *B. napus* L. cv Nugget resulted in an increase in total seed oil accumulation. Another investigation with MD embryos of *B. napus* demonstrated that there were genotypic differences on the effect of temperature on embryo oil content (Wilmer *et al*., 1996).

Although chloroplast GPAT probably does not participate in the synthesis of TAG in the ER, there appears to be different forms of the enzyme that are well suited to promote changes in the FA composition of membranes under different environmental conditions. The substrate specificity and selectivity properties of chloroplast GPAT suggest that the plastidial form of the enzyme has a role in controlling the unsaturated FA content of phosphatidylglycerol (Bertrams and Heinz, 1981; Frentzen *et al.*, 1983, 1987). Murata *et al.* (1992) were able to alter the chilling sensitivity of tobacco (*N. tabacum*) by transforming the plant with cDNAs encoding plastidial GPATs that differed in FA selectivity. The decreased chilling sensitivity of plants transformed with a cDNA from *A. thaliana* was linked to increased selectivity of the *A. thaliana* GPAT for *cis*-unsaturated FA moieties, which resulted in more unsaturation in the phosphatidylglycerol of chloroplast membranes. In contrast, transformation with squash (*Cucurbita moschata*) GPAT, which had increased selectivity for 16:0 moieties, resulted in increased chilling sensitivity.

High osmoticum and/or ABA treatment has been shown to induce TAG accumulation in a number of oil forming culture systems (Attree *et al.*, 1992; Radetzky and Langheinrich, 1994; Rodríguez-Sotres and Black, 1994; Pacheco-Moisés *et al.*, 1997). As well, DGAT activity was also shown to increase in response to increased osmoticum in immature maize embryos (Pacheco-Moisés *et al.*, 1997). As indicated in Section 6.5, treatment of MD embryos of *B. napus* with ABA or increased osmoticum has been shown to stimulate production of transcripts encoding oleosins (van Rooijen *et al.*, 1992) and of oleosin polypeptides (Holbrook *et al.*, 1991). Treatment of cell cultures of anise (*Pimpinella anisum* L.) with sorbitol, or ABA and sorbitol, has also been shown to induce oleosin production (Radetzky and Langheinrich, 1994). ABA-treated germinating embryo quarter seeds of rice have also been shown to have an enhanced production of mRNA encoding caleosin (Frandsen *et al.*, 1996). Interestingly, either high light intensity or drought stress has been shown to induce the production of fibrillin or plastid lipid-associated protein in tobacco leaves producing this lipid body protein (Rey *et al.*, 2000).

Introduction of increasing sucrose concentrations into MD cell suspension cultures of *B. napus* L. cv Jet Neuf have been shown to result in increased TAG accumulation after two weeks of culture, which may have been induced by both increased osmoticum and carbon supply (Weselake *et al.*, 1998). *BnDGAT1* transcript was positively associated with DGAT activity as the sucrose concentration was raised from 2 to 6% (w/v), but the level of transcript was lower following culture in 14% sucrose (Davoren *et al.*, 2002). Cells cultured in 14% sucrose may have formed TAG earlier in the culture period than cells cultured in 6% sucrose, or the temporal regulation of the transcript may have varied under the two sucrose regimes. Analysis of gene expression as a function of increasing sucrose concentration also revealed a sucrose-induced accumulation of transcript encoding a putative mitochondrial ACP, hinting at a possible relationship between mitochondrial FA biosynthesis and TAG biosynthesis in the ER (Davoren *et al.*, 2002). Indeed, mitochondria from higher plants have been shown to be capable of synthesizing FAs (Gueguen *et al.*, 2000; Focke *et al.*, 2003; see Chapter 5 of this volume). The time course of TAG accumulation in MD cell suspension cultures of *B. napus* following transfer from 6 to 14% sucrose concentration has also been investigated (Nykiforuk *et al.*, 2002). Exposure to higher sucrose concentration led to a burst of TAG accumulation and associated DGAT activity, which peaked at 6 h after transfer, but this was followed by a gradual increase in TAG accumulation and recovery of DGAT activity from 12 to 72 h after transfer. The disappearance of the initial TAG peak coincided with a temporal increase in TAG lipase activity. Maximum levels of *BnDGAT1* transcript occurred shortly after the peak of DGAT activity suggesting some post-transcriptional control of enzyme activity. Thus, the MD cell suspension system is capable of both TAG biosynthesis and mobilization within a short period of time, and may prove beneficial in examining associated mechanisms.

The nature of the carbon source may also exert an effect on TAG accumulation. Arabidopsis *mutants* deficient in DGAT1 have been shown to display altered carbohydrate metabolism, and increased sensitivity to ABA, glucose and osmotic stress during germination and seedling development (Lu and Hills, 2002). A recent study by Lu *et al.* (2003), using a *GUS-DGAT1* fusion as a reporter gene, has shown that DGAT1 expression in Arabidopsis seed is upregulated by glucose and associated with glucose-induced changes in seedling development.

Aspects of TAG mobilization and FA oxidation are also affected by changes in environmental conditions. Indeed, it was indicated above that MD cell suspension cultures of *B. napus* entered a short catabolic phase following an initial sucrose-induced burst of TAG formation (Nykiforuk *et al.*, 2002). In addition, decreasing the sucrose concentration in cell cultures of anise has been shown to lead to TAG mobilization and a decrease in the amount of oleosin (Radetzky and Langheinrich, 1994). Sambanthamurthi *et al.* (1991) have reported the activation of a lipase in mesocarp of oil palm (*E. guineensis*) by low temperature stress. The investigators suggested that the lipase activity was the result of differential effects on protein synthesis in response to the cold. In terms of FA β-oxidation, the *AtACX1* gene appears to respond to a number of environmental stresses (Graham and Eastmond, 2002). For example, studies have shown that the gene is induced in response to dehydration and drought stress.

Other environmental effects on plants have been shown to involve aspects of both lipid catabolism and TAG accumulation. Conversion of chloroplast MGDAG into TAG appears to function under normal environmental conditions in spinach leaves (see Figure 6.3) (Sakaki *et al.*, 1990a,b). Ozone treatment of spinach leaves, however, has been shown to enhance TAG accumulation in the leaves through this pathway. Ozone treatment stimulated the hydrolysis of this galactolipid to produce free FAs and *sn*-1,2-DAG catalyzed by galactolipase and galactolipid: galactosyltransferase activity, respectively. Following activation of free FA to acyl-CoA, the necessary substrates were in place for DGAT action to produce TAG. The investigators refer to a number of other environmental stresses that exert the same effect.

## 6.9 The role of lipid–protein particles and plasma membrane vesicles in membrane turnover

Lipid–protein particles and plasma membrane (PM) vesicles have been characterized and are involved in both normal membrane turnover and membrane degradation during senescence (Thompson *et al.*, 1998). These particles and vesicles serve as reservoirs for the removal of lipids and other components that might lead to membrane destabilization. Without having mechanisms for membrane repair, the functional integrity of membranes is lost and they become 'leaky'. Hudak and Thompson (1996) have isolated cytosolic lipid–protein particles from the leaves of carnation petals, which contain phospholipid, free FAs, steryl and wax esters, PA, DAG, TAG and a 17 kDa protein unrelated to oleosin. Smith *et al.* (2000) have isolated two distinct populations of lipid–protein particles from the chloroplasts of leaves of wax beans, which appear to act as reservoirs for catabolites of thylakoid membranes. One population of particles was buoyant and corresponded to plastoglobuli whereas the second population of particles was of higher density. Both populations of lipid–protein particles, however, contained remnants of cytochrome *f* from the thylakoid membrane suggesting that they were involved in removal of catabolite proteins as a part of normal membrane turnover. In another study, Madey *et al.* (2001) used antibodies raised against a recombinant polypeptide of PM $H^+$-ATPase to isolate a subpopulation of PM vesicles from carnation petals and *B. napus* leaves. The vesicles were enriched in free FAs, DAG, TAG, and steryl and wax esters. The vesicles were distinct, however, from the cytosolic lipid–protein particles observed by Hudak and Thompson (1996). It was interpreted that the vesicles arose by phase-separation of metabolites in the lipid bilayer, which formed domains in the PM that were released as vesicles into the cytosol, presumably by microvesiculation. The vesicles probably function in normal membrane turnover in the PM.

Both an ethylene-induced lipase (Hong *et al.*, 2000) and DGAT1 (Kaup *et al.*, 2002) appear to have important roles in senescence. The isolation of the cDNA for ethylene-induced lipase is first mentioned in Section 6.6.1. Expression of the equivalent cDNA from Arabidopsis in antisense orientation under the control of a constitutive 35S promoter resulted in delayed leaf senescence and increased seed yield (Thompson *et al.*, 2000). Kaup *et al.* (2002) have suggested that senescence-related enhancement of AtDGAT1 production and TAG accumulation may serve as a mechanism to sequester galactolipid FAs produced during senescence in a fashion similar to the TAG-generation mechanism in ozone-treated spinach leaves proposed by Sakaki *et al.* (1990a,b). The sequestering of FAs in plastidial TAG may represent an intermediate step in converting thylakoid FAs to phloem-mobile sucrose (Kaup *et al.*, 2002).

Inhibition of the sterol pathway in leek (*Allium porrum*) by fenpropimorph, a sterol biosynthesis inhibitor, has been shown to result in large increases in TAG content in both aerial parts and roots of seedlings of this plant (Hartmann *et al.*,

2002). Also, the fenpropimorph treatment led to the appearance of lipid droplets in the cytosol of cells from root and aerial parts, and enhanced DGAT activity. The sterol biosynthesis inhibitor resulted in the accumulation of 9 β19-cyclopropylsterols in place of the usual $\Delta^5$ sterols. It was presumed that retailoring of ER phospholipid was necessary to improve cyclopropylsterol-phospholipid interactions. Thus, it was suggested that DGAT had a role in diverting the non-utilized FAs into TAG.

## 6.10 Biosynthesis of liquid wax esters

The jojoba desert shrub of the Southwestern United States is the only angiosperm that is known to produce liquid wax esters as an energy store (Voelker and Kinney, 2001). The seed oil of the jojoba shrub is used in cosmetics and may have industrial applications as a lubricant, but the high price of the oil limits its use in the latter application (Lassner *et al.*, 1999). An example of a molecular species of a liquid wax ester is depicted in Figure 6.1B. Paul Stumpf and associates conducted the initial work on the biosynthesis of jojoba liquid wax (Ohlrogge *et al.*, 1978; Pollard *et al.*, 1979; Wu *et al.*, 1981). Oleic acid is produced in plastids of developing jojoba seeds in the same fashion as regular oilseed crops (Ohlrogge and Browse, 1995; Lassner *et al.*, 1999). Oleoyl-CoA is converted into liquid wax through the catalytic action of three enzymes, which include the FA elongase complex, acyl-CoA reductase (EC 1.2.1.42) and wax synthase (EC 2.3.1.75) (Pollard *et al.*, 1979; Lassner *et al.*, 1999) (Figure 6.7). β-Ketoacyl-CoA synthase catalyzes the condensation of malonyl-CoA with

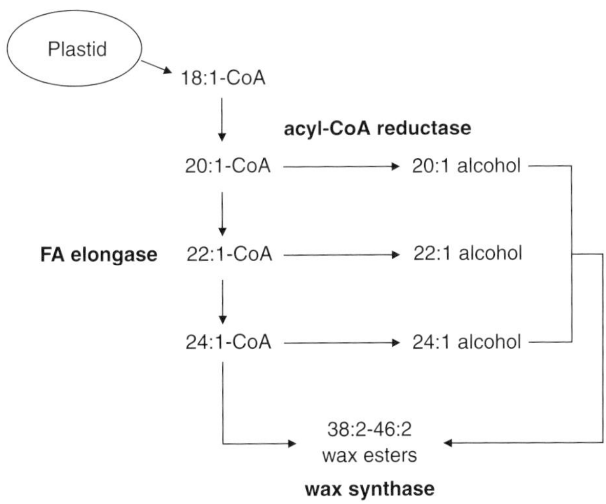


**Figure 6.7** Scheme for the synthesis of liquid wax esters in developing jojoba seed. Adapted from Pollard *et al.* (1979) and Lassner *et al.* (1999). FA, fatty acid.

long-chain acyl-CoA and is the initial step of the microsomal elongation pathway (Lassner *et al.*, 1996). This pathway converts 18:1 to eicosenoic acid (20:1), 22:1 and nervonic acid (24:1). Acyl-CoA reductase catalyzes the formation of fatty alcohols from 20:1-, 22:1 and 24:1-CoAs (Lassner *et al.*, 1999; Metz *et al.*, 2000). Wax synthase (fatty acyl-CoA:fatty alcohol acyltransferase) catalyzes the final step in the production of the linear esters from acyl-CoA substrates and fatty alcohols (Lardizabal *et al.*, 2000). cDNAs encoding β-ketoacyl-CoA synthase, acyl-CoA reductase and wax synthase have all been cloned (Lassner *et al.*, 1996; Lardizabal *et al.*, 2000; Metz *et al.*, 2000). Co-expression of jojoba acyl-CoA reductase and β-ketoacyl-CoA synthase from *Lunaria annua* during seed development in Arabidopsis resulted in seed oils with wax levels of up to 70% by weight (Lardizabal *et al.*, 2000). Also, expression of jojoba acyl-CoA reductase in high-erucic acid *B. napus* L. cv Reston resulted in the synthesis of a small amount of wax ester (Metz *et al.*, 2000). The *B. napus* embryos contained an endogenous wax synthase activity that could account for the wax synthesis.

Steinke *et al.* (2001) implemented an industrial approach to produce oils with more than 95% wax esters having compositions approaching that of jojoba oil. The procedure involved lipase-catalyzed esterification with FAs and alcohols from crambe (*Crambe abyssinica*) and camelina (*Camelina sativa*) oils. Recently, a gene encoding a novel bifunctional wax synthase/DGAT, which catalyzes both wax ester and TAG synthesis, was identified in *Acinetobacter calcoaceticus* ADP1 (Kalscheuer and Steinbüchel, 2003). The chemical structure of the wax ester produced by this microorganism is similar to jojoba wax ester. The bifunctional enzyme was not related to the DGAT1 and two families, PDAT or joboba wax synthase.

## 6.11 Do plants transport storage lipids?

A recent study by Madey *et al.* (2002) has identified spherical lipid particles in the phloem sap of stems of *B. napus* plants, which consisted of phospholipid, DAG, TAG, steryl and wax esters and a high proportion of free FAs. The unesterified FAs consisted mainly of medium-chain FAs, whose proportion increased when plants were stressed with sublethal doses of ultraviolet radiation. The investigators suggested that the phloem lipid particles might represent a long distance lipid transport system in plants, which is analogous to the function of lipoproteins in animals.

## 6.12 Conclusions and future directions

Storage lipid synthesis is a complex process that involves not only the donation of acyl chains from thioesters to a glycerol backbone, but also acyl-exchange reactions with the acyl-CoA pool and the movement of acyl chains between glycerolipids. Membrane repair appears to work in close association with TAG accumulation.

The cloning of most of the cDNAs for enzymes involved in TAG synthesis and degradation and conversion of lipid to carbohydrate has created a new era where a multitude of knockout and overexpression studies are being used to gain insight into the regulation of these processes. These studies will also set the foundation for developing new and improved strategies to genetically engineer oilseeds for enhanced seed oil content and desired FA composition of seed oil, especially FAs for nutraceutical and non-food applications. There are still, however, some key genes encoding TAG synthetic enzymes that have yet to be isolated, including type-1 PA phosphatase, DGTA and LPCAT. Detailed structural information is available on plastidial GPAT, but structural information on ER-resident acyltransferases is lacking and has largely been restricted to structural predictions based on the amino acid sequence. Hydrophobic proteins may also exist in the ER, which regulate the acyltransferases of TAG accumulation. In other cases, soluble factors might interact with segments of acyltransferases facing either the cytosol or lumen. The enzymology of storage lipid formation in developing pollen grains is also in need of investigation. An assortment of lipid-containing particles, in both cytoplasmic and plastidial cellular compartments, is intimately involved in TAG accumulation, membrane turnover and stress responses. The study of TAG mobilization in the germinating seed is also in need of investigation, in particular the identification of TAG lipase genes and the possible interaction of these acylhydrolases with constituents of lipid bodies including caleosins. It has become increasingly clear that TAG degradation can occur during seed development and TAG synthesis can occur during seed germination. Introduction of PHA synthases into plants can divert β-oxidation intermediates into PHA and thus provide information on the extent of this catabolic process under various conditions, especially during seed maturation. Studies on the effects of various environmental conditions on oilseeds and oil-forming culture systems have provided new insights into TAG synthesis, in particular the recent discovery (Vigeolas *et al.*, 2003) which has shown that low oxygen limits oil accumulation in developing seeds of *B. napus*. The genetic engineering of major oilseed crops to produce liquid wax esters holds the promise of an economical source of this lipid for applications such as biolubricants. The discovery of lipid particles in the phloem by Madey *et al.* (2002) will lead to experiments to assess their origin and the possible role of these particles in the transport of lipid in the plant.

## Acknowledgements

The author is grateful to the following funding agencies for supporting his research on TAG biosynthesis in oilseeds: Alberta Agricultural Research Institute, Alberta Crop Industry Development Fund, Alberta Network for Proteomics Innovation, Flax Council of Canada, Genome Prairie and Genome Canada, Imperial Oil Ltd., Natural Sciences and Engineering Research Council of Canada, Saskatchewan Agricultural Development Fund, Western Economic Diversification and The University of Lethbridge. The author also thanks Tara Furukawa-Stoffer, Phillip Huff and Brent Sorensen for their help with proofreading of the manuscript.

## References

Abell, B.M., High, S. and Moloney, M.M. (2002) Membrane protein topology of oleosin is constrained by its long hydrophobic domain. *The Journal of Biological Chemistry*, **277**, 8602–8610.

Abell, B.M., Holbrook, L.A., Abenes, M., Murphy, D.J., Hills, M.J. and Moloney, M.M. (1997) Role of the proline knot motif in oleosin endoplasmic reticulum topology and oil body targeting. *The Plant Cell*, **9**, 1481–1493.

Akermoun, M., Tester, E., Cassagne, C. and Bessoule, J.-J. (2000) Solubilization of the plastidial lysophosphatidylcholine acyltransferase from *Allium porrum* leaves: towards plants devoid of eukaryotic plastid lipids? *Biochemical Society Transactions*, **28**, 713–715.

Alexander, L.G., Sessions, R.B., Clarke, A.R., Tatham, A.S., Shewry, P.R. and Napier, J.A. (2002) Characterization and modeling of the hydrophobic domain of a sunflower oleosin. *Planta*, **214**, 546–551.

Attree, S.M., Pomeroy, M.K. and Fowke, L.C. (1992) Manipulation of conditions for the culture of somatic embryos of white spruce for improved triacylglycerol biosynthesis and desiccation tolerance. *Planta*, **187**, 395–404.

Bafor, M., Smith, M.A., Jonsson, L., Stobart, K. and Stymne, S. (1991) Ricinoleic acid biosynthesis and triacylglycerol assembly in microsomal preparations from developing castor-bean (*Ricinus communis*). *Biochemical Journal*, **280**, 507–514.

Bafor, M., Smith, M.A., Jonsson, L., Stobart, K. and Stymne, S. (1993) Biosynthesis of vernoleate (*cis*-12-epoxyoctadeca-*cis*-9-enoate) in microsomal preparations from developing endosperm of *Euphorbia lagascae*. *Archives of Biochemistry and Biophysics*, **303**, 145–151.

Bafor, M., Stobart, A.K. and Stymne, S. (1990) Properties of the glycerol acylating enzymes in microsomal preparations from the developing seeds of safflower (*Carthamus tinctorius*) and turnip rape (*Brassica campestris*) and their ability to assemble cocoa-butter type fats. *Journal of the American Oil Chemists' Society*, **67**, 217–225.

Bahn, S.C., Lee, H.Y., Kim, H.J., Ryu, S.B. and Shin, J.S. (2003) Characterization of Arabidopsis secretory phospholipase $A_2$-γ(cDNA) and its enzymatic properties. *FEBS Letters*, **553**, 113–118.

Baldo, A., Sniderman, A.D., St. Luce, S., Zhang, X.J. and Cianflone, K. (1995) Signal transduction pathway of acylation stimulating protein: involvement of protein kinase C. *Journal of Lipid Research*, **36**, 1415–1426.

Banaś, A., Dahlqvist A., Ståhl, U., Lenman, M. and Stymne S. (2000) The involvement of phospholipid:diacylglycerol acyltransferases in triacylglycerol production. *Biochemical Society Transactions*, **28**, 703–705.

Banaś, A., Johansson, I. and Stymne, S. (1992) Plant microsomal phospholipases exhibit preference for phosphatidylcholine with oxygenated acyl groups. *Plant Science*, **84**, 137–144.

Bao, X. and Ohlrogge, J. (1999) Supply of fatty acid is one limiting factor in the accumulation of triacylglycerol in developing embryos. *Plant Physiology*, **120**, 1057–1062.

Beaudoin, F. and Napier, J.A. (2000) The targeting and accumulation of ectopically expressed oleosin in non-seed tissues of *Arabidopsis thaliana*. *Planta*, **210**, 439–445.

Beaudoin, F. and Napier, J.A. (2002) Targeting and membrane-insertion of a sunflower oleosin *in vitro* and in *Saccharomyces cerevisiae*: the central hydrophobic domain contains more than one signal sequence, and directs oleosin insertion into the endoplasmic reticulum membrane using a signal anchor sequence mechanism. *Planta*, **215**, 293–303.

Beaudoin, F., Lacey, D.J. and Napier, J.A. (1999) The biogenesis of the plant seed oil body: oleosin protein is synthesized by ER-bound ribosomes. *Plant Physiology and Biochemistry*, **37**, 481–490.

Beaudoin, F., Wilkinson, B.M., Stirling, C.J. and Napier, J.A. (2000) *In vivo* targeting of a sunflower oil body protein in yeast secretory (*sec*) mutants. *The Plant Journal*, **23**, 159–170.

Beisson, F., Ferté, N., Bruley, S., Voultoury, R., Verger, R. and Arondel, V. (2001) Oil-bodies as substrates for lipolytic enzymes. *Biochimica et Biophysica Acta*, **1531**, 47–58.
Bergfeld, R. Hong, Y.-N., Kühnl, T. and Schopfer, P. (1978) Formation of oleosomes (storage lipid bodies) during embryogenesis and their breakdown during seedling development in cotyledons of *Sinapis alba* L. *Planta*, **143**, 297–307.
Bernerth, R. and Frentzen, M. (1990) Utilization of erucoyl-CoA by acyltransferases from developing seeds of *Brassica napus* (L.) involved in triacylglycerol biosynthesis. *Plant Science*, **67**, 21–28.
Bertrams, M. and Heinz, E. (1981) Positional specificity and fatty-acid selectivity of purified *sn*-glycerol 3-phosphate acyltransferases from chloroplasts. *Plant Physiology*, **68**, 653–657.
Bessoule, J.J., Testet, E. and Cassagne, C. (1995) Synthesis of phosphatidylcholine in the chloroplast envelope after import of lysophosphatidylcholine from endoplasmic reticulum membranes. *European Journal of Biochemistry*, **228**, 490–497.
Block, M.A., Dorne, A-J., Joyard, J. and Douce, R. (1983) The phosphatidic acid phosphatase of the chloroplast envelope is located on the inner envelope membrane. *FEBS Letters*, **164**, 111–115.
Bourgis, F., Kader, J-C., Barret, P., Renard, M., Robinson, D., Robinson, C., Delseny, M. and Roscoe, T.J. (1999) A plastidial lysophosphatidic acid acyltransferase from oilseed rape. *Plant Physiology*, **120**, 913–921.
Bouvier-Navé, P., Benveniste, P., Oelkers, P., Sturley, S.L. and Schaller, H. (2000) Expression in yeast and tobacco of plant cDNAs encoding acyl CoA:diacylglycerol acyltransferase. *European Journal of Biochemistry*, **267**, 85–96.
Brindley, D.N. (1984) Intracellular translocation of phosphatidate phosphohydrolase and its possible role in the control of glycerolipid synthesis. *Progress in Lipid Research*, **23**, 115–133.
Brockerhoff, H. (1971) Stereospecific analysis of triglycerides. *Lipids*, **6**, 942–956.
Brough, C.L., Coventry, J.M., Christie, W.W., Kroon, J.T.M., Brown, A.P., Barsby, T.L. and Slabas, A.L. (1996) Towards the genetic engineering of triacylglycerols of defined fatty acid composition: major changes in erucic acid content at the *sn*-2 position affected by the introduction of a 1-acyl-*sn*-glycerol-3-phosphate acyltransferase from *Limnanthes douglasii* into oil seed rape. *Molecular Breeding*, **2**, 133–142.
Broun, P., Gettner, S. and Somerville, C. (1999) Genetic engineering of plant lipids. *Annual Review of Nutrition*, **19**, 197–216.
Brown, A.P., Brough, C.L., Kroon, J.T.M. and Slabas, A.R. (1995) Identification of a cDNA that encodes a 1-acyl-*sn*-glycerol-3-phosphate acyltransferase from *Limnanthes douglasii*. *Plant Molecular Biology*, **29**, 267–278.
Brown, A.P., Carnaby, S., Brough, C., Brazier, M. and Slabas, A.R. (2002b) *Limnanthes douglasii* lysophosphatidic acid acyltransferases: immunological quantification, acyl selectivity and functional replacement of *Escherichia coli plsC* gene. *Biochemical Journal*, **364**, 795–805.
Brown, A.P., Coleman, J., Tommey, A.M., Watson, M.D. and Slabas, A.R. (1994) Isolation and characterization of a maize cDNA that complements a 1-acyl *sn*-glycerol-3-phosphate acyltransferase mutant of *Escherichia coli* and encodes a protein which has similarities to other acyltransferases. *Plant Molecular Biology*, **26**, 211–223.
Brown, A.P., Johnson, P., Rawsthorne, S. and Hills, M.J. (1998) Expression and properties of acyl-CoA binding protein from *Brassica napus*. *Plant Physiology and Biochemistry*, **36**, 629–635.
Brown, A.P., Slabas, A.R. and Denton, H. (2002a) Substrate selectivity of plant and microbial lysophosphatidic acid acyltransferases. *Phytochemistry*, **61**, 493–501.
Byers, S.D., Laroche, A., Smith, K.C. and Weselake, R.J. (1999) Factors enhancing diacylglycerol acyltransferase activity in microsomes from cell-suspension cultures of oilseed rape. *Lipids*, **34**, 1143–1149.
Cahoon, E.B. and Ohlrogge, J.B. (1994) Apparent role of phosphatidylcholine in the metabolism of petroselinic acid in developing *Umbelliferae* endosperm. *Plant Physiology*, **104**, 845–855.
Canvin, D.T. (1965) The effect of temperature on the oil content and fatty acid composition of the oils from several oil seed crops. *Canadian Journal of Botany*, **43**, 63–69.
Canvin, D.T. and Beevers, H.T. (1961) Sucrose synthesis from acetate in the germinating castor bean: kinetics and pathway. *The Journal of Biological Chemistry*, **236**, 988–995.

Cao, J.S., Lockwood, J., Burn, P. and Shi, Y.G. (2003) Cloning and functional characterization of a mouse intestinal acyl CoA-monoacylglycerol acyltransferase, MGAT2. *Journal of Biological Chemistry*, **278**, 38870–38876.

Cao, Y-Z. and Huang, A.H.C. (1986) Diacylglycerol acyltransferase in maturing oil seeds of maize and other species. *Plant Physiology*, **82**, 813–820.

Cao, Y-Z. and Huang, A.H.C. (1987) Acyl coenzyme A preference of diacylglycerol acyltransferase from the maturing seeds of *Cuphea*, maize, rapeseed, and canola. *Plant Physiology*, **84**, 762–765.

Cao, Y-Z., Oo, K-C. and Huang, A.H.C. (1990) Lysophosphatidate acyltransferase in the microsomes from maturing seeds of Meadowfoam (*Limnanthes alba*). *Plant Physiology*, **94**, 1199–1206.

Cases, S., Smith, S.J., Zheng, Y-W., Myers, H.M., Lear, S.R., Sande, E., Novak, S., Collins, C., Welch, C.B., Lusis, A.J., Erickson, S.K. and Farese, Jr., R.V. (1998) Identification of a gene encoding an acyl CoA:diacylglycerol acyltransferase, a key enzyme in triacylglycerol synthesis. *Proceedings of the National Academy of Sciences USA*, **95**, 13018–13023.

Cases, S., Stone, S.J., Zhou, P., Yen, E., Tow, B., Lardizabal, K.D., Voelker, T. and Farese, Jr., R.V. (2001) Cloning of DGAT2, a second mammalian diacylglycerol acyltransferase, and related family members. *The Journal of Biological Chemistry*, **276**, 38870–38876.

Chapman, K.D. and Trelease, R.N. (1991) Acquisition of membrane lipids by differentiating glyoxysomes: role of lipid bodies. *Journal of Cell Biology*, **115**, 995–1007.

Chen, E.C.F., Tai, S.S.K., Peng, C.-C. and Tzen, J.T.C. (1998) Identification of three novel unique proteins in seed oil bodies of sesame. *Plant and Cell Physiology*, **39**, 935–941.

Chen, J.C.F. and Tzen, J.T.C. (2001) An *in vitro* system to examine the effective phospholipids and structural domain for protein targeting to seed oil bodies. *Plant and Cell Physiology*, **42**, 1245–1252.

Chen, J.C.F., Tsai, C.C.Y. and Tzen, J.T.C. (1999) Cloning and secondary structure analysis of caleosin, a unique calcium-binding protein in oil bodies of plant seeds. *Plant and Cell Physiology*, **40**, 1079–1086.

Cheng, D., Meegalla, R.L., He, B., Cromley, D.A., Billheimer, J.T. and Young, P.R. (2001) Human acyl-CoA:diacylglycerol acyltransferase is a tetrameric protein. *Biochemical Journal*, **359**, 707–714.

Chia, T. and Rawsthorne, S. (2000) Fatty acid breakdown in developing embryos of *Brassica napus* L. *Biochemical Society Transactions*, **28**, 753–754.

Chuong, S.D.X., Mullen, R.T. and Muench, D.G. (2002) Identification of a rice RNA- and microtubule-binding protein as the multifunctional protein, a peroxisomal enzyme involved in the β-oxidation of fatty acids. *The Journal of Biological Chemistry*, **277**, 2419–1429.

Coleman, R.A. and Lee, D.P. (2004) Enzymes of triacylglycerol synthesis and their regulation. *Progress in Lipid Research*, **43**, 134–176.

Coopers, T.G. and Beevers, H.J. (1969) Beta oxidation in glyoxysomes from castor bean endosperm. *The Journal of Biological Chemistry*, **224**, 3514–3520.

Cramer, C.L., Boothe, J.G. and Oishi, K.K. (1999) Transgenic plants for therapeutic proteins: linking upstream and downstream strategies. *Current Topics in Microbiology and Immunology*, **240**, 95–118.

Dahlqvist, A., Ståhl, U., Lenman, M.A., Lee, M., Sandager, L., Ronne, H. and Stymne, S. (2000) Phospholipid:diacylglycerol acyltransferase: an enzyme that catalyzes the acyl-CoA-independent formation of triacylglycerol in yeast and plants. *Proceedings of the National Academy of Sciences USA*, **97**, 6487–6492.

Davies, M.H., Hawkins, D.H. and Nelsen, J.S. (1995) Lysophosphatidic acid acyltransferase from immature coconut endosperm having medium chain length substrate specificity. *Phytochemistry*, **39**, 989–996.

Davoren, J.D., Nykiforuk, C.L., Laroche, A. and Weselake, R.J. (2002) Sucrose-induced changes in the transcriptome of cell suspension cultures of oilseed rape reveal genes associated with lipid biosynthesis. *Plant Physiology and Biochemistry*, **40**, 719–725

Del Vecchio, A.J.D. (1996) High laurate canola. *Inform*, **7**, 230–242.

Dewey, R.E., Wilson, R.F., Novitzky W.P. and Goode, J.H. (1994) The *AAPT1* gene of soybean complements a cholinephosphotransferase-deficient mutant of yeast. *The Plant Cell*, **6**, 1495–1507.

Douady, D. and Dubacq, J.-P. (1987) Purification of acyl-CoA:glycerol-3-phosphate acyltransferase from pea leaves. *Biochimica et Biophysica Acta*, **921**, 615–619.

Eastmond, P.J., Hooks, M. and Graham, I.A. (2000) The *Arabidopsis* acyl-CoA oxidase gene family. *Biochemical Society Transactions*, **28**, 755–757.

Eccleston, V.S. and Harwood, J.L. (1995) Solubilisation, partial purification and properties of acyl-CoA:glycerol-3-phosphate acyltransferase from avocado (*Persea americana*) fruit mesocarp. *Biochimica et Biophysica Acta*, **1257**, 1–10.

Eccleston, V.S. and Ohlrogge, J.B. (1998) Expression of lauroyl-acyl carrier protein thioesterase in *Brassica napus* seeds induces pathways for both fatty acid oxidation and biosynthesis and implies a set point for triacylglycerol accumulation. *The Plant Cell*, **10**, 613–621.

Edqvist, J. and Farbos, I. (2002) Characterization of germination-specific lipid transfer proteins from *Euphorbia lagascae*. *Planta*, **215**, 41–50.

Evans, D.E., Rothnie, N.E., Palmer, M.V., Burke, D.G., Sang, J.P., Knox, R.B., Williams, E.G., Hilliard, E.P. and Salisbury, P.A. (1987) Comparative analysis of fatty acids in pollen and seed of rapeseed. *Phytochemistry*, **26**, 1895–1897.

Evans, D.E., Taylor, P.E., Singh, M.B. and Knox, R.B. (1992) The interrelationship between the accumulation of lipids, protein and the level of acyl carrier protein during the development of *Brassica napus* L. pollen. *Planta*, **186**, 343–354.

Fernandez, D.E. and Staehelin, L.A. (1987) Does gibberellic acid induce the transfer of lipase from protein bodies to lipid bodies in barley aleurone cells? *Plant Physiology*, **85**, 487–496.

Feussner, I., Kuhn, H. and Wasternack, C. (2001) Lipoxygenase-dependent degradation of storage lipids. *Trends in Plant Science*, **6**, 268–273.

Feussner, I., Wasternack, C., Kindl, H. and Kuhn, H. (1995) Lipoxygenase-catalyzed oxygenation of storage lipids is implicated in lipid mobilization during germination. *Proceedings of the National Academy of Sciences USA*, **92**, 11849–11853.

Finkelsten, R.R. and Somerville, C.R. (1990) Three classes of abscisic acid (ABA)-insensitive mutations of *Arabidopsis* define genes that control overlapping subsets of ABA responses. *Plant Physiology*, **94**, 1172–1179.

Finlayson, S.A. and Dennis, D.T. (1980) $NAD^+$-specific glycerol 3-phosphate dehydrogenase from developing castor bean endosperm. *Archives of Biochemistry and Biophysics*, **199**, 179–185.

Focke, M., Gieringer, E., Schwan, S., Jänsch, L., Binder, S. and Braun, H-P. (2003) Fatty acid biosynthesis in mitochondria of grasses: malonyl-Coenzyme A is generated by a mitochondrial-localized acetyl-Coenzyme A carboxylase. *Plant Physiology*, **133**, 875–884.

Foster, E., Schneiderman, D., Cloutier, M., Gleddie, S. and Robert, L.S. (2002) Modifying the pollen coat protein composition in *Brassica*. *The Plant Journal*, **31**, 477–486.

Francki, M.G., Whitaker, P., Smith, P.M. and Atkins, C.A. (2002) Differential expression of a novel gene during seed triacylglycerol accumulation in lupin species (*Lupinus angustifolius* L. and *L. mutabilis* L.). *Functional and Integrative Genomics*, **2**, 292–300.

Franco, L.O., de O. Manes, C.-L., Hamdi, S., Sachetto-Martins, G. and de Oliveira, D.E. (2002) Distal regulatory regions restrict the expression of *cis*-linked genes to the tapetal cells. *FEBS Letters*, **517**, 13–18.

Frandsen, G., Müller-Uri, F., Nielsen, M., Mundy, J. and Skriver, K. (1996) Novel plant $Ca^{+2}$-binding protein expressed in response to abscisic acid and osmotic stress. *The Journal of Biological Chemistry*, **271**, 343–348.

Frandsen, G.I., Mundy, J. and Tzen, J.T.C. (2001) Oil bodies and their associated proteins, oleosin and caleosin. *Physiologia Plantarum*, **112**, 301–307.

Fraser, T. and Stobart, K. (2000) Partial purification and photoaffinity labeling of sunflower acyl-CoA:lysophosphatidylcholine acyltransferase. *Biochemical Society Transactions*, **28**, 715–718.

Fraser, T., Waters, A., Chatrattanakunchai, S. and Stobart, K. (2000) Does triacylglycerol biosynthesis require diacylglycerol acyltransferase (DAGAT)? *Biochemical Society Transactions*, **28**, 698–700.

Frentzen, M. (1990) Comparison of certain properties of membrane bound and solubilized acyltransferase activities of plant microsomes. *Plant Science*, **69**, 39–48.

Frentzen, M. (1993) Acyltransferases and triacylglycerols, in *Lipid Metabolism in Plants* (ed. T.S. Moore, Jr.), CRC Press, Ann Arbor, Massachusetts, pp. 195–230.

Frentzen, M. and Wolter, F. P. (1998) Molecular biology of acyltransferases involved in glycerolipid synthesis, in *Plant Lipid Biosynthesis: Fundamentals and Agricultural Applications (Society of Experimental Biology Seminar Series, 67)* (ed. J.L. Harwood), Cambridge University Press, New York, pp. 247–272.

Frentzen, M., Heinz, E., McKeon, T.A. and Stumpf, P.K. (1983) Specificities and selectivities of glycerol-3-phosphate acyltransferase and monoacylglycerol-3-phosphate acyltransferase from pea and spinach chloroplasts. *European Journal of Biochemistry*, **129**, 629–636.

Frentzen, M., Nishida, I. and Murata, N. (1987) Properties of the plastidial acyl-(acyl-carrier-protein): glycerol-3-phosphate acyltransferase from the chilling-sensitive plant squash (*Cucurbita moschata*). *Plant and Cell Physiology*, **28**, 1195–1201.

Froese, C.D., Nowack, L., Cholewa, E. and Thompson, J.E. (2003) Molecular composition and surface properties of storage lipid particles in wax bean (*Phaseolus vulgaris*). *Journal of Plant Physiology*, **160**, 215–225.

Fulda, M., Shockey, J., Werber, M., Wolter, F.P. and Heinz, E. (2002) Two long-chain acyl-CoA synthetases from *Arabidopsis thaliana* involved in peroxisomal fatty acid β-oxidation. *The Plant Journal*, **32**, 93–103.

Furukawa-Stoffer, T.L., Boyle, R.M., Thomson, A.L., Sarna, M.A. and Weselake, R.J. (2003) Properties of lysophosphatidylcholine acyltransferase from *Brassica napus* cultures. *Lipids*, **38**, 651–656.

Furukawa-Stoffer, T.L., Byers, S.D., Hodges, D.M., Laroche, A. and Weselake, R.J. (1998) Identification of *N*-ethylmaleimide-sensitive and –insensitive phosphatidate phosphatase activity in microspore-derived cultures of oilseed rape. *Plant Science*, **131**, 139–147.

Gangar, A., Karande, A.A. and Rajasekharan, R. (2001) Isolation and localization of a cytosolic 10S triacylglycerol biosynthetic multienzyme complex from oleaginous yeast. *The Journal of Biological Chemistry*, **276**, 10290–10298.

García-Agustín, P., Benaches-Gastaldo, M.J. and Primo-Millo, E. (1992) Lipid mobilization in *Citrus* cotyledons during germination. *Journal of Plant Physiology*, **140**, 1–7.

Gerhardt, B. (1992) Fatty acid degradation in plants. *Progress in Lipid Research*, 31, 417–446.

Germain, V., Rylott, E.L., Larson, T.R., Sherson, S.M., Bechtold, N., Carde, J.-P., Bryce, J.H., Graham, I.A. and Smith, S.M. (2001) Requirement for 3-ketoacyl-CoA thiolase-2 in peroxisome development, fatty acid β-oxidation and breakdown of triacylglycerol in lipid bodies of *Arabidopsis* seedlings. *The Plant Journal*, **28**, 1–12.

Giannoulia, K., Haralampidis, K., Poghosyan, Z., Murphy, D.J. and Hatzopoulos, P. (2000) Differential expression of diacylglycerol acyltransferase (*DGAT*) genes in olive tissues. *Biochemical Society Transactions*, **28**, 695–697.

Gillikin, J.W., Zhang, F., Coleman, C.E., Bass, H.W., Larkins, B.A. and Boston, R.S. (1997) A defective signal peptide tethers the *floury-2* zein to the endoplasmic reticulum membrane. *Plant Physiology*, **114**, 345–352.

Graham, I.A. and Eastmond, P.J. (2002) Pathways of straight and branched chain fatty acid catabolism in higher plants. *Progress in Lipid Research*, **41**, 156–181.

Griffiths, G., Stobart, A.K. and Stymne, S. (1985) The acylation of *sn*-glycerol 3-phosphate and the metabolism of phosphatidate in microsomal preparations from the developing cotyledons of safflower (*Carthamus tinctorius* L.) seed. *Biochemical Journal*, **230**, 379–388.

Gueguen, V., Macherel, D., Jaquinod, M., Douce, R. and Bourguignon, J. (2000) Fatty acid and lipoic acid biosynthesis in higher plant mitochondria. *The Journal of Biological Chemistry*, **275**, 5016–5025.

Gunstone, F.D. (2001) Production and consumption of rapeseed oil on a global scale. *European Journal of Lipid Science and Technology*, **103**, 447–449.

Gurr, M.I. (1980) The biosynthesis of triacylglycerols, in *The Biochemistry of Plants, Vol. 4, Lipids: Structure and Function* (ed P.K. Stumpf), Academic Press, New York, pp. 205–248.

Hanke, C., Wolter, F-P, Coleman, J., Peterek, G. and Frentzen, M. (1995) A plant acyltransferase involved in triacylglycerol biosynthesis complements an *Escherichia coli sn*-1-acylglycerol-3-phosphate acyltransferase mutant. *European Journal of Biochemistry*, **232**, 806–810.

Hartmann, M.-A., Perret, A.M., Carde, J.-P., Cassagne, C. and Moreau, P. (2002) Inhibition of the sterol pathway in leek seedlings impairs phosphatidylserine and glucosylceramide synthesis but triggers an accumulation of triacylglycerols. *Biochimica et Biophysica Acta*, **1583**, 285–296.

Hayashi, H., De Bellis, L., Hayashi, Y., Nito, K., Kato, A., Hayashi, M., Hara-Nishimura, I. and Nishimura, M. (2002) Molecular characterization of an Arabidopsis acyl-Coenzyme A synthetase localized on glyoxysomal membranes. *Plant Physiology*, **130**, 2019–2026.

Hayman, M.W., Fawcett, T., Schierer, T.F. *et al.* (2000) Mutagenesis of squash (*Cucurbita moschata*) glycerol-3-phosphate acyltransferase (GPAT) to produce an enzyme with altered substrate selectivity. *Biochemical Society Transactions*, **28**, 680–681.

Hayman, M.W., Fawcett, T. and Slabas, A.R. (2002) Kinetic mechanism and order of substrate binding for *sn*-glycerol-3-phosphate acyltransferase from squash (*Cucurbita moschata*). *FEBS Letters*, **514**, 281–284.

He, X., Chen, G.Q., Lin, J.-T. and McKeon, T.A. (2003) Functional expression of *Ricinus communis* diacylglycerol acyltransferase in yeast. *Annual Meeting of the American Society of Plant Biology*, Hawaii.

Hellyer, S.A., Chandler, I.C. and Bosley, J.A. (1999) Can the fatty acid selectivity of plant lipases be predicted from the composition of the seed triglyceride? *Biochimica et Biophysica Acta*, **1440**, 215–224.

Herman, E.M. (1987) Immunogold-localization and synthesis of an oil-body membrane protein in developing soybean seeds. *Planta*, **172**, 336–345.

Hernández-Pinzón, I., Ross, J.H.E., Barnes, K.A., Damant, A.P. and Murphy, D.J. (1999) Composition and role of tapetal lipid bodies in the biogenesis of the pollen coat of *Brassica napus*. *Planta*, **208**, 588–598.

Hernández-Pinzón, I., Patel, K. and Murphy, D.J. (2001) The *Brassica napus* calcium-binding protein, caleosin, has distinct endoplasmic reticulum – and lipid body – associated isoforms. *Plant Physiology and Biochemistry*, **39**, 615–622.

Hershenson, S. and Ernst-Fonberg, M.L. (1983) Kinetic studies of the glycerophosphate acyltransferase from *Euglena* microsomes, including the effects of serum albumin. *Biochimica et Biophysica Acta*, **751**, 412–421.

Hills, M.J. and Beevers, H. (1987) $Ca^{2+}$ stimulated neutral lipase activity in castor bean lipid bodies. *Plant Physiology*, **84**, 272–276.

Hills, M.J. and Mukherjee, K.D. (1990) Triacylglycerol lipase from rape (*Brassica napus* L.) suitable for biotechnological purposes. *Applied Biochemistry and Biotechnology*, **26**, 1–10.

Hills, M.J. and Murphy, D.J. (1988) Characterization of lipases from lipid bodies and microsomal membranes of erucic acid-free oilseed rape (*Brassica napus*) cotyledons. *Biochemical Journal*, **249**, 687–693.

Hills, M.J., Kiewitt, I. and Mukherjee, K.D. (1990) Lipase from *Brassica napus* L. discriminates against *cis*-4 and *cis*-6 unsaturated fatty acids and secondary and tertiary alcohols. *Biochimica et Biophysica Acta*, **1042**, 237–240.

Hills, M.J., Watson, M.D. and Murphy, D.J. (1993) Targeting of oleosins to the oil bodies of oilseed rape (*Brassica napus* L.). *Planta*, **189**, 24–29.

Hiroyama, M. and Takenawa, T. (1998) Purification and characterization of a lysophosphatidic acid-specific phosphatase. *Biochemical Journal*, **336**, 483–489.

Hjelmstad, R.H. and Bell, R.M. (1990) The *sn*-1,2-diacylglycerol cholinephosphotransferase of *Saccharomyces cerevisiae*. Nucleotide sequence, transcriptional mapping, and gene product analysis of the *CPT1* gene. *The Journal of Biological Chemistry*, **265**, 1755–1764.

Hjelmstad, R.H. and Bell, R.M. (1991) *sn*-1,2-Diacylglycerol choline- and ethanolaminephosphotransferases in *Saccharomyces cerevisiae*. *The Journal of Biological Chemistry*, **266**, 5094–5103.

Hobbs, D.H. and Hills, M.J. (2000) Expression and characterization of diacylglycerol acyltransferase from *Arabidopsis thaliana* and insect cell cultures. *Biochemical Society Transactions*, **28**, 687–689.

Hobbs, D.H., Lu, C. and Hills, M.J. (1999) Cloning of a cDNA encoding diacylglycerol acyltransferase from *Arabidopsis thaliana* and its functional expression. *FEBS Letters*, **452**, 145–149.

Holbrook, L.A., van Rooijen, G.J.H., Wilen, R.W. and Moloney, M.M. (1991) Oil-body proteins in microspore-derived embryos of *Brassica napus*. *Plant Physiology*, **97**, 1051–1058.

Hong, Y., Wang, T.-W., Hudak, K.A., Schade, F., Froese, C.D. and Thompson, J.E. (2000) An ethylene-induced cDNA encoding a lipase expressed at the onset of senescence. *Proceedings of the National Academy of Sciences USA*, **97**, 8717–8722.

Hope, R.G., Murphy, D.J. and McLauchlan, J. (2002) The domains required to direct core proteins of hepatitis C virus and GB virus-B to lipid droplets share common features with plant oleosin proteins. *The Journal of Biological Chemistry*, **277**, 4261–4270.

Huang, A.H.C. (1987) Lipases, in *The Biochemistry of Plants, Vol. 9, Lipids: Structure and Function* (ed. P.K. Stumpf), Academic Press, New York, pp. 91–119.

Huang, A.H.C. (1992) Oil bodies and oleosins in seeds. *Annual Review of Plant Physiology and Plant Molecular Biology*, **43**, 177–200.

Huang, A.H.C. (1996) Oleosins and oil bodies in seeds and other organs. *Plant Physiology*, **110**, 1055–1061.

Huang, A.H.C. and Wang, S.M. (1992) Proteinaceous inhibitors of lipase activities in soybean and other oil seeds. *Modern Methods of Plant Analysis*, **14**, 263–271.

Hudak, K.A. and Thompson, J.E. (1996) Flotation of lipid-protein particles containing triacylglycerol and phospholipid from the cytosol of carnation petals. *Physiologia Plantarum*, **98**, 810–818.

Ichihara, K. (1984) *sn*-Glycerol-3-phosphate acyltransferase in a particulate fraction from maturing safflower seeds. *Archives of Biochemistry and Biophysics*, **232**, 685–698.

Ichihara, K. (1991) The action of phosphatidate phosphatase on the fatty-acid composition of safflower triacylglycerol and spinach glycerolipids. *Planta*, **183**, 353–358.

Ichihara, K. and Noda, M. (1982) Some properties of diacylglycerol acyltransferase in a particulate fraction from maturing safflower seeds. *Phytochemistry*, **21**, 1895–1901.

Ichihara, K., Norikura, S. and Fujii, S. (1989) Microsomal phosphatidate phosphatase in maturing safflower seeds. *Plant Physiology*, **90**, 413–419.

Ichihara, K. and Suda, Y. (2003) Lipid biosynthesis in developing perilla seeds. *Phytochemistry*, **63**, 139–143.

Ichihara, K., Mae, K., Sano, Y. and Tanaka, K. (1995) 1-Acylglycerophosphocholine *O*-acyltransferase in maturing safflower seeds. *Planta*, **196**, 551–557.

Ichihara, K., Murota, N. and Fujii, S. (1990) Intracellular translocation of phosphatidate phosphatase in maturing safflower seeds: a possible mechanism of feedforward control of triacylglycerol synthesis by fatty acids. *Biochimica et Biophysica Acta*, **1043**, 227–234.

Ichihara, K., Takahashi, T. and Fujii, S. (1988) Diacylglycerol acyltransferase in maturing safflower seeds: its influences on the fatty acid composition of triacylglycerol and on the rate of triacylglycerol synthesis. *Biochimica et Biophysica Acta*, **958**, 125–129.

Ishizaki, O., Inishida, I., Agata, K., Eguchi, G. and Murata, N. (1988) Cloning and nucleotide sequence of cDNA for the plastid glycerol-3-phosphate acyltransferase from squash. *FEBS Letters*, **238**, 424–430.

Imeson, H.C., Rood, S.B., Weselake, R.J. *et al.* (1993) Hormonal control of lipase activity in oilseed rape germinants. *Physiologia Plantarum*, **89**, 476–482.

IUPAC-IUB Commission on Biochemical Nomenclature. The nomenclature of lipids. (1967) *Biochemical Journal*, **105**, 897–902.

Jacks, T.J., Hensarling, T.P., Neucere, J.N., Yatsu, L.Y. and Barker, R.H. (1990) Isolation and physiochemical characterization of the half-unit membranes of oilseed lipid bodies. *Journal of the American Oil Chemists' Society*, **67**, 353–361.

Jain, R.K., Coffey, M., Lai, K., Kumar, A. and MacKenzie, S.L. (2000) Enhancement of seed oil content by expression of glycerol-3-phosphate acyltransferase genes. *Biochemical Society Transactions*, **28**, 958–961.

Jako, C., Kumar, A., Wei, Y. *et al.* (2001) Seed-specific over-expression of an Arabidopsis cDNA encoding a diacylglycerol acyltransferase enhances seed oil content and seed weight. *Plant Physiology*, **126**, 861–874.

Jamal, Z., Martin, A., Gomez-Munoz, A. and Brindley, D.N. (1991) Plasma membrane fractions from rat liver contain phosphatidate phosphohydrolase distinct from that in the endoplasmic reticulum and cytosol. *Journal of Biological Chemistry*, **266**, 2988–2996.

Jaworski, J. and Cahoon, E.B. (2003) Industrial oils from transgenic plants. *Current Opinions in Plant Biology*, **6**, 178–184.

Joyard, J., Marechal, E., Malherbe, A., Block, M.A. and Douce, R. (1994) Importance of diacylglycerol in glycerolipid biosynthesis by spinach chloroplast envelope membranes. *Progress in Lipid Research*, **33**, 105–118.

Jung, K.M. and Kim, D.K. (2000) Purification and characterization of a membrane-associated 48-kilodalton phospholipase $A_2$ in leaves of broad bean. *Plant Physiology*, **123**, 1057–1067.

Kalscheuer, R. and Steinbüchel, A. (2003) A novel bifunctional wax ester synthase/acyl-CoA:diacylglycerol acyltransferase mediates wax ester and triacylglycerol biosynthesis in *Acinetobacter calcoaceticus* ADP1. *The Journal of Biological Chemistry*, **278**, 8075–8082.

Kamisaka, Y., Mishra, S. and Nakahara, T. (1997) Purification and characterization of diacylglycerol acyltransferase from the lipid body fraction of an oleaginous fungus. *Journal of Biochemistry*, **121**, 1107–1114.

Kannan, L., Knudsen, J. and Jolly, C.A. (2003) Aging and acyl-CoA binding protein alter mitochondrial glycerol-3-phosphate acyltransferase activity. *Biochimica et Biophysica Acta*, **1631**, 12–16.

Katavic, V., Mietkiewska, E., Barton, D.L., Giblin, E.M., Reed, D.W. and Taylor, D.C. (2002) Restoring enzyme activity in nonfunctional low erucic acid *Brassica napus* fatty acid elongase 1 by a single amino acid substitution. *European Journal of Biochemistry*, **269**, 5625–5631.

Katavic, V., Reed, D.W., Taylor, D.C. *et al.* (1995) Alteration of seed fatty acid composition by an ethyl methanesulfonate-induced mutation in *Arabidopsis thaliana* affecting diacylglycerol acyltransferase activity. *Plant Physiology*, **108**, 399–409.

Kaup, M.T., Froese, C.D. and Thompson, J.E. (2002) A role for diacylglycerol acyltransferase during leaf senescence. *Plant Physiology*, **129**, 1616–1626.

Kennedy, E.P. (1961) Biosynthesis of complex lipids. *Federation Proceedings of the American Society for Experimental Biology*, **20**, 934–940.

Kessler, F., Schnell, D. and Blobel, G. (1999) Identification of proteins associated with plastoglobules isolated from pea (*Pisum sativum* L.) chloroplasts. *Planta*, **208**, 107–113.

Kim, H.U., Hsieh, K., Ratnayake, C. and Huang, A.H.C. (2002) A novel group of oleosins is present inside the pollen of *Arabidopsis*. *The Journal of Biological Chemistry*, **277**, 22677–22684.

Kim, J.Y., Chung, Y.S., Ok, S.H. *et al.* (1999) Characterization of the full-length sequences of phospholipase $A_2$ induced during flower development. *Biochimica et Biophysica Acta*, **1489**, 389–392.

Kleiter, A.E. and Gerhardt, B. (1998) Glyoxysomal β-oxidation of long chain fatty acids: completeness of degradation. *Planta*, **206**, 125–130.

Knutzon, D.S., Hayes, T.R., Wyrick, A., Xiong, H., Davies, H.M. and Voelker, T.A. (1999) Lysophosphatidic acid acyltransferase from coconut endosperm mediates the insertion of laurate at the *sn*-2 position of triacylglycerols in lauric rapeseed oil and can increase total laurate levels. *Plant Physiology*, **120**, 739–746.

Knutzon, D.S., Lardizabal, K.D., Nelsen, J.S., Bleibaum, J.L., Davies, H.M. and Metz, J.G. (1995) Cloning of a coconut endosperm cDNA encoding a 1-acyl-*sn*-glycerol-3-phosphate acyltransferase that accepts medium-chain-length substrates. *Plant Physiology*, **109**, 999–1006.

Kocsis, M.G. and Weselake, R.J. (1996) Phosphatidate phosphatases of mammals, yeast, and higher plants. *Lipids*, **31**, 785–802.

Kocsis, M.G., Weselake, R.J., Eng, J.A., Furukawa-Stoffer, T.L. and Pomeroy, M.K. (1996) Phosphatidate phosphatase from developing seeds and microspore-derived cultures of *Brassica napus*. *Phytochemistry* **41**, 353–363.

Kwanyuen, P. and Wilson, R.F. (1986) Isolation and purification of diacylglycerol acyltransferase from germinating soybean cotyledons. *Biochemica et Biophysica Acta*, **877**, 238–245.

Lacey, D.J. and Hills, M.J. (1996) Heterogeneity of the endoplasmic reticulum with respect to lipid synthesis in developing seeds of *Brassica napus* L. *Planta*, **199**, 545–551.

Lacey, D.J., Beaudoin, F., Dempsey, C.E., Shewry, P.R and Napier, J.A. (1999) The accumulation of triacylglycerols within the endoplasmic reticulum of developing seeds of *Helianthus annuus*. *The Plant Journal*, **17**, 397–405.

Lacey, D.J., Wellner, N., Beaudoin, F., Napier, J.A. and Shewry, P.R (1998) Secondary structure of oleosins in oil bodies isolated from seeds of safflower (*Carthamus tinctorius* L.) and sunflower (*Helianthus annuus* L.). *Biochemical Journal*, **334**, 469–477.

Lands, W.E.M. and Hart, P. (1965) Metabolism of glycerolipids. VI. Specificities of acyl Coenzyme A:phospholipid acyltransferases. *The Journal of Biological Chemistry*, **240**, 1905–1911.
Lardizabal, K.D., Mai, J.T., Wagner, N.W., Wyrick, A., Voelker, T. and Hawkins, D.J. (2001) DGAT2 is a new diacylglycerol acyltransferase gene family: purification, cloning and expression in insect cells of two polypeptides from *Mortierella ramanniana* with diacylglycerol acyltransferase activity. *The Journal of Biological Chemistry*, **276**, 38862–38869.
Lardizabal, K.D., Metz, J.G., Sakamoto, T., Hutton, W.C., Pollard, M.R. and Lassner, M.W. (2000) Purification of a jojoba embryo wax synthase, cloning of its cDNA, and production of high levels of wax in seeds of transgenic Arabidopsis. *Plant Physiology*, **122**, 645–655.
Larson, T.R., Edgell, T., Byrne, J., Dehesh, K. and Graham, I.A. (2002) Acyl CoA profiles of transgenic plants that accumulate medium-chain fatty acids indicate inefficient storage lipid synthesis in developing oilseeds. *The Plant Journal*, **32**, 519–527.
Lassner, M.W., Lardizabal, K. and Metz, J.G. (1996) A jojoba beta-ketoacyl-CoA synthase cDNA complements the canola fatty acid elongation mutation in transgenic plants. *The Plant Cell*, **8**, 281–292.
Lassner, M.W., Lardizabal, K. and Metz, J.G. (1999) Producing wax esters in transgenic plants by expression of genes derived from jojoba, in *New Crops and New Uses* (ed. J. Janick), ASHS Press, Alexandria, Virginia, pp. 220–224.
Lassner, M.W., Levering, C.K., Davies, H.M. and Knutzon, D.S. (1995) Lysophosphatidic acid acyltransferase from Meadowfoam mediates insertion of erucic acid at the *sn*-2 position of triacylglycerol in transgenic rapeseed oil. *Plant Physiology*, **109**, 1389–1394.
Laurent, P. and Huang, A.H.C. (1992) Organ- and development-specific acyl Coenzyme A lysophosphatidate acyltransferases in palm and meadowfoam. *Plant Physiology*, **99**, 1711–1715.
Lee, D.P., Deonarine, A.S., Kienetz, M. *et al.* (2001) A novel pathway for lipid biosynthesis: the direct acylation of glycerol. *Journal of Lipid Research*, **42**, 1979–1986.
Lehner, R. and Kuksis, A. (1993) Triacylglycerol synthesis by an *sn*-1,2(2,3)-diacylglycerol transacylase from rat intestinal microsomes. *The Journal of Biological Chemistry*, **268**, 8781–8786.
Leprince, O., van Aelst, A.C., Pritchard, H.W. and Murphy, D.J. (1998) Oleosins prevent oil-body coalescence during seed imbibition as suggested by a low-temperature scanning electron microscope study of desiccation-tolerant and sensitive oilseeds. *Planta*, **204**, 109–119.
Li, M., Murphy, D.J., Lee, K.-H.K. *et al.* (2002) Purification and structural characterization of the central hydrophobic domain of oleosin. *The Journal of Biological Chemistry*, **277**, 37888–37895.
Lin, J.-T. and McKeon, T.A. (2000) The biosynthesis of castor oil studied by high-performance liquid chromatography. *Recent Research Developments in Lipids Research*, **4**, 361–368.
Lin, J.-T., Chen, J.M., Chen, P., Liao, L.P. and McKeon, T.A. (2002a) Molecular species of PC and PE formed during castor oil biosynthesis. *Lipids*, **37**, 991–995.
Lin, J.-T., Lew, K.M., Chen, J.M., Iwasaki, Y. and McKeon, T.A. (2000) Metabolism of 1-acyl-2-oleoyl-*sn*-glycero-3-phosphoethanolamine in castor oil biosynthesis. *Lipids*, **35**, 481–486.
Lin, J.-T., Woodruff, C.L., Lagouche, O.J. *et al.* (1998) Biosynthesis of triacylglycerols containing ricinoleate in castor microsomes using 1-acyl-2-oleoyl-*sn*-glycero-3-phosphocholine as the substrate of oleoyl-12-hydroxylase. *Lipids*, **33**, 59–69.
Lin, L.-J., Tai, S.S.K., Peng, C.-C. and Tzen, J.T.C. (2002b) Steroleosin, a sterol-binding dehydrogenase in seed oil bodies. *Plant Physiology*, **128**, 1200–1211.
Lin, Y.-H. and Huang, A.H.C. (1983) Lipase in lipid bodies of cotyledons of rape and mustard seedlings. *Archives of Biochemistry and Biophysics*, **225**, 360–369.
Lin, Y.-H. and Huang, A.H.C. (1984) Purification and initial characterization of lipase from the scutella of corn seedlings. *Plant Physiology*, **76**, 719–722.
Lin, Y.-H., Yu, C. and Huang, A.H.C. (1986) Substrate specificities of lipases from corn and other seeds. *Archives of Biochemistry and Biophysics*, **244**, 346–356.
Little, D., Weselake, R., Pomeroy, K., Furukawa-Stoffer, T. and Bagu, J. (1994) Solubilization and characterization of diacylglycerol acyltransferase from microspore-derived cultures of oilseed rape, *Biochemical Journal*, **304**, 951–958.

Loer, D.S. and Herman, E.M. (1993) Co-translational integration of soybean (*Glycine max*) oil body membrane protein oleosin into microsomal membranes. *Plant Physiology*, **101**, 993–998.

Löhden, I. and Frentzen, M. (1992) Triacylglycerol biosynthesis in developing seeds of *Tropaeolum majus* L. and *Limnanthes douglasii* R. Br. *Planta*, **188**, 215–224.

Lu, C. and Hills, M.J. (2002) Arabidopsis mutants deficient in diacylglycerol acyltransferase display increased sensitivity to abscisic acid, sugars, and osmotic stress during germination and seedling development. *Plant Physiology*, **129**, 1352–1358.

Lu, C.L., de Noyer S.B., Hobbs, D.H. *et al.* (2003) Expression pattern of diacylglycerol acyltransferase-1, an enzyme involved in triacylglycerol biosynthesis, in *Arabidopsis thaliana*. *Plant Molecular Biology*, **52**, 31–41.

MacKintosh, R.W., Hardie, D.G. and Slabas, A.R. (1989) A new assay procedure to study the induction of beta-ketoacyl-ACP synthase-I and synthase-II, and the complete purification of beta-ketoacyl-ACP synthase-I from developing seeds of oilseed rape (*Brassica napus*). *Biochimica et Biophysica Acta*, **1002**, 114–124.

Madey, E., Nowack, L.M., Su, L., Hong, Y., Hudak, K.A. and Thompson, J.E. (2001) Characterization of plasma membrane domains enriched in lipid metabolites. *Journal of Experimental Botany*, **52**, 669–679.

Madey, E., Nowack, L.M. and Thompson, J.E. (2002) Isolation and characterization of lipid in phloem sap of canola. *Planta*, **214**, 625–634.

Maeshima, M. and Beevers, H. (1985) Purification and properties of glyoxysomal lipase from castor bean. *Plant Physiology*, **79**, 489–493.

Maisonneuve, S., Bessoule, J.-J., Lessire, R., Delseny, M. and Roscoe, T.J. (2000) Mutagenesis of a plastidial lysophosphatidic acid acyltransferase. *Biochemical Society Transactions*, **28**, 961–964.

Malherbe, A., Block, M.A., Douce, R. and Joyard, J. (1995) Solubilization and biochemical properties of phosphatidate phosphatase from spinach chloroplast envelope membranes. *Plant Physiology and Biochemistry*, **33**, 149–161.

Malherbe, A., Block, M.A., Joyard, J. and Douce, R. (1992) Feedback inhibition of phosphatidate phosphatase from spinach chloroplast envelope membranes by diacylglycerol. *Journal of Biological Chemistry*, **267**, 23546–23553.

Manaf, A.M. and Harwood, J.L. (2000) Purification and characterisation of acyl-CoA:glycerol 3-phosphate acyltransferase from oil palm (*Elaeis guineensis*) tissues. *Planta*, **210**, 318–328.

Mancha, M. and Stymne, S. (1997) Remodelling of triacylglycerols in microsomal preparations from developing castor bean (*Ricinus communis* L.) endosperm. *Planta*, **203**, 51–57.

Marcel, G.C.F., Matos, A., d'Arcy-Lameta, A., Kader, J-C., Zuily-Fodil, Y. and Pham-Thi, A. (2000) Two novel plant cDNAs homologous to animal type-2 phosphatidate phosphatase are expressed in cowpea leaves and are differentially regulated by water deficits. *Biochemical Society Transactions*, **28**, 915–917.

Martin, B.A. and Wilson, R.F. (1984) Subcellular localization of triacylglycerol synthesis in spinach leaves. *Lipids*, **19**, 117–121.

Martin, B.A., Horn, M.E., Widholm, J.M. and Rinne, R.W. (1984) Synthesis, composition and location of glycerolipids in photoautotrophic soybean cell cultures. *Biochimica et Biophysica Acta*, **796**, 146–154.

Matsui, K., Hijiya, K., Tabuchi, Y. and Kajiwara, T. (1999) Cucumber cotyledon lipoxygenase during postgerminative growth. Its expression and action on lipid bodies. *Plant Physiology*, **119**, 1279–1288.

Mayfield, J.A. and Preuss, D. (2000) Rapid initiation of *Arabidopsis* pollination requires the oleosin-domain protein GRP17. *Nature Cell Biology*, **2**, 128–130.

McMaster, C.R. and Bell, R.M. (1997) CDP-choline:1,2-diacylglycerol cholinephosphotransferase. *Biochimica et Biophysica Acta*, **1348**, 100–110.

Metz, J.G., Pollard, M.R., Anderson, L., Hayes, T.R. and Lassner, M.W. (2000) Purification of a jojoba embryo fatty acyl-Coenzyme A reductase and expression of its cDNA in high erucic acid rapeseed. *Plant Physiology*, **122**, 635–644.

Mikami, K. and Murata, N. (2003) Membrane fluidity and the perception of environmental signals in cyanobacteria and plants. *Progress in Lipid Research*, **42**, 527–543.

Millichip, M., Tatham, A.S., Jackson, F., Griffiths, G., Shewry, P.R. and Stobart, A.K. (1996) Purification and characterization of oil bodies (oleosomes) and oil-body boundary proteins (oleosins) from the developing cotyledons of sunflower (*Helianthus annuus* L.). *Biochemical Journal*, **314**, 333–337.

Mishra, S. and Kamisaka, Y. (2001) Purification and characterization of thiol-reagent-sensitive glycerol-3-phosphate acyltransferase from the membrane fraction of an oleaginous fungus. *Biochemical Journal*, **355**, 315–322.

Mittendorf, V., Bongcam, V., Allenbach, L., Coullerez, G, Martini, N. and Poirier, Y. (1999) Polyhydroxyalkanoate synthesis in transgenic plants as a new tool to study carbon flow through β-oxidation. *The Plant Journal*, **20**, 45–55.

Moreau, P., Bessoule, J.J., Mongrand, S., Testet, E., Vincent, P. and Cassagne, C. (1998) Lipid trafficking in plant cells. *Progress in Lipid Research*, **37**, 371–391.

Moreau, R.A. and Stumpf, P.K. (1982) Solubilization and characterization of an acyl-Coenzyme A. O-lysophospholipid acyltransferase from the microsomes of developing safflower seeds. *Plant Physiology*, **69**, 1293–1297.

Murata, N., Ishizaki-Nishizawa, O., Sigashi, S., Hayashi, H., Tasaka, Y. and Nishida I. (1992) Genetically engineered alteration in the chilling sensitivity of plants. *Nature*, **356**, 710–713.

Murata, N. and Tasaka, Y. (1997) Glycerol-3-phosphate acyltransferase in plants. *Biochimica et Biophysica Acta*, **1348**, 10–16.

Murphy, D.J. (1993) Structure, function and biogenesis of storage lipid bodies and oleosins in plants. *Progress in Lipid Research*, **32**, 247–280.

Murphy, D.J. (2001) The biogenesis and functions of lipid bodies in animals, plants and microorganisms. *Progress in Lipid Research*, **40**, 325–438.

Murphy, D.J., Cummins, I. and Kang, A.S. (1989) Synthesis of the major oil-body membrane protein in developing rapeseed (*Brassica napus*) embryos. Integration with storage-lipid and storage-protein synthesis and implications for the mechanism of oil-body formation. *Biochemical Journal*, **258**, 285–293.

Murphy, D.J., Hernández-Pinzón, I. and Patel, K. (2001) Role of lipid bodies and lipid-body proteins in seeds and other tissues. *Journal of Plant Physiology*, **158**, 471–478.

Murphy, D.J., Keen, J.N., O'Sullivan, J.N. *et al.* (1991) A class of amphipathic proteins associated with lipid storage bodies in plants. Possible similarities with animal serum apolipoproteins. *Biochimica et Biophysica Acta*, **1088**, 86–94.

Næsted, H., Frandsen, G.I., Jauh, G.-Y. *et al.* (2000) Caleosins: $Ca^{2+}$-binding proteins associated with lipid bodies. *Plant Molecular Biology*, **44**, 463–476.

Nagiec, M.M., Wells, G.B., Lester, R.L. and Dickson, R.C. (1993) A suppressor gene that enables *Saccharomyces cerevisiae* to grow without making sphingolipids encodes a protein that resembles an *Escherichia coli* fatty acyltransferase. *The Journal of Biological Chemistry*, **268**, 22156–22163.

Napier, J.A., Stobart, A.K. and Shewry, P.R. (1996) The structure and biogenesis of plant oil bodies: the role of the ER membrane and the oleosin class of proteins. *Plant Molecular Biology*, **31**, 945–956.

Nishida, I., Frentzen, M., Ishizaki, O. and Murata, N. (1987) Purification of isomeric forms of acyl-[acyl-carrier-protein]:glycerol-3-phosphate acyltransferase from greening squash cotyledons. *Plant and Cell Physiology*, **28**, 1071–1079.

Norton, G. and Harris, J. F. (1983) Triacylglycerols in oilseed rape during seed development. *Phytochemistry*, **22**, 2703–2707.

Nykiforuk, C.L., Furukawa-Stoffer, T.L., Huff, P.W. *et al.* (2002) Characterization of cDNAs encoding diacylglycerol acyltransferase from cultures of *Brassica napus* and sucrose-mediated induction of enzyme biosynthesis. *Biochimica et Biophysica Acta*, **1580**, 95–109.

Nykiforuk, C.L., Laroche, A. and Weselake, R. J. (1999a) Isolation and sequence analysis of a novel cDNA encoding a putative diacylglycerol acyltransferase from a microspore-derived cell suspension culture of *Brassica napus* (accession no. AF155224) (PGR99-123). *Plant Physiology*, **120**, 1207.

Nykiforuk, C.L., Laroche, A. and Weselake, R.J. (1999b) Isolation and characterization of a cDNA encoding a second putative diacylglycerol acyltransferase from a microspore-derived cell suspension culture of *Brassica napus* L. cv Jet Neuf (accession no. AF164434) (PGR99-158). *Plant Physiology*, **121**, 1053.

Oelkers, P., Tinkelenberg, A., Erdeniz, N., Cromley, D., Billheimer, J.T. and Sturley, S.L. (2000) A lecithin cholesterol acyltransferase-like gene mediates diacylglycerol esterification in yeast. *The Journal of Biological Chemistry*, **275**, 15609–15612.

Ohlrogge, J. and Browse, J. (1995) Lipid biosynthesis. *The Plant Cell*, **7**, 957–970.

Ohlrogge, J.B., Pollard, M.R. and Stumpf, P.K. (1978) Studies on biosynthesis of waxes by developing jojoba seed tissue. *Lipids*, **13**, 203–210.

Oo, K.-C. and Huang, A.H.C. (1989) Lysophosphatidate acyltransferase activities in the microsomes from palm endosperm, maize scutellum, and rapeseed cotyledon of maturing seeds. *Plant Physiology*, **91**, 1288–1295.

Pacheco-Moisés, F., Valencia-Turcotte, L., Altuzar-Martínez, M. and Rodríguez-Sotres, R. (1997) Regulation of acyltransferase activity in immature maize embryos by abscisic acid and the osmotic environment. *Plant Physiology*, **114**, 1095–1101.

Pacini, E., Franchi, G.G. and Hesse, M. (1985) The tapetum: its form, function, and possible phylogeny in *Embryophyta. Plant Systematics and Evolution*, **149**, 155–185.

Parmenter, D.L., Boothe, J.G., van Rooijen, G.J.H., Yeung, E.C. and Moloney, M.M. (1995) Production of biologically active hirudin in plant seeds using oleosin partitioning. *Plant Molecular Biology*, **29**, 1167–1180.

Pearce, M.L. and Slabas, A.R. (1998) Phosphatidate phosphatase from avocado (*Persea americana*) – purification, substrate specificity and possible metabolic implications for the Kennedy pathway and cell signalling in plants. *The Plant Journal*, **14**, 555–564.

Perry, H.J., Bligny, R., Gout, E. and Harwood, J.L. (1999) Changes in Kennedy pathway intermediates associated with increased triacylglycerol synthesis in oilseed rape. *Phytochemistry*, **52**, 799–804.

Perry, H.J. and Harwood, J.L. (1993a) Changes in the lipid content of developing seeds of *Brassica napus*. *Phytochemistry*, **32**, 1411–1415.

Perry, H.J. and Harwood, J.L. (1993b) Radiolabelling studies of acyl lipids in developing seeds of *Brassica napus*: use of [1-$^{14}$C]acetate precursor. *Phytochemistry*, **33**, 329–333.

Pierrugues, O., Brutesco, C., Oshiro, J. *et al.* (2001) Lipid phosphate phosphatases in *Arabidopsis*. Regulation of the *AtLPP1*gene in response to stress. *The Journal of Biological Chemistry*, **276**, 20300–20308.

Piffanelli, P. and Murphy, D.J. (1998) Novel organelles and targeting mechanisms in the anther tapetum. *Trends in Plant Science*, **3**, 250–253.

Piffanelli, P., Ross, J.H.E. and Murphy, D.J. (1997) Intra- and extracellular lipid composition and associated gene expression patterns during pollen development in *Brassica napus*. *The Plant Journal*, **11**, 549–562.

Piffanelli, P., Ross, J.H.E. and Murphy, D.J. (1998) Biogenesis and function of the lipidic structures of pollen grains. *Sexual Plant Reproduction*, **11**, 65–80.

Poirier, Y. (2002) Polyhydroxyalkanoate synthesis in plants as a tool for biotechnology and basic studies of lipid metabolism. *Progress in Lipid Research*, **41**, 131–155.

Poirier, Y., Ventre, G. and Caldelari, D. (1999) Increased flow of fatty acids toward β-oxidation in developing seeds of Arabidopsis deficient in diacylglycerol acyltransferase activity or synthesizing medium-chain-length fatty acids. *Plant Physiology*, **121**, 1359–1366.

Pollard, M.R., McKeon, T., Gupta, L.M. and Stumpf, P.K. (1979) Studies on biosynthesis of waxes by developing jojoba seed. II. The demonstration of wax biosynthesis by cell-free homogenates. *Lipids*, **14**, 651–662.

Pozueta-Romero, J., Rafia, F., Houlné, G. *et al.* (1997) A ubiquitous plant housekeeping gene, *PAP*, encodes a major protein component of bell pepper chromoplasts. *Plant Physiology*, **115**, 1185–1194.

Princen, L.H. and Rothfus, J.A. (1984) Development of new crops for industrial raw materials. *Journal of the American Oil Chemists' Society*, **61**, 281–289.

Pritchard, S.L., Charlton, W.L., Baker, A. and Graham, I.A. (2002) Germination and storage reserve mobilization are regulated independently in *Arabidopsis*. *The Plant Journal*, **31**, 639–647.

Radetzky, R. and Langheinrich, U. (1994) Induction of accumulation and degradation of the 18.4-kDa oleosin in a triacylglycerol-storing cell culture of anise (*Pimpinella anisum* L.). *Planta*, **194**, 1–8.
Rajasekharan, R. and Nachiappan, V. (1994) Use of photoreactive substrates for characterization of lysophosphatidate acyltransferases from developing soybean cotyledons. *Archives of Biochemistry and Biophysics*, **311**, 389–394.
Ramli, U.S., Baker, D.S., Quant, P.A. and Harwood, J.L. (2002a) Control mechanisms operating for lipid biosynthesis differ in oil-palm (*Elaeis guineensis* Jacq.) and olive (*Olea europaea* L.) callus cultures. *Biochemical Journal*, **364**, 385–391.
Ramli, U.S., Baker, D.S., Quant, P.A. and Harwood, J.L. (2002b) Control analysis of lipid biosynthesis in tissue cultures from oil crops shows that flux control is shared between fatty acid synthesis and lipid assembly. *Biochemical Journal*, **364**, 393–401.
Ranjan, R. and Lewak, S. (1992) Jasmonic acid promotes germination and lipase activity in non-stratified apple embryos. *Physiologia Plantarum*, **86**, 335–339.
Ranjan, R. and Lewak, S. (1995) Interaction of jasmonic acid and abscisic acid in the control of lipases and proteases in germinating apple embryos. *Physiologia Plantarum*, **93**, 421–426.
Rey, P., Gillet, B., Romer, S. *et al.* (2000) Over-expression of a pepper plastid lipid-associated protein in tobacco leads to changes in plastid ultrastructure and plant development upon stress. *The Plant Journal*, **21**, 483–494.
Reynolds, T.L. (1997) Pollen embryogenesis. *Plant Molecular Biology*, **33**, 1–10.
Richmond, T.A. and Bleecker, A.B. (1999) A defect in β-oxidation causes abnormal inflorescence development in Arabidopsis. *The Plant Cell*, **11**, 1911–1923.
Rodriguez, M.A., Dias, C. and Lau, T.E. (1992) Reversible ATP-dependent inactivation of adipose diacylglycerol acyltransferase. *Lipids*, **27**, 577–581.
Rodríguez-Sotres, R. and Black, M. (1994) Osmotic potential and abscisic acid regulate triacylglycerol synthesis in developing wheat embryos. *Planta*, **192**, 9–15.
Ross, J.H.E. and Murphy, D.J. (1996) Characterization of anther-expressed genes encoding a major class of extracellular oleosin-like proteins in the pollen coat of Brassicaceae. *The Plant Journal*, **9**, 625–637.
Rossell, J.B., King, B. and Downes, M.J. (1985) Composition of oil. *Journal of the American Oil Chemists' Society*, **62**, 221–230.
Routaboul, J-M., Benning, C., Bechtold, N., Caboche, M. and Lepiniec, L. (1999) The *TAG1* locus of *Arabidopsis* encodes for a diacylglycerol acyltransferase. *Plant Physiology and Biochemistry*, **37**, 831–840.
Rylott, E.L., Gilday, A.D. and Graham, I.A. (2003b) The gluconeogenic enzyme phosphoe*nol*pyruvate carboxykinase in Arabidopsis is essential for seedling establishment. *Plant Physiology*, **131**, 1834–1842.
Rylott, E.L., Hooks, M.A. and Graham, I.A. (2001) Co-ordinate regulation of genes involved in storage lipid mobilization in *Arabidopsis thaliana*. *Biochemical Society Transactions,* **29**, 283–287.
Rylott, E.L., Rogers, C.A., Gilday, A.D., Edgell, T., Larson, T.R. and Graham, I.A. (2003a) *Arabidopsis* mutants in short- and medium-chain acyl-CoA oxidase activities accumulate acyl-CoAs and reveal that fatty acid β-oxidation is essential for embryo development. *The Journal of Biological Chemistry*, **278**, 21370–21377.
Sadeghipour, H.R. and Bhatla, S.C. (2002) Differential sensitivity of oleosins in proteolysis during oil body mobilization in sunflower seedlings. *Plant and Cell Physiology*, **43**, 1117–1126.
Sakaki, T., Kondo, N. and Yamada, M. (1990b) Pathway for the synthesis of triacylglycerols from monogalactosyldiacylglycerols in ozone-fumigated spinach leaves. *Plant Physiology*, **94**, 773–780.
Sakaki, T., Saito, K., Kawaguchi, A., Kondo, N. and Yamada, M. (1990a) Conversion of monogalactosyldiacylglycerols to triacylglycerols in ozone-fumigated spinach leaves. *Plant Physiology*, **94**, 766–772.
Sambanthamurthi, R., Let, C.C., Cheang, O.K., Huat, Y.K. and Rajan, P. (1991) Chilling-induced lipid hydrolysis in the oil palm (*Elaeis guineensis*) mesocarp. *Journal of Experimental Botany*, **42**, 1199–1205.

Schultz, D.J. and Ohlrogge, J.B. (2000) Biosynthesis of triacylglycerol in *Thunbergia alata*: additional evidence for involvement of phosphatidylcholine in unusual monoenoic oil production. *Plant Physiology and Biochemistry*, **38**, 169–175.
Settlage, S.B., Kwanyuen, P. and Wilson, R.F. (1998) Relation between diacylglycerol acyltransferase activity and oil concentration in soybean. *Journal of the American Oil Chemists' Society*, **75**, 775–781.
Settlage, S.B., Wilson, R.F. and Kwanyuen, P. (1995) Localization of diacylglycerol acyltransferase to oil body associated endoplasmic reticulum. *Plant Physiology and Biochemistry*, **33**, 399–407.
Shank, K.J., Su, P., Brglez, I., Boss, W.F., Dewey, R.E. and Boston, R.S. (2001) Induction of lipid metabolic enzymes during the endoplasmic reticulum stress response in plants. *Plant Physiology*, **126**, 267–277.
Shekar, S., Tumaney, A.W., Sreenivasa Rao, T.J.V. and Rajasekharan, R. (2002) Isolation of lysophosphatidic acid phosphatase from developing peanut cotyledons. *Plant Physiology*, **128**, 988–996.
Shockey, J.M., Fulda, M.S. and Browse, J.A. (2002) Arabidopsis contains nine long-chain acyl-coenzyme A synthetase genes that participate in fatty acid and glycerolipid metabolism. *Plant Physiology*, **129**, 1710–1722.
Slabas, A.R., Harding, J., Hellyer, A., Roberts, P. and Bambridge, H.E. (1987) Induction, purification and characterization of acyl carrier protein from developing seeds of oil seed rape (*Brassica napus*). *Biochimica et Biophysica Acta*, **921**, 50–59.
Slabas, A.R., Sidebottom, C.M., Hellyer, A., Kessell, R.M.J. and Tombs, M.P. (1986) Induction, purification and characterization of NADH-specific enoyl acyl carrier protein reductase from developing seeds of oil seed rape (*Brassica napus*). *Biochimica et Biophysica Acta*, **877**, 271–280.
Slack, C.R., Campbell, L.C., Browse, J.A. and Roughan, P.G. (1983) Some evidence for the reversibility of the cholinephosphotransferase-catalysed reaction in developing linseed cotyledons *in vivo*. *Biochimica et Biophysica Acta*, **754**, 10–20.
Slack, C.R., Roughan, P.G., Browse, J.A. and Gardiner, S.E. (1985) Some properties of cholinephosphotransferase from developing safflower cotyledons. *Biochimica et Biophysica Acta*, **833**, 438–448.
Smith, M.D., Licatalosi, D.D. and Thompson, J.E. (2000) Co-association of cytochrome *f* catabolites and plastid-lipid-associated protein with chloroplast lipid particles. *Plant Physiology*, **124**, 211–221.
Somerville, C., Browse, J., Jaworski, J.G. and Ohlrogge, J.B. (2000) Lipids, in *Biochemistry & Molecular Biology of Plants* (eds B.B. Buchanan, W. Gruissem and R.L. Jones) American Society of Plant Physiologists, Rockville, MD, pp. 456–527.
Sorger, D. and Daum, G. (2002) Synthesis of triacylglycerols by the acyl-coenzyme A:diacylglycerol acyltransferase Dga1p in lipid particles of the yeast *Saccharomyces cerevisiae*. *Journal of Bacteriology*, **184**, 519–524.
Ståhl, U., Banas, A. and Stymne, S. (1995) Plant microsomal phospholipid acyl hydrolases have selectivities for uncommon fatty acids. *Plant Physiology*, **107**, 953–962.
Ståhl, U., Ek, B. and Stymne, S. (1998) Purification and characterization of a low-molecular-weight phospholipase $A_2$ from developing seeds of elm. *Plant Physiology*, **117**, 197–205.
Ståhl, U., Lee, M., Sjödahl, S. *et al.* (1999) Plant low-molecular-weight phospholipase $A_2$s ($PLA_2$s) are structurally related to the animal secretory $PLA_2$s and are present as a family of isoforms in rice (*Oryza sativa*). *Plant Molecular Biology*, **41**, 481–490.
Steinke, G., Weitkamp, P., Klein, E. and Mukherjee, K.D. (2001) High-yield preparation of wax esters via lipase-catalyzed esterification using fatty acids and alcohols from crambe and camelina oils. *Journal of Agriculture and Food Chemistry*, **49**, 647–651.
Stobart, A.K. and Stymne, S. (1985) The interconversion of diacylglycerol and phosphatidylcholine during triacylglycerol production in microsomal preparations of developing cotyledons of safflower (*Carthamus tinctorius* L.). *Biochemical Journal*, **232**, 217–221.
Stobart, A.K. and Stymne, S. (1990) Triacylglycerol biosynthesis, in *Methods in Plant Biochemistry*, Vol. 4 (eds J.L. Harwood and J.R. Bowyer), Academic Press, London, pp. 19–46.
Stobart, A.K., Stymne, S. and Höglund, S. (1986) Safflower microsomes catalyse oil accumulation *in vitro*: a model system. *Planta*, **169**, 33–37.

Stobart, K., Mancha, M., Lenman, M., Dahlqvist, A. and Stymne S. (1997) Triacylglycerols are synthesized and utilized by transacylation reactions in microsomal preparations of developing safflower (*Carthamus tinctorius* L.) seeds. *Planta*, **203**, 58–66.

Stymne, S. (1993) Biosynthesis of uncommon fatty acids and their incorporation into triacylglycerols, in *Biochemistry and Molecular Biology of Membrane and Storage Lipids of Plants* (eds N. Murata and C. Somerville), American Society of Plant Physiologists, Rockville, USA, pp. 150–158.

Stymne, S. and Glad, G. (1981) Acyl exchange between oleoyl-CoA and phosphatidylcholine in microsomes of developing soya bean cotyledons and its role in fatty acid desaturation. *Lipids*, **16**, 298–305.

Stymne, S. and Stobart, A.K. (1984a) Evidence for the reversibility of the acyl-CoA: lysophosphatidylcholine acyltransferase in microsomal preparations from developing safflower (*Carthamus tinctorius* L.) cotyledons and rat liver. *Biochemical Journal*, **223**, 305–314.

Stymne, S. and Stobart, A.K. (1984b) The biosynthesis of triacylglycerols in microsomal preparations of developing cotyledons of sunflower (*Helianthus annuus* L.). *Biochemical Journal*, **220**, 481–488.

Stymne, S. and Stobart, A. K. (1987) Triacylglycerol biosynthesis, in *The Biochemistry of Plants, Vol. 9, Lipids: Structure and Function* (ed. P.K. Stumpf), Academic Press, New York, pp. 175–214.

Stymne, S., Stobart, A.K. and Glad, G. (1983) The role of the acyl-CoA pool in the synthesis of polyunsaturated 18-carbon fatty acids and triacylglycerol production in the microsomes of developing safflower seeds. *Biochimica et Biophysica Acta*, **752**, 198–208.

Sukumar, V. and Sastry, P.S. (1987) Triacylglycerol synthesis in developing seeds of groundnut (*Arachis hypogaea*): studies on phosphatidic acid phosphatase and diacylglycerol acyltransferase during seed maturation. *Biochemistry International*, **14**, 1153–1158.

Sun, C., Cao, Y-Z. and Huang, A.H.C. (1988) Acyl coenzyme A preference of the glycerol phosphate pathway in the microsomes from the maturing seeds of palm, maize, and rapeseed. *Plant Physiology,* **88**, 56–60.

Taylor, D.C., Katavic, V., Zou, J. *et al.* (2001) Field testing of transgenic rapeseed cv Hero transformed with a yeast *sn*-2 acyltransferase results in increased oil content, erucic acid content and seed yield. *Molecular Breeding*, **8**, 317–322.

Taylor, D.C., MacKenzie, S.L., McCurdy, A.R. *et al.* (1994) Stereospecific analyses of seed triacylglycerols from high-erucic acid Brassicaceae: detection of erucic acid at the *sn*-2 position in *Brassica oleracea* L. Genotypes. *Journal of the American Oil Chemists' Society*, **71**, 163–167.

Taylor, D. C., Weber, N., Barton, D.L. *et al.* (1991) Triacylglycerol bioassembly in microspore-derived embryos of *Brassica napus* L. cv Reston. *Plant Physiology*, **97**, 65–79.

Taylor, D.C., Weber, N., Hogge, L.R., Underhill, E.W. and Pomeroy, M.K. (1992) Formation of trierucoylglycerol (trierucicn) from 1,2-dierucoylglycerol by a homogenate of microspore-derived embryos of *Brassica napus* L. *Journal of the American Oil Chemists' Society*, **69**, 355–358.

Taylor, D.C., Weber, N., Underhill, E.W. *et al.* (1990) Storage-protein regulation and lipid accumulation in microspore embryos of *Brassica napus* L. *Planta*, **181**, 18–26.

Thompson, J., Taylor, C. and Wang, T.-W. (2000) Altered membrane lipase expression delays leaf senescence. *Biochemical Society Transactions*, **28**, 775–777.

Thompson, J.E., Froese, C.D., Madey, E., Smith, M.D. and Hong, Y. (1998) Lipid metabolism during plant senescence. *Progress in Lipid Research*, **37**, 119–141.

Thoyts, P.J.E., Millichip, M.I., Stobart, A.K., Griffiths, W.T., Shewry, P.R. and Napier, J.A. (1995) Expression and *in vitro* targeting of a sunflower oleosin. *Plant Molecular Biology*, **29**, 403–410.

Ting, J.T.L., Balsamo, R., Ratnayake, C. and Huang, A.H.C. (1997) Oleosin of plant seed oil bodies is correctly targeted to the lipid bodies in transformed yeast. *The Journal of Biological Chemistry*, **272**, 3699–3706.

Ting, J.T.L., Lee, K., Ratnayake, C., Platt, K.A., Balsamo, R.A. and Huang, A.H.C. (1996) Oleosin genes in maize kernels having diverse oil contents are constitutively expressed independent of oil contents. Size and shape of intracellular oil bodies are determined by the oleosins/oils ratio. *Planta*, **199**, 158–165.

Ting, J.T.L., Wu, S.S.H., Ratnayake, C. and Huang, A.H.C. (1998) Constituents of the tapetosomes and elaioplasts in *Brassica campestris* tapetum and their degradation and retention during microsporogenesis. *The Plant Journal*, **16**, 541–551.

Tumaney, A.W. and Rajasekharan, R. (1999) Synthesis of azidophospholipids and labeling of lysophosphatidylcholine acyltransferase from developing soybean cotyledons. *Biochimica et Biophysica Acta*, **1439**, 47–56.

Tumaney, A.W., Shekar, S. and Rajasekharan, R. (2001) Identification, purification, and characterization of monoacylglycerol acyltransferase from developing peanut cotyledons. *The Journal of Biological Chemistry*, **276**, 10847–10852.

Turnbull, A.P., Rafferty, J.B., Sedelnikova, S.E. *et al.* (2001) Analysis of the structure, substrate selectivity, and mechanism of squash glycerol-3-phosphate (1)-acyltransferase. *Structure*, **9**, 347–353.

Tzen, J.T.C., Cao, Y-Z., Laurent, P., Ratnayake, C. and Huang, A.H.C. (1993) Lipids, proteins, and structure of seed oil bodies from diverse species. *Plant Physiology*, **101**, 267–276.

Tzen, J.T.C. and Huang, A.H.C. (1992) Surface structure and properties of plant seed oil bodies. *Journal of Cell Biology*, **117**, 327–335.

Valencia-Turcotte, L. and Rodríquez-Sotres, R. (2001) The treatment of purified maize oil bodies with organic solvents and exogenous diacylglycerol allows the detection and solubilization of diacylglycerol acyltransferase. *Biochimica et Biophysica Acta*, **1534**, 14–26.

van de Loo, F.J., Broun, P., Turner, S. and Somerville, C. (1995) An oleate 12-hydroxylase from *Ricinus communis* L. is a fatty acyl desaturase homolog. *Proceedings of the National Academy of Sciences USA*, **92**, 6743–6747.

van der Luit, A.H., Piatti, T., van Doorn, A., Musgrave, A., Felix G., Boller, T. and Munnik, T. (2000) Elicitation of suspension-cultured tomato cells triggers the formation of phosphatidic acid and diacylglycerol pyrophosphate. *Plant Physiology*, **123**, 1507–1515.

van Rooijen, G.J.H. and Moloney, M.M. (1995) Structural requirement of oleosin domains for subcellular targeting to the oil body. *Plant Physiology*, **109**, 1353–1361.

van Rooijen, G.J.H., Wilen, R.W., Holbrook, L.A. and Moloney, M.M. (1992) Regulation of accumulation of mRNAs encoding a 20-kDa oil-body protein in microspore-derived embryos of *Brassica napus*. *Canadian Journal of Botany*, **70**, 503–508.

Vigeolas, H., van Dongen, J.T., Waldeck, P., Hühn, D. and Geigenberger, P. (2003) Lipid storage metabolism is limited by the prevailing low oxygen concentrations within developing seeds of oilseed rape. *Plant Physiology*, **133**, 1–13.

Voelker, T. and Kinney, A.J. (2001) Variations in the biosynthesis of seed-storage lipids. *Annual Review of Plant Physiology and Plant Molecular Biology*, **52**, 335–361.

Voelker, T.A., Hayes, T.R., Cranmer, A.M., Turner, J.C. and Davies, H.M. (1996) Genetic engineering of a quantitative trait: metabolic and genetic parameters influencing the accumulation of laurate in rapeseed. *The Plant Journal*, **9**, 229–241.

Vogel, G. and Browse, J. (1995) Preparation of radioactively labeled synthetic *sn*-1,2-diacylglycerols for studies of lipid metabolism. *Analytical Biochemistry*, **224**, 61–67.

Vogel, G. and Browse, J. (1996) Cholinephosphotransferase and diacylglycerol acyltransferase. Substrate specificities at a key branch point in seed lipid metabolism. *Plant Physiology*, **110**, 923–931.

Waggoner, D.W., Gomez- Munõz, A., Dewald, J. and Brindley, D.N. (1996) Phosphatidate phosphohydrolase catalyzes the hydrolysis of ceraminde 1-phosphate, lysophosphatidate, and sphingosine 1-phosphate. *The Journal of Biological Chemistry*, **271**, 16506–16509.

Waggoner, D.W., Martin, A., Dewald, J., Gomez- Munõz, A. and Brindley, D.N. (1995) Purification and characterization of novel plasma membrane phosphatidate phosphohydrolase from rat liver. *The Journal of Biological Chemistry*, **270**, 19422–19429.

Wahlroos, T., Soukka, J., Denesyuk, A., Wahlroos, R., Korpela, T. and Kilby, N.J. (2003) Oleosin expression and trafficking during oil body biogenesis in tobacco leaf cells. *Genesis*, **35**, 125–132.

Wanasundara, P.K.J.P.D., Wanasundara, U.N. and Shahidi, F. (2001) Lipoytic activity of enzymes from germinating seeds of sesame (*Sesamum indicum* L.). *Journal of Food Lipids*, **8**, 75–84.

Wang, S.-M. and Huang, A.H.C. (1987) Biosynthesis of lipase in the scutellum of maize kernel. *The Journal of Biological Chemistry*, **262**, 2270–2274.

Wang, X. (2001) Plant phospholipases. *Annual Review of Plant Physiology and Plant Molecular Biology*, **52**, 211–231.
Wanner, G., Formanek, H. and Theimer, R.R. (1981) The ontogeny of lipid bodies (spherosomes) in plant cells. Ultrastructural evidence. *Planta*, **151**, 109–123.
Wanner, G. and Theimer, R.R. (1978) Membranous appendices of spherosomes (oleosomes). Possible role in fat utilization in germinating oilseeds. *Planta*, **140**, 163–169.
Weichert, H., Kolbe, A., Kraus, A., Wasternack, C. and Feussner, I. (2002) Metabolic profiling of oxylipins in germinating cucumber seedlings-lipoxygenase-dependent degradation of triacylglycerols and biosynthesis of volatile aldehydes. *Planta*, **215**, 612–619.
Weselake, R.J. (2000) Lipid biosynthesis in cultures of oilseed rape. *In vitro Cellular and Developmental Biology – Plant*, **36**, 338–348.
Weselake, R.J. (2002) Biochemistry and biotechnology of triacylglycerol accumulation in plants, in *Lipid Biotechnology* (eds T.M. Kuo and H.W. Gardner), Marcel Dekker, Inc., New York, pp. 27–56.
Weselake, R.J. (2003) Biochemistry and biotechnology of seed oil formation. *Annual Meeting & Expo of the American Oil Chemists' Society.* BIO5: New Industrial and Food Oils.
Weselake, R.J. and Taylor, D.C. (1999) The study of storage lipid biosynthesis using microspore-derived cultures of oilseed rape. *Progress in Lipid Research*, **38**, 401–460.
Weselake, R.J., Byers, S.D., Davoren, J.M., Laroche, A., Hodges, D.M., Pomeroy, M.K. and Furukawa-Stoffer, T.L. (1998) Triacylglycerol biosynthesis and gene expression in microspore-derived cell suspension cultures of oilseed rape. *Journal of Experimental Botany*, **49**, 33–39.
Weselake, R.J., Kazala, E.C., Cianflone, K., Boehr, D.D., Middleton, C.K., Rennie, C.D., Laroche A. and Recnik, I. (2000b) Human acylation stimulating protein enhances triacylglycerol biosynthesis in plant microsomes. *FEBS Letters*, **481**, 189–192.
Weselake, R.J., Nykiforuk, C.L., Laroche, A., Patterson, N.A., Wiehler, W.B., Szarka, S.J., Moloney, M.M., Tari, L.W. and Derekh, U. (2000a) Expression and properties of diacylglycerol acyltransferase from cell-suspension cultures of oilseed rape. *Biochemical Society Transactions*, **28**, 684–686.
Weselake, R.J., Pomeroy, M.K., Furukawa, T.L., Golden, J.L., Little, D.B. and Laroche, A. (1993) Developmental profile of diacylglycerol acyltransferase in maturing seeds of oilseed rape and safflower and microspore-derived cultures of oilseed rape. *Plant Physiology*, **102**, 565–571.
Weselake, R.J., Pomeroy, M.K., Furukawa-Stoffer, T., Little, D. and Rajasekharan, R. (1995) Interaction of photoreactive substrate analogs with diacyglycerol acyltransferase from microspore-derived embryos of oilseed rape, in *Plant Lipid Metabolism* (eds J.-C. Kader and P. Mazliak), Kluwer Academic Publishers, Boston, Massachucetts, pp. 518–529.
Weselake, R.J., Taylor, D.C., Pomeroy, M.K., Lawson, S.L. and Underhill, E.W. (1991) Properties of diacylglycerol acyltransferase from microspore-derived embryos of *Brassica napus*. *Phytochemistry*, **30**, 3533–3538.
Weselake, R.J., Thomson, L.W., Tenaschuk, D. and MacKenzie, S.L. (1989) Properties of solubilized microsomal lipase from germinating *Brassica napus*. *Plant Physiology*, **91**, 1303–1307.
Weselake, R.J., Wiehler, W.B., Patterson, N.A. *et al.* (2003) Transformation of *Brassica napus* with cDNAs encoding proteins that stimulate *in vitro* triacylglycerol biosynthesis, in *Plant Biotechnology 2002 and Beyond* (ed. I.K. Vasil), Kluwer Academic Publishers, Netherlands, pp. 321–322.
Wiberg, E., Banaś, A. and Stymne, S. (1997) Fatty acid distribution and lipid metabolism in developing seeds of laurate-producing rape (*Brassica napus* L.). *Planta*, **203**, 341–348.
Wiberg, E., Edwards, P., Byrne, J., Stymne, S. and Dehesh, K. (2000) The distribution of caprylate, caprate and laurate in lipids from developing and mature seeds of transgenic *Brassica napus* L. *Planta*, **212**, 33–40.
Wiberg, E., Tillberg, E. and Stymne, S. (1994) Substrates of diacylglycerol acyltransferase in microsomes from developing oil seeds. *Phytochemistry*, **36**, 573–577.
Wilmer, J.A., Brown, A.P., Forsyth, K., Carnaby, S., Barsby, T. and Slabas, A.R. (2000) *Limnanthes douglasii* erucic acid-specific lysophosphatidic acid acyltransferase activity in

oilseed rape, an analysis of biochemical effects. *Biochemical Society Transactions*, **28**, 964–966.
Wilmer, J.A., Helsper, J.P.F.G. and van der Plas, L.H.W. (1996) Effect of growth temperature on erucic acid levels in seeds and microspore-derived embryos of oilseed rape, *Brassica napus* L. *Journal of Plant Physiology*, **147**, 486–492.
Wilmer, J.A., Lessire, R., Helsper, J.P.F.G. and van der Plas, L.H.W. (1998) Regulation of elongase activity by abscisic acid and temperature in microspore-derived embryos of oilseed rape (*Brassica napus*). *Physiologia Plantarum*, **102**, 185–191.
Wilson, R.F. and Kwanyuen, P.K. (1986) Triacylglycerol synthesis and metabolism in germinating soybean cotyledons. *Biochimica et Biophysica Acta*, **877**, 231–237.
Wu, S.S.H., Moreau, R.A., Whitaker, B.D. and Huang, A.H.C. (1999) Steryl esters in the elaioplasts of the tapetum in developing *Brassica* anthers and their recovery on the pollen surface. *Lipids*, **34**, 517–523.
Wu, S.S.H., Platt, K.A., Ratnayake, C., Wang, T.-W., Ting, J.T.L. and Huang, A.H.C. (1997) Isolation and characterization of neutral-lipid-containing organelles and globuli-filled plastids from *Brassica napus* tapetum. *Proceedings of the National Academy of Sciences USA*, **94**, 12711–12716.
Wu, X.-Y., Moreau, R.A. and Stumpf, P.K. (1981) Studies of biosynthesis of waxes by developing jojoba seed: III. Biosynthesis of wax esters from acyl-CoA and long chain alcohols. *Lipids*, **16**, 897–902.
Xue, L., McCune, L.M., Kleppinger-Sparace, K.F., Brown, M.J., Pomeroy, M.K. and Sparace, S.A. (1997) Characterization of the glycerolipid composition and biosynthetic capacity of pea root plastids. *Plant Physiology*, **113**, 549–557.
Yasruel, Z., Cianflone, K., Sniderman, A.D., Rosenbloom, M., Walsh, M. and Rodriguez, M.A. (1991) Effect of acylation stimulating protein on the triacylglycerol synthetic pathway of human adipose tissue. *Lipids*, **26**, 495–499.
Yatsu, L.Y. and Jacks, T.J. (1972) Spherosome membranes. Half unit-membranes. *Plant Physiology*, **49**, 937–943.
Yen, C.-L.E. and Farese, Jr., R.V. (2003) MGAT2, a monoacylglycerol acyltransferase expressed in the small intestine.*The Journal of Biological Chemistry*, **278**, 18532–18537.
Yen, C.-L.E., Stone, S.J., Cases, S., Zhou, P. and Farese, Jr., R.V., (2002) Identification of a gene encoding MGAT1, a monoacylglycerol acyltransferase. *Proceedings of the National Academy of Sciences*, **99**, 8512–8517.
Żarska-Maciejewska, B. (1992) Lipolytic activity during dormancy removal in apple seeds. *Plant Physiology and Biochemistry*, **30**, 65–70.
Zheng, A. and Zou, J. (2001) The initial step of the glycerolipid pathway. Identification of glycerol 3-phosphate/dihydroxyacetone phosphate dual substrate acyltransferases in *Saccharomyces cerevisiae*. *The Journal of Biological Chemistry*, **276**, 41710–41716.
Zheng, Z., Xia, Q. Dauk, M., Shen, W., Selvaraj, G. and Zou, J. (2003) Arabidopsis *AtGPAT1*, a member of the membrane-bound glycerol-3-phosphate acyltransferase gene family, is essential for tapetum differentiation and male fertility. *The Plant Cell*, **15**, 1872–1887.
Zou, J., Katavic, V., Giblin, E.M. *et al.* (1997) Modification of seed oil content and acyl composition in the Brassicaceae by expression of a yeast *sn*-2 acyltransferase gene. *The Plant Cell*, **9**, 909–923.
Zou, J., Wei, Y., Jako, C., Kumar, A., Selvaraj, G. and Taylor, D.C. (1999) The *Arabidopsis thaliana TAG1* mutant has a mutation in a diacylglycerol acyltransferase gene. *The Plant Journal*, **19**, 645–653.
Zweytick, D., Athenstaedt, K. and Daum, G. (2000) Intracellular lipid particles of eukaryotic cells. *Biochimica et Biophysica Acta*, **1469**, 101–120.

# 7 Lipid-associated proteins

Denis J. Murphy

## 7.1 Introduction

Lipid-associated proteins are a relatively newly discovered class of proteins that are specifically associated with macromolecular lipid assemblies, other than bilayer membranes, in the cells of a wide range of organisms, from eubacteria to mammals. The most common type of non-bilayer lipid assembly in cells is a spherical organelle known as a lipid body or lipid droplet. These organelles, which are typically 0.5–2 μm in diameter, are made up of a neutral lipid core surrounded by an annulus comprising a phospholipid monolayer and a specific population of proteins. Lipid bodies (often called oil bodies in plants) have tended to be regarded as mere storage sites for carbon and energy. However, over the last decade, progress in elucidating the functions of intra- and extra-cellular lipid bodies – and especially their associated proteins – have revealed hitherto unsuspected dynamic roles for these organelles in processes, such as lipid import/export and the subcellular trafficking of both lipids and proteins.

Lipid-associated proteins have been particularly well characterised in plants, most notably in lipid-storing tissues like seeds and fruits where they are exemplified by the oleosins. Oleosins are tightly associated with storage lipid bodies in many, but not all, oil-accumulating plant tissues. Although oleosin genes appear to be ubiquitous components of the genomes of true plants, i.e. the Plantae, the levels of oleosin protein accumulation can vary enormously between different species. For example, some types of desiccation-intolerant oil-rich seeds many accumulate little or no oleosin protein (Leprince *et al*., 1998), although this has recently been questioned (Guilloteau *et al*., 2003). Proteins similar to seed oleosins may also be present within the cytosol of the cells of certain types of pollen grain that store lipid, e.g. entomophilous pollen like that of many Brassicaceae, including *Arabidopsis thaliana* (Kim *et al*., 2002). A separate class of oleosin-like proteins, called oleo-pollenins, has also been found in floral tissues including the tapetum and on the external surfaces of pollen grains, but so far reports of their occurrence have been restricted to the Brassicaceae and it is not known whether similar proteins are found in other plants (Murphy, 2001). Oleosins and oleosin-like proteins appear to be structural proteins with no discernible enzymatic motifs.

In the past few years, several additional classes of lipid-binding proteins have also been described in plants, including caleosins, steroleosins and protein kinases. Unlike oleosins, caleosins are not only associated with lipid bodies but have also been found on ER membranes. Since caleosins have a single putative membrane-spanning domain, as well as calcium-binding and protein kinase domains, they may have a role in signalling and also in oil-body assembly and mobilisation. Steroleosins are of unknown function, but may be analogous to the sterol-binding proteins that are principal components of fungal lipid bodies. The plastidial lipid-associated proteins (PLAPs) were originally found in the specialised plastids called chromoplasts, which are found in non-green pigmented tissues like coloured flower petals and fruits. Inside the chromoplasts, lipidic pigments and their associated PLAPs, such as carotenoids, are stored in a range of differently shaped structures – from long, thin fibrils to classic globular droplets similar to storage oil bodies. Plastidial lipid-associated proteins appear to be involved in several aspects of lipid assembly, sequestration and mobilisation in chromoplasts; however, these or similar proteins may also have wider roles in other types of plastid including chloroplasts where they are also associated with thylakoid membranes.

In this chapter, the major classes of lipid-associated proteins in plants will be described. These proteins will be compared with lipid-associated proteins in other organisms including eukaryotes like animals and yeasts as well as in prokaryotes like eubacteria and Archaea, and even in several viruses. Readers wishing to follow up the topic in greater detail are recommended to consult the comprehensive review by the author published in 2001 (Murphy, 2001). In the present chapter, some aspects of this earlier work will be summarised but it will also be extended considerably to discuss the impressive body of additional information on lipid bodies and their associated proteins (totalling more than 50 papers) that has emerged during the past 3–4 years.

## 7.2 Plant lipid-associated proteins

The vast majority of lipid-associated proteins in plants that have been described to date are intracellular. As with the first classes of lipid-associated proteins to be discovered in animals, the plant proteins appeared originally to be associated with the sequestration and long-term storage of neutral lipids, either in the cytosol or in plastids. Such proteins include oleosins and the plastid lipid-associated proteins. Since 2000, several new classes of lipid-associated proteins have been discovered, some of which can associate with bilayer membranes as well as with lipid bodies. For example, the caleosin family includes some isoforms that are major components of storage oil bodies in seeds while other very similar isoforms are membrane bound and are strongly responsive to stresses, such as drought and salinity. The emerging picture is of a diverse group of protein families, some of which, like oleosins, are obligate lipid-body components, while others, like plastid lipid-associated proteins and caleosins, can also associate with other types of lipid assembly, such as membrane bilayers or fibrils. This diversity in distribution is mirrored by an increasing recognition that lipid-associated proteins in plants play much more dynamic roles in cell biology than was realised hitherto.

### 7.2.1 Oleosins

Oleosins are a class of relatively low molecular weight alkaline proteins of about 15–24 kDa that accumulate on the surfaces of lipid bodies in desiccation-tolerant seeds. The structure, biosynthesis and function of oleosins have been the subject of several reviews (Moloney, 1999; Murphy and Vance, 1999; Murphy, 2001; Napier *et al.*, 1996, 2001). The roles of oleosins in neutral lipid formation/storage are also discussed in Chapter 6 of this volume. Readers interested in the historical perspective of oleosins are encouraged to consult the above reviews. In this section, the principal focus will be on the structural aspects of oleosins, especially in comparison with other lipid-associated proteins in plants and in other organisms.

The presence of proteins in isolated and washed lipid bodies from peanut seeds was recognised as long ago as 1967 (Jacks *et al.*, 1967). In 1972, the first amino acid analysis of this protein fraction was reported (Yatsu and Jacks, 1972): the close resemblance of this analysis to the known amino acid composition of oleosins means that it is now possible to say that these were probably the first oleosins to be isolated from seeds. Although it is often assumed that oleosins are present on the lipid bodies of all oil-storing seeds, this has only been demonstrated for those seeds that normally undergo desiccation as part of their maturation process. In contrast, many tropical or subtropical seeds are released from the parent plant in a partially or fully hydrated state and analysis of the lipid bodies of such seeds revealed that they contained either no detectable oleosin or much reduced levels (on a triacylglycerol basis) compared to desiccation-tolerant seeds (Leprince *et al.*, 1998). There was a recent report that oil bodies from cocoa seeds contain oleosins and that the cocoa genome encodes at least two olcosin genes (Guilloteau *et al.*, 2003). Although this is apparently at variance with the previous report of very low or undetectable levels of oleosins in such seeds (Leprince *et al.*, 1998), careful scrutiny of the data reveals that the oleosin: triacylglycerol ratio in cocoa is much lower than in orthodox, desiccation-tolerant plants and is not sufficient to cover the surfaces of the oil bodies. Oleosins are also absent or undetectably rare from oil-bearing fruit tissues, such as the mesocarp of olive, oil palm and avocado.

Oleosins are often stated as occurring only in seeds, and more recently inside pollen grains (Kim *et al.*, 2002). However, this view has been challenged by recent evidence from several species where oleosin expression was detected in root and in meristematic tissues. Oleosin proteins have been detected by immunofluorescence analysis of young roots from peas, rapeseed and Arabidopsis (Murphy *et al.*, 2000). In addition, transgenic rapeseed plants expressing the marker enzyme β-glucuronidase (GUS) under the control of an oleosin promoter were unexpectedly found to have significant GUS activity in the roots of the young seedlings, in addition to the normal activity in seeds (Murphy *et al.*, 2001). This suggests that the oleosin gene promoter can direct gene expression in non-seed tissues. There is also emerging evidence that some cells types in roots are able to express additional genes that are normally associated with storage product formation in seeds

(Baumlein *et al*., 1991; Murphy *et al*., 2001). Therefore, the distribution of oleosins in plants is rather more complex than was originally thought; oleosins appear to be present in all desiccation-tolerant seeds but may also be expressed in some cell types in roots, while they are absent from some other highly lipogenic tissues such as non-desiccating seeds and fruits.

To date, well over 100 seed oleosin sequences have been filed in public databases. The defining characteristic of oleosins is their unique central domain of about 70–80 non-polar residues that is uninterrupted by any charged residues. This hydrophobic domain is relatively well-conserved across most plants and includes a motif in the middle of the domain that contains three prolines as follows: AT**P**XXXXFS**P**XXX**P**A, where X = V, I, L or F. This motif, which is often termed the 'proline knot' (Abell *et al*., 1997), probably plays a key role in oleosin structure and targeting to lipid bodies. Another group of plant lipid-body associated proteins, the caleosins, was discovered in 2000. Caleosins also have a rather similar proline motif that is also highly conserved in different species, as described in more detail later. The proline knot motif in oleosins is flanked on each side by about 30 non-polar residues, some of which are well conserved and presumably have some structural significance. For example, there are two or three heptad leucine repeats plus several invariant Gly, Thr and Ser residues on both sides of the proline knot.

Although we now have primary sequence information on over 100 different oleosins from numerous plants, the elucidation of their secondary structures still presents formidable challenges. This is largely due to the extreme hydrophobicity of oleosins, which results in their aggregation, except in the presence of non-polar solvents or ionic detergents. Their *in vivo* localisation at a monolayer interface also makes it difficult to prepare paracrystalline arrays of the sort that have led to the elucidation of the structures of bilayer proteins like rhodopsin and the thylakoid photosystem complexes. Therefore, more indirect methods have been used to infer structural information about oleosins. For example, protease protection studies and differential availability to antibodies suggest that the central hydrophobic domain is embedded in the lipid-body matrix, while the more polar *C*- and *N*-terminal domains extend into the aqueous phase. However, the fine structure of the central domain remains the subject of some controversy. This is because it has been difficult to use conventional methods, such as circular dichroism (CD) or Fourier-transform infrared (FT-IR) spectroscopy, for the direct determination of oleosin secondary structure. Technical problems with these methods include the need to mimic the lipid–water interface during analysis (e.g. by CD spectroscopy) in order to preserve the native conformation of the protein, while at the same time trying to reduce effects such as light-scattering that are caused by particle emulsions. Equally, the use of solid-phase systems for FT-IR analysis may affect the conformation of this unusual type of protein. Some of the most recent structural studies of oleosins continue to give conflicting results suggestive that the central hydrophobic domain is mostly β-structure (Li *et al*., 2002) or is mostly α-helical structure (Alexander *et al*., 2002).

Some of the earliest structural studies of intact oleosins in solvents or on potassium bromide discs were consistent with the predicted β-strand structure for the central domain (Li *et al.*, 1992). In contrast, more recent structural studies of fragments corresponding to the central hydrophobic domain were consistent with a largely α-helical structure (Lacey *et al.*, 1998; Alexander *et al.*, 2002). These authors have now proposed a third model in which the central domain consists of an α-helical hairpin with a turn at the proline knot and with the two helical arms being connected to each other by inter-helix hydrogen bonding (Napier *et al.*, 2001). One of the difficulties with the second and third models of oleosin structure is that the hydrogen bonding requirements of both the proposed anti-parallel β-sheet and the α-helical hairpin are accommodated internally. This means that inter-molecular interactions between neighbouring oleosin molecules can only proceed via the surface-located *N*- and/or *C*-terminal domains. Several groups have noted that isolated oleosins can form very stable oligomeric associations, even in the presence of strong ionic detergents, including SDS (Li *et al.*, 1992; Pons *et al.*, 1998). While such interactions may not be the same as those found in the lipid-body environment, they do raise the possibility that native oleosins may interact with each other via their hydrophobic domains in order to form the kind of extremely stable surface coating that extends around lipid bodies in many seeds. We can now be fairly sure that the central hydrophobic domain of oleosin is firmly embedded in the lipid-body matrix. However, in the absence of high-resolution structural data, all of the detailed models of the conformations of this domain remain largely conjectural.

Oleosins have a net negative charge, which means that, in view of their steric and electrostatic repulsion, neighbouring lipid bodies would be unlikely to fuse with one another. The extreme inertness of lipid bodies from mature seeds is borne out by their exceptional stability, not only to lipases but also to the complete removal of water followed by a rapid rehydration in distilled water. It is probably only the fully mature lipid bodies of dehydrating seeds that contain a full complement of oleosins and therefore have the associated stability to lipase attack or changes in water content. We found that lipid bodies from younger seeds were relatively deficient in oleosins and were unstable to a cycle of dehydration/rehydration, while lipid bodies from seeds that had germinated for just 2–3 days were completely susceptible to lipase digestion (Murphy, 2001). These observations are also consistent with the known availability of lipid bodies in developing seeds to remodelling of their contents in association with ER-bound enzymes (Sarmiento *et al.*, 1998).

Although oleosins in many oilseeds are almost as abundant as seed storage proteins, the timing of their breakdown after germination is quite different. We have shown that, while storage proteins in rapeseed are mobilised within the first 2–3 days after germination, the breakdown of oleosins is much more protracted and occurs simultaneously with TAG mobilisation (Murphy and Cummins, 1989; Murphy *et al.*, 1989). More recently, we have also observed a partial cleavage of at least some of the oleosins that occurs prior to the full breakdown of storage lipid bodies in germinated rape seedlings (see Murphy, 2001). In this case, the oleosins were cleaved from 19 kDa to about 16 kDa but retained their recognition by antibodies and their binding to lipid bodies. This limited proteolytic processing of the oleosins does not immediately affect the integrity of the lipid bodies, but may render the latter more susceptible to digestion by lipases. This is analogous to the proposed role of the processing of perilipin (in this case by phosphorylation) in disrupting the integrity of lipid bodies in adipocytes and steroidogenic cells and hence allowing them to be accessed by lipases (see below). More substantial evidence to support the processing of oleosins prior to oil-body mobilisation has recently been reported in the case of germinating seedlings of sunflower (Sadeghipour and Bhatla, 2002). Here, a soluble protease of 65 kDa was found to cleave oleosins into successively smaller fragments that remained associated with the oil bodies. Only those oil bodies that contained the processed oleosins were susceptible to lipase mediated mobilisation of the storage triacylglycerol. The entire process of oleosin processing and triacylglycerol mobilisation was greatly enhanced in the light, relative to germination in the dark (Sadeghipour and Bhatla, 2003).

Surprisingly, over 15 years after the first oleosin gene was sequenced, the function of these proteins remains an enigma. One difficulty in studying oleosin function using molecular genetic approaches is that the proteins are encoded by relatively large gene families in all of the major plant model systems. Even in Arabidopsis, which has one of the smallest genomes of any higher plant, there are at least seven different genes encoding seed-expressed oleosins. These genes are sufficiently different from each other so that, despite repeated attempts in several labs including our own, gene knockouts using antisense approaches only succeeded in reducing the expression of a few members of the gene family, while the remaining oleosins were present in sufficient amounts to mask any resultant phenotype. In view of this lack of success to date, the function of oleosins remains the subject of conjecture.

Data from transgenic plants and cell lines show that oleosins can be expressed in a variety of lipid-body accumulating cells but cannot themselves induce the formation of lipid bodies in cell types where such organelles are not normally accumulated in any quantity. If oleosins are expressed ectopically in cells that do not accumulate storage lipid bodies, the proteins appear to accumulate in relatively small amounts on the ER (Beaudoin and Napier, 2000; Wahlroos *et al.*, 2003). This suggests that oleosins require the active formation of lipid bodies in order to accumulate further. In the absence of ongoing lipid-body biogenesis, the oleosins that are initially synthesised on the ER either cause a feedback inhibition of their further accumulation or may be actively turned over. The only difference between lipid bodies in oleosin-expressing tissues and those in non-expressing tissues is that the former tend to be smaller and are more stable to dehydration/rehydration. This may provide a clue to the true function of oleosins in seeds. We have observed that the oleosin-deficient lipid bodies of desiccation-sensitive seeds are unstable to drying and rehydration *in vivo* (Leprince *et al.*, 1998). Although the lipid bodies retained their integrity during the drying phase of the cycle, they fused with each other during rehydration of the seed tissues (Leprince *et al.*, 1998). Severe disruption of lipid-body integrity has also been reported in lettuce seeds subjected to cycles of freezing and thawing that would cause a combination of hydration and mechanical stresses (Jordan *et al.*, 1982). Therefore, a possible function of oleosins is to enable seeds to undergo extensive dehydration (that is often followed by a prolonged period of dormancy) and then allow them to suffer the rigours of rehydration while still being able to store over 50% of their weight as lipid bodies. This would allow oleosin-containing seeds to colonise new habitats away from the moist tropical and sub-tropical regions, i.e. to become desiccation-tolerant. The importance of lipid-body integrity in contributing to seed viability is demonstrated by observations that one of the earliest markers for deterioration in stored seeds is the coalescence of their lipid bodies (van Standen *et al.*, 1975; Dawidowicz-Grzegorzewska and Podstolski, 1992). This coalescence is presumably caused by damage to the oleosin layer that normally stabilises such lipid bodies, leading to the irreversible loss of seed viability.

### *7.2.2 Oleo-pollenins ('oleosin-like proteins')*

During the early- to mid-1990s, several groups noted the presence of genes encoding proteins containing an oleosin-like motif in members of the Brassicaceae, such as *Arabidopsis thaliana* and several Brassica spp. (Roberts *et al.*, 1993, 1995; Lee *et al.*, 1994; Robert *et al.*, 1994). In contrast to true oleosins, these genes are expressed specifically in floral tissues – most notably in the tapetum – which is a nutritive layer of cells that provides many of the precursors required for the development of microspores into mature pollen grains (Roberts *et al.*, 1991, 1993, 1995). Oleosin-like proteins contain a 70-residue domain that is homologous with the central hydrophobic lipid-binding domain of seed oleosins (Roberts *et al.*, 1993, 1995; Lee *et al.*, 1994; Robert *et al.*, 1994; Ross and Murphy, 1996). More recent evidence suggests that the oleosin-like region may simply be a novel type of cleavable targeting motif as it is not present in the mature form of these polypeptides (Ross and Murphy, 1996; Hernandez-Pinzon *et al.*, 1998, 1999; Piffanelli *et al.*, 1998a; Ting *et al.*, 1998). Since the mature forms of these proteins constitute the main protein component of the pollen coat in several plants, and since they have no resemblance to oleosins, it has been suggested that they be renamed as 'pollenins', while their precursor forms, which contain the oleosin-like domain may be termed 'oleo-pollenins' (Murphy, 2001; Franco *et al.*, 2002). The oleo-pollenins are an interesting class of diverse structural proteins that appear to use tapetal lipid bodies and oleosin-like domains as a non-secreted extracellular targeting mechanism. A comparison of the domain structures of true oleosins and oleo-pollenins is presented in Figure 7.1 (see also Piffanelli *et al.*, 1998b for review).

Oleo-pollenin (or oleosin-like) genes were first recognised as a large class of anther-specific transcripts in *Brassica napus* (Roberts, 1991, 1993, 1995) and it soon became apparent that they were not expressed in microspores or pollen grains but only in tapetal cells (Robert *et al.*, 1994; Ross and Murphy, 1996; Franco *et al.*, 2002). The three oleosin-like genes reported to be present in microspores

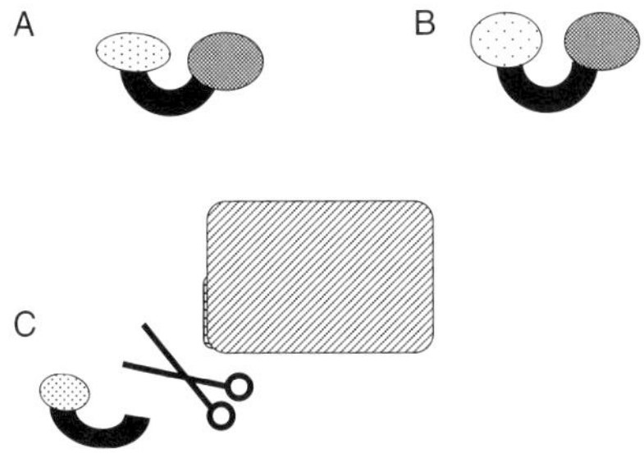


**Figure 7.1** Models of the domain organisations of oleosins and oleosin-like proteins. A. seed oleosins have three major domains consisting of a small hydrophilic *N*-terminal region (spotted), a hydrophobic central domain (dark shading) and an amphipathic *C*-terminal region (striped); B. pollen intracellular oleosins form a different family but are structurally similar to seed oleosins; C. anther oleo-pollenins contain very small *N*-terminal regions and a hydrophobic central domain that is quite similar to seed/pollen oleosins, but their *C*-terminal domains are very different and highly variable in size and composition. These *C*-terminal domains, termed pollenins, are cleaved from the rest of the protein following tapetal apoptosis and become relocated to the pollen coat where they play a key role in fertilisation.

of Arabidopsis (Kim *et al.*, 2002) are much more similar to seed-specific oleosins than to the six modular, tapetal-specific oleo-pollenin genes of Arabidopsis. Proteins encoded by oleo-pollenin genes contain an *N*-terminal oleosin-like domain followed by a *C*-terminal domain that is highly variable both in length and in amino acid composition (Ross and Murphy, 1996). The *C*-terminal domains in Brassica spp. range in length from 7–37 kDa and often contain repeating motifs characteristic of structural proteins (Ross and Murphy, 1996, 1998). A total of eight genes encoding glycine-rich proteins, similar to some of the oleo-pollenins, have been described in *Arabidopsis thaliana* (de Oliveira *et al.*, 1993), and some of these sequences provide clues as to the origin and possible function of pollenins. Many, but not all, oleo-pollenins are glycine-rich proteins that are predicted to adopt extended conformations, such as anti-parallel β-sheets or glycine loops (Sachetto-Martins *et al.*, 2000), and such motifs are often found in cell wall proteins (Ross and Murphy, 1996). In all the sequenced oleo-pollenin genes where introns are present, these are located at or near to the junction of the oleosin-like and *C*-terminal (pollenin) coding regions Therefore it is likely that oleo-pollenins arose by exon shuffling, whereby genes encoding a variety of structural polypeptides acquired additional exons encoding the hydrophobic central domain of proteins similar to the seed oleosins (Schein *et al.*, 2004).

Oleo-pollenins accumulate initially as full-length proteins (i.e. with the oleosin-like domain still present) in tapetal cells, where they are associated with the cytosolic lipid bodies, termed tapetosomes. There is some evidence that oleo-pollenins co-localise with the hexagonally arranged fibrils that are often seen inside tapetosomes viewed in thin sections (Owen and Makaroff, 1995; Wang *et al.*, 1997). Isolation of tapetosomes from rapeseed anthers has confirmed that the oleo-pollenins remain attached to these organelles as intact full-length proteins, including the oleosin-like domains, throughout tapetal development (Hernandez-Pinzon *et al.*, 1998, 1999; Ting *et al.*, 1998). However, once the tapetal cells undergo programmed cell death, the tapetosomes are released into the anther locule and the oleo-pollenins are proteolytically cleaved at the junction of the oleosin-like domain and the pollenin domain. This coincides with a reorganisation of the tapetosomes, together with the lipid-rich elaioplasts and other tapetal cell contents, to form the pollen coat that constitutes the extracellular outer coating of the pollen grains (Hernandez-Pinzon *et al.*, 1998, 1999; Piffanelli and Murphy, 1998; Piffanelli *et al.*, 1998; Ting *et al.*, 1998). Sequence analysis of the proteome of the pollen coat of rapeseed showed that all eight of the most abundant polypeptides were mature pollenins, i.e. the *C*-terminal domains of the original full-length precursor proteins (Murphy and Ross, 1998). Sequences of some of the oleosin-like domains were also detected in the pollen coat (Ross and Murphy, 1996) but in much lower abundance than the pollenin domains, indicating that these oleosin-like fragments are largely degraded after cleavage. Cleavage of precursor oleo-pollenins to yield pollenins is also consistent with the results of expressing a translational fusion of the Brassica oleo-pollenin (BnOlnB;4) with the marker protein, GUS (Foster *et al.*, 2002). In this case, the authors found that the full-length BnOlnB;4-GUS fusion protein of 125 kDa initially accumulated in the tapetum. Later on, in anther development the protein was found on the pollen coat, where it had been cleaved to an 115 kDa product. Such a cleavage is consistent with the loss of the oleosin-like *N*-terminal domain, as occurs with native oleo-pollenins. The mature, oleosin-free, pollenins have recently been shown to be required for normal rapid initiation of pollination in *Arabidopsis*, since a mutation that removed one of the major pollenins resulted in a severe impairment of pollen hydration and competitive ability after germination (Mayfield and Preuss, 2000).

To summarise, it seems likely that oleo-pollenins are hybrid proteins that consist of a 7–8 kDa lipid-binding domain, which is very similar to the central hydrophobic region of seed oleosins. However, this hydrophobic domain has now become fused to any one of a wide variety of structural protein domains that range in size from 7–37 kDa. The full-length oleo-pollenin proteins containing both domains, associate with, and possibly stabilise, the unique cytosolic lipid bodies of tapetal cells, namely the tapetosomes. It is only when tapetal cells undergo apoptosis at the latter stages of pollen development that the tapetosomes are released from the cells. In the lumen of the tapetum, the oleo-pollenins come into contact with proteases that remove the now-redundant oleosin-like domain to leave the mature pollenins to function as the major protein components of the pollen coat. The exact role of pollenins, although evidently important, is still unclear. It seems likely that they are involved in pollen rehydration, possibly by facilitating the creation of water channels through the otherwise relatively impermeable lipidic pollen coat. The reason for the diversity of pollenins may be that there are many alternative combinations of amino acid sequences that can form water channels and that some redundancy is desirable in such an important process as pollination, which is the basis of sexual reproduction in higher plants. In this case, the oleosin-like domains may effectively act as cleavable signal sequences that tether pollenins to the lipid bodies and ensure their delivery to the pollen wall at the appropriate stage of development.

### *7.2.3 Caleosins*

Caleosins were so named because they contain a conserved EF-hand, calcium-binding domain and because they were initially believed to be similar to oleosins in being uniquely associated with oil bodies in seeds (Chen *et al.*, 1999; Naested *et al.*, 2000). Similar proteins had been discovered earlier as gene products expressed in developing and germinating seeds of rice in response to abscisic acid or osmotic stress (Frandsen *et al.*, 1996). Similar proteins or the genes encoding them have now been found in a wide range of plants from maize, rice and barley to soya and sesame. A caleosin-like sequence is also present in the genome of the single-celled alga, *Auxenochlorella protothecoides*, which indicates that caleosins are probably ubiquitous in plants and algae (Partridge and Murphy, 2004). This is in contrast with oleosins, which are only found in true plants (so far they have not been reported in algae). Caleosin-like sequences are also present in at least two fungi, namely the lipid-accumulating fungus, *Neurospora crassa*, and the cereal pathogen, *Magnaporthe grisea*. Immunodetection assays indicated that the caleosin protein from rice, OsEFA27, was associated with cell membrane fractions (Frandsen *et al.*, 1996) although the caleosin from sesame was reported to be exclusively associated with lipid bodies (Chen *et al.*, 1999). More recently, we have reported evidence from immunofluorescence microscopy that caleosins in rapeseed were localised both on lipid bodies and in specific domains of the ER that may be associated with vesicle trafficking (Naested *et al.*, 2000). Very recent findings have emphasised that, although there are many intriguing similarities between caleosins and oleosins, there also important differences. In particular, analysis of caleosins from *A. thaliana* shows that they are members of a large family of as many as nine genes (Partridge and Murphy, 2004). Although one or more caleosin isoforms are tightly bound to lipid bodies during seed development, other isoforms accumulate in vegetative tissues and are probably integral membrane proteins of the ER (Murphy *et al.*, 2000; Naested *et al.*, 2000; Hernandez-Pinzon *et al.*, 2001).

The key structural features of caleosins are an $N$-terminal region with a single $Ca^{2+}$-binding EF hand domain, a central hydrophobic region able to form a single bilayer span and a $C$-terminal region with several putative protein kinase phosphorylation sites, as shown in Figure 7.2. So far, only two caleosin proteins have been shown experimentally to bind calcium (Frandsen *et al.*, 1996; Takahashi *et al.*, 2000). Most EF-hand proteins are either soluble in the cytosol or are on membranes facing the cytosol (Ikura, 1996). Some notable exceptions, such as the ER chaperone calreticulin and the Golgi resident protein nucleobindin (Lin *et al.*, 1998), are present within the lumen of organelles in the secretory pathway. Unlike these two proteins, caleosins lack an $N$-terminal signal peptide, but do have a central,

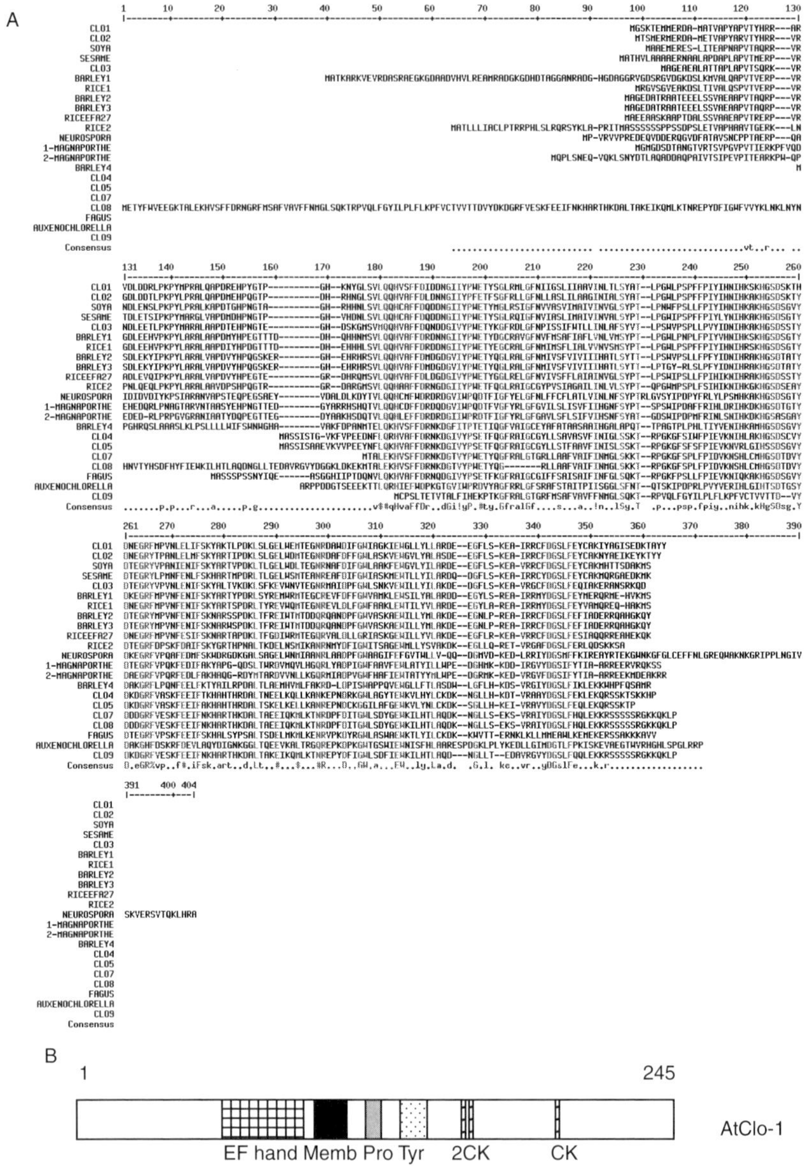


**Figure 7.2** Primary sequences and domain organisation of caleosins from plants and fungi. (A) Comparison of the 23 amino acid sequences of caleosins from plants and fungi that have been described to date. CLO1–9, the nine caleosin-like sequences from Arabidopsis; BARLEY1–3, three barley sequences; RICE1, 2, EFA27, three rice sequences; FAGUS, one sequence from fig; SOYA, one sequence from soya; SESAME, one sequence from sesame; NEUROSPORA, one sequence from the fungus *Neurospora crassa*; 1-,2-MAGNAPORTHE, two sequences from the fungus *Magnaporthe grisea*; AUXENOCHLORELLA, one sequence from the microalga *Auxenochlorella protothecoides*. Data were obtained from BLAST searches and aligned using the PRODOM database. (B) Domain organisation of a typical caleosin isoform that shows the major putative functional regions, namely the calcium-binding EF hand, membrane spanning region (Memb), proline-rich motif (Pro), tyrosine kinase site (Tyr) and three casein kinase II phosphorylation sites (CK).

hydrophobic region of more than 30 residues with the potential to form a transmembrane helix and amphipathic β-sheets (Frandsen *et al.*, 1996). This hydrophobic domain is much shorter than the analogous 70-residue domain of oleosins. Like oleosins, caleosins contain a proline-rich region with the potential to form a 'proline knot' motif of the type that appears to be so important in the lipid-body targeting of oleosins (Abell *et al.*, 1997, 2002). In addition to the hydrophobic and proline-rich domains, caleosins also possess an immediately adjacent potential amphipathic α-helical domain, which may play a role in their binding both to bilayer membranes and to lipid bodies. These properties have been used as the basis of structural models of the different forms of caleosins (Murphy, 2001).

Highly resolving Tricine-based SDS-PAGE gels have enabled us to distinguish physically between the caleosin isoforms from *A. thaliana* that bind respectively to lipid-bodies and the ER membrane (Naested *et al.*, 2000; Hernandez-Pinzon *et al.*, 2001). A 25-kDa isoform is only synthesised during the mid–late stage of seed development and is exclusively located on the surfaces of lipid-bodies. This 25-kDa isoform persists after seed desiccation and dispersal, along with oleosin, as a major lipid-associated protein and is then mobilised concurrently with the storage lipid bodies after seed germination. In contrast, the 27-kDa caleosin isoform is ER-associated and appears to be present in many tissues including roots, stems, young leaves and seeds. Using sections of rapeseed root tip cells, immunoblotting and immunolocalisation studies revealed that caleosin co-localised with the ER marker BiP and also with membranes labelled for α-TIP, a marker for protein storage vacuoles (Naested *et al.*, 2000). Parallel experiments indicated that immunodetectable oleosin is expressed in rapeseed root tip cells, and that caleosin is associated with it on what appeared to be lipid bodies. The presence of lipid bodies has been reported in root tips or root caps of rice, pea and maize and garden cress and these data indicate that root lipid bodies contain two proteins previously believed to be specific for seed lipid bodies, i.e. caleosin and oleosin. It has also been reported that lipid bodies in root tip cells from garden cress concentrate calcium (Busch *et al.*, 1993), which would be consistent with the presence of caleosin on their surfaces.

From the known primary structures of caleosins and the presence of conserved functional motifs, like EF hands and kinase domains, one can speculate about their possible functions in plants. For example, the $Ca^{2+}$-binding status of caleosins, and perhaps their phosphorylation status, could well modulate aspects of their function. Caleosins may be involved in processes such as membrane and lipid-body fusion. $Ca^{2+}$-mediated fusion has been shown to be involved in the maturation of microlipid bodies released from the ER to produce the large cytosolic lipid bodies found in milk secreting cells of mammary glands (Murphy and Vance, 1999). Likewise, in seeds and other storage-lipid accumulating plant tissues, nascent lipid-bodies are probably released as small droplets from the ER and then undergo several rounds of fusion to produce the mature 0.4–2.0 μm diameter lipid bodies characteristic of such tissues. We have also observed that the lipid-body caleosins persisted throughout seed desiccation, dormancy and for at least the first 6 days of post-germinative development. It is likely that lipid bodies need to dock with glyoxysomes to facilitate the concerted lipolysis, fatty acid oxidation and gluconeogenesis that occur during storage lipid mobilisation. Once again, caleosins may play a role in this lipid trafficking process. The microsomal caleosins found in very young embryos and in other plant tissues, such as roots and leaves, may be involved in other membrane-fission and/or fusion events relating to trafficking between the ER and storage or transport vesicles. The association of caleosins with either ER membranes or lipid bodies may be regulated by their binding of $Ca^{2+}$ in a manner similar to the $Ca^{2+}$-mediated association of lipocortin-1 with plasma membranes in human cell lines (Traverso *et al.*, 1998). The emerging dynamic role of lipid bodies in the cellular metabolism of most organisms requires a mechanism for trafficking of components between the lipid bodies and other organelles such as the ER, glyoxysomes and plasma membrane. This will involve protein mediators to regulate appropriate targeting of fusion/fission of the lipid bodies. The role, if any, of caleosins in such processes in plants (and perhaps in fungi) promises to be a fascinating topic for future investigations.

### *7.2.4 Plastid lipid-associated proteins*

Plastid lipid-associated proteins (PLAPs) are localised exclusively in plastids and, like caleosins but unlike oleosins, they are not restricted to the Plantae but are also found in unicellular photosynthetic eukaryotes (algae) and homologues have even been found in cyanobacteria. Originally regarded as having a storage/stabilisation role for lipidic pigments in chromoplasts, PLAPs are now recognised as having several additional roles in plastids. For example, PLAPs are implicated in thylakoid membrane assembly/turnover as well as participating in various stress responses. They were first described in 1976 when it was observed that the pigment-containing fibrillar lipoprotein assemblies of nasturtium flower petal chromoplasts had a distinctive protein composition dominated by a polypeptide of about 30 kDa (Winkenbach *et al.*, 1976). Similar lipid-associated proteins of 35–38 kDa were reported from higher plant chromoplasts (Knoth *et al.*, 1986; Smirra *et al.*, 1993; Vainstein *et al.*, 1994) and triacylglycerol/carotenoid globules of the alga, *Dunaliella bardawil* (Katz *et al.*, 1995). The algal protein was localised on the surface of plastoglobuli and, as with oleosin in seeds, its cleavage by trypsin led to coalescence of the globules, which suggested that the function of this PLAP might be to stabilise the plastid lipid bodies. In 1994, the gene encoding the 32-kDa bell pepper protein was sequenced and named fibrillin in view of the fibrillar nature of the chromoplast lipoprotein structures from which it was derived (Deruere *et al.*, 1994). However, the name fibrillin had already been assigned to a common class of structural proteins in animals (Handford, 2000) and it is also now clear that the plant protein is associated not only with fibrils and globules in various plastid types, but also with thylakoid membranes (Pozueta-Romero *et al.*, 1997; Hernández-Pinzón *et al.*, 1999). Therefore, a more general term such as plastid lipid-associated protein – PLAP – is probably more appropriate (Hernández-Pinzón *et al.*, 1999).

It is now clear that PLAPs belong to a large class of homologous proteins found throughout oxygenic photosynthetic organisms (Katz *et al.*, 1995; Vishnevetsky *et al.*, 1996; Pozueta-Romero *et al.*, 1997; Hernández-Pinzón *et al.*, 1999; Kim *et al.*, 2001). The discovery of a PLAP homologue in the cyanobacterium, *Synechocystis*, indicates the probable ancient origin of this protein in the endosymbiotic precursors of plastids (Kaneko *et al.*, 1996). In addition to forming the major protein component of triacylglycerol/carotenoid-rich fibrils and globules in chromoplasts, PLAPs are present in other plastid types such as elaioplasts and chloroplasts (Pozueta-Romero *et al.*, 1997; Ting *et al.*, 1998; Hernández-Pinzón *et al.*, 1999). The PLAPs of elaioplasts are located on globular lipid bodies that resemble the triacylglycerol/carotenoid globules of chromoplasts, except that their lipid components are mainly sterol esters and fatty aldehydes (Hernández-Pinzón *et al.*, 1999). By way of contrast, the PLAPs of chloroplasts are associated both with plastoglobuli and thylakoid membranes (Pruvot *et al.*, 1996a,b; Pozueta-Romero *et al.*, 1997; Kessler *et al.*, 1999). Indeed, it has been reported that plastids from *Brassica rapa* can contain up to three distinct PLAP isoforms, each of which is associated with globules containing a different mixture of neutral lipids (Kim *et al.*, 2001).

While the presence of PLAPs on neutral lipid bodies, including fibrils and globules, can be rationalised as providing a stabilising surface structure (Deruere *et al.*, 1994), their apparently ubiquitous distribution in plant tissues and their association with thylakoid membranes is more difficult to explain. A possible clue has come from studies of the induction of PLAP gene expression in response to various environmental factors, including drought stress, wounding or application of exogenous ABA (Pruvot *et al.*, 1996a; Chen *et al.*, 1998). Since one of the primary responses of plants to such stresses is often a re-arrangement of their photosynthetic membranes, it has been proposed that PLAP has a general role in the formation/disassembly/turnover of plastid membrane complexes (Chen *et al.*, 1998). Later, it was found that the PLAP homologue from potato was associated with photosystem II, which is one of the major multi-subunit pigment–protein complexes of thylakoid membranes (Monte *et al.*, 1999). Antisense-mediated reduction of PLAP accumulation in transgenic potato plants led to reduced photosynthetic efficiency and stunted growth, which reinforces the view that PLAPs play important role(s) in both plastid membranes and lipid bodies. It is likely that there are several classes of PLAP-like proteins in plants. Some of these may be expressed ubiquitously while others, like the ChrC protein of cucumber (Vishnevetsky *et al.*, 1999), may have a more restricted pattern of expression, e.g. in response to stress.

Those PLAPs that are expressed ubiquitously in plants are of particular interest since they appear to associate strongly, both with the monolayer surfaces of plastid lipid bodies and fibrils and with the bilayer membrane of the thylakoids. Analysis of the conserved regions of PLAPs does not reveal any obvious homology with other plant lipid-body proteins, such as oleosins or caleosins, although there are some interesting motifs in PLAPs that may be significant in their lipid-binding properties. For example, in the middle of the protein there are two non-polar regions of

16 and 22 residues, respectively, each of which is flanked by relatively polar regions, which could potentially form transbilayer- or monolayer-associating domains. Further structural studies are required to elucidate the mechanism(s) of lipid binding, and, hence, the biological functions of PLAPs. Finally, like seed lipid bodies, it is now emerging that plastid lipid assemblies probably contain several minor protein components in addition to the dominant oleosin or PLAP classes (Libal-Weksler *et al.*, 1997). In view of their lower abundance, these minor proteins are less likely to have structural roles, but may well be involved in other aspects of lipid-body function or possibly in more generalised intra-plastidial lipid trafficking. It is possible that some of these PLAPs may play a role in the conversion of thylakoid membranes into the triacylglycerol-rich globules found in senescing leaves (Thompson *et al.*, 1998a) and following exposure to a wide range of stresses, including ozone exposure, fungal infection, chilling, freezing and thawing (Murphy, 2001).

### *7.2.5 Minor lipid-associated proteins in plants*

Oleosins tend to be very abundant when they occur on storage oil bodies and caleosins are moderately abundant, i.e. caleosin bands can be readily discerned on SDS-PAGE gels of many oil-body extracts. However, there are also several relatively minor lipid-associated polypeptides that appear to be enzymes rather than structural proteins. It is still not known whether these are specific oil-body proteins or mere contaminants. Potential candidates include enzymes involved in triacylglycerol biosynthesis and, for many years, there have been reports of the presence of such enzymes on lipid bodies. This is mirrored by the well-documented findings of both triacylglycerol and sterol ester biosynthetic enzymes on fungal lipid bodies (see below). The apparent presence of triacylglycerol biosynthetic enzymes on seed lipid bodies may actually be due to the existence of membranous appendages of the ER remaining attached to the lipid bodies following their budding off from the ER proper. This has been proposed from ultrastructural studies (Wanner and Theimer, 1978; Wanner *et al.*, 1981) and would explain why membrane bilayer proteins can be associated with a non-bilayer structure like an oil body. The membranous appendages may facilitate re-fusion of oil bodies with the ER for the further metabolism of oil-body triacylglycerol, e.g. by desaturases or transacylases, as has been reported during sunflower seed development (Mancha and Stymne, 1997; Sarmiento *et al.*, 1998). Similar lipid-body appendages have been described in animal cells (Glaumann *et al.*, 1975; Howell and Palade, 1982; Gibbons *et al.*, 2000).

More recently, the apparently specific binding of a sterol-binding dehydrogenase to oil bodies has been reported in sesame (Lin *et al.*, 2002). This protein, tentatively named steroleosin, is similar in sequence to a protein encoded by a family of eight genes in Arabidopsis. It is not known whether all steroleosin-like proteins in plants are associated with oil bodies or whether, like caleosins, some isoforms associate instead with other cellular components. Although steroleosins are relatively minor components of oil bodies and are not yet confirmed as being ubiquitous in plants, their discovery is interesting in view of the finding that another sterol-binding protein (a sterol Δ24-methyltransferase) is the major protein associated with lipid bodies in yeast (Athenstaedt *et al.*, 1999). The importance of this class of enzyme in plants is shown by reports that sterol Δ24-methyltransferases in tobacco and rapeseed control the flux of carbon into sterols in seeds (Holmberg *et al.*, 2002; Harker *et al.*, 2003) and modulate growth in Arabidopsis (Schaeffer *et al.*, 2001; this topic is also discussed in Chapter 11). Another potentially relevant finding is the recent report of the presence of a sterol dehydrogenase on the surfaces of mammalian lipid bodies, with the implication that these organelles might play important roles in sterol metabolism and other aspects of lipid transport and membrane trafficking (Ohashi *et al.*, 2003).

A further putative oil-body protein described from developing sesame seeds is a calcium-dependent protein kinase of 55 kDa (Anil *et al.*, 2003). This oil-body-associated polypeptide was capable of calcium-dependent autophosphorylation *in vitro* and cross-reacted with an antibody to a soybean protein kinase. The detection of similar proteins in oil bodies of other plant species suggests that these lipid-associated kinases may be ubiquitous in oilseeds. It can be speculated that these kinases, which are specifically active during seed development, may be involved in oil-body ontogeny. A possible substrate would be the relatively abundant lipid-associated isoforms of caleosin, which are also present in developing seeds and possess several conserved phosphorylation sites.

In some species, seed oil bodies have been found to contain proteins that are associated with the mobilisation of the storage triacylglycerol that occurs after germination. An example is the lipoxygenase that is reportedly associated with oil bodies in cucumber seeds (Hause *et al.*, 2000). Using GST-fusion constructs, it was shown that this protein can be targeted both to oil bodies and to liposomes and that the targeting required the presence of an *N*-terminal beta-barrel domain (May *et al.*, 2000). Lipoxygenases are only active with polyunsaturated acyl substrates but many seeds do not store such fatty acids, and it is therefore unlikely that these proteins are present on storage oil bodies in all oilseeds. However, another class of cucumber oil-body protein described by the same group could well be ubiquitous in plants. This is a patatin-like protein that has phospholipase A2 activity, which was transiently expressed and associated with oil bodies coincidentally with lipid mobilisation (May *et al.*, 1998). Although the authors posit a role for this protein in storage lipid mobilisation, it could also have a signalling function. Phospholipases A2 have been shown to have such roles in both plants and animals (Holk *et al.*, 2002). Interestingly, a similar patatin-like protein has recently been found in the proteome of lipid bodies from human CHO K2 cells (Liu *et al.*, 2004). Patatins are the major family of storage proteins in potato tubers and, like some seed storage proteins, they appear to be derived from a group of enzymes that had esterase activities. If this is the case, then it may be misleading to refer to the motif common to the plant and animal lipid-body proteins as a patatin domain. Rather, these are all proteins with esterase-like domains, some of which have secondarily lost their enzymatic activity and become storage proteins in some higher plants. See Chapter 5 for more discussion of phospholipases and patatin-like esterases.

### *7.2.6 Lipid-associated proteins in non-storage tissues in plants*

Oleosins are mostly found in seeds and in some pollen grains, while oleo-pollenins are found in the anthers of some higher plants. Caleosins are cytoplasmic proteins that can be associated with lipid bodies and/or ER membranes in most plant tissues and PLAPs are plastidial proteins that can associate with lipid assemblies and/or thylakoid membranes in many plant tissues. In addition to these relatively well-known proteins, one can ask whether there are other proteins in other locations in plants that also associate with lipids; e.g. are there any plant analogues of the transport apolipoproteins of animals?

#### *7.2.6.1 Phloem*

The phloem is a long-distance transport system that ramifies throughout the plant, from roots to leaves. The role of the phloem in the transport of low molecular weight assimilates, such as sugars and amino acids, has long been established. It has emerged that the phloem may also participate in more complex processes involving transport of macromolecules including proteins, mRNA and even viruses (see Thompson, 1999 for review). Coincidentally, with this re-appraisal of phloem function, there has been a report that phloem exudates from stems of carnation and rapeseed contained lipid bodies of 90–160 nm and 60–240 nm diameters, respectively (Thompson *et al.*, 1998b; Madey *et al.*, 2002). These lipid bodies contained a variety of neutral phospholipids including sterol esters, waxes, free fatty acids and triacylglycerols. If this observation is confirmed, it may imply that the phloem could transport lipid bodies over long distances in plants, in a manner that would be somewhat analogous to the transport of lipoproteins in the circulatory systems of animals ranging from insects to mammals (Murphy, 2001). The analogy is not exact because, whereas animals transport lipids through an extracellular medium, namely the blood, plants and fungi would have to utilise an intracellular, symplastic system, such as the phloem or hyphae. A closer analogy is given by the intriguing recent observation made in 2002 that the hyphae of some mycorrhizal fungi are apparently able to transport lipids from their plant hosts all the way to their sporulating structures that may be well over a metre away. The lipid is transported in the form of 1–2-μm triacylglycerol-rich bodies that appear to move along an organised network through the hyphal cells (Bago *et al.*, 2002). This observation, which is also discussed below (Section 7.3.2), begs the question of whether there are proteins associated with these transport lipid bodies in the fungi. So far, no transport-related lipid-associated proteins have been discovered either in fungal hyphae or in the phloem of plants, but if macromolecular lipidic assemblies are really transported through either of these systems, it is difficult to imagine a mechanism that does not involve at least some involvement of lipid-binding proteins. This is an interesting area for future study.

#### *7.2.6.2 Roots and meristems*

There are numerous reports of intracellular lipid bodies in different cell types from roots (see Murphy, 2001 for review). A common location is in the young roots emerging from seeds shortly after germination, where the lipid bodies may be involved in the extensive lipid trafficking that is required to support the rapid expansion of this tissue. This is substantiated by reports that significant amounts of oleosins and lipid-body binding caleosins can be detected in the root tips of 2–3-day-old seedlings of rapeseed and Arabidopsis (Naested *et al.*, 2000; Hernandez-Pinzon *et al.*, 2001). However, lipid bodies may also play other roles in roots. For example, during their initial period of differentiation, root cap cells act as statocytes (gravity sensors). In root cap statocytes of cress there are abundant lipid bodies that appear to determine a preferential distribution of ER at the distal cell pole (Hensel, 1986). Hence, at least in cress, the lipid bodies may be one component of the positive orthogravitropic growth of roots. As well as oleosins, caleosins are found in young root tips where they may be located either on lipid bodies or on the ER membrane (Hernandez-Pinzon *et al.*, 2001). Cells from the shoot apical meristem in birch reportedly contain organelles similar to lipid bodies and the rearrangement of these structures may be involved in the breakage of bud cell dormancy (Rinne *et al.*, 2002). We have also observed the expression of the GUS reporter gene driven by an oleosin gene promoter in the meristems of seedlings of *B. napus*. This implies that oleosins may be present, albeit in small quantities, in such tissues (Murphy, unpublished results).

#### *7.2.6.3 Rubber*

There are very few examples of extracellular, lipid-associated proteins in plants; this contrasts with animals where there are numerous proteins involved in the transport of lipids via extracellular circulatory systems, such as the blood. Amongst the short list of plant proteins is the 14.5-kDa elongation factor associated with rubber bodies. Several species produce cytoplasmic rubber bodies in all tissues, where it remains until plant death, while others, like the commercial rubber tree, *Hevea brasiliensis*, secrete the rubber bodies as part of a complex mixture called latex (De Fay *et al.*, 1989; Siler *et al.*, 1997). Rubber bodies are similar to other cytosolic lipid bodies in consisting of 0.4–4.0-$\mu$m diameter droplets with a hydrophobic isoprenoid core that is surrounded by a lipid/protein annulus. The rubber lipid bodies are formed in the cytosol and may continue to accumulate there or may be released into the large vacuole (Backhaus and Walsh, 1983). Latex is produced by the disintegration of rubber-producing cells and the release of their contents into lactifers (Dennis *et al.*, 1989; Dennis and Light, 1989). The major component of the protein layer surrounding rubber bodies of *H. brasiliensis* is a 14.5-kDa rubber elongation factor polypeptide that may be involved in both the formation and stabilisation of these unusual lipid assemblies, although the mechanism of its binding to rubber bodies is not known (Dennis and Light, 1989).

## 7.3 Comparisons with non-plant systems

Although plant lipid-associated proteins like oleosins have been studied for over 15 years, the focus of much of the research has been on storage organelles, such as oil bodies from seeds. It is only since 2000 that other proteins, like caleosins, have been discovered where they may play a role in more dynamic events such as lipid trafficking. Over the past decade, there has also been a huge amount of work on lipid-associated proteins in non-plant organisms that range from animals to Archaea. The results of these investigations are opening up new horizons for the study of lipid assemblies, both within and without cells, and may suggest novel lines of study for plant cell biologists.

### *7.3.1 Animals*

There are two major classes of non-bilayer lipid assembly in animals. The first class is the intracellular, cytosolic lipid-bodies that function in storage and subcellular trafficking. Proteins associated with these intracellular lipid bodies include the recently discovered, structurally related family of so-called PAT proteins named after perilipin, adipophilin and TIP47/PP17 (TIP47, Tail-interacting 47-kDa protein; PP17, placental protein 17) (Londos *et al*., 1999; Wang *et al*., 2003). PAT-like proteins are widespread in Metazoans ranging from mammals to insects with similar proteins found in the plasmodial slime mold, Dictyostelium, which is normally classified in the Fungi (Miura *et al*., 2002). There are also some surprising recent findings that the Golgi/plasma membrane protein family of caveolins may be involved in the ontogeny of lipid bodies (Robenek *et al*., 2004). Similarly, other proteins, like vimentin and $P_{200}$, which have hitherto been regarded as cytoskeletal components, can associate with lipid bodies (Murphy, 2001; Wang *et al*., 2003). The second major class of non-bilayer lipid structure found in animals is the lipoproteins. These include various forms of extracellular, circulating lipid-bodies, which are involved in the long-range transport of lipids throughout the animal body.

Intra- and extra-cellular lipid assemblies are characterised by different populations of lipid-body proteins. Cytosolic lipid-body proteins include adipophilins, perilipins and related proteins like TIP47 or PP17, whereas several classes of apolipoproteins (A, B, C, E, etc.) are associated with the various types of extracellular, circulating lipoproteins. The likely dynamic intracellular roles of animal and plant lipid bodies and their proteins (i.e. apart from simple storage) has recently been shown by the proteomic analysis of lipid bodies in CHO K2 cells (Liu *et al*., 2004). In recognition of this role in membrane trafficking, these authors have proposed that these complex and active organelles, which may be a subset of all the lipid bodies within a given cell, be renamed 'adiposomes'. Meanwhile, there has been a fascinating series of recent findings that implicate proteins mistargeted to lipid bodies in the progression of some serious diseases. For example, the normally soluble Parkinson's disease protein, σ-synuclein, becomes bound to lipid bodies in affected patients, suggesting that relocation to lipid bodies may be an early and important stage in the onset of this killer disease (Cole *et al.*, 2002). Another protein, Nir2, which is essential for proper function of the retina, can, upon phosphorylation, become relocated to lipid *bodies* (Litvak *et al.*, 2002). The resulting disruption of intracellular lipid trafficking may be involved in the consequent degeneration of retinal function.

#### *7.3.1.1 The PAT family of cytosolic lipid-body proteins*

The presence of specific proteins on the surfaces of cytosolic lipid bodies in animals was discovered in 1991 for perilipin (Greenberg *et al.*, 1991), and in 1997 for adipophilin (Brasaemle *et al.*, 1997). Perilipin and adipophilin share several regions of substantial sequence homology with each other and are often grouped with two other proteins called TIP47 and PP17 (which are encoded by the same gene) to make up the so-called PAT family (for perlipin, adipophilin and TIP47). PAT proteins are characterised by their lipid-binding properties and by two conserved domains, termed PAT1 and PAT2. Whereas perlipin and adipophilin are exclusively bound to lipid bodies, TIP47/PP17 has a more complex subcellular distribution and has been variously reported as endosomal, soluble and lipid-body associated, as discussed below. It is now clear that as well as being ubiquitous in mammals, PAT proteins are present in organisms as diverse as Dictyostelium, Drosophila and Xenopus (Wolins *et al.*, 2003). In addition to perlipin, adipophilin and TIP47, a fourth PAT-like sequence, called S3–12 has been described (Scherer *et al.*, 1998). The S3–12 protein has recently been shown to bind lipid bodies in a manner similar to other members of the PAT family (Wolins *et al.*, 2003).

***Adipophilin.*** Adipophilin was discovered as one of the earliest markers of adipocyte development in mammals and was originally termed adipose differentiation-related protein (often abbreviated to ADRP), a name that is still used in some of the literature. However, it is now clear that this class of protein is not adipose-specific but is distributed ubiquitously in mammalian tissues, where it is localised on the surface of the cytosolic lipid bodies that can be found in almost all cell types at some stage of their development (Murphy, 2001). The mammalian adipophilins are proteins of about 400–440 residues with apparent masses of about 50 kDa on SDS-PAGE. Adipophilin has no obvious lipid-binding motifs, such as hydrophobic regions or amphipathic α-helices, of the sort found in apolipoproteins or the plant lipid-body proteins, oleosin and caleosin. Unlike perilipin, it is unlikely that adipophilin is phosphorylated *in vivo* but it can undergo post-translational modifications, including acylation (estimated at 5–6 mol fatty acid/mol protein), which may promote its binding to lipid bodies (Heid *et al.*, 1996). Several recent studies have addressed the lipid targeting and binding of adipophilins, by examining truncated variants, although a consensus view of such a mechanism has yet to emerge (McManaman *et al.*, 2003; Nakamura and Fujimoto, 2003). Adipophilin is probably more firmly bound to lipid bodies than other lipid-associated proteins. For example, in steroidogenic cells of the adrenal gland, the lipid-associated proteins perilipin and p160 were readily detached from lipid bodies following activation by protein kinase A, whereas adipophilin remained firmly attached (Fong *et al.*, 2002).

The ubiquitous distribution of adipophilin has encouraged speculation that it may play a key role in lipid-body biogenesis in mammals; a view encouraged by its apparent lipid-body localisation. Recent studies following the distribution of GFP-tagged adipophilin in human cell cultures are consistent with the existence of two populations of adipophilin-bound lipid bodies (Targett-Adams *et al.*, 2003). One population moved rapidly and apparently exchanged with other intracellular lipid pools, while the other population remained in the vicinity of the ER and was more static. It is possible that these two pools of lipid bodies are involved respectively in lipid trafficking and longer-term storage. Adipophilin may also be involved in several types of cellular dysfunctions of lipid metabolism. For example, it has been shown that adipophilin expression, and hence lipid-body accumulation, is strongly induced by oxidised low-density lipoprotein during foam cell formation in human macrophages, which is one of the earliest events in the aetiology of atherosclerosis (Wang *et al.*, 1999). Adipophilin is also induced when keratinocytes are exposed to skin irritants, resulting in the redistribution of lipids from intracellular membranes to form cytosolic lipid bodies (Corsini *et al.*, 2003).

***Perilipin.*** Perilipins are found in adipocytes and steroidogenic cells in mammals but do not appear to be present in the many other cell types that accumulate cytosolic lipid bodies. The accumulation of perilipin on lipid bodies is normally at the expense of endogenous adipophilin and therefore cells can contain either one of these two proteins, but not both together (Miura *et al.*, 2002). Perilipins occur as three isoforms, A, B and C, all of which are products of the same gene via differential mRNA splicing (Lu *et al.*, 2001). Perilipin A at 56-kDa is the largest and most abundant isoform in both adipose and steroidogenic cells while the B (46 kDa) and C (42 kDa) isoforms are found mainly in adipose and steroidogenic cells, respectively. The protein consists of an *N*-terminal domain of about 200 residues that contains several extensive regions of significant homology to TIP47/PP17 and adipophilin followed by about 300 residues of a few isolated regions of similarity to these proteins. Perilipins A and B contain six and three consensus PKA phosphorylation sites, respectively, and it has been shown that all three perilipin isoforms are phosphorylated by PKA in association with lipolytic stimulation (Londos *et al.*, 1996). Like adipophilin, perilipins contain no obvious lipid-binding motifs and studies of mutagenised and GFP-linked variants suggest that an extensive central region comprising 25% of the entire protein is required for lipid targeting and binding (Garcia *et al.*, 2003). These authors hypothesise that perilipins bind to lipid via three separate hydrophobic domains of 18, 23 and 16 residues, which are interspersed within more polar regions within the central portion of the protein molecule.

Immunocytochemical studies have shown that perilipin is associated with the surface layer of cytosolic lipid bodies in adipocytes and steroidogenic cells, but not with the ER, even when this membrane is in intimate contact with lipid bodies (Greenberg *et al.*, 1991). This raises the question of how perilipin is targeted to lipid bodies. It is known that oleosins in plants are first inserted into the ER via an SRP-dependent mechanism and then accumulate on lipid-body surfaces. This seems unlikely in the case of perilipin, which may be expressed initially as a soluble protein (it is not at all hydrophobic) that is later inserted directly onto the lipid-body surface, possibly following a post-translational modification, such as acylation. The hormonally induced phosphorylation of perilipin has been proposed to weaken its attachment to the surface of the lipid bodies and eventually to allow access of lipase(s), such as HSL, which will then mobilise their TAG or sterol ester contents (Brasaemle *et al.*, 2000). The stability of perilipin may be regulated by the presence of lipid bodies, which would have parallels with both oleosin and apolipoprotein B, both of which are rapidly turned over in the absence of suitable lipid-body acceptors. In the case of perilipin, an increase in lipid-body accumulation following fatty acid supplementation of the culture medium of steroidogenic cells led to a transient six-fold stimulation of mRNA levels (and even greater increases in protein levels) within 10–12 h (Brasaemle *et al.*, 1997). These data suggest a direct link between lipid-body formation/maturation and perilipin synthesis/accumulation, although whether perilipin stabilises lipid bodies (e.g. by shielding them from lipase attack) or the lipid bodies stabilise perilipin, or both, remains to be determined. A major role for perilipin in shielding lipid bodies from lipolysis by HSL is indirectly supported by the three- to six-fold higher activity of HSL in perilipin-knockout mice (Martinez-Botas *et al.*, 2000). These mice, while leaner than controls, developed normally and were apparently healthy, which suggests that the absence of perilipin in steroidogenic cells was somehow compensated for, as the mice suffered no obvious hormonal problems. While the central region of perilipin is required for lipid targeting and binding, the *C*- and *N*-terminal parts of the protein play a crucial role in facilitating the long-term storage of lipid bodies within cells (Garcia *et al.*, 2004).

***TIP47, PP17, S3–12 and other PAT family proteins.*** The homologous N-terminal regions of adipophilins and perilipins are very similar to a third class of proteins, variously referred to as TIP47 or PP17. These latter two proteins are, in fact, identical to one another, but were originally characterised in different cell systems and, hence, were given different names. Tail-interacting 47-kDa protein (or TIP47) is a mammalian protein that was originally reported to bind the cytosolic domains of mannose 6-phosphate receptors and to be involved in their transport from endosomes to the Golgi (Diaz and Pfeffer, 1998). In the past few years, there have been several reports that TIP47 can also be associated with lipid bodies (Wolins *et al.*, 2001; Miura *et al.*, 2002), although this has been challenged (Barbero *et al.*, 2001). However, several independent groups have now confirmed the presence of TIP47 on lipid bodies (Ohashi *et al.*, 2003; Than *et al.*, 2003). It looks increasingly likely, therefore, that this is not merely aberrant targeting, but rather a reflection of a more complex role of both TIP47 and lipid bodies in lipid trafficking in mammalian cells.

In view of the findings that adipophilin may be involved in the import of fatty acids, it is tempting to draw analogies with the suggested role of TIP47 in mediating protein import. However, current evidence suggests that TIP47 is only involved in trafficking between late endosomes and the Golgi and not in endocytosis or protein import *per se*. Also TIP47 appears to be extremely selective in the set of cargo proteins that it recognises (Krise *et al.*, 2000). Nevertheless, the degree of resemblance between TIP47/PP17 and adipophilins and perilipins encourages the view that there may be fundamental similarities in the processes involved in the trafficking of lipid bodies and bilayer membranes in cells. This view is further encouraged by reports (see below) that link adipophilin-coated lipid bodies with the vesicle coat protein, caveolin and cholesterol trafficking in mammalian cells.

Placental tissue proteins 17 (PP17, which are identical to TIP47) are members of a class of human pregnancy-related proteins that include four splicing variants of a single gene product (Than *et al.*, 2003). Although present during normal pregnancy, PP17 variants are also upregulated in patients with cervical carcinoma. In view of its expression pattern during oncogenesis, it has been suggested that PP17/TIP47 induction may be linked with infection by the herpes simplex virus, type 2 (HSV-2), which is one of the major pathological agents associated with the aetiology of cervical cancer. The PP17b variant has now been shown to be associated with lipid bodies in HeLa cells (Than *et al.*, 2003), which further suggests that PP17/TIP47 may have a dual role in both Golgi-lysosomal trafficking and as lipid-body components.

Like adipophilin, the S3-12 protein was originally found in a screen of differentiating mammalian adipocytes (Scherer *et al.*, 1998). Approximately two-thirds of the sequence is made up of tandem repeats of a 33 mer that is similar to the PAT1 domain. S3-12 has now been show to bind lipid bodies in a manner similar to other members of the PAT family but its function and interaction, if any, with other PAT proteins remains to be elucidated (Wolins *et al.*, 2003).

An adipophilin homologue, termed the fatvg protein, is present in oocytes of *Xenopus laevis* (Chan *et al.*, 1999). This protein, which is localised to the vegetal cortex, is 60% identical to adipophilin over its entire 408-residue sequence. There seems little doubt that this protein is an amphibian equivalent of adipophilin, although its exact function and whether it binds to any subcellular lipidic structures both remain to be determined. It has also been shown recently that lipid storage in the fruit fly, *Drosophila melanogaster*, is regulated by a 352-residue PAT-like protein, called Lsd2, which is similar to both adipophilin and perilipin from mammals and to the Xenopus fatvg protein (Gronke *et al.*, 2003; Teixeira *et al.*, 2003).

*7.3.1.2 Caveolins – the unexpected lipid-associated proteins*

Caveolins are small proteins with a subunit weight of about 21 kDa that contain a central hydrophobic domain of 33 amino acid residues, which is presumed to form a tight α-helical hairpin anchoring the protein in the plasma membrane of most mammalian cells. A caveolin-like gene has also been found in invertebrates– in the worm, *Caenorhabditis elegans* (Tang *et al.*, 1997). The *N*- and *C*-terminal domains of caveolin are both located on the cytosolic face of the membrane, which means that the membrane-embedded domain probably has a hairpin structure reminiscent of many models of oleosins (see Section 7.2.1). Caveolins are normally found as homooligomers that behave as complexes of about 350 kDa or more and are located on the surface of caveolae (Sargiacomo *et al.*, 1995). Caveolae are now generally regarded as a form of lipid raft that contains caveolin as its major structural component. Caveolae are found as 50–100 nm disk-shaped pits, or invaginations, in the plasma membrane of many mammalian cells. Caveolae give rise to a class of non-clathrin-coated vesicles that are involved in receptor-mediated endocytosis, cholesterol transport and signal transduction (Oh *et al.*, 1998; Fujimoto *et al.*, 2000, Nabi and Le, 2003). There are three isoforms of caveolin, caveolin 1 is the most common form, caveolin 2 is co-expressed with caveolin 1 and caveolin 3 is specific to muscle cells. Evidence of a role for lipid bodies in subcellular lipid trafficking originally came from reports that variant forms of caveolin were targeted to adipophilin-coated lipid bodies in fibroblasts (Ostermayer *et al.*, 2001; Pol *et al.*, 2001). The apparent mistargeting of caveolin to lipid bodies led to the accumulation of free cholesterol in late endosomes (Pol *et al.*, 2001), possibly at the expense of cholesterol ester formation. These studies pointed to a direct link in mammalian cells between cytosolic lipid bodies and elements of the vesicular trafficking system including plasma membrane-derived vesicles and the Golgi apparatus. Evidence is now emerging that a hydrophobic domain of caveolin 1 (which is predicted to have an α-helical structure) is required for binding of the protein to lipid bodies in animal cells (Ostermeyer *et al.*, 2004).

A link with cytoskeletal-mediated trafficking of lipid bodies was recently provided by a report that mammalian cell cultures and cell-free microsomal systems assemble lipid bodies that contain caveolin 1 and 2, vimentin, adipophilin and the 78-kDa glucose regulatory protein (Marchesan *et al.*, 2003). Caveolin 2 is also associated with lipid rafts and ER-associated lipid bodies that bind σ–1 receptors; these receptors in turn bind lipidic neurosteroids and psychotropic drugs, including neuroleptics and cocaine, which once more demonstrates the key role of lipid bodies in intracellular lipid trafficking (Hayashi and Su, 2003). There are now several reports of the involvement of caveolins in the formation of lipid bodies in specialised domains of the ER and of the co-localisation of caveolins with other lipid-associated proteins such as adipophilin (Liu *et al.*, 2004; Robenek *et al.*, 2004). Clearly, caveolins have a more complex and wider ranging role in cellular trafficking than was first thought and lipid bodies are intimately involved in at least some of these functions.

#### 7.3.1.3 *Extracellular lipid-body proteins*

***Apolipoproteins and related proteins.*** Apolipoproteins are the proteins that bind to the surfaces of the various extracellular lipid bodies, or lipoproteins, which are secreted into the circulatory system from mammalian organs, such as the liver or intestine. Exchangeable apolipoproteins are capable of moving from one lipoprotein particle to another. The best characterised structural motif that is involved in the binding of apolipoproteins to either lipid bilayers or lipid bodies is the amphipathic α-helix. The α-helices of apolipoproteins are distinctive in having an extensive non-polar face that interacts with acyl chains of the lipids and a polar face, the centre of which is enriched in the acidic residues Asp and Glu, while the basic residues Lys and Arg are at the edges. However, the most striking feature of such apolipoproteins is the presence of long runs of 11-residue repeats that in some cases have evolved into 22-mer tandem repeats. Based on sequence comparisons, an evolutionary tree has been proposed for all exchangeable apolipoproteins that traces their origin, via extensive gene duplication, from a single ancient gene (Luo *et al.*, 1986). The reason for the 11-mer organisation is straightforward; with a periodicity of 3.6 residues per turn, 11 residues correspond to three complete helical turns. Duplication to form a 22-mer will produce an amphipathic α-helix with a small twist of less than 20° between the polar and non-polar faces of the two 11-mer halves. The degree of amphipathicity of each α-helix is correlated with its affinity for lipid binding, while association with lipids tends to increase further the α-helical nature of the peptide.

Exchangeable apolipoproteins are able to interact with a range of macromolecular lipid assemblies, from relatively large globular VLDL droplets to tiny discoid HDL particles that are only 5–15 mn in diameter. The exchangeable apolipoproteins in non-mammalian species are similar to those of mammals in their physiological role, i.e. lipid transport, and in their predominant amphipathic α-helical structure, but they do not contain the characteristic 11- and 22-mer repeats of the mammalian apolipoproteins. The fine structure and dynamic behaviour of the locust apolipophorin, apoLp-III, have been determined to high resolution by X-ray and NMR methods (Breiter *et al.*, 1991; Fan *et al.*, 2003). Upon contact with diacylglycerol in a nascent lipoprotein (most insect lipoproteins preferentially transport diacylglycerol rather than triacylglycerol), the apoLp-III binds to the lipid, which triggers opening of the helix bundle about hinge loops located at opposite ends of the molecule. This exposes the non-polar faces of the α-helices, which then embed themselves into the lipid droplet contributing to its stabilisation and potential for further lipid loading. It is likely that the paradigm of the amphipathic α-helix is applicable to exchangeable apolipoproteins in most, if not all, animal species.

In contrast to exchangeable apolipoproteins, the non-exchangeable apolipoproteins, i.e. apoB48 and apoB100, remain on the same lipoprotein particle from biosynthesis to breakdown and their mechanism of lipid binding is much less well-understood. ApoB in mammals is a very large protein and each VLDL or chylomicron droplet contains just one molecule of either apoB48 or apoB100.

The apoB polypeptide has been proposed to loop in and out of the lipid droplet over its whole length. This model has been broadly substantiated by studies in 2001, which show that apoB is made up of two parts. These are a small *N*-terminal domain that adopts a compact globular structure, which folds independently and is probably not involved in bulk lipid binding, plus a much larger domain that forms a belt-like structure wrapped around the lipid droplet (reviewed in Murphy, 2001). The lipid-binding part of the apoB polypeptide contains extensive regions of amphipathic α-helices and β-sheets. The major lipid-binding regions of apoB are proposed to be two large amphipathic β-pleated sheet structures alternating with two much smaller amphipathic α-helical domains (Segrest *et al.*, 1998).

The *N*-terminal domain of apoB has significant sequence homology with both microsomal transport protein (MTP) and another ancient family of lipid transport and storage proteins, the vitellogenins, which are found in animals ranging from *C. elegans* and insects through to fish, amphibia and birds. These three proteins are evolutionarily related and are similar both structurally and, possibly to a more limited extent, functionally, although they differ significantly in the organisation of their *C*-terminal lipid-binding domains. The structural differences between these proteins are reflected in differences in their lipid-binding capacities, i.e. MTP binds lipid at 1–5 moles/mole protein, the lamprey vitellogenin binds ~35 moles/mole mostly of phospholipid, while the huge *C*-terminal domain of apoB (~390 kDa) can bind a neutral lipid droplet of 1 μm diameter containing millions of lipid molecules. The crystal structure of the lamprey lipovitellin (the mature form of vitellogenin) has been determined at 2.8 Å resolution (Timmins *et al.*, 1992; Anderson *et al.*, 1998). The lipid-binding domain of the lamprey vitellogenin is made up of an extensive β-sheet structure forming a cavity into which the lipids can partition, while the homologous domain in MTP is similar, albeit much reduced in size. The predicted structure of MTP, based on the lamprey lipovitellin structure, has been reported together with experimental data based on protein mutagenesis (Read *et al.*, 2000). For more details on apolipoproteins and their relatives, see the following reviews – Murphy, 2001; Narayanaswami and Ryan, 2000; Ursini *et al.*, 2002; Bolanos-Garcia and Miguel, 2003; Hussain *et al.*, 2003.

### *7.3.2 Microorganisms*

Many microorganisms accumulate cytosolic lipid bodies, including unicellular algae (see Section 7.2), protozoans, fungi and prokaryotes. One of the first indications that intracellular triacylglycerol may be involved in phospholipid metabolism was made with regard to the ciliated protozoan, *Tetrahymena pyriformis*, as long ago as 1976 (Borowitz and Blum, 1976). In this study, it was found that *T. pyriformis* contained a very labile triacylglycerol pool, which may be separate from other endogenous triacylglycerol pools, and that both the glycerol backbone and the acyl groups of this labile pool served as precursors for membrane phospholipid biosynthesis.

The results suggest the sort of intimate relationship between triacylglycerol and phospholipid metabolism that has subsequently been observed in many other types of organism including plants, animals and microorganisms.

*7.3.2.1 Fungi*

Cytosolic lipid bodies can be found in most fungal cells where they serve numerous different purposes that can vary according to species, developmental stage or environmental conditions. It is likely that the types of proteins associated with these different sorts of lipid bodies will also vary. Cytosolic lipid-body formation is a commonly observed feature of vegetative growth in saprophytic fungi and increases markedly during the formation of resting and reproductive structures (Beckett *et al.*, 1974; Mills and Cantino, 1977; Losel and Sancholle, 1996). The pathogenic fungus, *Plasmodiophora brassicae*, inserts itself into the cytoplasm of its Brassica plant host, whereupon it rapidly accumulates large numbers of lipid bodies in its own cytoplasm (Losel and Sancholle, 1996). Presumably these are temporary carbon stores synthesised from precursors extracted from the host plant. An unusual function of fungal lipid bodies is found in the sporangiophores of the unicellular fungus, *Phycomyces blakesleeanus*, which contain aggregates of several dozen 1–2-μm diameter lipid bodies that are proposed, by virtue of their buoyancy, to play a role in gravity sensing (Schimek *et al.*, 1999). These lipid bodies are clustered in a specific area about 30–50 μm below the cell apex, which implies that they are tethered together, possibly by microfilaments.

The proteome of yeast lipid bodies has been systematically identified using electrospray mass spectrometry followed by an attempt at functional characterisation by deletion of the open reading frames (ORFs) corresponding to each identified protein (Athenstaedt *et al.*, 1999). A total of 19 polypeptides ranging from 31–78 kDa were identified in the lipid-body fraction. With the possible exception of a small amount of contamination from the soluble enzyme, glyceraldehyde-3-phosphate dehydrogenase, the authors concluded that the polypeptides were genuine components of yeast lipid bodies. On the basis of Coomassie blue staining, the most abundant protein was a 43-kDa band that was identified as sterol Δ24-methyltransferase (SMT). Two other enzymes of ergosterol synthesis, i.e. squalene epoxidase and lanosterol synthetase, and three enzymes of long-chain fatty acid activation were also identified. However, several of these enzymes are also associated with the ER or plasma membranes (Stahmann *et al.*, 1994; Leber *et al.*, 1998), which either implies some cross-contamination of the fractions or, perhaps more likely, an intimate association of lipid bodies with these other membranes that may involve shared enzyme complements and/or direct physical continuities. Also, while the activity of the 1-acylglycerol-3-phosphate acyltransferase, *SLC1*, was previously detected in isolated lipid-body fractions, no trace of the protein was found in this mass spectrometry analysis.

Many of the sequenced polypeptides in the above study could only be assigned putative functions on the basis of similarity with other proteins and several more had no homology with any known proteins. The construction of deletion strains of each of the 19 lipid-body polypeptides yielded the surprising result that all but three of the deletions had no effect on either the viability of the yeast cells or the formation of lipid bodies. Deletion of squalene epoxidase and lanosterol synthetase resulted in the expected sterol auxotrophy but deletion of the major yeast lipid-body protein, sterol $\Delta$24-methyltransferase, had no effect. Only one deletion – that of an unknown ORF termed YDL193w – resulted in lethality and, since this could not be rescued by sterol or fatty acid supplementation, it seems unlikely that the gene is involved in neutral lipid biosynthesis. One interesting point is that only 2 out of the 19 lipid-body proteins contained putative transmembrane sequences and no other obvious lipid-targeting motifs were found. Since some of these proteins apparently associate with both lipid bodies and bilayer membranes, it is possible that the lipid binding does not occur by insertion of the bulk of the protein into the lipid matrix but rather by insertion of a small hydrophobic region, e.g. a short amphipathic $\alpha$-helix or a covalently-bound acyl group, or even by binding to a receptor protein that is itself integrally associated with the lipid body or bilayer membrane. Other unpublished data (cited in Zweytick *et al.*, 2000) imply that the deletion of *C*-terminal hydrophobic regions of squalene epoxidase and SMT results in the impairment of its targeting to lipid bodies and its accumulation instead on the ER membrane. If confirmed, these observations will be analogous to data from us and others showing that truncation of plant oleosins results in their failure to target to lipid bodies, and their accumulation on the ER membrane and/or their turnover (Abell *et al.*, 1997; Beaudoin and Napier, 2002).

One of the five homologues of the lipid-trafficking protein, Sec14p, in yeast has been shown to be associated with lipid bodies, whereas the other homologues were found in different locations such as the cytosol and microsomes (Schnabl *et al.*, 2003).

The lipid bodies of *Mortierella ramanniana* were originally reported to be enriched in three major polypeptide components of 24, 29 and 59 kDa (Kamisaka and Nakahara 1994). However, it was later reported that there were many more bands visible in a Triton X-100 extracts of purified lipid bodies, including a diacyglycerol acyltransferase of 53 kDa (Kamisaka *et al.*, 1997). Only one of the proteins, a 29-kDa polypeptide, was sufficiently prominent to be a putative structural protein analogous to oleosin from plants, phasins from prokaryotes or adipophilin from mammals. It will be interesting to determine whether this 29-kDa polypeptide is indeed exclusive to lipid bodies and, if so, has any similarity to any of the other structural proteins from other organisms. In an unpublished study (Hills and Murphy, unpublished data), we found that the principal polypeptides associated with lipid bodies of the oleogenic fungus, *Candida curvata*, were 30 and 55 kDa, which is similar to the major polypeptides of *M. ramanniana* lipid bodies. The presence of diacyglycerol acyltransferase in considerable quantities on lipid bodies of *M. ramanniana* is in sharp contrast to yeast where no such protein was found (Athenstaedt *et al.*, 1999). However, yeast accumulates a lot less lipid than *M. ramanniana* and about half of the yeast neutral lipid is sterol ester. The most abundant lipid-body protein in yeast is sterol Δ24-methyltransferase, the terminal enzyme of sterol ester biosynthesis, which is analogous to the presence of diacyglycerol acyltransferase on *M. ramanniana* lipid bodies. We have also found caleosin-like sequences (caleosins belong to a family of plant calcium-binding proteins that associate with both the ER membrane and lipid bodies) in the genomes of at least two fungi, namely the lipid-accumulating fungus, *Neurospora crassa*, and the cereal pathogen, *M. grisea* (Partridge and Murphy, 2004) (see Figure 7.2). It will be interesting to determine whether these proteins are distributed more widely in fungi, and whether they play any role in lipid storage and/or trafficking.

It is a surprising and little known fact that 90–95% of all land plants are colonised by a class of filamentous fungi, the Glomales, that form so-called mycorrhizal associations with the roots of their plant host (Bago *et al.*, 2000). Indeed, it can be argued that the major organs of nutrient uptake by many plants are not their roots, but rather their associated mycorrhizae. In mycorrhizal associations, the fungi derive assimilated carbon from the plant and much of this must be transported from the intraradical mycelium within the roots to the extraradical mycelium located in the rhizosphere outside the plant. This distance can exceed 1 m in many cases and the transport is wholly intracellular, namely, via the interconnected network of hyphal filaments.

The carbon that is transported to the extraradical mycelium is either stored or utilised either for anabolism or for the production of reproductive spores. Recent findings suggest that in some mycorrhizal species, such as *Glomus intraradices* and *Glomus margarita*, the majority of carbon is translocated as lipid bodies that can comprise as much as 16% of the hyphal volume (Bago *et al.*, 1999, 2002; Pffeffer *et al.*, 1999; Lammers *et al.*, 2001), although some carbon is also translocated as simple carbohydrates (Bago *et al.*, 2003). In an elegant series of time-lapse micrographs, the translocated lipid bodies can be seen moving along specific tracks within the hyphal cells at speeds of up to 11 μm/s (Bago *et al.*, 2002). It is calculated that, for each of the principal fungal hyphae, as much as 1.3 μg/h of triacylglycerol is transported in this way – a most impressive feat for such a tiny structure. The movement of the lipid bodies along specific tracks, and independently of cytoplasmic streaming, suggests some sort of organised transport, possibly via cytoskeletal elements as has been shown for some lipid bodies in animal cells (see Section 7.3.1 above). In such animal cells, lipid bodies are tethered to structural and motor proteins of the cytoskeleton via adaptor proteins, such as $P_{200}$, and it is very likely that the fungal lipid bodies also contain associated proteins to facilitate their movement and recognition, although such proteins have yet to be identified. In this regard, it is interesting that at least two species of fungus, including a plant parasite (the cereal pathogen, *M. grisea*, which also needs to transport carbon assimilates from its host), contain caleosin-like genes. It will be interesting to investigate the mechanism of intracellular lipid movement in such fungi and the role, if any, of their caleosin-like proteins.

*7.3.2.2 Prokaryotes*

***Phasins.*** Most eubacterial genera and many members of the family Halobactericeae from the Archaea are able to accumulate polyhydroxyalkanoates (PHAs) as insoluble cytosolic spheroidal bodies, normally termed 'granules' (Steinbuchel and Fuchtenbusch, 1998). Although solid and made up of a polymerised lipid, in contrast to the triacylglycerol-rich lipid bodies in most plants and animals, the PHA granules are also surrounded by a layer of specialised structural proteins (called phasins) that may have some analogies with oleosins from plants. The storage granules containing PHA in bacteria are morphologically and biochemically similar to the cytosolic lipid bodies of most eukaryotes. The granules, which typically have diameters ranging from 0.3–0.4 μm, consist of rounded droplets made up of a hydrophobic PHA core bounded by an outer monolayer of phospholipid and a discrete population of proteins. The protein composition of purified PHA granules is often dominated by a particularly abundant group of polypeptides of about 20–25 kDa (Wieczorek *et al.*, 1996). This class of proteins was first described in *Rhodococcus ruber* (Pieper-Furst *et al.*, 1994, 1995) and *Ralstonia eutropha* (formerly *Alcaligenes eutrophus*) (Wieczorek *et al.*, 1996), where they constitute up to 5% total cell protein during PHA accumulation. Since they are found exclusively associated with the boundary layer of PHA granules and may share some structural and/or functional attribute with plant oleosins, these prokaryotic proteins have been termed 'phasins' (Steinbuchel *et al.*, 1995).

Phasins are distinct from other PHA granule-associated proteins, such as PHA synthase or PHA depolymerase, in being very abundant and having no apparent enzymatic function. They are aligned at the interface of the hydrophobic PHA granule and the cytoplasm. Physical measurements and biochemical analyses suggest that the PHA boundary layer is mostly made up of phasin subunits with an incomplete phospholipid monolayer filling in the gaps between protein molecules (Griebel *et al.*, 1968; Pieper-Furst *et al.*, 1994). The functional importance of phasins in PHA granule formation in *R. eutropha* has been shown by gene disruption experiments (Wieczorek *et al.*, 1995). Such treatments resulted in a significant decrease in PHA synthase activity and the accumulation of a single large PHA granule, rather than many smaller granules as in wild-type cells. The effect of phasins on PHA granule size is also shown in heterologous hosts, such as *E. coli*, containing the PHA synthetic enzymes. In the absence of phasins, such cells form a single large PHA granule, but introduction of phasin genes results in the accumulation of many smaller granules, with granule size being inversely proportional to the amount of phasin protein.

There may also be some functional similarities between phasins and plant oleosins (and perhaps adipophilin in animals) in that the rate of synthesis of these boundary proteins seems to determine the size of their respective intracellular hydrophobic droplets. Phasins in *R. eutropha* are also necessary for optimal activity of PHA synthase (Wieczorek *et al.*, 1995), which implies that there is some sort of interaction between the two proteins. A 16-kDa PHA granule-associated protein from *Paracoccus denitrificans*, that is a putative phasin, has also been shown to stimulate the rate of PHA accumulation when expressed in *E. coli* (Maehara *et al.*, 1991). The effect of several phasins from different species in decreasing the molecular mass of PHA polymers in heterologous systems suggests that phasins may interact directly with the PHA synthase on the surface of the granule. Phasins may also enable more rapid mobilisation of PHA granules, not only by increasing the granule number and hence the surface area, but also by providing a docking point for the PHA depolymerase.

Initial studies of the major phasin from *R. ruber* identified two hydrophobic domains adjacent to the *C*-terminal that were suggested to form a lipid-binding motif with some similarity to that found in plant oleosins (Pieper-Furst *et al.*, 1994). The expression of phasin variants lacking this motif resulted in the failure of the protein to bind to PHA granules in *E. coli*, while a fusion protein containing the soluble protein, acetaldehyde dehydrogenase II, linked to the phasin *C*-terminal hydrophobic region was efficiently targeted to PHA granules (Pieper-Furst *et al.*, 1995). However, a subsequent re-evaluation of the sequence of the phasin from *R. eutropha* suggested that the protein does not contain the two *C*-terminal hydrophobic domains reported earlier (Hanley *et al.*, 1999). Instead, the authors propose that binding of this phasin to the PHA granule occurs via as many as four smaller domains that are mostly, but not exclusively, hydrophobic. Phasin-like proteins, as defined by size similarity and, in some cases, by immunological cross-reactivity, have been found associated with PHA granules in a wide range of prokaryotic species (Wieczorek *et al.*, 1996).

Despite these earlier indications that phasins might form a single class of structurally similar proteins, comparison of the *R. eutropha* sequence with the growing number of new phasin sequences now available from public databases shows that, unlike oleosins from plants, phasins are extremely diverse with respect to both their primary sequences and predicted secondary structures (Hanley *et al.*, 1999). This applies even in the location and extent of the putative PHA-binding domains, where perhaps some higher level of similarity may have been expected. It should also be noted that a predominant class of granule-bound protein that could be termed a phasin, could not always be detected in PHA-producing prokaryotes. This may simply be an analytical problem but it may also mean that some of these extremely diverse organisms have evolved alternative mechanisms for packaging PHA granules. The question of whether phasins really are a single class of functionally related interfacial proteins is unlikely to be resolved until the physical mechanism(s) of their interaction with lipoprotein assemblies has been elucidated.

#### *7.3.2.3 Viruses*

***Hepatitis C virus (HCV).*** This virus is a major pathogen that affects as many as 300 million people worldwide (McLauchlan, 2000). In many cases, acute infection leads to progressive liver disease culminating in cirrhosis or hepatocellular carcinoma. Hepatic steatosis, or the accumulation of intracellular fatty deposits (i.e. lipid bodies) is a commonly observed feature of chronic HCV infection (Bach *et al.*, 1992; Scheuer *et al.*,1992). This phenotype could be mimicked in two independent lines of transgenic mice expressing the gene encoding the HCV core protein specifically in the liver (Koike *et al.*, 1995), whereas expression of other HCV genes, such as those encoding envelope proteins, had no such effect (Moriya *et al.*, 1997). In addition, analysis of liver biopsies from HCV-infected chimpanzees showed that the HCV core protein was cytoplasmically located, mainly on the lipid bodies and ER membranes (Barba *et al.*, 1997). Parallel studies *in vitro*, using non-hepatic (CHO) and hepatic (HepG2) cell lines transfected with the HCV core gene revealed that in both cases, the HCV core protein was mainly localised on cytosolic lipid bodies of average diameter 0.8 μm (Barba *et al.*, 1997).

The core protein of the virus is cleaved from the *N*-terminal end of the polyprotein (McLauchlan, 2000) and is believed to comprise the viral capsid. The mature form of the core protein is a 21-kDa polypeptide. Other work has confirmed the predominantly lipid-body localisation of HCV core protein expressed in a variety of mammalian cell lines, including those derived from baby hamster kidney (BHK C13) and human liver (Huh 7). This work has also elucidated sequence motifs required for lipid-body targeting (Hope and McLauchlan, 2000; Hope *et al.*, 2002). These and other recent studies, involving the expression of a series of HCV core deletion *mutants* in BHK cells (Murphy *et al.*, 2000; Hope *et al.*, 2002), confirmed that most of the residues in the more hydrophobic proline-rich domain are necessary for lipid-body targeting. In contrast, the more *N*- and *C*-terminally located domains were required for association with lipid droplets. In the absence of lipid bodies, i.e. in BHK cells grown on delipidated serum, the HCV core protein was still produced and processed as normal but it now accumulated on ER membranes. This shows that HCV core synthesis, processing and accumulation do not require the presence of lipid bodies. However, under the right nutrient conditions, i.e. when a supply of lipid precursors is available, the presence of the HCV core protein in cells lacking lipid bodies can stimulate the formation of lipid bodies to which the protein will bind (Barba *et al.*, 1997).

Adipophilin is a ubiquitous lipid-body protein in mammalian cells, as discussed above. The apparent displacement of adipophilin by HCV core protein, when both proteins are co-expressed, has interesting parallels with the situation during the normal development of adipocytes and steroidogenic cells where the proliferation of lipid bodies is accompanied by the replacement of adipophilin with another lipid-binding protein, perilipin. The reason for this substitution of one lipid-body protein is unknown, but may be connected with a requirement for rapid neutral lipid accumulation in such cells. Although the accumulation of HCV core protein is readily observed in cultured cells, there does not appear to be a similar hyper-accumulation of HCV core protein in biopsy material from steatotic livers. However, it is possible that the presence of HCV core protein in infected liver cells somehow stimulates the accumulation of lipid bodies *de novo* (Barba *et al.*, 1997), such that they hyper-accumulate to form the large droplets characteristic of steatotic cells (Moriya *et al.*, 1997). Further investigations are clearly required to characterise in more detail the mode of interaction of HCV core protein with the various types of lipid bodies, lipid-binding proteins and with lipid biosynthetic pathways both *in vitro* and in HCV-infected individuals. Hopefully, such studies will not only shed light on the general mechanism of core protein binding to lipid bodies, but may also suggest novel therapeutic strategies for combating this extremely destructive human virus.

***Other lipid-associated viral proteins.*** Several other viruses are associated with lipid abnormalities, or lipodystrophies, during their infectious stages in animals and humans. Examples include cytomegaloviruses and other forms of herpesvirus that are associated with atherosclerosis (Berensci *et al.*, 1998; Lin *et al.*, 2000); the AIDS-causing human immunodeficiency virus that is often correlated with hyper-accumulation of lipids (Lo *et al.*, 1998; Safrin and Grunfeld, 1999); and the reports of the induction of obesity by avian viruses in animal models (Dhurandhar *et al.*, 1991). More recently, it has been reported that inoculation of mice or chickens with the human adenovirus, Ad-36, resulted in significantly increased adiposity, whereas inoculation with a control virus had no effect. Viral DNA could be detected in the adipose tissue of infected animals but was not detected in skeletal muscles (Dhurandhar *et al.*, 2000). Although these results are still controversial, they do at least hint at a possible implication of some human viruses in the aetiology of obesity, although whether the putative effect of such viruses is manifest by direct interactions with lipid bodies (as may be the case with the HCV) remains to be seen.

## 7.4 Conclusions

During the past few years, it has become increasingly evident that lipid bodies in many cells may be far more dynamic than was previously assumed according to the stereotypical view that these lipids were simply rather inert carbon stores. Our changing view of the nature and function of these hitherto misunderstood organelles has emerged largely thanks to the progress in the characterisation of the various classes of lipid-associated proteins that have been described in this chapter. Ironically, the first class of these proteins to be studied in detail, the oleosins, appears to be solely involved in lipid storage and mobilisation, especially in seeds, although even here there are hints that oleosins may sometimes be present in non-storage, meristematic tissues of shoots and roots where they may have other functions (Murphy *et al.*, 2000, 2001). Over the past 5 years, much of the progress in elucidating the non-storage roles of intracellular lipid bodies has come from studies in animal systems. Here, new and exciting discoveries are being made at a rapid rate. To give just a few noteworthy examples, the PAT family of lipid-associated proteins are now known to be present throughout the Metazoa and are also found in slime molds

(Miura *et al.*, 2002; Teixeira *et al.*, 2003); caveolins are true lipid-body proteins (Ostermeyer *et al.*, 2004; Robonek *et al.*, 2004); lipid bodies contain other proteins associated with lipid metabolism and trafficking (Liu *et al.*, 2004); and lipid-body proteins retain their targeting properties in ectopic systems (Hope *et al.*, 2002). It is also becoming apparent that lipid-associated proteins may be implicated in a host of serious human diseases and pathologies that include hepatitis C (Hope *et al.*, 2002), Parkinson's disease (Cole *et al.*, 2002), CHILD syndrome (Ohashi *et al.*, 2003), retinopathy (Litvak *et al.*, 2002) and even skin irritation (Corsini *et al.*, 2003).

For the first time, improved imaging techniques, such as multi-photon, laser-assisted confocal microscopy, coupled with the use of more powerful reporter reagents like high-output fluorophores, have allowed for the real-time analysis of the behaviour of lipid bodies and their associated proteins in living cells (Bago *et al.*, 2002). This has enabled investigators to begin to dissect out the various populations of lipid bodies, some in rapid flux in cells while others are less dynamic (Targett-Adams *et al.*, 2003). Such direct real-time observational studies form a vital link with other more 'snapshot' approaches and are beginning to allow us to assert with some confidence that lipid bodies are more than mere storage entities. There is, therefore, an emerging consensus that cytosolic lipid bodies in animals and fungi are complex, multifunctional organelles that participate in a host of cellular processes including membrane trafficking, lipid-based signalling and sterol homeostasis (Bago *et al.*, 2002; Ohashi *et al.*, 2003; Liu *et al.*, 2004). Of course, in some cases, lipid bodies may also sometimes act as carbon stores!

Are these results from animal and microbial systems of relevance to lipid-body function in plants? Before 2000, it would have been difficult to answer this question with any confidence. Nevertheless, the overall similarity of many fundamental aspects of cell biology in the higher Eukarya has encouraged the view that we would eventually also uncover evidence of a dynamic role of lipid bodies, and their proteins, in plants. In the past few years, we have started to make some significant progress in this direction. First, the caleosins were discovered and shown to be both membrane and lipid-body associated (Chen *et al.*, 1999; Naested *et al.*, 2000; Hernandez-Pinzon *et al.*, 2001). With their putative calcium-binding and protein kinase A domains, caleosins are ideal candidates for signalling and trafficking functions, although we have yet to demonstrate such functions experimentally. There is also good evidence of additional proteins, including phospholipase A2 (May *et al.*, 1998), lipoxygenase (Hause *et al.*, 2000), sterol dehydrogenase (Lin *et al.*, 2002) and a calcium-dependent protein kinase (Anil *et al.*, 2003). We have also pointed out the non-random arrangement of lipid bodies, even in storage tissues (Murphy, 2001), but also in other tissues like roots (Rinne *et al.*, 2002). This implies an ordered mechanism, possibly linked to the cytoskeleton, for the organisation of lipid bodies in specific spatial domains in plant cells, as recently described in human CHO K2 cells (Liu *et al.*, 2004). With the availability of improved cytological techniques and the vast resources of molecular biology, it should prove possible to make good progress in elucidating the manifold roles of lipid bodies and their associated proteins in plant cells over the next few years.

## References


Abell, B.M., High, S. and Moloney, M.M. (2002) Membrane protein topology of oleosin is constrained by its long hydrophobic domain. *Journal of Biological Chemistry*, **277**, 8602–8610

Abell, B.M., Holbrook L.A., Abenes, M., Murphy D.J., Hills M.J. and Moloney, M.M. (1997) Role of the proline knot motif in oleosin endoplasmic reticulum topology and oil body targeting. *Plant Cell*, **9**, 1481–1493.

Alexander, L.G., Sessions R.B., Clarke A.R., Tatham, A.S., Shewry, P.R., and Napier, J.A. (2002) Characterization and modelling of the hydrophobic domain of a sunflower oleosin, *Planta*, **214**, 546–551

Anderson, T.A., Levitt, D.G. and Banaszak, L. (1998) The structural basis of lipid interactions in lipovitellin, a soluble lipoprotein. *Structure*, **6**, 895–909.

Anil, V.S., Harmon, A.C. and Rao, K.S. (2003) Temporal association of Ca(2+)-dependent protein kinase with oil bodies during seed development in *Santalum album* L.: its biochemical characterization and significance. *Plant and Cell Physiology*, **44**, 367–376

Athenstaedt, K., Zweytick, D., Jandrositz, A., Kohlwein, S.D. and Daum, G. (1999) Identification and characterization of major lipid particle proteins of the yeast *Saccharomyces cerevisiae*. *Journal of Bacteriology*, **181**, 1458–1463

Bach, N., Thung, S.N. and Schaffner, F. (1992) The histological features of chronic hepatitis C and autoimmune chronic hepatitis: a comparative analysis. *Hepatology*, **15**, 572–577.

Backhaus, R.A. and Walsh, S. (1983) Ontogeny of rubber formation in guayule, *Parthenium argentatum Gray. Botanical Gazette*, **144**, 391–400.

Bago, B., Pfeffer, P.E., Abubaker, J. *et al.* (2003) Carbon export from arbuscular mycorrhizal roots involves the translocation of carbohydrate as well as lipid. *Plant Physiology*, **131**, 1496–1507.

Bago, B., Pfeffer, P.E., Douds, D.D. Jr, Brouillette, J., Becard, G. and Shachar-Hill, Y. (1999) Carbon metabolism in spores of the arbuscular mycorrhizal fungus *Glomus intraradices* as revealed by nuclear magnetic resonance spectroscopy. *Plant Physiology*, **121**, 263–272.

Bago, B., Pfeffer, P.E. and Shachar-Hill, Y. (2000) Carbon metabolism and transport in arbuscular mycorrhizas. *Plant Physiology*, **124**, 949–958.

Bago, B., Zipfel, W., Williams, R.M. *et al.* (2002) Translocation and utilization of fungal storage lipid in the arbuscular mycorrhizal symbiosis. *Plant Physiology*, **128**, 108–124.

Barba, G., Harper, F., Harada, T. *et al.* (1997) Hepatitis C virus core protein shows a cytoplasmic localization and associates to cellular lipid storage droplets. *Proceedings of the National Academy of Sciences USA*, **94**, 1200–1205.

Barbero, P., Buell, E., Zulley, S. and Pfeffer, S.R. (2001) TIP47 is not a component of lipid droplets. *Journal of Biological Chemistry*, **277**, 24348–24351.

Baumlein, H., Boerjan, W., Nagy, I., Panitz, R., Inze, D. and Wobus, U. (1991) Upstream sequences regulating legumin gene expression in heterologous transgenic plants. *Molecular and General Genetics*, **225**, 459–467.

Beaudoin, F. and Napier, J.A. (2000) The targeting and accumulation of ectopically expressed oleosin in non-seed tissues of Arabidopsis thaliana. *Planta*, **210**, 439–445.

Beaudoin, F. and Napier, J.A. (2002) Targeting and membrane-insertion of a sunflower oleosin *in vitro* and in *Saccharomyces cerevisiae*: the central hydrophobic domain contains more than one signal sequence, and directs oleosin insertion into the endoplasmic reticulum membrane using a signal anchor sequence mechanism. *Planta*, **215**, 293–303.

Beaudoin, F., Wilkinson, B.M., Stirling, C.J., and Napier, J.A. (2000) *In vivo* targeting of a sunflower oil body protein in yeast secretory (sec) mutants. *Plant Journal*, **23**, 159–170.

Beckett, A., Heath, I.B. and McLaughlin, D.J., eds (1974) *An Atlas of Fungal Ultrastructure*, Longman, London.

Berencsi, K., Endrész, V., Klurfeld, D., Kari, L., Kritchevsky, D. and Gönczöl, É. (1998) Early atherosclerotic plaques in the aorta following cytomegalovirus infection of mice. *Cell Adhesion and Communication*, **5**, 39–47.

Bolanos-Garcia, V.M. and Miguel, R.N. (2003) On the structure and function of apolipoproteins: more than a family of lipid-binding proteins. *Progress in Biophysics and Molecular biology*, **83**, 47–68.

Borowitz, M.J. and Blum, J.J. (1976) Triacylglycerol turnover in *Tetrahymena pyriformis*. Relation to phospholipid synthesis. *Biochimica et Biophysica Acta*, **424**, 114–124.

Brasaemle, D.L., Barber, T., Wolins, N.E., Serrero, G., Blanchette-Mackie, E.J. and Londos, C. (1997) Adipose differentiation-related protein is an ubiquitously expressed lipid storage droplet-associated protein. *Journal of Lipid Research*, **38**, 2249–2263.

Brasaemle, D.L., Rubin, B., Harten, I.A., Gruia-Gray, J., Kimmel, A.R., and Londos, C. (2000) Perilipin A increases triacylglycerol storage by decreasing the rate of triacylglycerol hydrolysis. *Journal of Biological Chemistry*, **275**, 38486–38493.

Breiter, D.R., Kanost, M.R., Benning, M.M. *et al.* (1991) Molecular structure of an apolipoprotein determined at 2.5-A resolution., *Biochemistry* **30**, 603–608

Busch, M.B., Körtje, K.H., Rahmann, H. and Sievers, A. (1993) Characteristic and differential calcium signals from cell structures of the root cap detected by energy-filtering electron microscopy. *Eur. Journal of Cell Biology*, **60**, 88–100.

Chan, A.P., Kloc, M. and Etkin, L.D. (1999) fatvg encodes a new localized RNA that uses a 25-nucleotide element (FVLE1) to localize to the vegetal cortex of Xenopus oocytes. *Development*, **126**, 4943–4953.

Chen, H., Klein, A., Xiang, M., Backhaus, R.A. and Kuntz, M. (1998) Drought and wound induced expression in leaves of a gene encoding a chromoplast carotenoid-associated protein. *Plant Journal*, **14**, 317–326.

Chen, J.C.F., Tsai, C.C.Y. and Tzen, J.T.C. (1999) Cloning and secondary structure analysis of caleosin, a unique calcium-binding protein in oil bodies of plant seeds. *Plant and Cell Physiology*, **40**, 1079–1086.

Cole, N.B., Murphy, D.D., Grider, T., Rueter, S., Brasaemle, D. and Nussbaum, R.L. (2002) Lipid droplet binding and oligerization properties of the Parkinson's Disease Protein σ-Synuclein. *Journal of Biological Chemistry*, **277**, 6344–6352

Corsini, E., Viviani, B., Zancanella, O. *et al.* (2003) Induction of adipose differentiation related protein and neutral lipid droplet accumulation in keratinocytes by skin irritants. *The Journal of Investigative Dermatology*, **121**, 337–344.

Dawidowicz-Grzegorzewska, A. and Podstolski, A. (1992) Degradation of protein and lipid bodies during dormancy in apple seeds. *Annals of Botany*, **69**, 39–46.

De Fay, E., Hebant, C. and Jacob, J.L. (1989) (eds D'auzac, J. J.L. Jacob and H. Chrestin) *Physiology of Rubber Tree Latex*, CDC Press, Boca Raton, pp. 15–29

de Oliveira, D., Seurinck, J., Inzé, D., van Montagu, M. and Botterman, J. (1990) Differential expression of five Glycine-Rich Proteins encoding genes in *Arabidopsis thaliana. Plant Cell*, **2**, 427–436.

Dennis, M.S., Henzel, W.J., Bell, J., Kohr, W. and Light, D.R.J. (1989) Amino acid sequence of rubber elongation factor protein associated with rubber particles in Hevea latex. *Journal of Biological Chemistry*, **264**, 18618–18626.

Dennis, M.S. and Light, D.R.J. (1989) Rubber elongation factor from *Hevea brasiliensis*. Identification, characterization, and role in rubber biosynthesis. *Journal of Biological Chemistry*, **264**, 18608–18817

Deruere, J., Romer, S., D'Harlingue, A., Backhaus, R.A., Kuntz, M. and Camara, B. (1994) Fibril assembly and carotenoid over accumulation in chromoplasts: a model for supramolecular lipoprotein structures. *Plant Cell*, **6**, 119–133.

Dhurandhar, N.V., Israel, B.A., Kolesar, J.M., Mayhew, G.F., Cook, M.E. and Atkinson, R.L. (2000) Increased adiposity in animals due to a human virus. *International Journal of Obesity and Related Metabolic Disorders*, **9**, 89–96

Dhurandhar, N.V., Kulkarni, P., Ajinkya, S.M. and Sherikar, A. (1991) Avian adenovirus leading to pathogenic obesity in chickens. *Veterinary Microbiology*, **31**, 101–107.

Diaz, E. and Pfeffer, S.R. (1998) TIP47: a cargo selection device for mannose 6-phosphate receptor trafficking. *Cell*, **93**, 433–443.

Fan, D., Zheng, Y., Yang, D. and Wang, J. (2003) NMR solution structure and dynamics of an exchangeable apolipoprotein, *Locusta migratoria* apolipophorin III. *Journal of Biological Chemistry*, **278**, 21212–21220.

Fong, T.H., Yang, C.C., Greenberg, A.S. and Wang, S.M. (2002) Immunocytochemical studies on lipid droplet-surface proteins in adrenal glands. *Journal of Cellular Biochemistry*, **86**, 432–439.

Foster, E., Schneiderman, D., Cloutier, M., Gleddie, S., Robert, L.S. (2002) Modifying the pollen coat protein composition in Brassica. *Plant Journal*, **31**, 477–486

Franco, L.O., de O Manes, C.L., Hamdi, S., Sachetto-Martins, G. and de Oliveira, D.E. (2002) Distal regulatory regions restrict the expression of cis-linked genes to the tapetal cells. *FEBS Letters*, **517**, 13–18.

Frandsen, G., Müller-Uri, F., Nielsen, M., Mundy, J. and Skriver, K. (1996) Novel plant Ca(2+)-binding protein expressed in response to abscisic acid and osmotic stress. *Journal of Biological Chemistry*, **271**, 343–348.

Fujimoto, T., Kogo, H., Nomura, R. and Une, T. (2000) Isoforms of caveolin 1 and caveolar structure. *Journal of Cell Science*, **113**, 3509–3517.

Garcia, A., Sekowski, A., Subramanian, V., Brasaemle, D.L. (2003) The central domain is required to target and anchor perilipin A to lipid droplets. *Journal of Biological Chemistry*, **278**, 625–635.

Garcia, A., Subramanian, V., Sekowski, A., Bhattacharyya, S., Love, M.W. and Brasaemle, D.L. (2004) The amino and carboxyl termini of perilipin a facilitate the storage of triacylglycerols. *Journal of Biological Chemistry*, **279**, 8409–8416.

Gibbons, G.F., Islam, K. and Pease, R.J. (2000) Mobilisation of triacylglycerol stores. *Biochimica et Biophysica Acta*, **1483**, 37–57.

Glaumann, H., Bergstrand, A. and Ericsson, J.L.E. (1975) Studies on the synthesis and intracellular transport of lipoprotein particles in rat liver. *Journal of Cell Biology*, **64**, 356–377.

Greenberg, A.S., Egan, J.J., Wek, S.A., Garty, N.B., Blanchette-Mackie, E.J. and Londos, C. (1991) Perilipin, a major hormonally regulated adipocyte-specific phosphoprotein associated with the periphery of lipid storage droplets. *Journal of Biological Chemistry*, **266**, 11341–11346.

Griebel, R., Smith, Z. and Merrick, J.M. (1968) Metabolism of poly-beta-hydroxybutyrate. I. Purification, composition, and properties of native poly-beta-hydroxybutyrate granules from *Bacillus megaterium*. *Biochemistry*, **7**, 3676–3681.

Gronke, S., Beller, M., Fellert, S., Ramakrishnan, H., Jackle, H. and Kuhnlein, R.P. (2003) Control of fat storage by a Drosophila PAT domain protein. *Current Biology*, **13**, 603–606.

Guilloteau, M., Laloi, M., Blais, D., Crouzillat, D. and McCarthy, J. (2003) Oil bodies in *Theobroma cacao* seeds: cloning and characterization of cDNA encoding the 15.8 and 16.9 kDa oleosins. *Plant Science*, **164**, 597–606.

Handford, P.A. (2000) Fibrillin-1, a calcium binding protein of extracellular matrix. *Biochimica et Biophysica Acta*, **1498**, 84–90.

Hanley, S.Z., Pappin, D.J.C., Rahman, D., White, A.J., Elborough, K.M. and Slabas, A.R. (1999) Re-evaluation of the primary structure of *Ralstonia eutropha* phasin and implications for polyhydroxyalkanoic acid granule binding. *FEBS Letters*, **447**, 99–105.

Harker, M., Hellyer, A., Clayton, J.C., Duvoix, A., Lanot, A. and Safford, R. (2002) Co-ordinate regulation of sterol biosynthesis enzyme activity during accumulation of sterols in developing rape and tobacco seed. *Planta*, **216**, 707–715.

Hause, B., Weichert, H., Hohne, M., Kindl, H. and Feussner, I. (2000) Expression of cucumber lipid-body lipoxygenase in transgenic tobacco: lipid-body lipoxygenase is correctly targeted to seed lipid bodies. *Planta*, **210**, 708–714.

Hayashi, T. and Su, T.P. (2003) Sigma-1 receptors (sigma(1) binding sites) form raft-like microdomains and target lipid droplets on the endoplasmic reticulum: roles in endoplasmic reticulum lipid compartmentalization and export. *The Journal of Pharmacology and Experimental Therapeutics*, **306**, 718–725.

Heid, H.W., Schnolzer, M. and Keenan, T.W. (1996) Adipocyte differentiation-related protein is secreted into milk as a constituent of milk lipid globule membrane. *Biochemical Journal*, **320**, 1025–1030.

Hensel, W (1986) Cytodifferentiation of polar plant cells. *Planta*, **169**, 293–303.

Hernandez-Pinzon, I., Patel, K. and Murphy, D.J. (2001) The novel calcium-binding protein, caleosin, has distinct endoplasmic reticulum and lipid-body associated isoforms. *Plant physiology and Biochemistry*, **39**, 1–9.

Hernández-Pinzón, I., Ross, J.H.E., Barnes, K.A., Damant, A.P. and Murphy, D.J. (1999) Composition and role of tapetal lipid bodies in the biogenesis of the pollen coat of *Brassica napus*. *Planta*, **208**, 588–598.

Hernández-Pinzón, I., Ross, J.H.E. and Murphy, D.J. (1998) Characterisation of fibrillin-like and oleosin-like proteins in anther lipid bodies of *Brassica napus*, in *Advances in Plant Lipid Research* (eds J. Sanchez E. Cerdá-Olmedo and E. Martinez-Force.), Universidad de Sevilla, pp. 689–691.

Holk, A., Rietz, S., Zahn, M., Quader, H. and Scherer, G.F. (2002) Molecular identification of cytosolic, patatin-related phospholipases A from Arabidopsis with potential functions in plant signal transduction. *Plant Physiology*, **130**, 90–101.

Holmberg, N., Harker, M., Gibbard, C.L. *et al.* (2002) Sterol C-24 methyltransferase type 1 controls the flux of carbon into sterol biosynthesis in tobacco seed. *Plant Physiology*, **130**, 303–311.

Hope, R.G. and McLauchlan, J. (2000) Sequence motifs required for lipid droplet association and protein stability are unique to the hepatitis C virus core protein. *The Journal of General Virology*, **81**, 1913–1925.

Hope, R.G., Murphy, D.J. and McLauchlan, J. (2002) The domains required to direct core proteins of hepatitis C virus and GB virus-B to lipid droplets share common features with plant oleosin proteins. *Journal of Biological Chemistry*, **277**, 4261–4270.

Howell, K.E. and Palade, G.E. (1982) Heterogeneity of lipoprotein particles in hepatic Golgi fractions. *Journal of Cell Biology*, **92**, 833–845.

Hussain, M.M., Shi, J. and Dreizen, P. (2003) Microsomal triglyceride transfer protein and its role in apoB-lipoprotein assembly. *Journal of Lipid Research*, **44**, 22–32.

Ikura, M. (1996) Calcium binding and conformational response by EF-hand proteins. *Trends in Biochemical Sciences*, **21**, 14–17.

Jacks, T.J., Yatsu, L.Y. and Altschul, A.M. (1967) Isolation and characterisation of peanut spherosomes. *Plant Physiology*, **42**, 585–597.

Jordan, J.L., Jordan, L.S. and Jordan, C.M. (1982) Ultrastructure of lettuce seeds subjected to cycles of freezing and thawing. *Cryobiology*, **19**, 435–442.

Kamisaka, Y. and Nakahara, T. (1994) Characterization of the diacylglycerol acyltransferase activity in the lipid body fraction from an oleaginous fungus. *Journal of Biochemistry*, **116**, 1295–1301.

Kamisaka, Y., Mishra, S. and Nakahara, T. (1997) Purification and characterization of diacylglycerol acyltransferase from the lipid body fraction of an oleaginous fungus. *Journal of Biochemistry*, **121**, 1107–1114.

Kaneko, T., Sato, S., Kotani, H. *et al.* (1996) Sequence analysis of the genome of the unicellular cyanobacterium sp. strain PCC6803. II. Sequence determination of the entire genome and assignment of potential protein-coding regions. *DNA Research*, **3**, 109–136.

Katz, A., Jimenez, C. and Pick, U. (1995) Isolation and characterization of a protein associated with carotene globules in the alga *Dunaliella bardawil*. *Plant Physiology*, **108**, 1657–1664.

Kessler, F., Schnell, D. and Blobel, G. (1999) Identification of proteins associated with plastoglobules isolated from pea (*Pisum sativum* L.) chloroplasts. *Planta*, **208**, 107–113.

Kim, H.U., Hsieh, K., Ratnayake, C. and Huang, A.H. (2002) A novel group of oleosins is present inside the pollen of Arabidopsis. *Journal of Biological Chemistry*, **277**, 22677–22684.

Kim, H.U., Wu, S.S.H., Ratnayake, C. and Huang, A.H.C. (2001) *Brassica rapa* has three genes that encode proteins associated with different neutral lipids in plastids of specific tissues. *Plant Physiology*, **126**, 330–341.

Knoth, R., Hansmann, P. and Sitte, P. (1986) Chromoplasts of *Palisota barteri*, and the molecular structure of chromoplast tubules. *Planta*, **168**, 167–174.

Koike, K., Moriya, K., Ishibashi, K. *et al.* (1995) Expression of hepatitis C virus envelope proteins in transgenic mice. *The Journal of General Virology*, **76**, 3031–3038.

Krise, J.P., Orsel, J.G., Sincock, P.M. and Pfeffer, S.R. (2000) Quantitative analysis of TIP47-receptor cytoplasmic domain interactions: implications for endosome-to-trans Golgi network trafficking. *Journal of Biological Chemistry*, **275**, 25188–25193.

Lacey, D.J., Wellner, N., Beaudoin, F., Napier, J.A. and Shewry, P.R. (1998) Secondary structure of oleosins in oil bodies isolated from seeds of safflower (*Carthamus tinctorius* L.) and sunflower (*Helianthus annuus* L.). *Biochemical Journal*, **334**, 469–477.

Lammers, P.J., Jun, J. Abubaker, J. *et al.* (2001) The glyoxylate cycle in an arbuscular mycorrhizal fungus. Carbon flux and gene expression. *Plant Physiology*, **127**, 1287–1298.

Leber, R., Landl, K., Zinser, E. *et al.* (1998) Dual localization of squalene epoxidase, erg1p, in yeast reflects a relationship between the endoplasmic reticulum and lipid particles. *Molecular Biology of the Cell*, **9**, 375–386.

Lee, K., Bih, F.Y., Learn, G.H., Ting, J.T., Sellers, C. and Huang, A.H. (1994) Oleosins in the gametophytes of Pinus and Brassica and their phylogenetic relationship with those in the sporophytes of various species. *Planta*, **193**, 461–469.

Leprince, O., van Aelst, A.C., Pritchard, H.W. and Murphy, D.J. (1998) Oleosins prevent oil-body coalescence during seed imbibition as suggested by a low-temperature scanning electron microscope study of desiccation-tolerant and -sensitive oilseeds. *Planta*, **204**, 109–119.

Li, M., Murphy, D.J., Lee, K.H. *et al.* (2002) Purification and structural characterization of the central hydrophobic domain of oleosin. *Journal of Biological Chemistry*, **277**, 37888–37895.

Li, M., Smith, L.J., Clark, D.C., Wilson, R. and Murphy, D.J. (1992) Secondary structures of a new class of lipid body proteins from oilseeds. *Journal of Biological Chemistry*, **267**, 8245–8253.

Libal-Weksler, Y., Vishnevetsky, M., Ovadis, M. and Vainstein, A. (1997) Isolation and regulation of accumulation of a minor chromoplast-specific protein from cucumber corollas. *Plant Physiology*, **113**, 59–63.

Lin, L.-J., Tai, S.S.K., Peng, C.-C. and Tzen, J.T.C. (2002) Steroleosin, a sterol-binding dehydrogenase in seed oil bodies. *Plant Physiology*, **128**, 1200–1211.

Lin, P., Le-Niculescu, H., Hofmeister, R. *et al.* (1998) The mammalian calcium-binding protein, Nucleobindin (CALNUC), is a Golgi resident protein. *Journal of Cell Biology*, **141**, 1515–1528.

Lin, T.-M., Jiang, M.-J., Eng, H.-L. *et al.* (2000) Experimental infection with bovine herpesvirus-4 enhances atherosclerotic process in rabbits. *Laboratory Investigation*, **80**, 3–11.

Litvak, V., Shaul, Y.D., Shulewitz, M., Amarilio, R., Carmon, S. and Lev, S. (2002) Targeting of Nir2 to lipid droplets is regulated by a specific threonine residue within its PI-transfer domain. *Current Biology*, **12**, 1513–1518.

Liu, P., Ying, Y., Zhao, Y., Mundy, D.I., Zhu, M. and Anderson, R.G. (2004) CHO K2 cell lipid droplets appear to be metabolic organelles involved in membrane traffic. *Journal of Biological Chemistry*, **279**, 3787–3792.

Lo, J.C., Mulligan, K., Tai, V.W., Algren, H. and Schambelan, M. (1998) 'Buffalo hump' in men with HIV-1 infection. *The Lancet*, **351**, 867–870.

Londos, C., Brasaemle, D.L., Schultz, C.J., Segrest, J.P. and Kimmel, A.R. (1999) Perilipins, ADRP, and other proteins that associate with intracellular neutral lipid droplets in animal cells. *Seminars in Cell and Developmental Biology*, **10**, 51–58.

Londos, C., Gruia-Gray, J., Brasaemle, D.L. *et al.* (1996) Perilipin: possible roles in structure and metabolism of intracellular neutral lipids in adipocytes and steroidogenic cells. *International Journal of Obesity*, **20** (Suppl. 3), S97–S101.

Losel, D.M. and Sancholle, M. (1996) in *Lipids of Pathogenic Fungi* (eds R. Prasad and M.A. Ghannoum), CRC Press, Boca Raton, USA, pp. 27–62.

Lu, X., Grui-Grey, J., Copeland, N.G., Gilbert, D.J., Londos, C. and Kimmel, A.R. (2001) The murine perilipin gene: the lipid droplet-associated perilipins derive from tissue-specific, mRNA splice variants and define a gene family of ancient origin. *Genome*, **12**, 741–749.

Luo, C.C., Li, W.H., Moore, M.N. and Chan, L. (1986) Structure and evolution of the apolipoprotein multigene family. *Journal of Molecular Biology*, **187**, 325–340.

Maehara, A., Ueda, S., Nakano, H. and Yamane, T. (1991) Analyses of a polyhydroxyalkanoic acid granule-associated 16-kilodalton protein and its putative regulator in the pha locus of *Paracoccus denitrificans*. *Journal of Bacteriology*, **181**, 2914–2921.

Madey, E., Nowack, L.M., Thompson, J.E. (2002) Isolation and characterization of lipid in phloem sap of canola. *Planta*, **214**, 625–634.

Mancha, M. and Stymne, S. (1997) Remodelling of triacylglycerols in microsomal preparations from developing castor bean (*Ricinus communis* L.) endosperm. *Planta*, **203**, 51–57.

Marchesan D., Rutberg, M.O., Andersson, L. *et al.* (2003) A phospholipase D-dependent process forms lipid droplets containing caveolin, adipocyte differentiation-related protein, and vimentin in a cell-free system. *Journal of Biological Chemistry*, **278**, 27293–27300.

Martinez-Botas J., Anderson, J.B., Tessier, D. *et al.* (2000) Absence of perilipin results in leanness and reverses obesity in Lepr(db/db) mice. *Nature Genetics*, **26**, 474–479.

May, C., Hohne, M., Gnau, P., Schwennesen, K. and Kindl, H. (2000) The *N*-terminal beta-barrel structure of lipid body lipoxygenase mediates its binding to liposomes and lipid bodies. *European Journal of Biochemistry/FEBS*, **267**, 1100–1109.

May, C., Preisig-Muller, R., Hohne, M., Gnau, P. and Kindl, H. (1998) A phospholipase A2 is transiently synthesized during seed germination and localized to lipid bodies. *Biochimica et Biophysica Acta*, **1393**, 267–276.

Mayfield, J.A. and Preuss, D. (2000) Rapid initiation of *Arabidopsis* pollination requires the oleosin-domain protein GRP17. *Nature Cell Biology*, **2**, 128–130.

McLauchlan, J. (2000) Properties of the hepatitis C virus core protein: a structural protein that modulates cellular processes. *Journal of Viral Hepatitis*, **7**, 2–14.

McManaman, J.L., Zabaronick, W., Schaack, J. and Orlicky, D.J. (2003) Lipid droplet targeting domains of adipophilin. *Journal of Lipid Research*, **44**, 668–673.

Mills, G.L. and Cantino, E.C. (1977) Isolation and characterization of lipid globules from the zoospores of *Blastocladiella emersonii*. *Archives of Microbiology*, **112**, 195–200.

Miura, S., Gan, J.W., Brzostowski, J. *et al.* (2002) Functional conservation for lipid storage droplet association among Perilipin, ADRP, and TIP47 (PAT)-related proteins in mammals, Drosophila, and Dictyostelium. *Journal of Biological Chemistry*, **277**, 32253–32257.

Moloney, M.M. (1999) Seed Oleosins. In *Seed Proteins* (eds P.R. Shewry and R. Casey) Kluwer, Dordrecht, pp. 781–806.

Monte, E., Ludevid, D. and Prat, S. (1999) Leaf C40.4: a carotenoid-associated protein involved in the modulation of photosynthetic efficiency? *Plant Journal*, **19**, 399–410.

Moriya, K., Yotsuyanagi, H., Shintani, Y. *et al.* (1997) Hepatitis C virus core protein induces hepatic steatosis in transgenic mice. *The Journal of General Virology*, **78**, 1527–1531.

Murphy, D.J. (2001) Biogenesis and functions of lipid bodies in animals, plants and microorganisms. *Progress in Lipid Research*, **40**, 325–438.

Murphy, D.J. and Cummins, I. (1989) Purification and immunogold localisation of the major oil-body membrane protein of oilseed rape. *Plant Science*, **60**, 47–54.

Murphy, D.J. and Ross, J.H.E. (1998) Biosynthesis, targeting and processing of oleosin-like proteins, which are major pollen coat components in *Brassica napus*. *Plant Journal*, **13**, 1–16.

Murphy, D.J. and Vance, J. (1999) Mechanisms of lipid body formation. *Trends in Biochemical Sciences*, **24**, 109–115.

Murphy, D.J., Hernandez-Pinzon, I. and Patel, K. (2001) Roles of lipid bodies and lipid-body proteins in seeds and other tissues *Journal of Plant Physiology*, **158**, 471–478.

Murphy, D.J., Hernendez-Pinzon, I., Patel, K., Hope, R.G. and McLauchlan, J. (2000) New insights into the mechanisms of lipid-body biogenesis in plants and other organisms. *Biochemical Society Transactions*, **28**, 710–711.

Nabi, I.R. and Le, P.U. (2003) Caveolae/raft-dependent endocytosis. *Journal of Cell Biology*, **161**, 673–677.

Naested, H., Frandsen, G.I., Jauh, G.Y. *et al.* (2000) Caleosins: Ca2+-binding proteins associated with lipid bodies. *Journal of Plant Molecular Biology*, **44**, 463–476.

Nakamura, N. and Fujimoto, T. (2003) Adipose differentiation-related protein has two independent domains for targeting to lipid droplets. *Biochemical and Biophysical Research Communications*, **306**, 333–338.

Napier, J.A., Beaudoin, F., Tatham, A.S., Alexander, L.G. and Shewry, P.R. (2001) The seed oleosins: structure, properties and biological role. *Advances in Botanical Research*, **35**, 111–138.

Napier, J.A., Stobart, A.K. and Shewry, P.R. (1996) The structure and biogenesis of plant oil bodies: the role of the ER membrane and the oleosin class of proteins. *Plant Molecular Biology*, **31**, 945–956.

Narayanaswami, V. and Ryan, R.O. (2000) Molecular basis of exchangeable apolipoprotein function. *Biochimica et Biophysica Acta*, **1483**, 15–36.

Oh, P., McIntosh, D.P. and Schnitzer, J.E. (1998) Dynamin at the neck of caveolae mediates their budding to form transport vesicles by GTP-driven fission from the plasma membrane of endothelium. *Journal of Cell Biology*, **14**, 101–114.

Ohashi, M., Mizushima, N., Kabeya, Y. and Yoshimori, T. (2003) Localization of mammalian NAD(P)H steroid dehydrogenase-like protein on lipid droplets. *Journal of Biological Chemistry*, **278**, 36819–36829.

Ostermeyer, A.G., Paci, J.M., Zeng Y., Lublin, D.M., Munro, S. and Brown, D.A. (2001) Accumulation of caveolin in the endoplasmic reticulum redirects the protein to lipid storage droplets. *Journal of Cell Biology*, **152**, 1071–1078.

Ostermeyer, A.G., Ramcharan, L.T., Zeng, Y., Lublin, D.M. and Brown, D.A. (2004) Role of the hydrophobic domain in targeting caveolin-1 to lipid droplets. *Journal of Cell Biology*, **164**, 69–78.

Owen, H.A. and Makaroff, C.A. (1995) Ultrastructure of microsporogenesis and microgametogenesis in *Arabidopsis thaliana* (L.) Heynh. ecotype Wassilewskija (Brassicaceae). *Protoplasma*, **185**, 7–21.

Partridge, M. and Murphy, D.J. (2004) Roles of caleosins in plants. *Biochemical Society Transactions*, **32**, C106.

Pfeffer, P.E., Douds, D.D., Bécard, G. and Shachar-Hill, Y. (1999) Carbon uptake and the metabolism and transport of lipids in an arbuscular mycorrhiza. *Plant Physiology*, **120**, 587–598.

Pieper-Furst, U., Madkour, M.H., Mayer, F. and Steinbuchel, A. (1994) Purification and characterization of a 14-kilodalton protein that is bound to the surface of polyhydroxyalkanoic acid granules in *Rhodococcus ruber. Journal of Bacteriology*, **176**, 4328–4337.

Pieper-Furst, U., Madkour, M.H., Mayer, F. and Steinbuchel, A. (1995) Identification of the region of a 14-kilodalton protein of *Rhodococcus ruber* that is responsible for the binding of this phasin to polyhydroxyalkanoic acid granules. *Journal of Bacteriology*, **177**, 2513–2523.

Piffanelli, P. and Murphy, D.J. (1998a) Novel organelles and targeting mechanisms in the anther tapetum. *Trends in Plant Science*, **3**, 250–253.

Piffanelli, P., Ross, J.H.E. and Murphy, D.J (1998b) Biogenesis and function of the lipidic structures of pollen grains. *Sexual Plant Reproduction*, **11**, 65–80.

Pol, A., Luetterforst, R., Lindsay, M., Heino, S., Ikonen, E. and Parton, R.G. (2001) A caveolin dominant negative mutant associates with lipid bodies and induces intracellular cholesterol imbalance. *Journal of Cell Biology*, **152**, 1057–1070.

Pons, L., Olszewski, A. and Gueant, J.-L. (1998) Characterization of the oligermeric behavior of a 16.5 kDa peanut oleosin protein by chromatography and electrophoresis of the iodinated form. *Journal of Chromatography B.*, **706**, 131–140.

Pozueta-Romero, J., Rafia, F., Houlne, G. *et al.* (1997) A ubiquitous plant housekeeping gene, PAP, encodes a major protein component of bell Pepper Chromoplasts. *Plant Physiology*, **115**, 1185–1194.

Pruvot, G., Cuine, S., Peltier, G. and Rey, P. (1996) Characterization of a novel drought-induced 34-kDa protein located in the thylakoids of *Solanum tuberosum* L. plants. *Planta*, **198**, 471–479.

Pruvot, G., Massimino, J., Peltier, G. and Rey, P. (1996b) Effects of low temperature, high salinity and exogenous ABA on the synthesis of two chloroplastic drought-induced proteins in *Solanum tuberosum. Physiologia Plantarum*, **97**, 123–131.

Read, J., Anderson, T.A., Ritchie, P.J. *et al.* (2000) A mechanism of membrane neutral lipid acquisition by the microsomal triglyceride transfer protein. *Journal of Biological Chemistry*, **275**, 30372–30377.

Rinne, P.L.H., Kaikuranta, P.M. and van der Schoot, C. (2002) The shoot apical meristem restores its symplasmic organization during chilling-induced release from dormancy. *Plant Journal*, **26**, 249–264.
Robenek, M.J., Severs, N.J., Schlattmann, K., Plenz, G., Zimmer, K.-P., Troyer, D. and Robenek, H. (2004) Lipids partition caveolin-1 from ER membranes into lipid droplets: updating the model of lipid droplet biogenesis. *The FASEB Journal*, **18**, 866–868.
Robert, L.S., Gerster, J., Allard, S., Cass, L. and Simmonds, J. (1994) Molecular characterization of two *Brassica napus* genes related to oleosins which are highly expressed in the tapetum. *Plant Journal*, **6**, 927–933.
Roberts, M.R., Hodge, R., Ross, J.H.E. *et al.* (1993) Characterisation of a new class of oleosins indicates a male gametophyte-specific lipid storage pathway. *Plant Journal*, **3**, 629–636.
Roberts, M.R., Hodge, R. and Scott, R. (1995) *Brassica napus* pollen oleosins possess a characteristic *C*-terminal domain. *Planta*, **195**, 469–470.
Roberts, M.R., Robson, F., Foster, G.D., Draper, J. and Scott, R.J. (1991) A *Brassica napus* mRNA expressed specifically in developing microspores. *Plant Molecular Biology*, **17**, 295–299.
Ross, J.H.E. and Murphy, D.J. (1996) Characterisation of anther-expressed genes encoding a major class of extracellular oleosin-like proteins in the pollen coat of *Brassicaceae*. *Plant Journal*, **9**, 625–637.
Sachetto-Martins, G., Franco, L.O. and de Oliveira, D.E. (2000) Plant glycine-rich proteins: a family or just proteins with a common motif? *Biochimica et Biophysica Acta*, **1492**, 1–14.
Sadeghipour, H.Z. and Bhatla, S.C. (2002) Differential sensitivity of oleosins to proteolysis during oil body mobilization in sunflower seedlings. *Plant & Cell Physiology*, **43**, 1117–1126.
Sadeghipour, H.Z. and Bhatla, S.C. (2003) Light-enhanced oil body mobilization in sunflower seedlings accompanies faster protease action on oleosins. *Plant Physiology and Biochemistry*, **41**, 309–316.
Safrin, S. and Grunfeld, C. (1999) Fat distribution and metabolic changes in patients with HIV infection. *Aids*, **13**, 2493–2505.
Sargiacomo, M., Scherer, P.E., Tang, Z. *et al.* (1995) Oligomeric structure of caveolin: implications for caveolae membrane organization. *Proceedings of the National Academy of Sciences USA*, **92**, 9407–9411.
Sarmiento, C., Garces, R. and Mancha, M. (1998) Oleate desaturation and acyl turnover in sunflower (*Helianthus annuus* L.) seed lipids during rapid temperature adaptation. *Planta*, **205**, 595–600.
Schaeffer, A., Bronner, R., Benveniste, P. and Schaller, H. (2001) The ratio of campesterol to sitosterol that modulates growth in Arabidopsis is controlled by STEROL METHYLTRANSFERASE 2;1. *Plant Journal*, **25**, 605–615.
Schein, M., Yang, Z., Mitchell-Olds, T. and Schmid, K., (2004) Rapid evolution of a pollen-specific oleosin-like gene family from *Arabidopsis thaliana* and closely related species. *Molecular Biology and Evolution*.
Scherer, P.E., Bickel, P.E., Kotler, M. and Londish, H.F. (1998) Cloning of cell-specific secreted and surface proteins by subtractive antibody screening. *Nature Biotechnology*, **16**, 581–586.
Scheuer, P.J., Ashrafzadeh, P., Sferloch, S., Brown, D. and Dusheiko, G.M. (1992) The pathology of hepatitis C. *Hepatology*, **15**, 567–571.
Schimek, C., Eibel, P., Grolig, F., Horie, T., Ootaki, T. and Galland, P. (1999) Gravitropism in *Phycomyces*: a role for sedimenting protein crystals and floating lipid globules. *Planta*, **210**, 132–142.
Schnabl, M., Oskolkova, O.V., Holic, R. *et al.* (2003) Subcellular localization of yeast Sec14 homologues and their involvement in regulation of phospholipid turnover. *European Journal of Biochemistry/FEBS*, **270**, 3133–3145.
Segrest, J.P., Jones, M.K., Mishra, V.K. *et al.* (1998) Apolipoprotein B-100: conservation of lipid-associating amphipathic secondary structural motifs in nine species of vertebrates. *Journal of Lipid Research*, **39**, 85–102.
Siler, D.J., Goodrich-Tanrikulu, M., Cornish, K., Stafford, A.E. and Mckeon, T. (1997) Composition of rubber particles of *Hevea brasiliensis, Parthenium argentatum, Ficus elastica,*

and *Euphorbia lactiflua* indicates unconventional surface structure. *Plant physiology and Biochemistry*, **35**, 881–889.

Smirra, I., Halevy, A.H. and Vainstein, A. (1993) Isolation and characterization of a chromoplast-specific carotenoid-associated protein from *Cucumis sativus* corollas. *Plant Physiology*, **102**, 491–496.

Stahmann, K.-P., Kupp, C., Feldmann, S.D. and Sahm, H. (1994) Formation and degradation of lipid bodies found in the riboflavin-producing fungus *Ashbya gossypi.Applied Microbiology and Biotechnology*, **42**, 121–127.

Steinbuchel, A. and Fuchtenbusch, B. (1998) Bacterial and other biological systems for polyester production. *Trends in Biotechnology*, **16**, 419–427.

Steinbüchel, A., Aerts, K., Babel, W. *et al.* (1995). Considerations on the structure and biochemistry of bacterial polyhydroxyalkanoic acid inclusions. *Canadian Journal of Microbiology*, **41**, 94–105.

Takahashi, S., Katagiri, T., Yamaguchi-Shinozaki, K. and Shinozaki, K. (2000) An *Arabidopsis* gene encoding a $Ca^{2+}$-binding protein is induced by abscisic acid during dehydration. *Plant and Cell Physiology*, **41**, 898–903.

Tang, Z., Okamoto, T., Boontrakulpoontawee, P., Katada, T., Otsuka, A.J. and Lisanti, M.P. (1997) Identification, sequence and expression of an invertebrate caveolin gene family from the nematode *Caenorhabditis elegans*. Implications for the molecular evolution of mammalian caveolin genes. *Journal of Biological Chemistry*, **272**, 2437–2445.

Targett-Adams, P., Chambers, D., Gledhill, S., Hope, R.G., Coy, J.F., Girod, A. and McLauchlan, J. (2003) Live cell analysis of the lipid droplet binding protein ADRP. *Journal of Biological Chemistry*, **278**, 15998–16007.

Teixeira, L., Rabouille, C., Rorth, P., Ephrussi, A. and Vanzo, N.F. (2003) Drosophila Perilipin/ADRP homologue Lsd2 regulates lipid metabolism. *Mechanisms of Development*, **120**, 1071–1081.

Than, N.G., Sumegi, B., Bellyei, S. *et al.* (2003) Lipid droplet and milk lipid globule membrane associated placental protein 17b (PP17b) is involved in apoptotic and differentiation processes of human epithelial cervical carcinoma cells. *European Journal of Biochemistry*, **270**, 1176–1188.

Thompson, G.A. and Schulz, A. (1999) Macromolecular trafficking in the phloem. *Trends in Plant Science*, **4**, 354–360.

Thompson, J.E., Froese, C.D., Madey, E., Smith, M.D. and Hong, Y. (1998a) Lipid metabolism during plant senescence. *Progress in Lipid Research*, **37**, 119–141.

Thompson, J.E., Madey, E., Nowack, L. and Froese, C. (1998b) Distinguishable populations of lipid particles of membrane origin in petal and stems, in *Advances in Lipid Research* (eds J. Sanchez, E. Cerda-Olmedo and E. Martinez-Force), University of Sevilla, Sevilla, pp. 377–379.

Timmins, P.A., Poliks, B. and Banaszak, L. (1992) The location of bound lipid in the lipovitellin complex. *Science*, **257**, 652–655.

Ting, J.T.L., Wu, S.S.H., Ratnayake, C. and Huang, A.H.C. (1998) Constituents of the tapetosomes and elaioplasts in *Brassica campestris* and their degradation and retention during microsporogenesis. *Plant Journal*, **16**, 541–551.

Traverso, V., Morris, J.F., Flower, R.J. and Buckingham, J. (1998) Lipocortin 1 (annexin 1) in patches associated with the membrane of a lung adenocarcinoma cell line and in the cell cytoplasm. *Journal of Cell Science*, **111**, 1405–1418.

Ursini, F., Davies, K.J., Maiorino, M., Parasassi, T. and Sevanian, A. (2002) Atherosclerosis: another protein misfolding disease? *Trends in Molecular Medicine*, **8**, 370–374.

Vainstein, A., Halevy, A.H., Smirra, I. and Vishnevetsky, M. (1994). Chromoplast biogenesis in *Cucumis sativus* corollas. *Plant Physiology*, **104**, 321–326.

van Standen, J., Gilliland, M.G. and Brown, N.A.C. (1975) Ultrastructure of dry viable and non-viable *Protea compacta* embryos. *Zeitschrift fur Pflanzenphysiologie*, **76**, 28–35.

Vishnevetsky, M., Ovadis, M., Itzhaki, H. *et al.* (1996) Molecular cloning of a carotenoid-associated protein from *Cucumis sativa* corollas: homologous genes involved in carotenoid biosynthesis. *Plant Journal*, **10**, 1111–1118.

Vishnevetsky, M., Ovadis, M., Zuker, A. and Vainstein, A. (1999) Molecular mechanism underlying carotenogenesis in the chromoplast: multilevel regulation of carotenoid associated genes. *Plant Journal*, **20**, 423–431.

Wahlroos, T., Soukka, J., Denesyuk, A., Wahlroos, R., Korpela, T. and Kilby, N.J. (2003) Oleosin expression and trafficking during oil body biogenesis in tobacco leaf cells. *Genesis*, **35**, 125–132.

Wang, S.M., Hwang, R.D., Greenberg, A.S. and Yeo, H.L. (2003) Temporal and spatial assembly of lipid droplet-associated proteins in 3T3-L1 preadipocytes. *Histochemistry and Cell Biology*, **120**, 285–292.

Wang, T.W., Balsamo, R.A., Ratnayake, C., Platt, K.A., Ting, J.T. and Huang, A.H. (1997) Identification, subcellular localization, and developmental studies of oleosins in the anther of *Brassica napus*. *Plant Journal*, **11**, 475–487.

Wang, X., Reape, T.J., Li, X. *et al.* (1999) Induced expression of adipophilin mRNA in human macrophages stimulated with oxidized low-density lipoprotein and in atherosclerotic lesions. *FEBS Letters*, **462**, 445–450.

Wanner, G., Formanek, H. and Theimer, R.R. (1981) The ontogeny of lipid bodies (spherosomes) in plant cells. Ultrastructural evidence. *Planta*, **151**, 109–123.

Wanner, G. and Theimer, R.R. (1978) Membranous appendices of spherosomes (oleosomes). *Planta*, **140**, 163–169.

Winkenbach, F., Falk, H., Liedvogel, B. and Sitte, P. (1976). Chromoplasts of *Tropaeolum majus* L.: isolation and characterisation of lipoprotein elements. *Planta*, **128**, 23–28.

Wieczorek, R., Pries, A., Steinbuchel, A. and Mayer, F. (1995) Analysis of a 24-kilodalton protein associated with the polyhydroxyalkanoic acid granules in *Alcaligenes eutrophus*. *Journal of Bacteriology*, **177**, 2425–2435.

Wieczorek, R., Steinbuchel, A. and Schmidt, B. (1996) Occurrence of polyhydroxyalkanoic acid granule-associated proteins related to the *Alcaligenes eutrophus* H16 GA24 protein in other bacteria. *FEMS Microbiology Letters*, **135**, 23–30.

Wolins, N.E., Rubin, B. and Brasaemle, D.L. (2001) TIP47 associates with lipid droplets. *Journal of Biological Chemistry*, **276**, 5101–5108.

Wolins, N.E., Skinner, J.R., Schoenfish, M.J., Tzekov, A., Bensch, K.G. and Bickel, P.E. (2003) Adipocyte protein S3–12 coats nascent lipid droplets. *Journal of Biological Chemistry*, **278**, 37713–37721.

Yatsu, L.Y. and Jacks, T.J. (1972) Spherosome membranes. *Plant Physiology*, **49**, 937–943.

Zweytick, D., Athenstaedt, K. and Daum, G. (2000) Intracellular lipid particles of eukaryotic cells. *Biochimica et Biophysica Acta*, **1469**, 101–120.

# 8 The plant cuticle: formation and structure of epidermal surfaces

Ljerka Kunst, A.L. Samuels and Reinhard Jetter

## 8.1 Introduction

All land plants are coated with a waxy cuticle that is essential for their protection and interaction with the environment. The main functions of the cuticle are to limit non-stomatal water loss and to repel water on the surface of the plant, minimizing deposition of dust, pollen and air pollutants. The cuticle also plays important roles in plant defense against bacterial and fungal pathogens, and in mediating a variety of plant-insect interactions (Kerstiens, 1996). The structure and chemical composition of the cuticle differ among plant species, and among individual organs of a single species. Even within one organ, dynamic changes in the cuticle can be observed during development.

The major component of the cuticle is cutin (Figure 8.1), comprising between 40% and 80% of cuticular weight in various plant organs (Holloway, 1982). Cutin is a polyester that is insoluble in organic solvents and consists of oxygenated fatty acids with a chain length of 16 or 18 carbons. Embedded in the cutin matrix are cuticular waxes, which are readily removed by dipping the plant in non-polar solvents. The cuticular waxes are complex mixtures of very-long-chain fatty acid (VLCFA) derivatives, but they may also include triterpenoids and minor secondary metabolites, such as sterols, alkaloids, phenylpropanoids and flavonoids. These waxes are arranged into an intracuticular layer in close association with the cutin matrix, as well as an epicuticular film exterior to this, which may include epicuticular wax crystals. The whitish bloom on many plant surfaces is due to the light-scattering epicuticular crystals, which have a fantastic variety of forms in different plant species. In contrast, some plant surfaces are glossy if the epicuticular wax is present in a smooth film on the epidermal cell surface. The outer purely lipidic portion of the cuticle comprising epicuticular wax, intracuticular wax and cutin has been called the 'cuticle proper' (Jeffree, 1996). A transition zone between the cuticle proper and the underlying cell wall is the 'cuticular layer', consisting of a mixture of polysaccharides, intracuticular waxes and cutin (Jeffree, 1996).

The cuticle is synthesized and secreted by the epidermis during plant development. Early in seed development, the cuticle can be demonstrated experimentally by fluorescent staining of cuticular lipids in the developing Arabidopsis embryo at the globular stage, where the embryo consists of a few dozen cells (Szczuka and

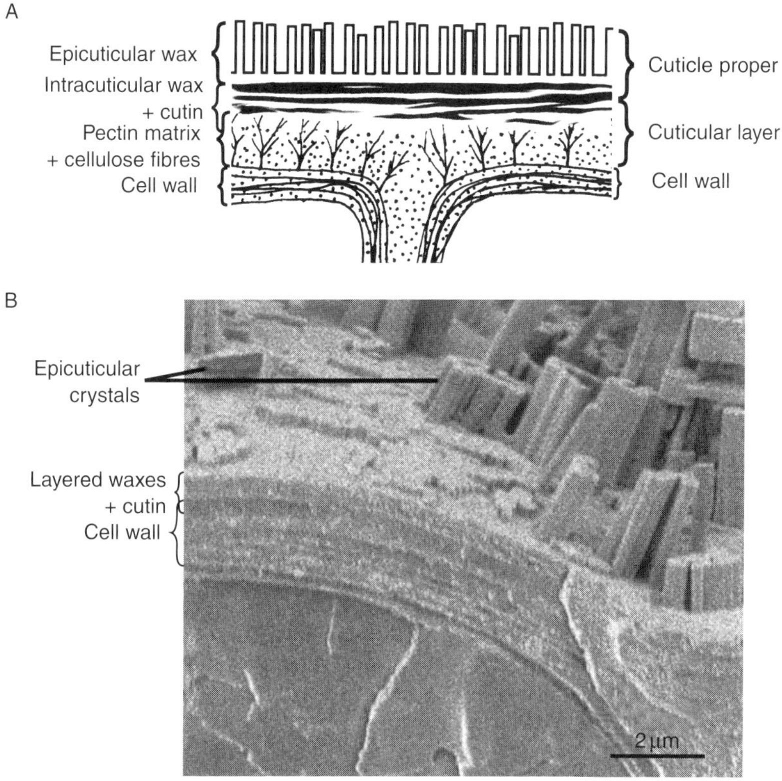


**Figure 8.1** A. Schematic drawing of the cuticle, showing the outer lipidic region (cuticle proper), as well as the cuticular layer that connects the cuticle to the cell wall. The cuticle proper consists of epicuticular wax film and crystals, intracuticular wax and cutin. The cuticular layer is made up of polysaccharides, intracuticular wax and the cutin matrix. B. Cryo-SEM of Arabidopsis stem cuticle, fractured and viewed while frozen. Epicuticular crystals on the plant surface, as well as the layered cuticle, cover the cell wall of an epidermal cell.

Szczuka, 2003). Only a few specialized cells of the embryo produce these lipids, which are secreted asymmetrically to the periclinal exterior surface. During further development of plant aerial organs, the epidermis arises from the L1 layer covering the shoot apical meristem and undergoes a series of predominantly anticlinal divisions, resulting in a seamless two-dimensional array of cells. The developing cuticle is connected to the procuticle on the meristematic surface. Within the epidermal sheet, many specialized cell types differentiate, such as guard cells, trichomes and their supporting socket cells, silicified cells, stylar secretory cells, as well as the surrounding pavement cells. Despite their structural and functional complexities, all epidermal cell types secrete and are covered by the cuticle, which has been called 'the universal marker of epidermal character' (Lolle and Pruitt, 1999). The continuity of the cuticle is maintained during division, differentiation and expansion of individual epidermal cell types. Individual cells must also coordinate cuticular lipid secretion to achieve the evenly thickened, expansible cuticle that protects the growing plant. The mechanisms allowing the controlled expansion of the cutin matrix and accumulation of waxes during development remain to be elucidated.

In this chapter, we will review our current understanding of the cuticular lipid biosynthesis, export and self-arrangement within the cuticle, with the goal to integrate the available biochemical, morphological and molecular data into a contemporary model of the plant cuticle. The physiology and ecology of plant cuticles will not be covered, as these topics have been extensively reviewed in Kerstiens (1996).

## 8.2 Biosynthesis of cuticle components

Generation of cuticular components in epidermal cells involves two types of pathways: those for the synthesis of carbon skeletons and those for modifying them into diverse aliphatic lipid classes. As with membrane and storage lipids, 16- and 18-carbon skeletons are supplied by a plastid-localized elongation pathway. These are substrates for cutin synthesis, or may be further extended into VLCFA wax precursors by extraplastidial elongases (Figure 8.2). VLCFAs are used by the decarbonylation and acyl-reduction pathways to yield major wax constituents, including aldehydes, alcohols, esters, hydrocarbons and ketones, present in all plant species (Figure 8.3). In addition, a unique membrane β-ketoacyl-elongation pathway with features of a polyketide synthase is responsible for the production of β-diketones (Figures 8.2 and 8.3), an important wax component in some plant species (von Wettstein-Knowles, 1993).

### *8.2.1* De novo *fatty acid synthesis*

Fatty acyl precursors for the synthesis of cuticular lipids are produced in plastids of epidermal cells by the soluble, dissociable enzymes of the fatty acid synthase (FAS) (Ohlrogge *et al.*, 1993). Fatty acid synthesis is initiated by condensation of a $C_2$ moiety originating from malonyl-ACP to acetyl-CoA, followed by the reduction of β-ketoacyl-ACP, the dehydration of β-hydroxyacyl-ACP and the reduction of *trans*-$\Delta^2$-enoyl-ACP (Jaworski *et al.*, 1993). During synthesis, growing acyl chains are attached to acyl carrier protein (ACP), an essential protein cofactor, and elongated by two carbons per cycle, supplied by malonyl-ACP. Three different types of FAS complexes are required for the synthesis of 18-carbon fatty acids in the plastid. They differ in their condensing enzymes, which have strict acyl length specificities: KAS III ($C_2$–$C_4$) (Clough *et al.*, 1992), KAS I ($C_4$–$C_{16}$) and KAS II ($C_{16}$–$C_{18}$) (Shimakata and Stumpf, 1982). In contrast, the two reductases and the dehydratase apparently have no particular acyl chain length specificity and are shared by all three plastidial elongation complexes (Stumpf, 1984).

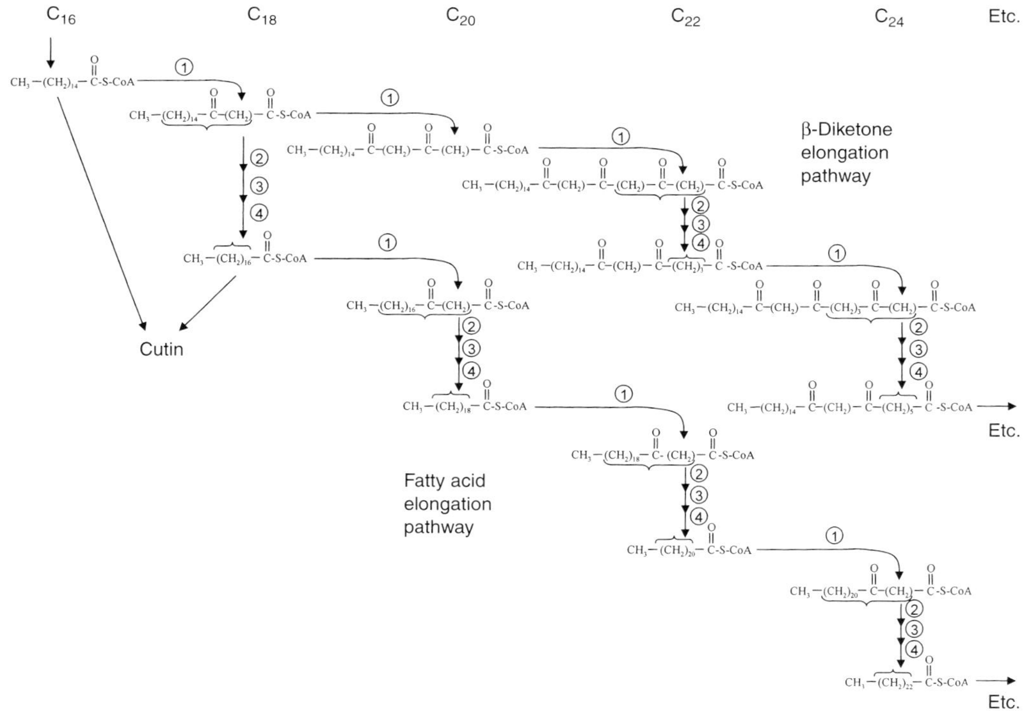


**Figure 8.2** Pathways for elongation of long-chain fatty acids, generating carbon skeletons for cuticular lipids. Cutin monomers are derived from $C_{16}$ and $C_{18}$ fatty acids without further elongation. The fatty acid elongation pathway produces VLCFAs ($C_{20}$–$C_{34}$), which are the substrates of modifying wax biosynthetic enzymes (shown in Figure 8.4). The β-diketone elongation pathway produces β-keto acyl intermediates en route to β-diketones. Sequential two-carbon additions to the growing chains are made by enzymes of the FAE: (1) β-ketoacyl-CoA synthase (KAS), (2) β-keto-acyl-CoA reductase, (3) β-hydroxyacyl-CoA dehydratase and (4) enoyl-CoA reductase. Pairs of brackets indicate corresponding parts of the molecules before and after reactions (2), (3), and (4).

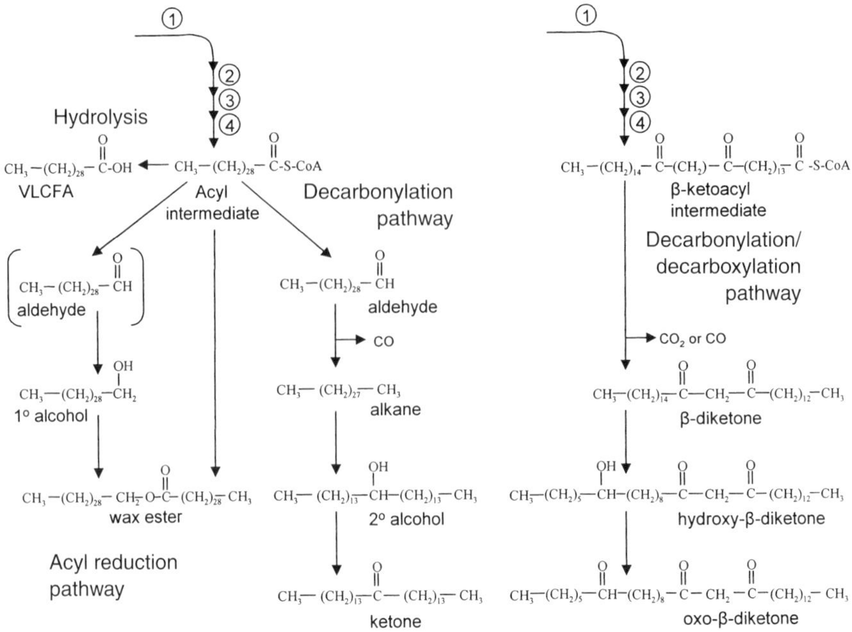


**Figure 8.3** Pathways for modification of VLCFA precursors into aliphatic wax constituents. The acyl reduction pathway produces primary alcohols and esters, the decarbonylation pathway generates aldehydes, alkanes, secondary alcohols and ketones, and the β-diketone pathway yields β-diketones, hydroxy- and oxo-β-diketones. Chemical formulas depicting representative species for each lipid class are shown.

Once they reach $C_{16}$ or $C_{18}$ in length, fatty acids must be partitioned between membrane glycerolipid synthesis and pathways for cutin and wax biosynthesis. This is an important control point in epidermal cells, which results in a massive flow of fatty acids into cutin and wax production. It is not clear, however, how the proportion of fatty acid precursors destined for different biosynthetic pathways is regulated. Activities and specificities of enzymes such as acyl-ACP acyltransferases, thioesterases and the 18:0 desaturase in the plastid, as well as extraplastidial acyltransferases, fatty acid elongase enzymes and the FAD2 desaturase, are all likely determining factors. In addition, the availability of fatty acids due to compartmentation or metabolic channeling may affect their distribution *in vivo* (Post-Beittenmiller, 1996).

For incorporation into cutin and wax, acyl chains are released from the ACP by a thioesterase and exported from the plastid to the ER. The 16:0- and 18:1-CoAs are then used as precursors for cutin monomer formation (Kolattukudy, 1980), whereas wax biosynthesis requires further rounds of elongation of 16:0 or 18:0 acyl groups to generate VLCFA wax precursors ranging between $C_{26}$ and $C_{34}$ in length. Two different classes of acyl-ACP thioesterases, FATA and FATB, have been identified in plants on the basis of gene sequence analysis and substrate preferences (Voelker *et al.*, 1997). The ubiquitously expressed FATA class, with 18:1-ACP as its preferred substrate, could be the key acyltransferase involved in export of 18:1 acyl groups for $C_{18}$ cutin monomer production. FATB thioesterases, with preference for saturated acyl-ACP substrates, could be involved in supplying 16:0 for the production of $C_{16}$ cutin monomers. The FATB type thioesterase is also likely involved in providing fatty acyl substrates for wax biosynthesis, as Arabidopsis *mutants* deficient in FATB activity exhibit a 20% reduction in total wax load in leaves and a 50% reduction in stems (Bonaventure *et al.*, 2003).

### *8.2.2 Cutin biosynthesis*

Cutin forms a three-dimensional network associated with polysaccharides and intracuticular wax (Holloway, 1982). Chemically, it is a polyester polymer insoluble in chloroform, composed of bifunctional oxygenated fatty acids, such as $C_{16}$ and $C_{18}$ ω-hydroxyacids, with hydroxyl or epoxy groups in mid-chain positions. Ester-breaking reactions, followed by analysis of released products by gas chromatography/mass spectrometry (GC/MS) revealed that the most abundant monomers of the $C_{16}$ family are dihydroxypalmitic acids containing one hydroxyl group at the ω-carbon and the other at a mid-chain carbon ($C_{10}$, $C_9$, $C_8$ or $C_7$), whereas diagnostic $C_{18}$ family cutin monomers are oleic acid, ω-hydroxyoleic acid, 9,10,18-trihydroxystearic acid and 9,10-epoxy-18-hydroxystearic acid (Walton and Kolattukudy, 1972; Kolattukudy, 1980). The $C_{16}$ family of monomers is predominantly found in rapidly growing plant organs, while slower-expanding plant organs usually contain a mixture of $C_{16}$ and $C_{18}$ monomers. In addition to fatty acid derivatives, small amounts of phenolic compounds, such as *p*-coumarate and ferulate are also esterified to cutin.

Compositional data, together with evidence suggesting that most of the primary hydroxyl and carboxyl groups of cutin monomers were involved in ester linkage formation, while the secondary hydroxyls were only partially esterified (Kolattukudy, 1977; Daes and Holloway, 1980), led to the initial models of cutin structure (Kolattukudy, 1977, 1980; Zlotnik-Mazori and Stark, 1988). These models depict cutin monomers linked head to tail via their primary functional groups in a linear fashion, and partially cross-linked through their secondary hydroxyl groups. A major revision of our view of cutin structure may be required following recent reports that glycerol is also one of the major constituents of cutin in diverse species, accounting for up to 14% of the total monomers (Moire *et al.*, 1999; Graça *et al.*, 2002). Glycerol was not detected in greater than trace amounts in earlier cutin analyses because the solvent extraction step of cutin monomers after depolymerization left small polar molecules, such as glycerol in the aqueous phase, which was discarded (Moire *et al.*, 1999). Nevertheless, the presence of glycerol in cutin, which can serve as an attachment point for ω-hydroxyacids, offers new possibilities for polymer expansion in different dimensions (Graça *et al.*, 2002). Recent structural studies by Fourier-transform infrared spectroscopy (FTIR), X-ray diffraction and NMR suggest that cutin is a moderately flexible, amorphous network with molecular spacing of 0.4–0.5 nm between polymer chains (Heredia, 2003).

Despite the abundance and structural importance of cutin, relatively little is known about its biosynthesis, transport to the plant surface and monomer polymerization or the enzymes involved in these processes. The first pathway of cutin biosynthesis was proposed by Kolattukudy and coworkers in the 1970s (reviewed by Kolattukudy, 1981). Precursors of $C_{16}$ and $C_{18}$ cutin monomers are 16:0-CoA and 18:1-CoA, respectively. These acyl groups then undergo hydroxylation and epoxidation reactions catalyzed by a variety of cytochrome P450-dependent enzymes (Figure 8.4). Even though cytochrome P450 enzymes capable of carrying out in-chain and ω-hydroxylation have been characterized (Benveniste *et al.*, 1998; Cabello-Hurtado *et al.*, 1998; Pinot *et al.*, 1998; Tijet *et al.*, 1998), biological confirmation of their involvement in cutin formation is lacking. An alternative lipoxygenase-peroxygenase-epoxide hydrolase pathway for the formation of $C_{18}$ cutin monomers has recently been demonstrated (Blee and Schuber, 1993; Lequeu *et al.*, 2003). According to this pathway (Figure 8.4), a peroxygenase catalyzes the epoxydation of oleic acid in the presence of unsaturated fatty acid hydroperoxides that act as co-substrates. These hydroperoxides are likely generated by a lipoxygenase from linoleic or linolenic acid.

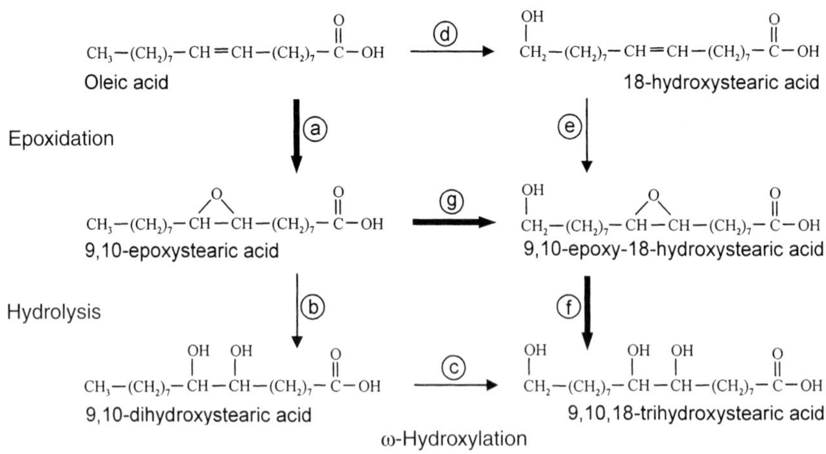


**Figure 8.4** Biosynthetic pathways for the formation of $C_{18}$ cutin monomers. The initial steps leading to the formation of 9,10-epoxy-8-hydroxystearic acid are pathway a, g (major route) or pathway d, e (minor route). The final steps in the synthesis of 9,10,18-trihydrostearic acid are either step f, or pathway b, c, depending on activities of epoxide hydrolase f or cytochrome P450-dependent ω-hydroxylase (c). Lipoxygenase is required for the formation of fatty acid hydroperoxides, the co-substrates of the peroxygenase. Modified from Blée *et al.* (1993).

The mode of transport of cutin monomers to their site of polymerization is currently unknown. How these monomers traverse the cell wall is also unclear, even though lipid transfer proteins (LTPs) have been proposed as possible carriers (Sterk *et al.*, 1991). Polymerization of cutin monomers into a cutin mesh has not been characterized to date. A single enzyme activity involved in this process, an ATP-dependent hydroxyl-CoA:cutin transacylase has been demonstrated in crude leaf extracts (Croteau and Kolattukudy, 1973, 1975).

Further progress in dissection of the cutin biosynthetic pathway, identification of transporters of cutin monomers and enzymes involved in cutin polymerization awaits isolation and characterization of *mutants* defective in these processes. Until recently, this was not possible, because the viability and morphological phenotypes of such *mutants* were not known. An elegant experiment carried out by Sieber *et al.* (2000), who expressed a fungal cutin-degrading enzyme in the cuticle, revealed that the disruption of the cutin polymer leads to organ fusions during plant development. This result suggested that *mutants* with altered cutin properties may also exhibit organ fusion phenotypes. A number of Arabidopsis *mutants* with major growth abnormalities due to organ fusion are known. Two fusion *mutants*, *lacerata (lcr)* and *fiddlehead (fdh)* have been characterized, and genes identified by mutation have been cloned (Yephremov *et al.*, 1999; Pruitt *et al.*, 2000; Wellesen *et al.*, 2001). The *LCR* gene may be a candidate for a role in cutin biosynthesis, since it encodes a cytochrome P450-dependent monooxygenase capable of catalyzing ω-hydroxylation of fatty acids from $C_{12}$ to $C_{18}$ in length. Further biochemical work on this mutant is necessary to demonstrate the role of LCR in cutin formation. The *FDH* gene encodes a protein with sequence similarity to β-ketoacyl-CoA synthases involved in extraplastidial fatty acid elongation, and its involvement in cutin production is not obvious. Characterization of additional *mutants* with organ fusion phenotypes, such as *wax1* (Jenks 1996), *wax2* (Chen *et al.*, 2003), *cer10* and *cer13* (Yephremov *et al.*, 1999), *ded* and others (Lolle and Pruitt, 1998) in Arabidopsis, as well as *bm2* in *Sorghum* and *cr4* in maize promises to unravel some of the complexities of cutin biosynthesis and deposition.

### *8.2.3 Synthesis of very-long-chain fatty acid wax precursors*

The predominant cuticular wax components, such as primary and secondary alcohols, aldehydes, alkanes, ketones, esters and β-diketones are derived from saturated VLCFAs with chain lengths between $C_{20}$ and $C_{34}$ (von Wettstein-Knowles, 1995). Thus, the first step in wax biosynthesis is the elongation of $C_{16:0}$ and $C_{18:0}$ fatty acids produced in the plastid to generate VLCFA wax precursors up to 34 carbons in length (Lemieux *et al.*, 1994; Post-Beittenmiller, 1996). Fatty acid elongation is catalyzed by extraplastidial membrane-associated, multienzyme complexes known as fatty acid elongases (FAE) (von Wettstein-Knowles, 1982) and involves four enzymatic reactions that are analogous to those of FAS. The initial condensation reaction (which adds two carbon units from malonyl-CoA to a CoA-esterified fatty acid substrate) is catalyzed by a β-ketoacyl-CoA synthase (Figure 8.2). Experimental evidence suggests that this is the substrate-specific step and the rate-limiting step of fatty acid elongation (Lassner *et al.*, 1996; Millar and Kunst, 1997). Subsequent reactions include a reduction of the β-ketoacyl-CoA to β-hydroxy-CoA, dehydration to an enoyl-CoA, and a second reduction to yield acyl-CoA extended by two carbons. Enzymes catalyzing these last three reactions are thought to be constitutively expressed throughout the plant and used with the diverse condensing enzymes present in the same cell (Millar and Kunst, 1997).

The differential effects of inhibitors on incorporation of radiolabeled precursors into wax components of various chain lengths, and analyses of *mutants* with defects in fatty acid elongation, demonstrated that multiple FAEs (parallel and sequential) are required for the production of VLCFA precursors (von Wettstein-Knowles, 1993). Consistent with the requirement for fatty acyl precursors of various chain lengths, in various plant tissues, at different times during plant development, a large family of 21 elongase condensing enzyme-like sequences has been identified in the Arabidopsis genome. It is not known how many of these putative condensing enzymes are wax-specific and how many different condensing enzymes are needed for the elongation of a $C_{18}$ to $C_{34}$ fatty acid, because elongase condensing enzymes can catalyze multiple elongation steps. A detailed analysis of the substrate specificity and expression patterns of individual condensing enzymes will therefore be necessary to determine their exact metabolic roles.

Of the four enzymes comprising the FAE, condensing enzymes are by far the best characterized. The condensing enzymes were first studied from several species of the Brassicaceae and jojoba (*Simmondsia chinensis*), because their seed storage components contain agronomically important VLCFAs (Kunst *et al.*, 1992; James *et al.*, 1995; Lassner *et al.*, 1996; Barett *et al.*, 1998; Ghanevati and Jaworski, 2001; Blacklock and Jaworski, 2002). Three additional condensing enzymes have been characterized from Arabidopsis and implicated in the synthesis of VLCFA precursors for wax production in shoots: KCS1 (Todd *et al.*, 1999), FDH (Yephremov *et al.*, 1999; Pruitt *et al.*, 2000) and CER6 (Millar *et al.*, 1999; Fiebig *et al.*, 2000). Interestingly, a major reduction of CER6 activity in *cer6 mutants* and sense suppressed *CER6* plants nearly abolished stem wax accumulation, and resulted in conditional male sterlity (Millar *et al.*, 1999), suggesting that there is no significant functional overlap of CER6 with KCS1 and FDH activities in the stem and anther of Arabidopsis. Furthermore, a recently identified VLCFA condensing enzyme CER60 (Fiebig *et al.*, 2000) with high amino acid sequence identity to CER6 does not appear to significantly contribute to the synthesis of stem and pollen surface lipids. This could be due to a very low level of *CER60* expression (Hooker *et al.*, 2002) or a different tissue specificity of CER60. Taken together, these data implicate CER6 as the major condensing enzyme for stem wax and pollen coat lipid biosynthesis in Arabidopsis.

Recent cloning of the β-ketoacyl reductase gene from *Saccharomyces cerevisiae* allowed identification and characterization of homologous genes from Arabidopsis (Beaudoin *et al.*, 2002). In maize, a mutation at the *GL8* locus results in decreased levels of wax ester components longer than $C_{24}$ (Bianchi *et al.*, 1979, 1985). The *GL8* gene was isolated by transposon tagging and suggested to encode a reductase involved in fatty acid elongation (Xu *et al.*, 1997). Based on evidence presented by Beaudoin *et al.* (2002) and further characterization of the Glossy8 mutant (Xu *et al.*, 2002), it is clear that the maize GL8 functions as a β-ketoacyl reductase of the FAE involved in wax production. Similarly, the isolation of the *TSC13* gene encoding the yeast enoyl reductase (Kohlwein *et al.*, 2001) allowed the identification of a similar sequence in the Arabidopsis genome. Functional complementation of the yeast mutant with the Arabidopsis sequence and reverse genetic approaches in Arabidopsis should establish the role of this gene as enoyl reductase in fatty acid elongation.

### *8.2.4 Synthesis of aliphatic cuticular wax components*

In most plants, there are two major pathways producing aliphatic wax components from the carbon skeletons generated by fatty acid elongation. An acyl reduction pathway gives rise to primary alcohols and wax esters, and a decarbonylation pathway is involved in the formation of aldehydes, alkanes, secondary alcohols and ketones (Figure 8.3).

#### *8.2.4.1 The acyl reduction pathway*

Partial purification of reducing activities from *Brassica oleracea* initially led to a proposal that primary alcohol production is a two-step process carried out by two separate enzymes. These are an NADH-dependent acyl-CoA reductase required for a reduction of VLCFAs to aldehydes, and an NADPH-dependent aldehyde reductase required for a further reduction of aldehydes to primary alcohols (Kolattukudy, 1971). Subsequent work of Pollard *et al.* (1979) in jojoba and Vioque and Kolattukudy (1997) in pea (*Pisum sativum*), demonstrated that alcohol formation from VLCFA precursors is carried out by a single fatty acyl-CoA reductase (FAR). Cloning of the cDNA encoding the FAR from developing embryos of jojoba, and expression in *Escherichia coli* and *B. napus* verified its biochemical role and resulted in the accumulation of fatty alcohols (Metz *et al.*, 2000). The pea and the jojoba alcohol-forming FARs are both integral membrane proteins, which appear to be associated with the ER (Metz *et al.*, 2000) and have similar molecular masses in the range of 56–58 kDa. FAR-related sequences from maize (*Zea mays*), rice (*Oryza sativa*), cotton (*Gossypium hirsutum*), oilseed rape (*B. napus*) and Arabidopsis found in the public databases, suggest that alcohol-generating reductases are ubiquitous in plants.

The final step of the acyl reduction pathway is the synthesis of wax esters, a reaction catalyzed by a fatty acyl-CoA: fatty alcohol acyltransferase (wax synthase, WS). WS activity from several different sources has been characterized biochemically (Kolattukudy, 1967; Khan and Kolattukudy, 1973; Wu *et al.*, 1981). Partial purification from jojoba embryos (Lardizabal *et al.*, 2000) resulted in identification of a cDNA encoding the WS enzyme. The identity of the cloned cDNA was demonstrated by expression in Arabidopsis, which resulted in high levels of WS activity and accumulation of wax esters in seeds of transgenic plants. Hydropathy analysis of the deduced protein sequence revealed seven to nine transmembrane domains indicating that jojoba WS is an integral membrane protein.

#### *8.2.4.2 The decarbonylation pathway*

The initial step of the decarbonylation pathway is the production of aldehydes from VLCFA precursors by a membrane-bound FAR. This enzyme has been purified to homogeneity from pea, and shown to be an enzyme with distinct molecular weight (28 kDa) and biochemical properties from the primary alcohol-forming reductase (Vioque and Kolattukudy, 1997). However, a gene encoding this enzyme has not yet been isolated. The generated aldehydes are then decarbonylated to odd-chain alkanes with a release of carbon monoxide. This reaction is catalyzed by an aldehyde decarbonylase (Cheesbrough and Kolattukudy, 1984). Two plant aldehyde decarbonylases have been studied in some detail, one from pea and one from the green colonial alga *Botryococcus braunii* (Cheesbrough and Kolattukudy, 1984; Dennis and Kolattukudy, 1992; Schneider-Belhaddad and Kolatukudy, 2000). Both enzymes need metal ions for their function and are integral membrane proteins. However, technical difficulties encountered during solubilization of decarbonylases have delayed molecular characterization of these enzymes and elucidation of the catalytic mechanism of decarbonylation.

A molecular genetic approach to identifying and characterizing a plant decarbonylase is an attractive alternative to conventional biochemical approaches. In Arabidopsis stems, products of the decarbonylation pathway account for approximately 90% of thc total wax (McNevin *et al.*, 1993; Millar *et al.*, 1999). Thus, a lesion in a key biosynthetic step of this pathway, such as decarbonylation, is likely to result in a significantly reduced wax load, and such a mutant could be easily identified on the basis of its increased glossiness. Interestingly, of the 23 different genetic loci identified by mutation to date, only *CER1* and *WAX2* were reported to specifically affect the decarbonylation pathway. Aarts *et al.* (1995) cloned the *CER1* gene, but the *CER1* sequence was not useful in assigning a biochemical function to its gene product. Similarly, the exact role of WAX2 in wax production remains to be established (Chen *et al.*, 2003).

Subsequent steps of the decarbonylation pathway, the hydroxylation of alkanes to secondary alcohols and oxidation of secondary alcohols to ketones, have been demonstrated in broccoli (*B. oleracea*) using radiolabeled alkanes (Kolattukudy and Liu, 1970; Kolattukudy *et al.*, 1973). Our knowledge of the enzymes catalyzing these reactions is, however, extremely limited. Progress in this area awaits identification and characterization of genes encoding the enzymes of this pathway.

### *8.2.5 The β-diketone pathway*

In some plants, including barley, carnation and species of the genera *Eucalyptus* and *Rhododendron*, β-diketones, hydroxy- and oxo-β-diketones and alkan-2-ols are important wax components. These lipids are synthesized by a unique polyketide-like elongation system elucidated in barley through elegant analyses of *cer-cqu mutants* using a combination of radioactive labeling and inhibitors (von Wettstein-Knowles, 1993). A comprehensive review describing the β-diketone pathway is available (von Wettstein-Knowles, 1995). Therefore, we will only describe the main features of this pathway.

Unlike acyl chain elongation carried out by FAEs, in which a β-keto group introduced by a condensing enzyme is removed by subsequent enzymes (β-keto reductase, β-hydroxy dehydratase and enoyl reductase), production of polyketides is characterized by successive condensations carried out by β-ketoacyl elongases, without the requirement to remove the β-keto groups before the next condensation takes place (Figure 8.2) (Hopwood and Sherman, 1990). In barley, β-ketoacyl elongases generate a pool of β-oxoacyl intermediates that can be converted to methylketones by a decarboxylase, followed by a methylketone reductase to give alkan-2-ols that can then be esterified. Alternatively, intermediates with two oxo-groups can be generated by successive condensations, followed by several chain elongations and decarboxylation or decarbonylation resulting in formation of β-diketones. A hydroxyl group can then be introduced and oxidized to yield hydroxy-β-diketones and oxo-β-diketones respectively (Figure 8.3) (von Wettstein-Knowles, 1993).

## 8.3 Cuticle biosynthesis in the context of the epidermal cell

Cuticular components are unusual biological products since many of them, particularly the saturated VLCFAs and derivatives, are solids at physiological temperatures. Yet, epidermal cells produce massive quantities of these high molecular weight lipids and arrange them into a highly ordered, functional cuticle. Some cells have much higher capacity for synthesis than their adjacent neighbors, as evidenced by some monocots and succulents where massive epicuticular wax 'stomatal chimneys' are secreted from epidermal cells adjacent to guard cells (Barthlott *et al.*, 1998). As the biochemical pathways of cutin and VLCFA production are characterized, the cellular environment in which cuticle biosynthesis occurs must be considered. The typical wax-secreting epidermal cell contains a large central vacuole surrounded by a very thin layer of cortical cytoplasm, with small leucoplasts, abundant endoplasmic reticulum (ER) and a basal nucleus.

### *8.3.1 Saturated long-chain fatty acids are exported from the plastid to the ER for elongation*

For the production of intra- and epicuticular waxes, plastidial fatty acids must be diverted to the ER, where elongation of saturated long-chain precursors ($C_{16}$–$C_{18}$) to very long chains ($C_{20}$–$C_{34}$) by the multi-enzyme FAE complexes occurs. Following cleavage from ACP by FATB thioesterases in the plastid, saturated fatty acids are esterified to CoA in the chloroplast envelope (Joyard and Stumpf, 1981; Schnurr *et al.*, 2002; Shockey *et al.*, 2003). Domains of the ER are in close physical proximity to the plastid, without apparent fusion or mixing of the bilayers (Kaneko and Keegstra, 1996; Staehelin, 1997). This proximity may facilitate fatty acid transfer to the ER. The non-vesicular fatty acid transfer is not well characterized in plants. In mammals, however, protein-mediated transfer and selection of long-chain fatty acids in the cytoplasm have been described (Storch and Thumser, 2000; Stremmel *et al.*, 2001; Schaffer, 2002). In *S. cerevisiae* (Zou *et al.*, 2002, 2003), a putative transport protein (Fat1p) and long-chain acyl-CoA synthetase are believed to act together in the 'vectorial acylation' or metabolic trapping of exogenous fatty acids into the cytoplasm. Acyl-CoA binding proteins (ACBP) found in mammals, yeast and plants (Rasmussen *et al.*, 1994; Engeseth *et al.*, 1996; Johnson *et al.*, 2002) may also be involved. For example, in *S. cerevisiae*, a mutant lacking ACBP accumulated $C_{18:0}$ and had reduced $C_{26:0}$, a VLCFA usually associated with sphingolipids, suggesting that ACBP is involved in transport of acyl-CoAs to the FAE (Gaigg *et al.*, 2001).

Experimental data from a variety of plants support the view that the ER is the site of fatty acid elongation. In a TEM study of *Sorghum* cork cells, there was a quantifiable increase in ER membranes in the cytoplasm that correlated with light-induced wax accumulation (Jenks *et al.*, 1994). Fatty acid elongation activities found in the microsomal fraction of leek (*Allium porrum*) co-localize with ER markers (Lessire *et al.*, 1985; Bessoule *et al.*, 1989). In maize, the β-ketoacyl reductase was shown to be associated with ER membranes by subcellular fractionation (Xu *et al.*, 2002). Finally, the Arabidopsis CER6 condensing enzyme involved in cuticular wax production, when fused to GFP, showed a fluorescent localization pattern typical of the ER (Kunst and Samuels, 2003; Zhang and Kunst, unpublished observations).

When elongation is complete, the acyl chains reach lengths of up to 34 carbons, which makes them highly hydrophobic. They are likely positioned within membranes and, due to their molecular geometry, might locally disrupt the membrane bilayer (Hamilton, 1998). The epidermal cell must control the flux of these VLCFA through the cell to avoid detrimental effects on housekeeping ER membrane domains. It is possible that specialized domains of ER are present in plant cells. A special domain of ER at the nuclear–vacuolar junction has been demonstrated in *S. cerevisiae* with GFP fusions to the TSC13 enoyl reductase of the FAE (Kohlwein *et al.*, 2001).

### *8.3.2 VLCFA modification and delivery of wax constituents to the plasma membrane*

Subcellular compartments in which VLCFAs are further metabolized, i.e. the location of the two possible pathways that the VLCFA may enter (acyl-reduction or decarbonylation pathways) in the epidermal cells has not been established. Interestingly, in the biochemical studies of the decarbonylation pathway, decarbonylase activity was found in the 'cell wall-cuticle' subcellular fraction of pea cells (Cheesbrough and Kolattukudy, 1984). This may reflect a plasma membrane (PM) location for this enzyme, since domains of the PM are strongly adherent to the cell wall (Wyat and Carpita, 1993; Oparka *et al.*, 1994; Lang-Pauluzzi and Gunning, 2000). In addition, the pH optimum of the enzyme is 7, instead of acidic pH found in the cell wall. Therefore, a PM location for the decarbonylation pathway (as postulated by von Wettstein-Knowles (1993)) is an attractive and viable hypothesis, which needs to be further explored.

One of the most intriguing questions of plant cuticular lipid deposition is how VLCFAs and/or VLCFA derivatives get moved from their sites of synthesis to the cell surface. Recently, two models for their transport between the ER to the PM, have been proposed (Kunst and Samuels, 2003): (1) direct molecular transfer from specialized ER domains to the PM, and (2) Golgi-mediated sorting and vesicular transfer in lipid-rafts. The first model suggests that wax constituents could be transferred directly to the PM via sites of close apposition of ER domains with the protoplasmic face of the PM. Analogous sites have been directly visualized in all plant cell types examined by TEM under conditions where the osmotic balance of the cell was preserved intact (Craig and Staehelin, 1988; Lichtscheidl *et al.*, 1990; Staehelin, 1997). No membrane fusion occurs in these sites, but studies with fluorescently labeled lipids showed their movement from PM into the ER (Grabsky *et al.*, 1993). In epidermal cells, similar sites of close ER/PM apposition may allow anterograde traffic of VLCFAs or their derivatives from the ER to the PM. Some support for the direct ER to PM transfer comes from the secretion of lipidic essential oils in glandular trichomes of peppermint (*Mentha x piperita*) (Gershenzon *et al.*, 2000; Lange *et al.*, 2000). Essential oil production in this system begins in the plastid where key enzymes such as limonene synthase have been localized (Turner *et al.*, 1999). Subsequent cytochrome P450-mediated hydroxylation reactions are carried out by microsomal enzymes, and final metabolism to menthol by cytoplasmic reductases (Wise and Croteau, 1999). Advanced microscopy techniques, such as microwave fixation and cryo-fixation of mechanically isolated peltate trichomes, demonstrated that terpene synthase activity was localized in non-photosynthetic plastids (leucoplasts) and that hydroxylation occurred on abundant smooth ER, which was closely associated with the PM. There was no evidence of Golgi-mediated secretion of essential oils in these cells, and a direct transfer mechanism from SER to PM was proposed.

The alternative hypothesis states that wax components are carried from the ER to the PM by Golgi-mediated vesicle traffic. This model fits with current paradigms of membrane lipid trafficking between compartments in plant cells. The kinetics of transfer between compartments has been measured by cell fractionation following pulse-chase of $^{14}C$-acetate (reviewed by Moreau *et al.*, 1998). Incorporated radioactivity follows a pattern typical of macromolecules moving through the secretory pathway, i.e. ER then Golgi and then PM-enriched fractions are labeled. In addition, sphingolipids containing VLCFA tails have been shown to combine with sterols to form raft-like structures in yeast and mammalian cells (Bagnat *et al.*, 2000; Brown and London, 2000). Association of the VLCFAs into lipid-rafts with sterols could be a mechanism for preventing disruption of the lipid bilayer structure by VLCFAs.

### *8.3.3 Export of wax components from the epidermal cell to the cuticle*

Once wax components reach the PM, they must be transferred across the membrane into the aqueous apoplast. Lipid export from mammalian cells is mediated by ATP-binding cassette (ABC) transporters (Schmitz *et al.*, 2000; Dean *et al.*, 2001). ABC transporters are found in all sequenced genomes (Sanchez-Fernandez *et al.*, 2001) and roles for these proteins are just beginning to emerge in plants (Martinoia *et al.*, 2002; Goossens *et al.*, 2003). For example, a plasma membrane ABC transporter upregulated in the presence of a toxic diterpene in *Nicotiana plumbaginifolia* cell culture, was recently cloned and characterized (Jasinski *et al.*, 2001). Under normal conditions, this transporter was expressed in the leaf epidermis, and may have a role in diterpene export into the cell wall. More significantly, a wax-deficient mutant of Arabidopsis, *cer5*, has a lesion in a gene encoding an ABC transporter (Pighin, Kunst and Samuels, unpublished results). Thus, ABC transporters are likely candidates for moving cuticular waxes across the PM.

The cell wall is the final barrier that cuticular lipids must traverse to be assembled in the cuticle. The transport across the cell wall is often attributed to LTPs (Sterk *et al.*, 1991; Jenks *et al.*, 2002). LTPs are a conserved family of small (9 kDa) , basic cell wall proteins (reviewed by Kader, 1996) found in all eukaryotic cells. In plants, the most abundant LTPs have a signal sequence, which targets them to the endomembrane system (Bernhard *et al.*, 1991; Thoma *et al.*, 1993). The role of LTPs in cuticular lipid transport has never been experimentally verified, but there is abundant experimental evidence in support of this possibility. LTPs accumulate to high levels in the wax in *B. oleracea* leaves. LTP WAX9, which shares high homology with Arabidopsis LTP1, makes up more than 90% of the protein present in the wax fraction (Pyee *et al.*, 1994). *LTP* transcript levels are one of the highest in Arabidopsis (Beisson *et al.*, 2003). There are *LTPs* that are exclusively expressed in the epidermis, based on *in situ* hybridization and promoter-GUS fusion experiments (Clark and Bohnert, 1999; Sohal *et al.*, 1999; Arondel *et al.*, 2001), although others are also expressed in internal tissues. Upregulation of LTP genes has been correlated with conditions that stimulate wax production, such as drought (Trevino and O'Connell, 1998), cadmium exposure (Hollenbach *et al.*, 1997) and light (Sohal *et al.*, 1999).

Biochemical characterization of LTPs also provides important information for assessing the potential function of these proteins in cuticular lipid transport across the cell wall. The LTP sequence contains eight conserved cysteines that form disulfide bonds believed to stabilize a hydrophobic tunnel, which can accommodate fatty acyl tails of a variety of chain lengths (Shin *et al.*, 1995; Gomar *et al.*, 1998; Tassin-Moindrot *et al.*, 2000). High-resolution X-ray crystallography of maize

LTPs showed the hydrophobic tunnel could accommodate ligands from $C_{10}$–$C_{18}$ (Han *et al.*, 2001), which allows binding of $C_{16}$ and $C_{18}$ cutin precursors, but not VLCFAs in the $C_{24}$–$C_{30}$ range (Keresztessy, 1998). In competition experiments, maize LTP showed substrate preference for $C_{18}$ over $C_{20}$, $C_{22}$ or hydroxylated $C_{18}$ acyl chains (Zachowski, 1998). These *in vitro* data are consistent with a role for LTPs in cutin monomer transport ($C_{16}$ or $C_{18}$), but not transport of VLCFA-derived wax constituents. An alternative route for VLCFA derivatives to the cell surface may be through the hydrophobic subdomains present within predominantly hydrophilic cell wall. Hydrophobic passages could be created by proteins, such as glycine-rich proteins (GRP) that have been localized in the epidermis of maize and Arabidopsis. The GRPs have been predicted to form β-pleated sheets with a large hydrophobic surface (Sachetto-Martins *et al.*, 2000), and could facilitate the flow of a narrow 'river' of cuticular lipids through the cell wall. Since the typical sample preparations for TEM are 70 nm thick, these 'rivers', if they were in the size range of membranes (10 nm wide), would not be visible with conventional techniques, especially since saturated fatty acids react poorly with EM fixatives.

## 8.4 Cuticle composition and structure

The biosynthesis and secretion of cuticular components described above lead to their accumulation outside the epidermal cell walls during epidermis differentiation and expansion. These constituents were analyzed, and the overall composition of cuticles is known for many different plant species. In the past ten years, additional progress has been made by combining chemical and structural information so that we can now clearly distinguish three layers within the cuticle defined, respectively, by: intracuticular wax, epicuticular wax film and epicuticular wax crystals protruding from this film. Available evidence points to the fact that molecules of cuticular components spontaneously self-arrange, and then self-sort among these layers. In the following section, we attempt to summarize our current knowledge on (1) the composition of epicuticular wax crystals, (2) the processes leading to separation of epicuticular film and intracuticular wax and (3) the crystalline arrangement of molecules within all three cuticular compartments.

### *8.4.1 Formation and composition of epicuticular crystals*

Scanning electron microscopy (SEM) surveys of various plants reveal an enormous diversity in the shape, size and arrangement of three-dimensional structures protruding from the cuticle surface (Jeffree, 1996; Barthlott *et al.*, 1998). Amongst other terms, these structures have been classified as platelets, tubules, threads, and found to be characteristic for certain plant species, organs and in some instances, even individual organ surfaces. Based on these surveys, questions arose regarding the composition, organization, biogenesis and function of epicuticular crystals.

**Table 8.1** Compilation of chemical and crystallographic references for six selected types of epicuticular wax crystals. Scanning electron micrographs depict the typical shape of crystals on an exemplary plant species (indicated by *). Evidence for crystal-forming compounds in all examples comes from comparative SEM and analytical studies, but for some species additionally from recrystallization experiments with neat compounds (indicated by ¥) or direct probing of crystals (indicated by ‡).

| Crystal shape | Species | Selected references for wax analyses | Main crystal component | References for crystalline order |
|---|---|---|---|---|
| Platelets 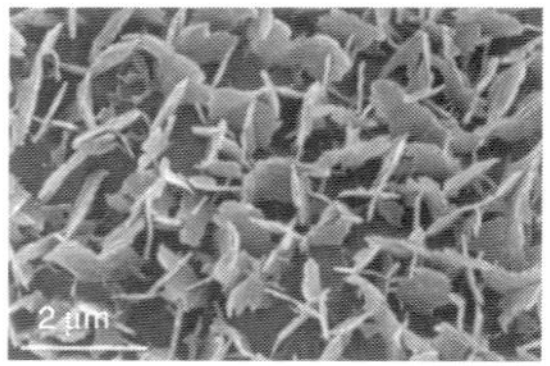 | *Pisum sativum** | Macey and Barber, 1970b | $CH_3-(CH_2)_{24}-CH_2-OH$ Hexacosanol | Kreger, 1948 |
| | *Iris germanica* | Gonzalez *et al.*, 1996 | | Kreger, 1948 |
| | *Hordeum vulgare* | Giese,1975 | | Kreger, 1948 |
| | *Allium porrum* | Rhee *et al.*, 1998 | $CH_3-(CH_2)_{14}-C(=O)-(CH_2)_{14}-CH_3$ Hentriacontan-16-one | Kreger, 1948 |
| | *Nepenthes alata* | Riedel *et al.*, 2003 ‡ | $CH_3-(CH_2)_{28}-C(=O)-H$ Triacontanal | N/A |
| Tubules 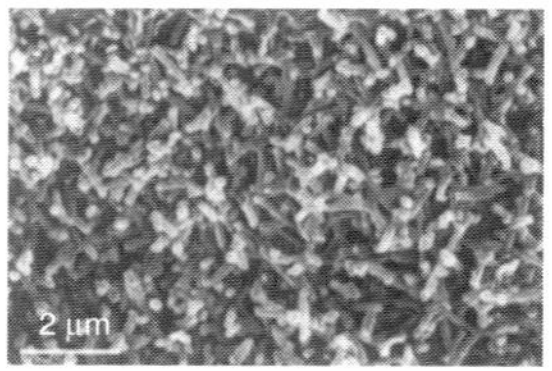 | *Papaver somniferum* | Jeffree *et al.*, 1975 ¥ | $CH_3-(CH_2)_8-CH(OH)-(CH_2)_{18}-CH_3$ Nonacosan-10-ol | Kreger, 1948 |
| | *Picea pungens** | Jetter and Riederer, 1994 ¥ | | Kreger, 1948 |
| | *Nelumbo nucifera* | Barthlott *et al.*, 1996 | | N/A |
| | *Pogonatum urnigerum* | Neinhuis and Jetter, 1995 | | N/A |
| | *Eucalyptus globulus* | Garrec *et al.*, 1995 | $CH_3-(CH_2)_{12}-C(=O)-(CH_2)-C(=O)-(CH_2)_{14}-CH_3$ Hentriacontane-14,16-dione | Kreger, 1948 |
| | *Secale cereale* | Dierickx, 1974 | | Kreger, 1948 |
| | *Hordeum vulgare* | von Wettstein-Knowles, 1972 | | Kreger, 1948 |
| Threads 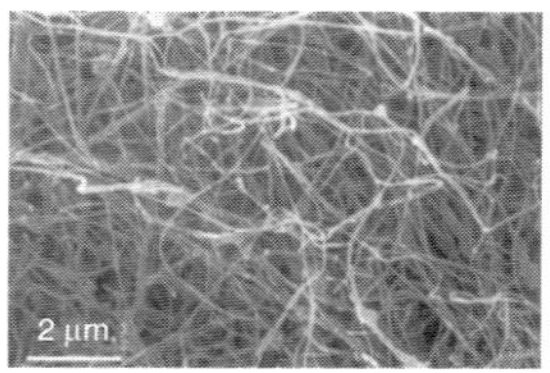 | *Dudleya brittonii* | Manheim and Mulroy, 1978 | 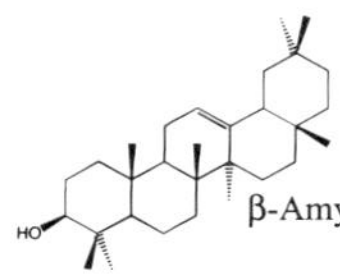 β-Amyrin 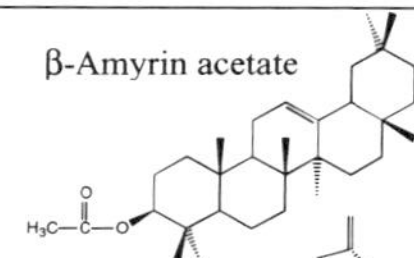 β-Amyrin acetate | Kreger, 1948 |
| | *Macaranga tanarius* | Markstädter *et al.*, 2000 | | N/A |
| | *Ricinus communis** | Hobl, Guhling and Jetter (unpublished) ‡ | 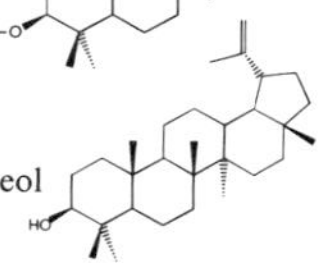 Lupeol | Kreger, 1948 |

**Table 8.1** *(continued)*

| | | | | |
|---|---|---|---|---|
| Dendrites 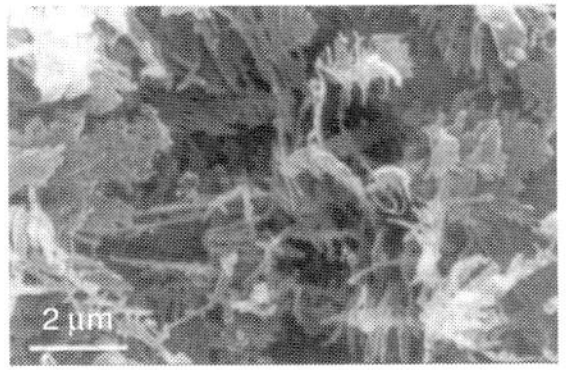 | *Brassica oleracea** | Macey and Barber, 1970a<br>Baker, 1974<br>Holloway *et al.*, 1977 | $CH_3-(CH_2)_{27}-CH_3$<br>Nonacosane<br>$CH_3-(CH_2)_{14}-C(=O)-(CH_2)_{14}-CH_3$<br>Nonacosan-15-one<br>$CH_3-(CH_2)_{14}-CH(OH)-(CH_2)_{14}-CH_3$<br>Nonacosan-15-ol | Kreger, 1948 |
| Longitudinal bundles of rodlets 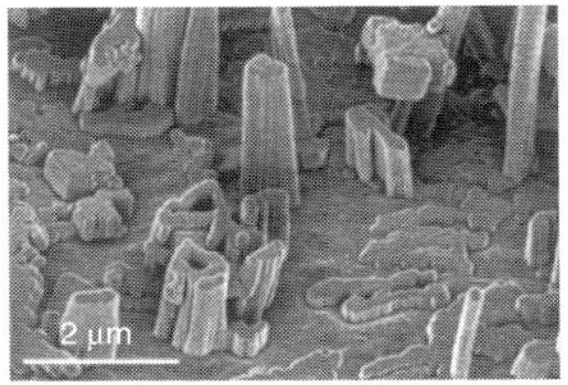 | *Arabidopsis thaliana** | McNevin *et al.*, 1993<br>Jenks *et al.*, 1995<br>Jenks *et al.*, 1996<br>Rashotte *et al.*, 1997<br>Rashotte *et al.*, 2001 | $CH_3-(CH_2)_{27}-CH_3$<br>Nonacosane<br>$CH_3-(CH_2)_{14}-C(=O)-(CH_2)_{14}-CH_3$<br>Nonacosan-15-one<br>$CH_3-(CH_2)_{14}-CH(OH)-(CH_2)_{14}-CH_3$<br>Nonacosan-15-ol | N/A |
| Transversely ridged rodlets  | *Aristolochia gigantea*<br>*Liriodendron tulipifera* | Meusel *et al.*, 1999¥<br>Gülz *et al.*, 1992 | $CH_3-(CH_2)_{14}-C(=O)-(CH_2)_{14}-CH_3$<br>Hentriacontan-16-one | Meusel *et al.*, 2000<br>Meusel *et al.*, 2000 |
| | *Gypsophila acuitfolia* | Meusel *et al.*, 1999¥ | $CH_3-(CH_2)_{25}-CH_3$<br>Hentriacontane | Meusel *et al.*, 2000 |
| | *Osmunda regalis** | Jetter and Riederer, 2000 | $CH_3-(CH_2)_8-C(=O)-(CH_2)_{18}-CH_3$<br>Nonacosan-10-one | N/A |

Compilation of chemical and crystallographic references for six selected types of epicuticular wax crystals. Scanning electron micrographs depict the typical shape of crystals on an exemplary plant species (indicated by *). Evidence for crystal-forming compounds in all examples comes from comparative SEM and analytical studies, but for some species, additionally, from recrystallization experiments with neat compounds (indicated by ¥) or direct probing of crystals (indicated by ‡).

The composition of the cuticular surface structures could not initially be assessed directly, but was inferred from the total wax composition. The wax mixtures were found to be very complex, containing diverse compound classes and chain lengths. Comparative studies performed over more than three decades revealed good correlations between characteristic shapes and the predominance of certain compounds in the wax mixture, e.g. for platelets and primary alcohols, tubules and nonacosan-10-ol, or threads and triterpenoids (Table 8.1). *In vitro* crystallization of pure compounds, or wax fractions containing them, yielded structures that closely matched those on the plant surfaces in shape, size and arrangement (Jeffree *et al.*, 1975; Jetter and Riederer, 1994; Meusel *et al.*, 1999), confirming that the presence of these components in high concentration was the sole and sufficient prerequisite for formation of the epicuticular wax crystals. Thus, the surface chemistry for plants carrying characteristic three-dimensional structures could be described, though indirectly, for the first time. Alternative hypotheses, claiming that surface shapes are created by extrusion of wax through pores of characteristic geometries (Hall and Donaldson, 1962) were ruled out.

Direct confirmation of the predominance of characteristic individual compounds in the surface crystals required the development of methods for the selective sampling of epicuticular wax. In most of the previous experiments, the total wax mixtures had been prepared by superficial extraction of the intact tissue. The solvents used to extract cuticular wax, however, not only mobilized epicuticular wax, but also entered into the cuticle and released intracuticular wax (Jetter *et al.*, 2000). Thus, it was not clear whether the composition of the epicuticular wax structures differed from the wax mixtures from which they protrude, and from the intracuticular layer underneath. Jeffree (1996) and later Ensikat *et al.* (2000) showed that frozen droplets of hydrophilic liquids can be used to mechanically lift epicuticular wax crystals from plant surfaces and transfer them onto artificial substrates, perfectly preserving their shape and arrangement. This new method was further developed to allow separate sampling and quantification of all epi- and intracuticular wax constituents (Jetter *et al.*, 2000).

The mechanical sampling technique has been applied in studies on the composition of epicuticular wax crystals of *Nepenthes alata* pitchers (Riedel *et al.*, 2003). A broad zone of the inner wall of these pitchers is densely covered with wax platelets, creating a microscopically rough surface that hampers adhesion of insect feet (Knoll, 1914). Previous studies, based on extraction of total waxes, did not allow the identification of the wax constituents that are responsible for the formation of these special surface features. The selective removal of epicuticular waxes, employing droplets of frozen water as a cryo-adhesive, demonstrated that the wax platelets consisted predominantly of triacontanal, i.e. $C_{30}$ aldehyde (Riedel *et al.*, 2003). Small amounts of primary alcohols with a broad chain length distribution (maxima for $C_{24}$ and $C_{30}$) and the $C_{24}$ fatty acid homolog were also detected. In contrast, the crystal-forming aldehydes were found in much lower concentration in the intracuticular wax of *N. alata*. These data represent the first direct evidence that the accumulation of a single compound is responsible for the formation of unique surface wax structures. The results also showed that wax platelets that are virtually identical in shape are formed on different plant species either by primary alcohols or by aldehydes (Table 8.1).

For plant systems with epicuticular wax crystals that contain individual constituents at high concentrations, crystal formation has been described as a spontaneous physical process based on phase separation within the cuticular wax mixture (Jetter and Riederer, 1994). The crystal-forming compound has to be accumulated above a critical concentration in the mixture before it starts to form a separate solid phase. From comparative studies of related plant species with or without epicuticular wax crystals, the threshold concentrations for some crystal-forming compounds can be inferred. Thus, Jetter and Riederer (1996) estimated that nonacosan-10-ol must be accumulated above 40% of the wax mixture before tubular crystals start to form. Beyond that threshold composition, crystal-forming compounds will accumulate in the crystals while the surrounding mixture is deprived of them. The crystal composition, consequently, differs drastically from that of the bulk wax.

Single chemical constituents, however, are not likely to be the sole determinants of the shape of epicuticular wax crystals, as there are instances where a clear correlation between individual dominating wax constituents and shapes could not be established. For example, there exists conflicting chemical evidence for dendrites, longitudinal bundles of rodlets and transversely ridged rodlets (Table 8.1): (1) a combination of several compounds, in varying relative amounts is likely involved in the formation of surface structures on *B. oleracea* and Arabidopsis; (2) although both these species share similar wax composition, they have very distinct surface features; (3) while transversely ridged rodlets on the surfaces of diverse plant species look very similar, they consist of at least three different compounds; (4) one of these compounds, hentriacontan-16-one probably causes the formation of platelet-shaped aggregates (instead of rodlets) on *Allium porrum*. All three types of structures are further characterized by the relatively complex geometry consisting of aggregated substructures that point to fluctuations in the crystallization process. Based on this information, it seems likely that these shapes are a product of local crystallization conditions rather than chemical composition.

Epicuticular wax crystals cause diffuse light scattering, giving respective plant surfaces a whitish (glaucous) appearance, while smooth surfaces, devoid of crystals, are shiny (glossy). All the wax biosynthetic *mutants* to date have been identified in visual screening approaches that relied on the macroscopic manifestation of epicuticular wax crystals (Lemieux *et al.*, 1994; von Wettstein-Knowles, 1995; Post-Beittenmiller, 1996). As discussed above, epicuticular crystals are usually formed by single wax constituents that accumulate to high levels in the wax mixture. This can be accomplished by high flux of metabolites through all early steps of the pathway and its abrupt termination at a maximum chain length and/or at a final modification step. Therefore, epicuticular crystals often consist of aliphatic compounds that are end products of respective biosynthetic pathways, such as β-diketones in *Hordeum vulgare* or alkanes plus secondary alcohols plus ketones in Arabidopsis. Consequently, the easiest way to abolish wax crystal accumulation is by downregulating the pathway, either entirely or at least in early steps. This is one of the reasons why previous mutant screens, based on changes in crystal coverage, recovered predominantly regulatory *mutants* and why relatively few *mutants* with lesions in structural pathway genes have been found.

### *8.4.2 Physical and chemical distinction between epicuticular film and intracuticular wax*

Another important question is whether spontaneous phase separation also gives rise to chemical gradients between the intracuticular wax and the overlying epicuticular wax film. To answer this question, it was initially attempted to distinguish between intracuticular waxes and the epicuticular wax film using variations of solvent extraction methods (Silva Fernandes *et al.*, 1964; Baker and Procopiou, 1975). Wax released by very brief (surface) extraction of the intact tissue was considered as epicuticular, while thorough extraction of isolated cuticular membranes was assumed to yield the intracuticular wax constituents. Alternatively, films of cellulose acetate or nitrocellulose were employed to mechanically remove surface waxes (Baker *et al.*, 1983; Haas and Rentschler, 1984). As these polymer films have to be applied in the presence of substantial amounts of organic solvents, intra- and epicuticular waxes could be mixed in the course of probing. Hence, none of these methods had sufficient selectivity and the individual constituents could not be localized unambiguously. Nonetheless, in all studies, gradients in the percentages of individual compounds were detected, suggesting chemical differences between the intracuticular wax and the epicuticular film.

Development of the mechanical sampling technique (see Section 8.4.1) allowed for the first time the selective removal of epicuticular wax films from plant surfaces. This new method was employed to probe the smooth epicuticular wax film on *Prunus laurocerasus* leaves (Jetter *et al.*, 2000) and to quantify its constituents. As with the epicuticular crystals, repeated cryo-adhesive stripping of the same tissue surface yielded rapidly decreasing amounts of wax. After mechanically removing the epicuticular wax, the remaining cuticle was extracted, which gave high amounts of intracuticular wax. Steep gradients between both layers of cuticular waxes were detected, proving that the mechanical sampling technique has much higher selectivity than the previously used extractive protocols. Based on these quantitative results, the selectivity of other mechanical sampling techniques could be tested, and finally aqueous gum arabic was established as a second adhesive for the selective removal of epicuticular waxes (Jetter and Schäffer, 2001).

Employing cryo-adhesive sampling, adaxial cuticles of mature *Prunus* leaves were found to have 13 $\mu g/cm^2$ of epicuticular waxes (Jetter *et al.*, 2000). As the total wax coverage was 28 $\mu g/cm^2$ and the cutin matrix had been characterized with 333 $\mu g/cm^2$ (Schreiber and Riederer, 1996), the epicuticular wax film accounted for less than 4% of the cuticular material. Based on the amount of material removed, the thickness of the epicuticular layer could be predicted, and the calculated value agreed well with the thickness measured using SEM. The epicuticular wax mixture consisted exclusively of aliphatic constituents in a characteristic pattern of $C_{25}$–$C_{33}$ alkanes, $C_{22}$–$C_{36}$ alcohols, $C_{24}$–$C_{32}$ aldehydes, $C_{20}$–$C_{34}$ fatty acids and $C_{22}$–$C_{36}$ alcohol acetates. In sharp contrast, the intracuticular wax contained large quantities of alicyclic triterpenoids (63% of the fraction) in addition to those aliphatic compounds. With the long-chain molecules located outside the cutin matrix and the cyclic terpenoids located inside the cutin matrix, it is clear that two chemically and morphologically distinct layers are present within the *Prunus* leaf cuticle. In several other species, a similar segregation of triterpenoids within the intracuticular layer has been found, while epicuticular wax mixtures again consisted entirely of aliphatics (Jetter, unpublished results).

This layered structure of cuticular waxes could be due to the sequential formation and/or deposition of the different compounds. In this case, the aliphatic compounds found on the surface of mature leaves should appear first, i.e. already in immature leaves, while the triterpenoids would be deposited in a lower layer during maturation. To test this hypothesis, the ontogenetic development of *Prunus* leaf surfaces was monitored over 60 days, during expansive epidermal cell growth and maturation (Jetter and Schäffer, 2001). Epicuticular wax coverage increased from 0 to 6 $\mu g/cm^2$ during this interval, in good accordance with a corresponding film thickness of 25 nm at day 13 of the time course to more than 100 nm at day 60 (Figure 8.5). In the course of leaf development, the epicuticular wax composition changed continuously going through phases dominated by very-long-chain alcohol acetates, free alcohols and, finally, alkanes. The composition of the intracuticular wax, however, did not change, the triterpenoids being present in very high amounts during all stages of leaf development. These results showed that chemical differences between epi- and intracuticular waxes are not due to sequential addition of layers during development, but must arise from diffusion of compounds and spontaneous phase separation. The diffusion of very-long-chain aliphatic molecules in wax mixtures (Schreiber and Schönherr, 1993) is fast enough to account for the developmental changes observed in the *Prunus* leaf cuticle.

### *8.4.3 Crystalline arrangement of epi- and intracuticular wax molecules*

Many paraffinic mixtures, e.g. synthetic Fischer-Tropsch waxes, contain wide ranges of aliphatic chain lengths. FTIR, NMR, X-ray and Differential Scanning Calorimetry (DSC) studies have revealed that a part of the methylene groups in such mixtures is packed in a dense lattice, while chain ends of the hydrocarbon molecules are in a much less ordered, amorphous state (Le Roux, 1969; Basson and Reynhardt, 1992). Accordingly, depending on the chain length distribution and the temperature, only a partial volume of the solid is crystalline. Typical wax characteristics, e.g. softness and broad melting ranges, are due to the amorphous

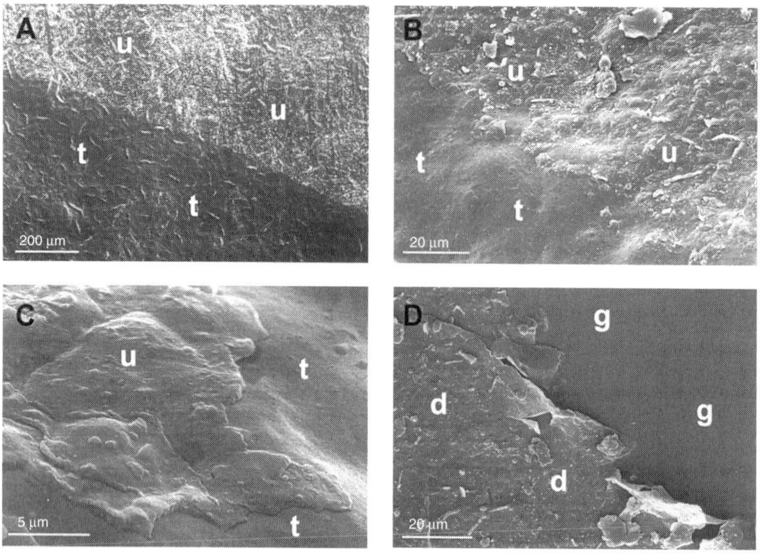


**Figure 8.5** Scanning electron micrographs of adaxial *Prunus laurocerasus* cuticles. A, B, C. Outer segment of the leaf area that had been treated with frozen glycerol (t) and the adjacent area that had been frozen but not treated with glycerol (u). D. Film of epicuticular waxes (d) transferred onto glass (g) using frozen glycerol.

portions of the solid. Merely based on the general chemical composition of plant cuticular waxes, consisting of very-long-chain aliphatic mixtures with (in some cases) extremely broad chain length distributions, it is frequently assumed that they should be amorphous throughout and exhibit properties of soft, 'waxy' material. Physical data on the solid state structure of these lipids that allow us to assess their lattice order, and to quantify the relative portions of crystalline and amorphous zones of these mixtures at physiological temperatures are available.

Early work by Chibnall *et al.*(1934) and Kreger (1948) investigated the molecular packing arrangements in plant wax using X-ray powder diffractograms. A systematic comparison of various plant waxes with synthetic compounds showed that the cuticular aliphatics are in a linear conformation and aligned in parallel (Piper *et al.*, 1931, 1934; Kreger and Schamhart, 1956). All investigated plant waxes yielded sharp signals, indicating lattices that are highly ordered in all three directions of space (Kreger, 1948). Thus, the diffractograms confirmed the principal crystalline nature of plant cuticular waxes. However, amorphous zones in the wax mixtures were neglected, and, hence, the relative volumes of crystalline and amorphous domains remained unclear.

This gap can now be filled by integrating the available crystallographic results with chemical information on epicuticular waxes (see Section 8.4.1). Overlapping sets of compositional data and X-ray diffractograms have been published for several plant species representing various types of surface shapes, e.g. platelets, tubules and threads (Table 8.1). It can be assumed that in these cases, the epicuticular wax layer contained, by far, the largest portion of the wax mixture, and, therefore, the crystallographic data largely reflect the molecular arrangement in respective surface wax structures. The results, without exception, confirm the view that these epicuticular structures contain predominant single components and exhibit relatively little chain length variation. The structures have high lattice order even in the regions of chain termini and consequently relatively large crystalline domains. Comparatively small amorphous domains are due to the co-crystallizing admixtures that are present at low percentages. Therefore, to appropriately describe the high degree of molecular order in all these surface structures, they should be called 'epicuticular wax crystals'.

Sometimes inhomogeneities in the aggregation medium that cause irregular crystal outlines, might also considerably disturb the solid state order of some structures, casting doubts on their crystalline nature. X-ray data suggest, however, that at least the dendrites on leaves of *B. oleracea* (Kreger, 1948) have substantial lattice order. In addition, Meusel *et al.* (2000) investigated original wax and recrystallized mixtures of two species characterized by transversely ridged rodlets. All samples exhibited high lattice order, demonstrating that these aggregates are also true crystals.

Finally, the arrangement of molecules within solid phases of intracuticular wax can be evaluated. Respective wax mixtures often contain diverse compound classes with relatively broad chain length distribution, and should, therefore, have relatively small crystalline domains restricted to the central portions of hydrocarbon chains. DSC, X-ray diffraction, solid state NMR and FTIR techniques have been employed to distinguish between amorphous and crystalline domains in the wax and to quantify their relative amounts for species that are devoid of surface crystals. The wax, allegedly largely intracuticular, showed crystallinities of aliphatic components between 20% (*Citrus aurantium*), 30% (*Fagus sylvatica*), 58% (*Juglans regia*) and 74% (*Hedera helix*) (Reynhardt and Riederer, 1991, 1994; Merk *et al.*, 1998). Triterpenoids and aliphatic molecules were found to form separate solid phases (Casado and Heredia, 1999; Merk *et al.*, 1998). Thus, the intracuticular wax has been described as a mosaic of submicroscopic solid domains, a relatively large part of which is amorphous. Diffusion of molecules and spontaneous phase separation suffice to explain the segregation into these crystalline aliphatic, amorphous aliphatic and (partially?) crystalline triterpenoid domains. Currently nothing is known about the geometry and relative arrangement of these zones within the intracuticular layer.

In summary, spontaneous self-arrangement of wax molecules leads to similar crystalline order within both the epicuticular crystals and the intracuticular compartment. This raises the question whether continuous crystalline zones exist between both layers. To date, no studies that address the crystalline nature of intracuticular wax, epicuticular film and surface crystals separately in one species have been performed. Future investigations should focus on the molecular arrangement in these layers, possibly employing wax *mutants*, to judge the continuity between crystalline domains in the different compartments.

## 8.5 Conclusions

The biochemical pathways for the production of cuticular lipids involve synthesis of carbon skeletons and their modification to produce diverse aliphatic structures. These basic pathways have been defined by biochemical and genetic approaches. New information concerning the enzymes involved in these pathways and their regulation is currently emerging from molecular and genomic studies in Arabidopsis and maize. In the near future, similar information will be available from other plant species, and should greatly contribute to our overall understanding of the diversity of plant cuticles.

Trafficking and export of cuticular lipids remain enigmatic. We still do not know how hydrophobic wax components are moved intracellularly, how they are exported out of the cell or translocated through the hydrophilic cell wall. Biochemical, genetic and genomic approaches in Arabidopsis promise to be particularly useful in identifying and characterizing gene products involved in these processes.

In contrast, good experimental evidence has been accumulating for the self-arrangement of cuticular lipids to create an intracuticular wax layer, an epicuticular wax film and, in many cases, also epicuticular wax crystals, as chemically and morphologically distinct compartments. Depending on their concentration and solubility in the mixture, single compounds or groups of compounds can segregate into separate solid phases. All the resulting phases may have crystalline order, and those exposed at the cuticle surface will be visible as epicuticular wax crystals with characteristic shapes. Depending on their chemical composition, the individual solid domains might have different affinities for the carbohydrate fibrils and cutin matrix present in the cuticular layer and cuticle proper, respectively. Accordingly, they will be arranged predominantly in one of these layers, or in the epicuticular wax to form the protective coating we know as the cuticle.

## Acknowledgements

We thank Ortwin Guhling, Jamie Pighin, Markus Riederer, Owen Rowland and Cornelia Vermeer for stimulating discussions, and Gail Kenner for critical reading of the manuscript. We also thank Franka Gniwotta for providing SEMs of *Pisum sativum*. The Natural Sciences and Engineering Research Council of Canada and the Deutsche Forschungsgemeinschaft are acknowledged for financial support.

## References

Aarts, M.G., Keijzer, C.J., Stiekema, W.J. and Pereira, A. (1995) Molecular characterization of the *CER1* gene of *Arabidopsis* involved in epicuticular wax biosynthesis and pollen fertility. *Plant Cell*, **7**, 2115–2127.

Arondel, V., Vergnolle, C., Cantrel, C. and Kader, J.-C. (2000) Lipid transfer proteins are encoded by a small multigene family in *Arabidopsis thaliana*. *Plant Science*, **157**, 1–12.

Bagnat, M., Kernen, S., Shevchenko, A., Shevchenko, A. and Simons, K. (2000) Lipid rafts function in biosynthetic delivery of proteins to the cell surface in yeast. *Proceedings of the National Academy of Sciences USA*, **97**, 3254 -3259.

Baker, E.A. (1974) The influence of environment on leaf wax development in *Brassica oleracea* var. *gemmifera*. *The New Phytologist*, **73**, 955–966.

Baker, E.A. and Procopiou, J. (1975) The cuticles of *Citrus* species. Composition of the intracuticular lipids of leaves and fruits. *Journal of the Science of Food and Agriculture*, **26**, 1347–1352.

Baker, E.A., Hunt, G.M. and Stevens, P.J.G. (1983) Studies of plant cuticle and spray droplet interactions: a fresh approach. *Pesticide Science*, **14**, 645–658.

Barret, P., Delourme, R., Renard, M. *et al.* (1998) A rapeseed *FAE1* gene is linked to the E1 locus associated with variation in the content of erucic acid. *Theoretical and Applied Genetics. Theoretische und Angewandte Genetik*, **96**, 177–186.

Barthlott, W., Neinhuis, C., Cutler, D. *et al.* (1998) Classification and terminology of plant epicuticular waxes. *Botanical Journal of the Linnean Society*, **126**, 237–260.

Barthlott, W., Neinhuis, C., Jetter, R., Bourauel, T. and Riederer, M. (1996) Waterlily, poppy, or sycamore: on the systematic position of *Nelumbo*. *Flora*, **191**, 169–174.

Basson, I. and Reynhardt, E.C. (1992) The structure and melting of paraffinic Fischer-Tropsch waxes. *Chemical Physics Letters*, **198**, 367–372.

Beaudoin, F., Gable, K., Sayanova, O., Dunn, T. and Napier, J.A. (2002). A *Saccharomyces cerevisiae* gene required for heterologous fatty acid elongase activity encodes a microsomal beta-keto-reductase. *Journal of Biological Chemistry*, **277**, 11481–11488.

Beisson, F., Koo, J.J.K., Ruuska, S. *et al.* (2003) *Arabidopsis* genes involved in acyl lipid metabolism. A 2003 census of the candidates, a study of the distribution of ESTs in organs, and a web-based database. *Plant Physiology*, **132**, 681–697.

Benveniste, I., Tijet, N., Adas, F., Philipps, G., Salaun, J.P. and Durst, F. (1998) CYP86A1 from *Arabidopsis thaliana* encodes a cytochrome P450-dependent fatty acid omega-hydroxylase. *Biochemical and Biophysical Research Communications*, **243**, 688–693.

Bernhard, W.R., Thoma, S., Botella, J. and Somerville, C.R. (1991) Isolation of a cDNA clone for spinach lipid transfer protein and evidence that the protein is synthesized by the secretory pathway. *Plant Physiology*, **95**, 164–170.

Bessoule, J.J., Lessire, R. and Cassagne, C. (1989) Partial purification of the acyl-CoA elongase of *Allium porrum* leaves. *Archives of Biochemistry and Biophysics*, **268**, 475–484.

Bianchi, A., Bianchi, G., Avato, P. and Salamini, F. (1985) Biosynthetic pathways of epicuticular wax in maize as assessed by mutation, light, plant age and inhibitor studies. *Maydica*, **30**, 179–198.

Bianchi, G., Avato, P. and Salamini, F. (1979) Glossy mutants of maize. IX. Chemistry of *glossy4, glossy8, glossy15, and glossy18* surface waxes. *Heredity*, **42**, 391–395.

Blacklock, B. and Jaworski, J.G. (2002) Studies into factors contributing to substrate specificity of membrane-bound 3-ketoacyl-CoA synthases. *European Journal of Biochemistry*, **269**, 4789–4798.

Blee, E. and Schuber, F. (1993) Biosynthesis of cutin monomers: involvement of a lipoxygenase/peroxygenase pathway. *Plant Journal*, **4**, 113–123.

Bonaventure, G., Salas, J.J., Pollard, M.R. and Ohlrogge, J.B. (2003) Disruption of the *FATB* gene in *Arabidopsis* demonstrates an essential role of saturated fatty acids in plant growth. *Plant Cell*, **15**, 1020–1033.

Brown, D.A. and London, E. (2000) Structure and function of sphingolipid- and cholesterol-rich membrane rafts. *Journal of Biological Chemistry*, **275**, 17221–17224.

Cabello-Hurtado, F., Batard, Y., Salaun, J.P., Durst, F., Pinot, F. and Werck-Reichart, D. (1998) Cloning, expression in yeast, and functional characterization of CYP81B1, a plant cytochrome P450 that catalyzes in-chain hydroxylation of fatty acids. *Journal of Biological Chemistry*, **273**, 7260–7267.

Casado, C.G. and Heredia, A. (1999) Structure and dynamics of reconstituted cuticular waxes of grape berry cuticle (*Vitis vinifera* L.). *Journal of Experimental Botany*, **50**, 175–182.

Cheesbrough, T.M. and Kolattukudy, P.E. (1984) Alkane biosynthesis by decarbonylation of aldehydes catalyzed by a particulate preparation from *Pisum sativum*. *Proceedings of the National Academy of Sciences USA,* **81**, 6613–6617.

Chen, X., Goodwin, M., Boroff, V.L., Liu, X. and Jenks, M.A. (2003) Cloning and characterization of the *WAX2* gene of *Arabidopsis* involved in cuticle membrane and wax production. *Plant Cell*, **15**, 1170–1185.

Chibnall, A.C., Piper, S.H., Pollard, A., Williams, E.F. and Sahai, P.N. (1934) The constitution of the primary alcohols, fatty acids and paraffins present in plant and insect waxes. *Biochemical Journal*, **28**, 2189–2208.

Clark, A.M. and Bohnert, H.J. (1999) Cell-specific expression of genes of the lipid transfer protein family from *Arabidopsis thaliana*. *Plant & Cell Physiology*, **40**, 69–76.

Clough R.C., Matthis A.L., Barnum S.R. and Jaworski J.G. (1992) Purification and characterization of 3-ketoacyl-acyl carrier protein synthase from spinach: a condensing enzyme utilizing acetyl-CoA to initiate fatty acid synthesis. *Journal of Biological Chemistry*, **267**, 20992–20998.

Croteau, R. and Kolattukudy, P.E. (1973) Enzymatic biosynthesis of a hydroxyl fatty acid polymer, cutin, by a particulate preparation from *Vicia faba* epidermis. *Biochemical and Biophysical Research Communications*, **52**, 863–869.

Croteau, R. and Kolattukudy, P.E. (1975) Biosynthesis of hydroxyl fatty acid polymers. Enzymatic epoxidation of 18-hydroxyoleic acid to 18-hydrohy-cis-9,10-epoxystearic acid by a particular preparation from spinach (*Spinacea oleracea*). *Archives of Biochemistry and Biophysics*, **170**, 61–72.

Craig, S. and Staehelin, L.A. (1988) High pressure freezing of intact plant tissues. Evaluation and characterization of novel features of the endoplasmic reticulum and associated membrane systems. *European Journal of Cell Biology*, **46**, 80–93.

Daes, A. and Holloway, P. (1980) The intermolecular structure of some plant cutins, in *Lipids and Lipid Polymers in Higher Plants* (eds M. Tevini and H. Lichtenthaler), Springer Verlag, Berlin, pp. 293–299.

Dean, M., Rzhetsky, A. and Allikmets, R. (2001) The human ATP-binding cassette (ABC) transporter superfamily. *Genome Research,* **11**, 1156–1166.

Dennis, M. and Kolattukudy, P.E. (1992) A cobalt-porphyrin enzyme converts a fatty aldehyde to a hydrocarbon and CO. *Proceedings of the National Academy of Sciences USA,* **89**, 5306–5310.

Dierickx, P. (1974) Waxy beta-diketones from *Secale cereale*. *Phytochemistry*, **13**, 682–683.

Engeseth, N.J., Pacovsky, R.S., Newman, T. and Ohlrogge, J. (1996) Characterization of an acyl-CoA-binding protein from *Arabidopsis thaliana*. *Archives of Biochemistry and Biophysics*, **331**, 55–62.

Ensikat, H.J., Neinhuis, C. and Barthlott, W. (2000) Direct access to plant epicuticular wax crystals by a new mechanical isolation method. *International Journal of Plant Sciences*, **161**, 143–148.

Fiebig, A., Mayfield, J.A., Miley, N.L., Chau, S., Fischer, R.L. and Preuss, D. (2000) Alterations in *CER6*, a gene identical to *CUT1*, differentially affect long-chain lipid content on the surface of pollen and stems. *Plant Cell*, **12**, 2001–2008.

Gaigg, B., Neergaard, T.B.F., Schneiter, R. *et al.* (2001) Depletion of acyl-CoA-binding protein affects sphingolipid synthesis and causes vesicle accumulation and membrane defects in *Saccharomyces cerevisiae*. *Molecular Biology of the Cell*, **12**, 1147–1160.

Garrec, J.P., Henry, C. and Le Maout, L. (1995) Cires epi- et intracuticulaires: etude de leur separation, de leurs caracteristiques chimiques et de leurs roles respectifs dans la permeabilite cuticulaire. *Environmental and Experimental Botany*, **35**, 399–409.

Gershenzon, J., McConkey, M.E. and Croteau, R.B. (2000) Regulation of monoterpene accumulation in leaves of peppermint. *Plant Physiology*, **122**, 205–213.

Ghanevati, M. and Jaworski, J.G. (2001) Active-site residues of a plant membrane-bound fatty acid elongase beta-ketoacyl-CoA synthase, FAE1. *Biochimica et Biophysica Acta*, **1530**, 77–85.

Giese, B.N. (1975) Effects of light and temperature on the composition of epicuticular wax of barley leaves. *Phytochemistry,* **14**, 921–929.

Gomar, J., Sodano, P., Sy, D. *et al.* (1998) Comparison of solution and crystal structures of maize nonspecific protein: a potential model for *in vivo* lipid carrier protein. *Proteins*, **31**, 160–171.

Gonzalez, R., Paul, N.D., Percy, K. *et al.* (1996) Responses to ultraviolet-B radiation (280–315 nm) of pea (*Pisum sativum*) lines differing in leaf surface wax. *Physiologia plantarum*, **98**, 852–860.

Goossens, A., Hakkinen, S.T., Laakso, I., Oksman-Caldentey, K.-M. and Inzé, D. (2003) Secretion of secondary metabolites by ATP-binding cassette transporters in plant cell suspension cultures. *Plant Physiology*, **131**, 1161–1164.

Grabsky, S., de Feijter, A.W. and Schindler, M. (1993) Endoplasmic reticulum forms a dynamic continuum for lipid diffusion between contiguous soybean root cells. *Plant Cell*, **5**, 25–38.

Graça, J., Schreiber, L., Rodrigues, L. and Pereira, H. (2002) Glycerol and glyceryl esters of ω-hydroxyacids in cutins. *Phytochemistry*, **61**, 205–215.

Gülz, P.-G., Müller, E., Schmitz, K., Marner, F.-J. and Güth, S. (1992) Chemical composition and surface structures of epicuticular leaf waxes of *Ginkgo biloba, Magnolia grandiflora* and *Liriodendron tulipifera. Zeitrschrift für Naturforschung C*, **47**, 516–526.

Haas, K. and Rentschler, I. (1984) Discrimination between epicuticular and intracuticular wax in blackberry leaves: ultrastructural and chemical evidence. *Plant Science Letters*, **36**, 143–147.

Hall, D.M. and Donaldson, L.A. (1962) Secretion from pores of surface wax on plant leaves. *Nature*, **194**, 1196.

Hamilton, J.A. (1998) Fatty acid transport: difficult or easy? *Journal of Lipid Research*, **39**, 467–481.

Heredia, A. (2003) Biophysical and biochemical characteristics of cutin, a plant barrier biopolymer. *Biochimica et Biophysica Acta*, **1620**, 1–7.

Hollenbach, B., Schreiber, L., Hartung, W. and Dietz, K.-J. (1997) Cadmium leads to stimulated expression of the lipid transfer protein genes in barley: implications for the involvement of lipid transfer proteins in wax assembly. *Planta*, **203**, 9–19.

Holloway, P.J. (1982) The chemical constitution of plant cutins, in *The Plant Cuticle* (eds D. Cutler, K. Alvin and C. Price), Linnean Society Symposium Series, Academic Press, London, pp. 45–85.

Holloway, P.J., Brown, G.A., Baker, E.A. and Macey, M.J.K. (1977) Chemical composition and ultrastructure of the epicuticular wax in three lines of *Brassica napus* (L.). *Chemical Physics Letters*, **19**, 114–127.

Hooker, T.S., Millar, A.A. and L. Kunst (2002) Significance of the expression of the CER6 (=CUT1) condensing enzyme for epicuticular wax production in *Arabidopsis*. *Plant Physiology*, **129**, 1568–1580.

Hopwood, D.A. and Sherman, D.H. (1990) Molecular genetics of polyketides and its comparison to fatty acid biosynthesis. *Annual Review of Genetics*, **4**, 37–66.

James, D.W., Lim, E., Keller, J., Plooy, I., Ralston, E. and Dooner, H.K. (1995) Directed tagging of the *Arabidopsis FATTY ACID ELONGATION1 (FAE1)* gene with the maize transposon activator. *Plant Cell*, **7**, 309–319.

Jasinski, M., Stukkens, Y., Degand, H., Purnelle, B., Marchand-Brynaert, J. and Boutry, M. (2001) A plant plasma membrane ATP Binding Cassette-type transporter is involved in antifungal terpenoid secretion. *Plant Cell,* **13**, 1095–1107.

Jaworski, J.G., Post-Beittenmiller, D. and Ohlrogge, J.B. (1993) Acetyl-acyl carrier protein is not a major intermediate in fatty acid biosynthesis in spinach. *European Journal of Biochemistry*, **213**, 981–987.

Jeffree, C.E. (1996) Structure and ontogeny of plant cuticles, in *Plant Cuticles. An Integrated Approach* (ed. G. Kerstiens), BIOS Scientific, Oxford, pp. 33–82.

Jeffree, C.E., Baker, E.A. and Holloway, P.J. (1975) Ultrastructure and recrystallisation of plant epicuticular waxes. *The New Phytologist*, **75**, 539–549.

Jenks, M.A., Eigenbrode, S.D. and Lemieux, B. (2002) Cuticular waxes of arabidopsis, in *The Arabidopsis Book* (eds C.R. Somerville and E.M. Meyerowitz), American Society of Plant Biologists, Rockville, MD, doi/10.1199/tab.0016.

Jenks, M.A., Rashotte, A.M., Tuttle, H.A. and Feldmann, K.A. (1996) Mutants in *Arabidopsis thaliana* altered in epicuticular wax and leaf morphology. *Plant Physiology*, **110**, 377–385.

Jenks, M.A., Rich, P.J. and Ashworth, E.N. (1994) Involvement of cork cells in the secretion of epicuticular wax filaments on *Sorghum bicolor* (L.) Moench.*International Journal of Plant Sciences*, **155**, 506–518.

Jenks, M.A., Tuttle, H.A., Eigenbrode, S.D. and Feldmann, K.A. (1995) Leaf epicuticular waxes of the *eceriferum* mutants in *Arabidopsis*. *Plant Physiology*, **108**, 369–377.

Jetter, R. and Riederer, M. (1994) Epicuticular crystals of nonacosan-10-ol:*in vitro* reconstitution and factors influencing crystal habits. *Planta*, **195**, 257–270.

Jetter, R. and Riederer, M. (1996) Cuticular waxes from the leaves and fruit capsules of eight Papaveraceae species. *Canadian Journal of Botany*, **74**, 419–430.

Jetter, R. and Riederer, M. (2000) Composition of cuticular waxes on *Osmunda regalis* L. fronds. *Journal of Chemical Ecology*, **26**, 399–412.

Jetter, R. and Schäffer, S. (2001) Chemical composition of the *Prunus laurocerasus* leaf surface. Dynamic changes of the epicuticular wax film during leaf development. *Plant Physiology*, **126**, 1725–1737.

Jetter, R., Schäffer, S. and Riederer, M. (2000) Leaf cuticular waxes are arranged in chemically and mechanically distinct layers: evidence from *Prunus laurocerasus* L. *Plant, Cell & Environment*, **23**, 619–628.

Johnson, P.E., Rawsthorne S. and Hills M.J. (2002) Export of acyl chains from plastids isolated from embryos of *Brassica napus* (L.). *Planta,* **215**, 515–517.

Joyard, J. and Stumpf, P.K. (1981) Synthesis of long chain acyl coenzyme A in chloroplast envelopes. *Plant Physiology*, **67**, 250–256.

Kader, J.-P. (1996) Lipid transfer proteins in plants. *Annual Review of Plant Physiology and Plant Molecular Biology*, **47**, 627–654.

Kaneko, Y. and Keegstra, K. (1996) Plastid biogenesis in embryonic pea leaf cells during early germination. *Protoplasma*, **195**, 59–67.

Keresztessy, Z. and Hughes, M.A. (1998) Homology modelling and molecular dynamics aided analysis of ligand complexes demonstrates functional properties of lipid-transfer proteins encoded by the barley low-temperature inducible gene family, blt4. *Plant Journal*, **14**, 523–533.

Kerstiens, G., ed. (1996) *Plant Cuticles*, BIOS Scientific Publishers, Oxford.

Khan, A.A. and Kolattukudy, P.E. (1973) Control of synthesis and distribution of acyl moieties in etiolated *Euglena gracilis. Biochemistry*, **12**, 1939–1948.

Knoll, F. (1914) Über die Ursache des Ausgleitens der Insektenbeine an wachsbedeckten Pflanzenteilen. *Jahrbücher für Wissenschaftliche Botanik*, **54**, 448–459.

Kohlwein, S.D., Eder, S., Oh, C.S., Martin, C.E., Gable, K., Bacikova, D. and Dunn, T. (2001) Tsc13p is required for fatty acid elongation and localizes to a novel structure at the nuclear-vacuolar interface in *Saccharomyces cerevisiae. Molecular and Cellular Biology*, **21**, 109–125.

Kolattukudy, P.E. (1967) Mechanisms of synthesis of waxy esters in broccoli (*Brassica oleracea*). *Biochemistry*, **6**, 2705–2717.

Kolattukudy, P.E. (1971) Enzymatic synthesis of fatty alcohols in *Brassica oleracea. Archives of Biochemistry and Biophysics*, **142**, 701–709.

Kolattukudy, P.E. (1977) Lipid polymers and associated phenols, their chemistry, biosynthesis and role in pathogenesis. *Recent Advances in Phytochemistry*, **77**, 185–246.

Kolattukudy, P.E. (1980) Biopolyester membranes of plants: cutin and suberin. *Science*, **208**, 990–1000.

Kolattukudy, P.E. (1981) Structure, biosynthesis, and biodegradation of cutin and suberin. *Annual Review of Plant Physiology*, **32**, 539–567.

Kolattukudy, P.E., Buckner, J.S. and Liu T.-Y.J. (1973) Biosynthesis of secondary alcohols and ketones from alkanes. *Archives of Biochemistry and Biophysics*, **156**, 613–620.

Kreger, D.R. (1948) An X-ray study of waxy coatings from plants. *Recueil des Travaux Botaniques Neerlandais*, **41**, 606–736.

Kreger, D.R. and Schamhart, C. (1956) On the long crystal-spacings in wax esters and their value in micro-analysis of plant cuticle waxes. *Biochimica et Biophysica Acta*, **19**, 22–44.

Kunst, L., Taylor, D.C. and Underhill, E.W. (1992) Fatty acid elongation in developing seeds of *Arabidopsis thaliana. Plant Physiology and Biochemistry*, **30**, 425–434.

Kunst, L. and Samuels, A.L. (2003) Biosynthesis and secretion of plant cuticular wax. *Progress in Lipid Rresearch*, **42**, 51–80.

Lang-Pauluzzi, I. and Gunning, B.E.S. (2000) A plasmolytic cycle: the fate of cytoskeletal elements. *Protoplasma*, **212**, 174–185.

Lange, B.M., Wildung, M.R., Stauber, E.J., Sanchez, C., Puchnik, D. and Croteau, R. (2000) Probing essential oil biosynthesis and secretion by functional evaluation of expressed sequence

tags from mint glandular trichomes. *Proceedings of the National Academy of Sciences USA*, **97**, 2934–2939.
Lardizabal, K.D., Metz, J.G., Sakamoto, T., Hutton, W.C., Pollard, M.R. and Lassner M.W. (2000) Purification of a jojoba embryo wax synthase, cloning of its cDNA, and production of high levels of wax in seeds of transgenic *Arabidopsis*. *Plant Physiology*, **122**, 645–655.
Lassner, M.W., Lardizabal, K. and Metz, J.G. (1996) A jojoba β-ketoacyl-CoA synthase cDNA complements the canola fatty acid elongation mutation in transgenic plants. *Plant Cell*, **8**, 281–292.
Lemieux, B., Koornneef, M. and Feldmann, K.A. (1994) Epicuticular wax and *eceriferum* mutants, in *Arabidopsis* (eds E.M. Meyerowitz and C.R. Somerville), Cold Spring Harbor Press, New York, pp. 1031–1047.
Lequeu, J., Fauconnier, M.-L., Chammai, A., Bronner, R. and Blee, E. (2003) Formation of plant cuticle: evidence for the occurrence of the peroxygenase pathway. *Plant Journal*, **36**, 155–164.
Le Roux, J.H. (1969) Fischer-Tropsch waxes. II. Crystallinity and physical properties. *Journal of Applied Chemistry*, **19**, 86–88.
Lessire, R., Bessoule, J.J. and Cassagne, C. (1985) Solubilization of C18-CoA and C20-CoA elongases from *Allium porrum* L. epidermal cell microsomes. *FEBS Letters*, **187**, 314–320.
Lichtscheidl, I.K., Lancelle, S.A. and Hepler, P.K. (1990) Actin-endoplasmic reticulum complexes in *Drosera*. Their structural relationship with the plasmalemma, nucleus, and organelles in cells prepared by high pressure freezing. *Protoplasma*, **155**, 116–126.
Lolle, S.J. and Pruitt, R.E. (1998) Genetic analysis of organ fusion in *Arabidopsis thaliana*. *Genetics*, **149**, 607–619.
Lolle, S.J. and Pruitt, R.E. (1999) Epidermal cell interactions: a case for local talk. *Trends in Plant Science*, **4**, 14–20.
Macey, M.J.K. and Barber, H.N. (1970) Chemical genetics of wax formation on leaves of *Pisum sativum*. *Phytochemistry*, **9**, 5–12.
Macey, M.J.K. and Barber, H.N. (1970) Chemical genetics of wax formation on leaves of *Brassica oleracea*. *Phytochemistry*, **9**, 13–23.
Manheim, B.S. and Mulroy, T.W. (1978) Triterpenoids in epicuticular waxes of *Dudleya* species. *Phytochemistry*, **17**, 1799–1800.
Markstädter, C., Federle, W., Jetter, R., Riederer, M. and Hölldobler, B. (2000) Chemical composition of the slippery epicuticular wax blooms on *Macaranga* (Euphorbiaceae) ant-plants. *Chemoecology*, **10**, 33–40.
Martinoia, E., Lein, M., Geisler, M. *et al.* (2002) Multifunctionality of plant ABC transporters-more than just detoxifiers. *Planta*, **214**, 345–355.
McNevin, J.P., Woodward, W., Hannoufa, A., Feldmann, K.A. and Lemieux, B. (1993) Isolation and characterization of *eceriferum* (*cer*) mutants induced by T-DNA insertions in *Arabidopsis thaliana*. *Genome*, **36**, 610–618.
Merk, S., Blume, A. and Riederer, M. (1998) Phase behaviour and crystallinity of plant cuticular waxes studied by Fourier transform infrared spectroscopy. *Planta*, **204**, 44–53.
Metz, J.G., Pollard, M.R., Anderson, L., Hayes, T.R. and Lassner, M.W. (2000) Purification of a jojoba embryo fatty acyl-Coenzyme A reductase and expression of its cDNA in high erucic acid rapeseed. *Plant Physiology*, **122**, 635–644.
Meusel, I., Barthlott, W., Kutzke, H. and Barbier, B. (2000) Crystallographic studies of plant waxes. *Powder Diffraction*, **15**, 123–129.
Meusel, I., Neinhuis, C., Markstädter, C. and Barthlott, W. (1999) Ultrastructure, chemical composition, and recrystallization of epicuticular waxes: transversely ridged rodlets. *Canadian Journal of Botany*, **77**, 706–720.
Millar, A.A. and Kunst, L. (1997) Very-long-chain fatty acid biosynthesis is controlled through the expression and specificity of the condensing enzyme. *Plant Journal*, **12**, 121–131.
Millar, A.A., Clemens, S., Zachgo, S., Giblin, E.M., Taylor, D.C. and Kunst, L. (1999) *CUT1*, an *Arabidopsis* gene required for cuticular wax biosynthesis and pollen fertility, encodes a very-long-chain fatty acid condensing enzyme. *Plant Cell*, **11**, 825–838.
Moire, L., Schmuts, A., Buchala, A., Stark, R.E. and Ryser, U. (1999) Glycerol is a suberin monomer. New experimental evidence for an old hypothesis. *Plant Physiology*, **119**, 1137–1146.

Moreau P., Bessoule S., Mongrand E., Testet E., Vincent P. and Cassagne C. (1998) Lipid trafficking in plant cells. *Progress in Lipid Research*, **37**, 371–391.
Neinhuis, C. and Jetter, R. (1995) Ultrastructure and chemistry of epicuticular wax crystals in Polytrichales sporophytes. *Journal of Bryology*, **18**, 399–406.
Ohlrogge, J.B., Jaworski, J.G. and Post-Beittenmiller, D. (1993) *De novo* fatty acid biosynthesis, in *Lipid Metabolism in Plants* (ed. T.S. Moore), CRC Press, Inc., Boca Raton, pp. 3–32.
Oparka, K.J., Prior, D.A.M. and Crawford J.W. (1994) Behaviour of plasma membrane, cortical ER and plasmodesmata during plasmolysis of onion epidermal cells. *Plant, Cell & Environment*, **17**, 163–171.
Pinot, F., Benveniste, I., Salaun, J.P. and Durst, F. (1998) Methyl jasmonate induces lauric acid omega-hydroxylase activity and accumulation of CYP94A1 transcripts but does not affect epoxide hydrolase activities in *Vicia sativa* seedlings. *Plant Physiology*, **118**, 1481–1486.
Piper, S.H., Chibnall, A.C., Hopkins, S.J., Pollard, A., Smith, J.A.B. and Williams, E.F. (1931) Synthesis and crystal spacings of certain long-chain paraffins, ketones and secondary alcohols. *Biochemical Journal*, **25**, 2072–2094.
Piper, S.H., Chibnall, A.C. and Williams, E.F. (1934) Melting-points and long crystal spacings of the higher primary alcohols and n-fatty acids. *Biochemical Journal*, **28**, 2175–2188.
Pollard, M.R., McKeon, T., Gupta, L.M. and Stumpf, P.K. (1979) Studies on biosynthesis of waxes by developing jojoba seed. II. The demonstration of wax biosynthesis by cell-free homogenates. *Lipids*, **14**, 651–662.
Post-Beittenmiller, D. (1996) Biochemistry and molecular biology of wax production in plants. *Annual Review of Plant Physiology and Plant Molecular Biology*, **47**, 405–430.
Pruitt, R.E., Vielle-Calzada, J.P., Ploense, S.E., Grossniklaus, U. and Lolle, S.J. (2000) *FIDDLE-HEAD*, a gene required to suppress epidermal cell interactions in *Arabidopsis*, encodes a putative lipid biosynthetic enzyme. *Proceedings of the National Academy of Sciences USA*, **97**, 1311–1316.
Pyee, J., Yu, H. and Kolattukudy P.E. (1994) Identification of a lipid transfer protein as the major protein in the surface wax of broccoli (*Brassica oleracea*) leaves. *Archives of Biochemistry and Biophysics*, **311**, 460–468.
Rashotte, A.M., Jenks, M.A. and Feldmann, K.A. (2001) Cuticular waxes on *eceriferum* mutants of *Arabidopsis thaliana*. *Phytochemistry*, **57**, 115–123.
Rashotte, A.M., Jenks, M.A., Nguyen, T.D. and Feldmann, K.A. (1997) Epicuticular wax variation in ecotypes of *Arabidopsis thaliana*. *Phytochemistry*, **45**, 251–255.
Rasmussen, J.T., Faergeman, N.J., Kristiansen, K. and Kundsen, J. (1994) Acyl-CoA binding protein can mediate intermembrane acyl-CoA transport and donate acyl-CoA for beta-oxidation and glycerolipid synthesis. *Biochemical Journal*, **299**, 165–170.
Reynhardt, E.C. and Riederer, M. (1991) Structure and molecular dynamics of the cuticular wax from leaves of *Citrus aurantium*. *Journal of Physics D: Applied Physics*, **24**, 478–486.
Reynhardt, E.C. and Riederer, M. (1994) Structures and molecular dynamics of plant waxes. II. Cuticular waxes from leaves of *Fagus sylvatica* L. and *Hordeum vulgare* L. *European Biophysics Journal*, **23**, 59–70.
Rhee, Y., Hlousek-Radojcic, A., Ponsamuel, J., Liu, D. and Post-Beittenmiller, D. (1998) Epicuticular wax accumulation and fatty acid elongation activities are induced during leaf development of leeks. *Plant Physiology*, **116**, 901–911.
Riedel, M., Eichner, A. and Jetter, R. (2003) Slippery surfaces of carnivorous plants: composition of epicuticular wax crystals in *Nepenthes alata* Blanco pitchers. *Planta*, **218**, 87–97.
Sachetto-Martins, G., Franco, L.O. and de Oliveira, D.E. (2000) Plant glycine-rich proteins: a family or just proteins with a common motif? *Biochimica et Biophysica Acta*, **1492**, 1–14.
Sánchez-Fernández, R., Emyr-Davies, T.G., Coleman, J.O.D. and Rea, P.A. (2001) The *Arabidopsis thaliana* ABC protein superfamily, a complete inventory. *Journal of Biological Chemistry*, **276**, 30231–30244.
Schaffer, J.E. (2002) Fatty acid transport: the roads taken. *American Journal of Physiology. Endocrinology and Metabolism*, **282**, E239–E246.
Schneider-Belhaddad, F. and Kolattukudy, P.E. (2000) Solubilization, partial purification and characterization of a fatty aldehyde decarbonylase from a higher plant, *Pisum sativum*. *Archives of Biochemistry and Biophysics*, **377**, 341–349.

Schmitz, G., Kaminski, W.E. and Orso, E. (2000) ABC transporters in cellular lipid trafficking. *Current Opinion in Lipidology,* **11**, 493–501.

Schnurr, J.A., Shockey, J.M., de Boer, G.-J. and Browse, J.A. (2002) Fatty acid export from the chloroplast. Molecular characterization of a major plastidial acyl-Coenzyme A synthetase from *Arabidopsis. Plant Physiology*, **129**, 1700–1709.

Schreiber, L. and Riederer, M. (1996) Ecophysiology of cuticular transpiration: comparative investigation of cuticular water permeability of plant species from different habitats. *Oecologia*, **107**, 426–432.

Schreiber, L. and Schönherr, J. (1993) Mobilities of organic compounds in reconstituted cuticular wax of barley leaves: determination of diffusion coefficients. *Pesticide Science*, **38**, 353–361.

Shimakata, T. and Stumpf, P.K. (1982) Isolation and function of spinach leaf β-ketoacyl-[acyl-carrier-protein] synthases.*Proceedings of the National Academy of Sciences USA,* **79**, 5808–5812.

Shin, D.H., Lee, J.Y., Qwang, K.Y., Kim, K.K. and Suh, S.W. (1995) High-resolution crystal structure of the non-specific lipid-transfer protein from maize seedlings. *Structure*, **3**, 189–199.

Shockey, J.M., Fulda, M.S. and Browse, J. (2003) *Arabidopsis* contains a large superfamily of acyl-activating enzymes. Phylogenetic and biochemical analysis reveals a new class of acyl-coenzyme A synthetases. *Plant Physiology*, **132**, 1065–1076.

Sieber, P., Schorderet, M., Ryser, U., Buchala, A., Kolattukudy, P.E., Metraux, J.-P. and Nawrath, C. (2000) Transgenic *Arabidopsis* plants expressing a fungal cutinase show alterations in structure and properties of the cuticle and postgenital organ fusions. *Plant Cell,* **12**, 721–737.

Silva Fernandes, A.M.S., Baker, E.A. and Martin, J.T. (1964) Studies on the plant cuticle. VI. The isolation and fractionation of cuticular waxes. *The Annals of Applied Biology*, **53**, 43–58.

Sohal, A.K., Pallas, J.A. and Jenkins, G.I. (1999) The promoter of a *Brassica napus* lipid transfer protein gene is active in a range of tissues and stimulated by light and viral infection in transgenic *Arabidopsis. Plant Molecular Biology*, **41**, 75–87.

Staehelin, L.A. (1997) The plant ER: a dynamic organelle composed of a large number of discrete functional domains. *Plant Journal*, **11**, 1151–1165.

Sterk, P., Booij, H., Schellekens, G.A., Van Kammen, A. and De Vries, S.C. (1991) Cell-specific expression of the carrot *EP2* lipid transfer protein gene. *Plant Cell,* **3**, 907–921.

Storch, J. and Thumser, A.E.A. (2000) The fatty acid transport functions of fatty acid binding proteins. *Biochimica et Biophysica Acta,***1486**, 28–44.

Stremmel, W., Pohl, J., Ring, A. and Herrman, T. (2001) A new concept of cellular uptake and intracellular trafficking of long-chain fatty acids. *Lipids,* **36**, 981–989.

Stumpf, P.K. (1984) Fatty acid biosynthesis in higher plants, in *Fatty acid Metabolism and its Regulation* (ed. S. Numa), Elsevier Science Publishers B.V., Amsterdam, pp. 155–179.

Szczuka, E. and Szczuka A. (2003) Cuticle fluorescence during embryogenesis of *Arabidopsis thaliana* (L.) Heynh. *Acta Biologica Cracoviensia Series Botanica*, **45**, 63–67.

Tassin-Moindrot, S., Caille, A., Douliez, J.-P., Marion, D. and Vovelle, F. (2000) The wide binding properties of a wheat nonspecific lipid binding protein. *European Journal of Biochemistry*, **267**, 1117–1124.

Thoma, S., Kaneko, Y. and Somerville, C. (1993) A non-specific lipid transfer protein from *Arabidopsis* is a cell wall protein. *Plant Journal*, **3**, 427–436.

Tijet, N., Helvig, C., Pinot, F. *et al.* (1998) Functional expression in yeast and characterization of a clofibrate-inducible plant cytochrome P450 (CYP94A1) involved in cutin monomer synthesis. *Biochemical Journal,* **332**, 583–589.

Todd, J., Post-Beittenmiller, D. and Jaworski, J.G. (1999) *KCS1* encodes a fatty acid elongase 3-ketoacyl-CoA synthase affecting wax biosynthesis in *Arabidopsis thaliana. Plant Journal*, **17**, 119–130.

Trevino, M.B. and O'Connell, M.A. (1998) Three drought-responsive members of the nonspecific lipid-transfer protein gene family in *Lycopersicon pennillii* show different developmental patterns of expression. *Plant Physiology*, **116**, 1461–1468.

Turner, G.W., Gershenzon, J., Nielson, E.E., Froehlich, J.E. and Croteau, R.B. (1999) Limonene synthase, the enzyme responsible for monoterpene biosynthesis in peppermint, is localized to leucoplasts of oil gland secretory cells. *Plant Physiology,* **120**, 879–886.

Vioque, J. and Kolattukudy, P.E. (1997) Resolution and purification of an aldehyde-generating and an alcohol generating fatty acyl-CoA reductase from pea leaves (*Pisum sativum* L.). *Archives of Biochemistry and Biophysics*, **340**, 64–72.

Voelker, T.A., Jones, A., Cranmer, A.M., Davies, H.M. and Knutzon, D.S. (1997) Broad-range and binary-range acyl-acyl-carrier protein thioesterases suggest an alternative mechanism for medium chain production in seeds. *Plant Physiology*, **114**, 669–677.

von Wettstein-Knowles, P. (1982) Elongase and epicuticular wax biosynthesis. *Physiologie vegetale*, **20**, 797–809.

von Wettstein-Knowles, P. (1993) Waxes, cutin, and suberin, in *Lipid Metabolism in Plants* (ed. T.S. Moore), CRC Press, Inc., Boca Raton, pp. 127–166.

von Wettstein-Knowles, P. (1995) Biosynthesis and genetics of waxes, in *Waxes: Chemistry, Molecular Biology and Functions* (ed. R.J. Hamilton), The Oily Press, Dundee, pp. 91–129.

Walton, T.J. and Kolattukudy, P.E. (1972) Demonstration of the structures of cutin monomers by a novel depolymerization procedure and combined gas-chromatography and mass spectrometry. *Biochemistry*, **11**, 1885–1897.

Wellesen, K., Durst, F., Pinot, F. *et al.* (2001) Functional analysis of the *LACERATA* gene of *Arabidopsis* provides evidence for different roles of fatty acid omega-hydroxylation in development. *Proceedings of the National Academy of Sciences USA,* **98**, 9694–9699.

Wise, M.L. and Croteau, R. (1999) Monoterpene synthesis, in *Comprehensive Natural Products Chemistry* (ed. D.E. Cane), Elsevier Science, Oxford, pp. 97–153.

Wu, W.-Y., Moreau, R.A. and Stumpf, P.K. (1981) Studies of biosynthesis of waxes by developing jojoba seed. III. Biosynthesis of wax esters from acyl-CoA and long chain alcohols. *Lipids,* **6**, 897–902.

Wyatt S.E. and Carpita N.C. (1993) The plant cytoskeleton-cell wall continuum. *Trends in Cell Biology*, **3**, 413–417.

Xu, X., Dietrich, C.R., Delledonne, M. *et al.* (1997) Sequence analysis of the cloned *glossy8* gene of maize suggests that it may code for a β-ketoacyl reductase required for the biosynthesis of cuticular waxes. *Plant Physiology*, **115**, 501–510.

Xu, X., Dietrich, C.R., Lessire, R., Nikolau, B.J. and Schnable, P.S. (2002) The endoplasmic reticulum-associated maize gl8 protein is a component of the acyl-CoA elongase involved in the production of cuticular waxes. *Plant Physiology*, **128**, 924–934.

Yephremov, A., Wisman, E., Huijser, P., Huijser, C., Wellesen, K. and Saedler, H. (1999) Characterization of the *FIDDLEHEAD* gene of *Arabidopsis* reveals a link between adhesion response and cell differentiation in the epidermis. *Plant Cell*, **11**, 2187–2201.

Zachowski A., Guerbette F., Grosbois M., Jolliot-Croquin A. and Kader J.-C. (1998) Characterization of acyl binding by a plant lipid-transfer protein. *European Journal of Biochemistry*, **257**, 443–448.

Zlotnik-Mazori, T. and Stark, R. (1988) Nuclear magnetic resonance studies of cutin, an insoluble plant polyester. *Macromolecules*, **21**, 2412–2417.

Zou, Z., Tong, F., Faergeman, N.J., Borsting, C., Black, P.N. and DiRusso, C.C. (2003) Vectorial acylation in *Saccharomyces cerevisiae. Journal of Biological Chemistry*, **278**, 16414–16422.

Zou, Z.Y., DiRusso, C.C., Ctrnacta, V. and Black, P.N. (2002) Fatty acid transport in *Saccharomyces cerevisiae* – Directed mutagenesis of Fat1 distinguishes the biochemical activities associated with Fat1p. *Journal of Biological Chemistry*, **277**, 31062–31071.

# 9 Inositol-containing lipids: roles in cellular signalling

Bjørn K. Drøbak

## 9.1 Introduction

Inositol-containing lipids belong to a large and heterogeneous family of lipids. Although tremendous progress has been made in the study of this class of lipids, many members of the family still remain to be fully characterised. The first inositol-containing lipids were identified around 1930 by R.J. Anderson, who discovered that inositol could be isolated from a 'phosphatide' fraction from tubercle bacilli. The presence of inositol in brain phospholipids was first reported by Folch and Woolley in 1942.

Phosphatidylinositol (PtdIns) is a common membrane phospholipid in eukaryotic cells. In plants, it is exceeded in abundance in membranes from non-photosynthetic tissues only by phosphatidylethanolamine and phosphatidylcholine (Harwood, 1980). It is now nearly 20 years since it was first realised that PtdIns in plants was similar to that in mammalian and other eukaryotic cells in the following respects. PtdIns can function as a structural membrane lipid, but it also acts as a precursor for a number of lipid molecules, which can act either as second messengers or as precursors for second messenger production. In spite of the very diverse functions of these PtdIns derivatives, they differ simply by the phosphorylation status and isomeric structure of their myo-inositol headgroups. This group of lipids is collectively known as phosphoinositides and the phosphorylated forms as polyphosphoinositides.

The enormous interest in phosphoinositides over the past 20 years is due primarily to the findings that these lipids play key roles in a vast array of cell signalling events in eukaryotic cells. A number of features of phosphoinositides make them particularly well suited to act as messenger molecules or precursors for messenger molecules. First, the highly polar inositol phosphate headgroup, which can rapidly have its phosphorylation status altered, provides an excellent and dynamic locus for protein targeting events within discrete microdomains in membranes. Second, the ratio between phosphoinositides and their interacting ligands can often be relatively large, which makes it possible to target a significant number of distinct proteins to a specific membrane site without saturating a relatively limited number of binding sites. Third, the specific isomeric conformation of the phosphorylated myo-inositol headgroup, combined with a rapid monoester phosphate turnover rate, provides the opportunity for highly flexible and dynamic interactions to occur without the loss of specificity between the polyphosphoinositide headgroup moieties and their multitude of cognate ligands. However, it was the discovery that activated forms of phospholipase C (PLC) were capable of hydrolysing phosphatidylinositol(4,5)bisphosphate [PtdIns(4,5)$P_2$] with the result being the production of the two second messenger molecules, inositol(1,4,5)trisphosphate [Ins(1,4,5)$P_3$] and diacylglycerol (DAG) that started the new revolution in phosphoinositide research. Ins(1,4,5)$P_3$ was found to be capable of specifically inducing $Ca^{2+}$-release from intracellular stores in a wide variety of eukaryotic cells (including plant cells), whilst DAG in many cell types was shown to modulate the activity of a group of enzymes known collectively as protein kinase(s) C. Thus, the concomitant increase in cytosolic $Ca^{2+}$ and the switch-on of protein kinase activity results in the triggering of bifurcated signalling pathways that lead to a multitude of down-stream events. This scenario can now rightfully be referred to as the 'classic PI-response'; but as is so often the case in biological research, matters are somewhat more complicated than first assumed. First, the syntheses and potential function of currently recognised plant phosphoinositides will be briefly described.

## 9.2 Phosphoinositides: synthesis, turnover and function

### *9.2.1 Biosynthesis of phosphatidylinositol*

The structure of phosphatidylinositol is shown in Figure 9.1.

*De novo* biosynthesis of PtdIns in higher plants, as in animals, occurs via the CDP-DAG:inositol-3-phosphatidyltransferase (EC 2.7.8.11) pathway (for review see Moore, 1990). The gene encoding CDP-DAG:inositol-3-phosphatidyltransferase, also called PtdIns synthase, has been cloned in yeast (Nikawa *et al.*, 1987), rats (Tanaka *et al.*, 1996) and humans (Lykidis *et al.*, 1997). Another reaction, initially described in plants, is the PtdIns:Ins exchange reaction where the inositol head-group is exchanged with free inositol in a CMP-dependent manner (Sexton and Moore, 1981; Sandelius and Morre, 1987; for review see Moore, 1990). Using the recombinant enzyme, Lykidis *et al.* (1997) has shown that both the CDP-DAG:inositol-3-phosphatidyltransferase and PtdIns:Ins reactions are

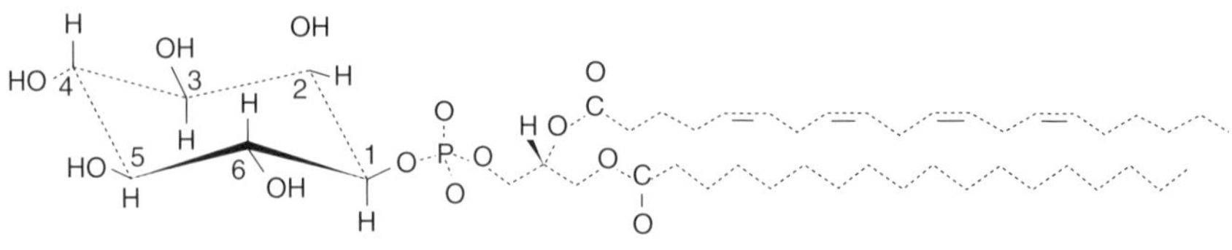

**Figure 9.1** Structure of phosphatidylinositol (PtdIns). The numbering of the inositol headgroup follows the current IUPAC convention.

catalysed by PtdIns synthase. PtdIns synthase is located primarily on the cytoplasmic face of the ER (Ballas and Bell, 1981). In addition to being an important component of biological membranes, PtdIns can be metabolised to form inositol sphingolipids in both yeast and plants. In this reaction, the inositol phosphate head-group of PtdIns is transferred to a hydroxyl of phytoceramide by inositolphosphorylceramide synthase producing DAG and an inositolphosphorylphytoceramide (Kearns *et al.*, 1997). Inositolphosphorylphytoceramide biosynthesis also takes place in the ER. The regulation of inositolphosphophytoceramide biosynthesis has been suggested to be a critical factor in membrane biogenesis (Kearns *et al.*, 1997).

## 9.3 Phosphorylation of phosphatidylinositol and other phosphoinositides

It is now known that PtdIns can be phosphorylated in the D-3, D-4 and D-5 position of the inositol headgroup in all eukaryotes studied so far. Phosphorylation of the PtdIns D-2 and D-6 positions has not been reported although D-2 and D-6 phosphorylation is well known to occur in water-soluble inositol phosphates. A schematic overview of the currently recognised pathways leading to polyphosphoinositide synthesis in plants is given in Figure 9.2.

Although PtdIns and polyphosphoinositide phosphorylation has been extensively studied in both higher eukaryotic cells as well as in e.g. yeast and plants, surprisingly little is known about the polyphosphoinositide phosphatases. Therefore,

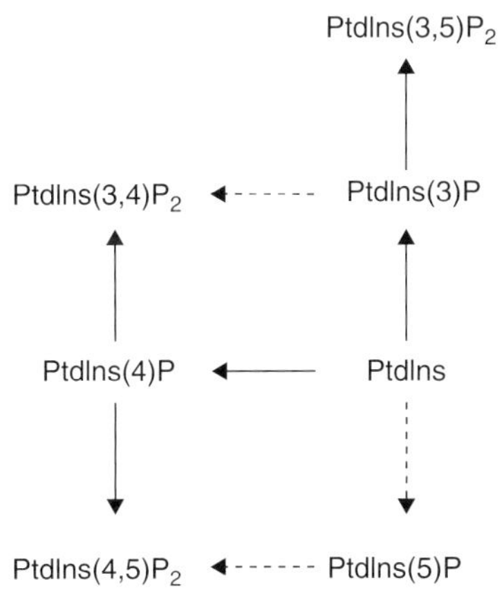


**Figure 9.2** Pathways for PtdIns phosphorylation in plant cells. Phosphorylation pathways, which are established to exist in a number of plant cell types or have been supported by direct *in vivo* biochemical data or cloning of the relevant kinases, are indicated by solid arrows. Stippled arrows indicate phosphorylation pathways that have some experimental support but still need to be verified. (Routes of synthesis involving phosphatase activity have not been included in the figure but are mentioned in the main text.)

the following sections will primarily concentrate on the phosphoinositide kinases and PLC isozymes, which have since the 1990s been identified in plant cells. In addition, a brief description of potential roles of individual phosphoinositides will be given.

### *9.3.1 Phosphatidylinositol 3-kinases*

In the literature, some confusion often occurs when PtdIns 3-kinases are discussed. This is due to the fact that there are at least two routes leading to the synthesis of polyphosphoinositides with a D-3 phosphomonoester. These two main synthetic pathways appear to be evolutionarily separate and the reactions are catalysed by quite different kinases. The first route involves phosphorylation of PtdIns(4,5)$P_2$ by 3-kinases, which leads to the formation of PtdIns(3,4,5)$P_3$. Dephosphorylation of PtdIns(3,4,5)$P_3$ is believed to proceed via sequential dephosphorylation by 5- and 3-phosphatases, resulting in the production of PtdIns(3,4)$P_2$ and PtdIns(4)P (Stephens *et al.*, 1991). As this pathway does not appear to exist in plant (or yeast) cells, it will not be described further in the current context. The other route utilises direct 3-phosphorylation of PtdIns by PtdIns 3-kinases. In some cells, there may be a certain 'overlap' between these two pathways and there is some evidence that smaller amounts of PtdIns(3,4)$P_2$ may be formed by 4-phosphorylation of PtdIns(3)P and/or 3-phosphorylation of PtdIns(4)P and that PtdIns(3)P may also, in certain cases, be generated by the action of a 4-phosphatase upon PtdIns(3,4)$P_2$.

In yeast and *Neurospora*, as much as 50% of the PtdInsP pool can be made up by PtdIns(3)P but early studies in plants suggested that if 3-phosphoinositides were present, they were likely only to be rather minor components of the polyphosphoinositide pool (Irvine *et al.*, 1989). Later work has confirmed this assumption but has also clearly demonstrated that both PtdIns(3)P and PtdIns(3,4)$P_2$ are normal constituents of the plant phosphoinositide pool (Brearley and Hanke, 1993; Munnik *et al.*, 1994a,b). Approximate estimates indicate that PtdIns(3)P in plant cells may constitute 2–20% of the PtdInsP pool whereas PtdIns(3,4)$P_2$, in most cases, constitutes approximately 5–10% of the PtdIns$P_2$ pool. However, in a few cases, PtdIns(3,4)$P_2$ may approach the levels of PtdIns(4,5)$P_2$ but the physiological relevance and frequency of this occurrence remain to be investigated further.

Only one class of PtdIns 3-kinases appears to be present in plant cells. These kinases exclusively utilise PtdIns as substrate and belong to the class III of eukaryotic PtdIns 3-kinases (Mueller-Rober and Pical, 2002). These kinases are all closely related to the *VPS34*-gene product from yeast, vps34, which was first identified in a screen of *mutants* defective in vesicle sorting. Since this kinase is the only form of PtdIns 3-kinase present in yeast and higher plants it perhaps represents the primordial form of PtdIns 3-kinase. However, in a manner similar to the 3-kinases belonging to the Group I and Group II enzymes, the activity of type III PtdIns 3-kinases is also dependent on the interaction with adaptor proteins. The activation partner of vps34 in yeast is a ser/thr kinase named vps15. It is believed that vps15 recruits vps34 to membranes where its lipid kinase activity is switched on. In humans a homologue to vps15 has also been identified and has been named p150. Both vps34 and p150 have *N*-terminal myristylation sites, a ser/thr kinase domain and a set of HEAT-repeats and WD-motifs towards the *C*-terminal. Experiments in yeast have shown that vps34 and vps15 can form tight complexes *in vivo* and that the formation of such complexes depends on the activation of vps15 protein kinase activity. It is thus reasonable to assume that vps15 is an upstream regulator of vps34 function. That the level of PtdIns 3-kinase activity is dependent on the vps34/vps15 interaction has been demonstrated by deletion experiments where levels of PtdIns(3)P in vps15 yeast strains (strains where the gene encoding vps15 has been deleted) were found to be below the level of detection. Virtually nothing is currently known about the regulation of PtdIns 3-kinase activity in plants.

The first structural information about plant PtdIns 3-kinases came from the studies of Hong and Verma (1994) and Welters *et al.* (1994). These groups cloned and sequenced very similar cDNAs encoding vps34-related enzymes from soybean and Arabidopsis, respectively.

The soybean PtdIns 3-kinase gene identified by Hong and Verma (1994) encodes a polypeptide (SPI3K-5) of 814 amino acids and a predicted molecular mass of 93 kDa; a second PtdIns 3-kinase gene encodes SPI3K-1, a polypeptide of 812 amino acids. Both kinases were found to have catalytic PtdIns 3-kinase activity when expressed in *Escherichia coli* and assayed using exogenous PtdIns liposomal substrates. One of the interesting features of the study by Hong and Verma (1994) was the discovery that the two PtdIns 3-kinase genes were differentially expressed during the organogenesis of root nodules induced by Rhizobium. Thus, the induction of expression of the *SPI3K-1* gene, encoding the nodule form of PtdIns 3-kinase was found to repress the expression of *SPI3K-5*, the root form of PtdIns 3-kinase. Only after nodule maturation had been completed did the expression levels of *SPI3K-5* return to pre-nodulation levels. These induction and repression events were further found to be directly correlated with the degree of membrane proliferation in the root nodules. These data lend support to the hypothesis that the plant PtdIns 3-kinase – like its yeast and mammalian counterparts – is involved in the trafficking and/or sorting of membrane and plays a central role in membrane biogenesis and the development of endosymbiotic compartments. Another plant PtdIns 3-kinase gene was identified in *Arabidopsis thaliana* by Welters *et al.* (1994). The gene product, AtVPS34 is a polypeptide of 814 amino acids and shares 40% identity with the yeast vps34 and is 25% identical to bovine p110. It is interesting to note that the *N*-terminal part of AtVPS34 contains a domain with strong homology to the CaLB-domain of rabphilin 3a, and Welters *et al.* (1994) suggest that this domain may be involved in the binding of phospholipids. A chimeric gene in which the coding sequence for the *C*-terminal of VPS34 was replaced with the corresponding sequence from the plant gene was able to rescue a yeast-deletion mutant. Perhaps of more significance was the finding that the expression of *AtVPS34* antisense constructs gave rise to second generation transformed plants which were severely inhibited in both their growth rate and development – indicating a significant role for PtdIns 3-kinase in these processes.

Two compounds have been extensively used in the characterisation of mammalian and other PtdIns 3-kinase enzymes, namely, the fungal metabolite wortmannin and the synthetic molecule LY 294002. Matsuoka *et al.* (1995) used wortmannin to investigate whether PtdIns 3-kinase had the potential to affect vacuolar-sorting processes in tobacco BY-2 cells. Wortmannin was found to inhibit both PtdIns 3-kinase and PtdIns 4-kinase activity as well as being highly inhibitory to at least one type of vacuolar sorting. These studies point to a link between PtdIns 3-kinase activity and vesicle trafficking in plant cells, but it must be stressed that it can often be difficult to unequivocally prove causal links merely by inhibitor studies, particularly when the inhibitor may have more than one target.

Little is known about the subcellular localisation of plant PtdIns 3-kinase(s) but Dove *et al.* (1994) found that the highest specific activity of plant PtdIns 3-kinase was associated with a detergent-resistant nucleo-cytoskeletal compartment. It is unclear whether this location is the true intracellular site for PtdIns 3-kinase as the methods used for preparing the nucleo-cytoskeletal fractions involve a number of steps known to upregulate and induce translocation of certain enzymes of the plant phosphoinositide signalling system. Little is still known about the function of PtdIns(3)P in plant cells. PtdIns(3)P has a turnover rate in plant cells, which is greater than that of most structural lipids. This suggests that the 3-phosphorylated phosphoinositides, like their 4-phosphorylated counterparts, have the potential to participate in rapid processes, such as those encountered during cell signalling events. There is now little doubt that PtdIns(3)P in plant cells, like in yeast and higher eukaryotes, plays an important role in the regulation of vesicle translocation and other processes associated with membrane biogenesis and function. Recent studies by Kim *et al.* (2001) and Jung *et al.* (2002) indicate that PtdIns 3-kinase activity and PtdIns(3)P are present in several intracellular pools including locations in the Golgi apparatus and vacuolar network and, in addition, PtdIns 3-kinase has been suggested to play an important role for normal function of stomata. Data which show that PtdIns 3-kinase activity is associated with active nuclear transcription sites in plants and that isolated nuclei possess PtdIns 3-kinase activity (Bunney *et al.*, 2000) are also available. The exact molecular basis underlying these functions remain unclear but one possibility is that the phosphorylation of membrane PtdIns by a type III PtdIns 3-kinase leads to local changes in the curvature of membrane bilayers, which in turn may lead to the initiation of transport vesicle formation. Another possibility is that PtdIns(3)P acts as a target for vesicle docking onto target membranes. A third proposition is that PtdIns(3)P during signalling events recruits otherwise soluble proteins to specific membranes. This latter hypothesis is supported by data since 2000, which show that proteins containing a FYVE-finger domain interact, with high specificity and affinity, with PtdIns(3)P in both mammalian and plant cells. A further description of plant FYVE-domain containing proteins and their potential roles in signalling events can be found in Section 9.4.

### *9.3.2 Phosphatidylinositol 4-kinases*

In contrast to membranes from mammalian cells, PtdOH (phosphatidic acid or PA) in plant cell membranes constitutes approximately 70% of the phosphorylated lipid products, and PtdIns(4)P and PtdIns(4,5)$P_2$, typically, constitute no more than maximum 20% and 3% (Sandelius and Sommarin, 1990; Gross and Boss, 1992). Under equal assay conditions, the specific activity of the PtdIns 4-kinase from rat liver is in the order of 10- to 20-fold greater than that of higher plants (Sandelius and Sommarin, 1990).

PtdIns 4-kinases appear to be present in low abundance in plant cells and often co-purify with heat shock proteins. As a result, they eluded researchers for many years and their characterisation was limited to biochemical traits of partially purified proteins. For example, PtdIns 4-kinases inhibited by adenosine ($K_i$: 20–100 μM) and having a low $K_m$ for ATP (30–100 μM) were denoted type II; whereas, PtdIns 4-kinases insensitive to 100 μM adenosine ($K_i$: 1520 μM) and having a high $K_m$ for ATP (150–750 μM) were denoted type III (Carpenter and Cantley, 1990; Pike, 1992). It was not until Flanagan and Thorner (1992) took advantage of yeast genetics by using a heat shock null mutant that the first PtdIns 4-kinase was cloned (Flanagan *et al.*, 1993). Subsequently, PtdIns 4-kinases have been cloned and sequenced from a number of higher eukaryotes (Wong and Cantley, 1994; Gehrmann *et al.*, 1996; Nakagawa *et al.*, 1996).

Based on the predicted amino acid sequences, the PtdIns 4-kinases fall into two major families. One family comprises the large molecular weight proteins (approximately 200–230 kDa), which have a *C*-terminal lipid kinase domain and a lipid kinase unique domain separated by a pleckstrin homology domain (PH-domain, for further details see Section 9.4). Members of this family are generally type III, adenosine-insensitive PtdIns 4-kinases (Nakagawa *et al.*, 1996; Balla *et al.*, 1997). The other family contains smaller molecular weight proteins, which lack a PH-domain. They have the lipid kinase domain at the *C*-terminus, whilst the unique lipid kinase domain is closer to the *N*-terminus. These enzymes are generally type II kinases. However, sequence data clearly indicate that several subclasses of the PtdIns 4-kinases exist within these two families (Balla *et al.*, 1997; Wong *et al.*, 1997).

Although several groups have purified enzymes with PtdIns 4-kinase activity from a number of plant tissues and cells, the first cloning and more detailed study of a plant PtdIns 4-kinase was described by Stevenson *et al.* (1998). This enzyme, named AtPIKα1, contains a functional PH-domain and has a high degree of homology to the large molecular weight, type III enzymes found in mammalian cells. This is consistent with earlier biochemical data. *In vitro* studies using *E. coli*-expressed PH-domain of AtPIK$\alpha$1 showed affinity for PtdIns(4)P, but not for PtdIns(3)P, suggesting that the binding is biologically relevant. In addition, based on Western blot analysis using antibodies raised to the *C*-terminus containing the conserved lipid kinase domain and the PH-domain, AtPIK$\alpha$1 was found in the F-actin fraction isolated from microsomes. In a subsequent study, Xue *et al.* (1999) isolated and functionally expressed a full-length plant PtdIns 4-kinase cDNA. The corresponding protein is 126 kD and was named AtPI4K$\beta$1. AtPI4K$\beta$1 is similar in size to Beta-isoforms identified in yeast and mammalian cells and also displays a high degree of similarity in primary structure (Xue *et al.*, 1999). A number of additional potential PtdIns 4-kinase genes have been identified by homology searches of the Arabidopsis databases but currently little is known about their potential physiological roles. For further biochemical and molecular/genetic details, the reader is referred to recent reviews (Drøbak *et al.*, 1999; Mueller-Roeber and Pical, 2002).

Apart from acting as a precursor for PtdIns(4,5)$P_2$ (and in some cases PtdIns(3,4)$P_2$), no 'independent' physiological role for PtdIns(4)P has so far been defined in eukaryotic cells, although some reports have indicated that PtdIns(4)P may be involved in the regulation of certain enzymes, such as DNA-polymerase $\alpha$ (Sylvia *et al.*, 1988). In many cases PtdIns(4)P has been found to share functionality with PtdIns(4,5)$P_2$ e.g. in its ability to bind to actin-binding proteins but, in all cases, PtdIns(4,5)$P_2$ has been shown to possess the highest affinity. However, it should be remembered that such experiments often are carried out using equimolar concentrations of polyphosphoinositides. Since PtdIns(4)P by far is the dominant polyphosphoinositide in plant cells, the possibility that this lipid may be the important physiological target for a number of *in vivo* processes in plant cells should not be discounted, although from *in vitro* experiments PtdIns(4,5)$P_2$ appears to be the favoured candidate.

The presence of PtdIns(3,4)$P_2$ in plant cells appears to be ubiquitous but so far no physiological role has been ascribed to this member of the plant phosphoinositide family. Until the late 1990s, PtdIns(3,4)$P_2$ was also viewed in mammalian cells simply as the first hydrolysis product following PtdIns(3,4,5)$P_3$ dephosphorylation. However, this view has changed significantly and it has been suggested that PtdIns(3,4)$P_2$ may well fulfil cellular functions, which are equal in importance to those of e.g. PtdIns(4,5)$P_2$ and PtdIns(3,4,5)$P_3$. One important target for PtdIns(3,4)$P_2$ is the *Akt* proto-oncogene product, Akt. Akt (also referred to as PKBa or Raca) is a serine-threonine kinase, which participates in the activation of the p70 ribosomal protein S6 kinase ($p70^{S6K}$) and also plays a role in proliferative and anti-apoptotic cell responses (Franke *et al.*, 1995; Dudek *et al.*, 1997). It is known that Akt activation by growth factors requires PtdIns 3-kinase activity but there are also alternative pathways, which can lead to Akt activation. There are, thus, good reasons to suspect that one (or more) of the phosphoinositide products resulting from 3-kinase activation could be a potential regulator of Akt activity and hence $p70^{S6K}$

activation. Franke *et al.* (1997) investigated the effect of various phosphoinositides on the activity of Akt and found, somewhat surprisingly, that PtdIns(3,4)$P_2$ was capable of activating Akt both *in vitro* and *in vivo*. The ability to activate Akt was not shared by any of the other phosphoinositides; so there is some reason to believe that this interaction is of physiological relevance and not just due to the strong negative charges found in e.g. PtdIns$P_2$ molecules. An interesting question is of course whether such a function for PtdIns(3,4)$P_2$ extends beyond mammalian cells. As mentioned above, there is no evidence for the presence of PtdIns(3,4,5)$P_3$ in plant cells. Hence, the production of PtdIns(3,4)$P_2$ by a dephosphorylation pathway continues to remain doubtful. As the yeast vps34p and its plant equivalents appear to be the only specific phosphoinositide 3-kinases in these organisms, the question of how PtdIns(3,4)$P_2$ is synthesised remains a point of debate. A gene encoding a functional plant PtdInsP kinase was identified by Mikami *et al.* (1998) and named AtPIP5K1. When this enzyme was heterologously expressed in *E. coli*, it was found to be capable of phosphorylating both PtdIns(3)P and PtdIns(4)P resulting in the synthesis of PtdIns(3,4)$P_2$ and PtdIns(4,5)$P_2$ (Westergren *et al.*, 2001). However, in studies by Elge *et al.* (2001) using insect cells to express AtPIP5K1, it was found that the recombinant enzyme preferentially catalysed the synthesis of PtdIns((4,5)$P_2$ (and surprisingly, PtdIns(3,4,5)$P_3$) whereas it was reluctant to synthesise PtdIns(3,4)$P_2$. So, at present it must be said that the jury is still out when it comes to the exact route of synthesis of PtdIns(3,4)$P_2$ in plant cells.

Nevertheless, all available evidence points to PtdIns(3,4)$P_2$ being a common constituent of the plant polyphosphoinositide pool, and it has the potential to fulfil physiological roles similar to its mammalian counterpart. There are several poignant examples of signalling molecules fulfilling similar functions in different organisms in spite of being produced by different synthetic routes. A more thorough investigation of the pattern of PtdIns(3,4)$P_2$ synthesis and turnover in plant cells and possible responses to environmental stress exposure seems an obvious target for future investigation.

### *9.3.3 Phosphatidylinositol 5-kinases*

Whereas direct PtdIns 3- and 4-phosphorylation is established to occur *in vivo* in plants the synthetic route(s) leading to PtdIns(5)P formation remains less well-defined. However, direct biochemical evidence for the presence of PtdIns(5)P in plant cells was reported by Meijer *et al.* (2001). It has been demonstrated that PtdIns(5)P can be produced in higher eukaryotes by direct D-5 PtdIns phosphorylation *in vitro*, but at present there is a paucity of evidence to suggest that this also happens *in vivo*. In addition, pulse-chase experiments in both mammalian and plant cells suggest that it is unlikely that PtdIns(5)P is formed by 4-dephosphorylation of PtdIns(4,5)$P_2$ (see, e.g. Meijer *et al.*, 2001 and references therein). Data from osmotic stress experiments in *Chlamydomonas* show that PtdIns(5)P levels are elevated in concert with PtdIns(3,5)$P_2$ levels whereas increased turnover of

PtdIns(4,5)$P_2$ did not appear to affect PtdIns(5)P levels. These findings would be consonant with a role for PtdIns(3,5)$P_2$ in PtdIns(5)P synthesis, but as yet there is insufficient evidence to assess this hypothesis with confidence. However, recent data from mammalian research indicates that proteins belonging to the myotubularin family are involved in the synthesis of PtdIns(5)P via 3-dephosphorylation of PtdIns(3,5)$P_2$. The physiological role of PtdIns(5)P remains to be investigated further but Rameh *et al.* (1997), surprisingly, found that type II PtdInsP kinases in mammalian cells phosphorylate PtdIns(5)P in the D-4 position rather than phosphorylating PtdIns(4)P in the D-5 position as it had been assumed for decades. The authors explain how this previous error in interpretation is likely to be due to small (but significant) contaminating amounts of PtdIns(5)P in commercial preparations of PtdIns(4)P. The fact that PtdIns(5)P is present, albeit in small quantities, in many types of eukaryotic cells, including plant cells, suggests that at least a portion of cellular PtdIns(4,5)$P_2$ is likely to be formed by sequential 5- and 4-phosphorylation of PtdIns.

### *9.3.4 Phosphatidylinositol 3-monophosphate 5-kinases*

The presence of PtdIns(3,5)$P_2$ in mouse fibroblasts was first reported by Whiteford *et al.* (1997), and findings by Dove *et al.* (1997) suggest that rapid formation of PtdIns(3,5)$P_2$, following the activation of a PtdIns(3)P 5-kinase is a conserved response to osmotic stress in a wide range of cell types. PtdIns(3,5)$P_2$ has also been identified in plant cells (Dove *et al.*, 1997). In Arabidopsis four putative PtdIns(3)P 5-kinase genes have been identified. Two of these genes have been cloned and appear to be active PtdIns(3)P 5-kinase (deVos, Dove and Drøbak, manuscript submitted). Fur further details of PtdIns(3)P 5-kinases in mammalian and plant cells, see McEwen *et al.*, 1999, Cooke, 2002 and Mueller-Roeber and Pical, 2002. In addition to its role as a potential second messenger in its own right, PtdIns(3,5)$P_2$ may also, as mentioned above, act as a precursor for the production of PtdIns(5)P.

### *9.3.5 Phosphatidylinositol 4-monophosphate 5-kinases*

The PtdIns(4)P 5-kinases form a unique family of lipid kinases, which lack sequence homology to the kinase domains of the phosphatidylinositol and inositol phosphate kinases (Boronenkov and Anderson, 1995; Loijens and Anderson, 1996). The absence of sequence homology implies that their catalytic mechanism is distinct – a notion that is consistent with the observation that some PtdIns(4)P 5-kinases can utilise both GTP and ATP as phosphoryl donors. The PtdIns(4)P 5-kinases, like the PtdIns 4-kinases, are classified into two families based on their biochemical characteristics. The activity of type I PtdIns(4)P 5-kinases is stimulated by PtdOH and small G-proteins whereas the type II enzymes are not. Both types I and II are rather promiscuous with regard to substrates (Zhang *et al.*, 1997).

Thus, Type I, in addition to phosphorylate PtdIns(4)P in the D-5 position, can also phosphorylate PtdIns(3)P and PtdIns(3,4)$P_2$ in the D-4 and D-5 positions, respectively. Type II will phosphorylate PtdIns(3)P and PtdIns(4)P about equally well, but does not phosphorylate PtdIns(3,4)$P_2$. The type II enzyme has also been shown to phosphorylate PtdIns(5)P in the D-4 position (Rameh *et al.*, 1997a). The general lack of substrate specificity led Zhang *et al.* (1997) to suggest that PtdIns kinases should be renamed as PtdInsP 4/5-kinases.

These unique lipid kinases, which catalyse the last step in the biosynthesis of PtdIns(4,5)$P_2$, appear to hold one of the keys to the understanding of how the Rho family of GTP-binding proteins transduce extracellular signals to the actin cytoskeleton. When PtdIns(4,5)$P_2$ binds to actin-capping proteins and monomeric actin-binding proteins, the result is an increase in filament formation. GTP-Rho binds and activates PtdIns(4)P 5-kinase *in vitro* (Chong *et al.*, 1994; Tolias, 1995), and *in vivo* studies of permeabilised and transfected cells indicate that GTP-activated Rac indeed promote increases in actin filament formation by activating PtdIns(4)P 5-kinase (Hartwig *et al.*, 1995; Shibasaki *et al.*, 1997). Overexpression of PtdIns(4)P 5-kinase alone also results in increase in actin filament formation, suggesting that PtdIns(4)P 5-kinase acts downstream of GTP-Rac, and it has also been found that not all Rac-related changes in actin polymerisation are mediated by PtdIns(4,5)$P_2$. Importantly, studies of PtdIns(4)P 5-kinases in animal cells indicate that these enzymes have other functions than merely replenishing phosphoinositide pools, and they suggest that the rate of biosynthesis of PtdIns(4,5)$P_2$ may act as regulator of several cytoskeletal and membrane-associated processes.

In spite of their obvious physiological importance, information about plant PtdIns(4)P 5-kinases is limited. Eleven putative PtdIns(4)P 5-kinases have been identified in the Arabidopsis genome, but apart from AtPIP5K1 none of these have been characterised in any detail. At the amino acid level, several of the Arabidopsis PtdIns(4)P 5-kinases show an almost equal degree of homology to both the type I and type II isoforms of animal cells. At least two cDNAs are predicted to encode large molecular weight proteins and one has 47% amino acid homology to the yeast Mss4p kinase (Yoshida *et al.*, 1994), which is an essential enzyme classified as a type II PtdIns(4)P 5-kinase (Zhang *et al.*, 1997). Further details of other putative PtdIns(4)P 5-kinases can be found in a recent review by Mueller-Roeber and Pical (2002).

It is clear that PtdIns(4,5)$P_2$ can act as a precursor for the production of the two second messengers: inositol(1,4,5)trisphosphate and DAG in plant cells. Whereas the messenger role of DAG still needs to be clearly defined, the role of PtdIns(4,5)$P_2$ as a precursor for Ins(1,4,5)$P_3$ production in plant cells is now well established. Moreover, Ins(1,4,5)$P_3$ has been shown to be produced in response to a wide variety of abiotic and biotic stimuli (see e.g. Drøbak, 1992, 1996; Cote and Crain, 1993; Munnik *et al.*, 1998). Being a precursor for at least two major second messengers seems justification enough for the existence of any molecule, but recent research suggests that PtdIns(4,5)$P_2$ plays several other important roles in both mammalian and plant in cells. One such function of PtdIns(4,5)$P_2$ is linked to the regulation of cytoskeletal dynamics. This aspect of PtdIns(4,5)P2 function will be described in more detail in Section 9.4.

PtdIns(4,5)$P_2$ has also been found to act as a specific activator of several enzymes including protein kinases (Palmer *et al.*, 1995), GTPase-activating proteins (Liscovitch and Cantley, 1995) and guanine nucleotide exchange factors (Zheng *et al.*, 1996); PtdIns(4,5)$P_2$ has also been shown to mediate protein–protein interactions as exemplified e.g. by the PtdIns(4,5)$P_2$-driven interaction between ADP-ribosylation factor 1 (Arf) and the Arf-GTPase activating protein (Arf-GAP; Randazzo, 1997). In plant cells, PtdIns(4,5)$P_2$ has been reported to modulate the activity of a number of enzymes including vanadate-sensitive plasma membrane $H^+$-ATPase (Memon and Boss, 1990), protein kinases (Schaller *et al.*, 1992) and DAG kinase (Lundberg and Sommarin, 1992). Other PtdIns(4,5)$P_2$-regulated enzymes include the mammalian and plant Phospholipase D (PLD) isozymes (for a general review see Exton, 1997). Pappan *et al.* (1997a,b) further identified a novel member of the plant PLD family in *A. thaliana* and this enzyme, named PLDβ, has been found to be dependent on both PtdIns(4,5)$P_2$ and submicromolar concentrations of $Ca^{2+}$ for activity. As such, PLDβ would appear to be eminently well-suited to carry out a role as a downstream effector enzyme for signals transduced through the plant phosphoinositide pathway.

Yet another role for PtdIns(4,5)$P_2$ is as an essential regulatory component in the 'priming' of exocytotic vesicles. Vesicle 'priming' is an ATP-dependent process, which occurs after vesicle docking but before fusion (Hay and Martin, 1993; Banerjee *et al.*, 1996). Hay *et al.* (1995) thus showed that three cytosolic factors are essential for the successful 'priming' of exocytotic vesicles; two are PtdIns transfer proteins, and the third is a type I PtdIns(4)P 5-kinase. Together with a vesicular PtdIns 4-kinase, the two PtdIns transfer proteins and a PtdIns(4)P 5-kinase catalyse the synthesis of PtdIns(4,5)$P_2$ in the vesicle membrane. Although the precise function of PtdIns(4,5)$P_2$ in this type of priming process is still under investigation, the most likely possibilities are that PtdIns(4,5)$P_2$ either functions to recruit PtdIns(4,5)$P_2$-binding proteins to vesicular membranes or perhaps activate enzymes within the vesicle membrane. At the moment there is limited direct evidence for a role for PtdIns(4,5)$P_2$ in vesicle transport in plant cells, although experiments by Matsuoka *et al.* (1995) indicate that inhibition of phosphoinositide kinase activity severely affect vesicle transport processes in BY-2 cells.

## 9.4 Phosphoinositide–protein interactions

In addition to acting as second messenger precursors and kinase and phosphatase substrates, phosphoinositides can, as mentioned previously, also act as specific ligands in intracellular protein-targeting events and in the assembly of signal-induced multi-protein–lipid complexes. In many of these processes, the specific recognition of the phosphoinositide headgroup by specialised intraprotein phosphoinositide-recognition domains have been found to play a key role. Although relatively little is known about the specific molecular aspects of protein–phosphoinositide interactions in plant cells, there are, nevertheless, a number of studies which show that they indeed occur and several phosphoinositide-interacting plant proteins have been identified and characterised in some detail. In the sections below, a brief overview of some of the currently known phosphoinositide-interacting proteins in plants will be given and, in addition, a number of specific phosphoinositide-recognition domains will be described.

### *9.4.1 Profilin*

Profilin is a low molecular weight (12–15 kDa), ubiquitous, cytoplasmic protein. It is essential for the viability of fission yeast (Lu and Pollard, 2001), flies (Verheyen and Cooley, 1994) and mice (Witke *et al.*, 2001), whereas loss of profilin function severely impacts on the growth of budding yeast (Haarer *et al.*, 1990) and *Dictyostelium* (Haugwitz *et al.*, 1994). Profilin was originally isolated from calf spleen and identified as a protein that forms high affinity 1 : 1 complexes with G-actin (Carlsson *et al.*, 1977). It is now known that profilin has complex effects on actin dynamics in a wide variety of cell types. In the presence of capped ends, profilin acts like a simple sequestering protein by binding to G-actin and preventing its polymerisation but when filament ends are free, profilin–actin complexes can add to F-actin and promote assembly (Pantaloni and Carlier, 1993; Kang *et al.*, 1999). Which activity predominates appears to be dictated by the size and activity of the G-actin pool, the presence of other actin-binding proteins and the stoichiometry between profilin and G-actin. In addition to binding G-actin and certain actin-related proteins (Machesky *et al.*, 1994; Zhao *et al.*, 1998), profilin interacts with a number of other protein and lipid ligands. One example is the ability of profilin to bind contiguous stretches of proline (Tanaka and Shibata, 1985), a process which is thought to be important for its interaction with a number of proline-rich proteins. Vertebrate profilin has a high affinity for phosphoinositide lipids, and shows a marked preference for D-3 phosphoinositides, with relative binding affinities for PtdIns(3,4)$P_2$ > PtdIns(3,4,5)$P_3$ > PtdIns(4,5)$P_2$ (Lu *et al.*, 1996). The observation that PtdIns(4,5)$P_2$ binding releases G-actin from profilin–actin complexes (Lassing and Lindberg, 1985) and a comparison of the amino acid homology between two biochemically-distinct isoforms of *Acanthamoeba* profilin (Fedorov *et al.*, 1994) led to the view that the lipid-binding site overlaps the actin-binding site within the profilin molecule. Positively charged side groups from lysine and arginine residues on β-strands 4, 5 and 6 are likely candidates for interactions with the polar headgroup of phosphoinositides and mutagenesis of specific residues in yeast and vertebrate profilins often reduce the ability of the recombinant protein to bind both PtdIns(4,5)$P_2$ and G-actin (Haarer *et al.*, 1993; Sohn *et al.*, 1995; Lambrechts *et al.*, 2002; Skare and Karlsson, 2002). Such competitive-binding would greatly reduce the actin-binding activity of profilins in the presence of lipids. Lassing and Lindberg (1985, 1988) interpreted this as a potential mechanism whereby a population of profilin at the plasma membrane might interact with PtdIns(4,5)$P_2$ and prevent interactions with actin until a signal transduction cascade is initiated. Goldschmidt-Clermont and coworkers provided direct biochemical evidence supporting this model by showing that profilin binding to PtdIns(4,5)$P_2$ inhibited membrane-associated polyphosphoinositide PLC activity (Goldschmidt-Clermont *et al.*, 1990). Moreover, the inhibitory effect of profilin could be overcome by phosphorylation of PLC-$\gamma$ (Goldschmidt-Clermont *et al.*, 1991), an event known to occur during the activation of certain membrane receptor tyrosine kinases. A number of profilin isoforms have now been identified in plant cells (Valenta *et al.*, 1991; Staiger *et al.*, 1993) and it has been demonstrated that they specifically interact with phosphoinositides (Drøbak *et al.*, 1994). It is now clear that profilin, in addition to acting as sensors for the phosphoinositide status in specific membranes, also appear to have a multitude of other functions. For reviews, see Staiger (2000), Staiger and Hussey (2003), Drøbak *et al.* (2004).

### *9.4.2 ADF/cofilin*

The actin-depolymerizing-protein (ADF)/cofilin group of proteins bind G- and F-actin and modulate actin dynamics. They do this by both severing actin filaments thereby providing more ends for polymerisation, and by increasing the rate of dissociation of actin monomer from the pointed end (Carlier 1998; Maciver 1998; Bamburg 1999; Chen *et al.*, 2000). Many ADFs are pH sensitive, can be inhibited by phosphorylation and, more recently, their activity has been shown to be enhanced by an associated protein known as actin-interacting protein 1 (AIP1). In plants, some ADFs are vegetatively expressed (e.g. maize ADF 3, ZmADF3) whereas others are expressed in reproductive tissues such as pollen (Gungabissoon *et al.*, 1998; Smertenko *et al.*, 1998; Allwood *et al.*, 2001). Several ADF/cofilins interact specifically with phosphatidylinositol 4,5-bisphosphate (PtdIns(4,5)$P_2$) and phosphatidylinositol 4-phosphate (PtdIns(4)P) (Yonezawa *et al.*, 1990; Quirk *et al.*, 1993; Gungabissoon *et al.*, 1998). ADF/cofilins reciprocally inhibit PLC-mediated hydrolysis of PtdIns(4,5)$P_2$ (Yonezawa *et al.*, 1991; Gungabissoon *et al.*, 1998). So whilst PtdIns(4,5)$P_2$ may participate in the processes that localise ADF/cofilins to membranes and affect actin filament dynamics, the ADF/cofilins may also have direct effects upon PtdIns(4,5)$P_2$ metabolism (Takenawa and Itoh, 2001).

In plants, both ZmADF3 and LlADF1 have been experimentally shown to interact with PtdIns(4)P and PtdIns(4,5)$P_2$, and this interaction inhibits the actin binding and depolymerisation activity of ADF (Gungabissoon *et al.*, 1998; Allwood *et al.*, 2002). When maize and lily ADF were tested for binding to a variety of phospholipids *in vitro*, the binding was found to be specific for phosphoinositides. It was found that ADF/cofilin does not bind to a number of other phospholipids, such as phosphatidylcholine, phosphatidylserine, phosphatidylethanolamine, 1-oleoyl-2-acetylglycerol or the water-soluble messenger molecule inositol 1,4,5-trisphosphate

(Gungabissoon *et al.*, 1998). Two sites for PtdIns(4,5)$P_2$ binding have been identified in chicken ADF, and although phospholipid binding has only been experimentally determined in two plant ADFs, comparison of these sites in a range of plant and animal ADFs suggests that phospholipid binding, and hence regulation of activity by phospholipids is likely to be a common feature of plant ADFs. This raises the possibility that the localised turnover of the actin cytoskeleton is regulated, at least in part, by the phosphoinositide composition of adjacent membranes.

### *9.4.3 PARF and other FYVE-finger domain proteins*

The Fab1, YOTB/ZK632.12, Vac1, EEA1 (FYVE) finger is a specific phosphoinositide-interacting domain, which was first identified by Stenmark *et al.* (1996) and was named after four of the proteins in which it was first found: Fab1, YOTB/ZK632.12, Vac1 and EEA1. The FYVE-domain is a zinc finger-like domain, which is predominantly found in proteins involved in membrane trafficking and signal transduction and more than 60 eukaryotic proteins containing FYVE domains have so far been identified. One of the features that distinguish the FYVE domain from e.g. the PH-domains is that they exhibit a very high degree of specificity for a single species of polyphosphoinsitides, namely PtdIns(3)P, although recent research has shown that PtdIns(5)P may also, in some cases, act as a ligand. The structure of several eukaryotic FYVE-domains as well as residues believed to be essential for specific interactions have been identified. The basic FYVE motif consists of eight cysteines (or in a certain cases, seven cysteines and one histidine). Two of the cysteines are part of the core motif R+HHC+XCG (where '+' is a positively charged residue and 'X' is any amino acid). Two $Zn^{2+}$ ions are bound by the FYVE-domains and the conserved Cys/(His) residues are located in pairs such that the first and the third pair bind one $Zn^{2+}$ ion, while the second and fourth pairs bind the other $Zn^{2+}$ ion. The crystal structure of the FYVE domain from Vps27p has revealed that it contains two double-stranded, antiparallel β-sheets and a *C*-terminal α-helix. β1 and β2 form the first sheet and β3 and β4 the second (Misra and Hurley, 1999). Although the FYVE domains are highly conserved, there may still be significant difference in the specificity and affinity of individual FYVE-domains towards phosphoinositides. Thus, EEA1 has been shown to have affinity for PtdIns(5)P as well as PtdIns(3)P, perhaps due to the fact that both PtdIns(3)P and PtdIns(5)P have their phosphomonoesters in the meta-position. Additional data suggests that the EEA1-FYVE and Vps27-FYVE domains may also differ in other aspects of their phosphoinositide-binding mechanisms (Driscoll, 2001; Kutateladze and Overduin, 2001).

Several FYVE-domain proteins have been identified in plant cells and have, on the basis of domain structure and amino acid sequence homology, been tentatively been divided into three groups. Group I contains proteins with a canonical FYVE-domain and a domain classified as a 'phosphatidylinositol(4)phosphate 5-kinase domain'. These proteins have a high degree of homology to the yeast and mammalian Fab proteins. The yeast and human Fab proteins have been found not to be phosphatidylinositol(4)phosphate 5-kinases but rather: phosphatidylinositol(3)phosphate 5-kinases (Dove *et al.*, 1997; McEwen *et al.*, 1999). It is believed that these kinases play an important role in the transduction of osmotic shock signals and in the regulation of the invagination of endosomal membranes. The high degree of similarity between the AtFab and their eukaryotic homologues suggest that AtFab could also be involved in similar processes in plant cells. The second group of plant proteins containing the FYVE-domain (Group II) contains large proteins with rather complex domain structures. A representative of this family, named TMVr1 (hypothetical TMV-resistance protein), in addition to the FYVE domain also contains a number of other recognisable domains. One such domain is the TIR (Toll/Interleukin 1R) domain, which is an intracellular signalling domain found in several eukaryotic proteins such as MyD88, the interleukin 1 receptor and the Toll receptor. Members of the Toll family are thought to play important roles in innate antibacterial and antifungal immunity in insects as well as in mammals. The similarity between Toll-like receptors (LTRs) and IL-1R appears not to be restricted to sequence homology since these proteins both induce the activation of a Rel type transcription factor via an adaptor protein and a protein kinase (Bonnert *et al.*, 1997). The potential involvement of TMVrs in disease resistance signalling is emphasised by the presence of a NB-ARC domain, which is a motif shared by many plant resistance gene products and regulators of cell death in animals (Van der Biezen and Jones, 1998). In addition to a short Leucine Rich Repeat, the final recognisable domain in TMVr1 is a set of RCC1-repeats (Regulator of Chromosome Condensation 1).

Although members of the plant FYVE-domain group III are significantly smaller than the members of the TMVr group (and several proteins contain PH-domains; collectively known as PARF or PARF-Like proteins from **PH** **A**nd **R**CC1 **F**YVE), these two groups have one distinguishing feature in common, namely that in addition to the FYVE-domain they also contain the RCC1 tandem repeats. RCC1 is a nuclear chromatin-bound protein, which acts as a guanine nucleotide exchange factor (GEF) for the small nuclear GTPase, ran (Dasso, 1993; Roepmen *et al.*, 1996). RCC1 has been found to be essential for the co-ordination of the onset of mitosis with S-phase completion in higher eukaryotic cells and has also recently been recognised to be involved in nucleo-cytoplasmic transport events through its control of the guanine-nucleotide bound status of ran. The function of ran (like other members of the ras superfamily) is dependent upon the type of guanine nucleotide with which it is associated. RCC1 promotes the release of GDP from ran and promotes ran–GTP association. Subcellular studies have shown that ran in its GTP bound form is primarily present within the nucleus whereas GDP-ran is mainly cytoplasmic. This difference in the nucleotide-bound state of ran is known to act as an indicator of compartmental identity between the cytoplasm and intranuclear space. Therefore, the transnuclear ran–GTP/ran–GDP gradient provides ran with the opportunity to act as switch determining the vectorial transport of protein cargoes across the nuclear envelope. Part of this function is thought to be mediated via its interaction with transnuclear transport factors (i.e. exportins and importins; for further details see Moore, 2001). Whether the FYVE-RCC1 proteins in plant have similar functions remains to be determined but several sets of data suggest that this indeed could be the case (for further information see Drøbak and Heras, 2002; Heras and Drøbak, 2002).

### *9.4.4 Proteins containing PH-domains*

The PH-domain is a protein module of around 120 amino acids, which is found in many proteins involved in intracellular signalling and cytoskeletal organisation. The PH-domain(s) is often found alongside other domains that participate in intermolecular interactions. More that 500 PH-domains have so far been identified in regulatory proteins and they have been found to have a wide range of ligand affinity and specificity. Although the amino acid structure of PH-domain is highly divergent (some 14 different PH-domain structures have been identified to date), they all retain a highly conserved 3-dimentional structure. The core structure is a β-sandwich of two nearly orthogonal β-sheets consisting of three and four strands, respectively. The sheets are tightly packed at the so-called 'close corners' of which one is spanned by the β1 strand while the other is completed by a loop connecting strands β4 and β5. Overall, there are six loops connecting the β-strands. Three of these (i.e. β1/ β2, β3/β4 and β6/β7) have been termed the 'variable loops' as these structures show a high degree of variance at the amino acid level. The opposite edge of the PH-domain structure is capped by an amphiphatic *C*-terminal α-helix. These latter regions have been termed 'splayed corners' and appear to be important determinants in the strong electrostatic polarisation of the PH-domain. When the electrostatic potential around e.g. the β-spectrin domain is calculated, clearly defined regions of positive potential are observed around the β1/ β2 loop whereas the remainder of the protein is surrounded by an overall negative electrochemical potential.

This electrostatic polarisation appears to be characteristic for all the PH-domain so far investigated and the majority of PH-domain containing proteins appear to have stringent requirements for membrane association in order to exert their functions. In several cases, PH-domain proteins also act as membrane 'docking-partners' for a variety of associated proteins. Although it has been known for sometime that PH-domains in certain cases can interact with membrane-associated phosphoinositides with high specificity and high affinity, it is now clear that only around 10–20% of the currently recognised PH-domain proteins utilise phosphoinositides as their primary binding partners. However, in the cases where PH-domains interact with phosphoinositides with high affinity and specificity, this has been found to be necessary and sufficient for selective membrane targeting of PH-domain containing protein.

Although a large number of PH-domains have been found in mammalian (and other) cells, relatively little is currently known about proteins containing PH-domains in plant cell. The first evidence for the presence of PH-domains in plant proteins came from work by Stevenson *et al.* (1998) who, as mentioned earlier, demonstrated that a plant phosphatidylinositol 4-kinase contained a PH-domain that was able to interact with PtdIns(4)P, PtdIns(4,5)$P_2$ and phosphatidic acid. Other plant proteins, which have subsequently been found to contain PH-domains, include a putative pleckstrin homologue (Mikami *et al.*, 1999) and a dynamin-like protein named ADL3 (Mikami *et al.*, 2000). A plant homologue of the human 3-phosphoinositide-dependent protein kinase-1 (PDK1) has been identified in *Arabidopsis* and rice (Deak *et al.*, 1999) and, like the mammalian PDK1, the plant kinase possesses a canonical kinase domain at the *N*-terminus and a PH-domain at the *C*-terminus. The Arabidopsis PDK1 is able to rescue lethality in *Saccharomyces cerevisiae* caused by disruption of the genes encoding yeast PDK1 homologues and interacts via its PH-domain with phosphatidic acid, PtdIns(3)P, PtdIns(3,4,5)$P_3$ and PtdIns(3,4)$P_2$ and to a lesser extent with PtdIns(4,5)P2 and PtdIns(4)P. Although the Arabidopsis PDK1 was found to be catalytically active and was able to activate human protein kinase B alpha (PKB/AKT) in the presence of PtdIns(3,4,5)P3, the absence of this phosphoinositide in plant cells currently makes it difficult to draw exact parallels between the function of the mammalian and the plant enzyme.

### 9.4.5 *Proteins containing PX-domains*

Recent research has revealed that the phagocyte NADPH oxidase (Phox)-homology (PX) domain also interacts specifically with phosphoinositides (Cheever *et al.*, 2001; Ellson *et al.*, 2001; Kanai *et al.*, 2001; Song *et al.*, 2001; Xu *et al.*, 2001). The Phox complex is a multi-subunit complex responsible for the generation of reactive oxygen species (ROS) and thus is central in the defence against a wide range of infectious agents. When phagocytes encounter microorganisms, the assembly of membrane associated Phox-complexes is triggered and ROS production is dramatically upregulated. Two of the Phox-subunits, p40phox and p47phox, have been found to contain PX-domains with specificity for PtdIns(3)P and PtdIns(3)P/PtdIns(3,4)P2, respectively and it has been demonstrated how their PX-domain-phosphoinositide interaction plays important roles in the assembly of Phox-complexes in activated neutrophils. The PX-domain is approximately 100–140 amino acids long and it contains two highly conserved arginine residues. So far it has been identified in more than 150 eukaryotic proteins, which in addition to the members of the Phox-complex family include certain PLD isoforms, members of the nexin family, a number of yeast vacuolar-sorting proteins and the cytokine-independent survival kinase, CISK. Although the highest specificity and affinity of PX-domains appear to be for PtdIns(3)P, some controversy still exists in this area as e.g. Xu *et al.* (2001) have shown that the CISK PX-domain appears also to have a high degree of specificity for PtdIns(3,4,5)P3 and PtdIns(3,5)P2. Very little is currently known about the potential presence and function of PX-domains in plant proteins but a recent database survey has revealed the existence of several proteins containing distinct PX-domains in *A. thaliana* (B.K. Drøbak, unpublished).

### *9.4.6 Proteins containing ENTH-, VHS- and FERM-domains*

The recent rapid progress in proteomics has led to the discovery of several new potential phosphoinositide-interacting domains. One of these is the epsin NH2-terminal homology, ENTH-domain. ENTH-domains are structural modules of approximately 140 amino acids found in mammalian epsin 1 and 2, AP180 and Hip1R as well as in their yeast homologues Ent1p, Ent4p, yAP180 and Sla2p (Itoh *et al.*, 2001). Mammalian epsin is known to play a central role in clathrin-mediated endocytosis and the yeast Ent1p and Ent2p are essential for endocytosis and proper actin function. It was recently demonstrated that ENTH-domains display a high specificity and affinity for PtdIns(4,5)$P_2$ suggesting potential links between PtdIns(4,5)$P_2$ metabolism, ENTH-domain containing proteins and events such as endocytosis and reorganisation of the actin cytoskeleton. Another domain to be mentioned in the current context is the Vps27, Hrs and STAM (VHS) domain. VHS-domains are like ENTH-domains found at the amino terminus of several proteins involved in endocytosis (Hurley and Meyer, 2001). Although the sequence homology between VHS-domains and ENTH-domains is relatively low, the three-dimensional structures display a surprisingly high degree of similarity as they both form eight-helical bundles and have a common conserved basic face. The potential of phosphoinositide–VHS interactions have until now received limited attention but, based on structural analyses, it would not be surprising if VHS-domains also had some affinity for specific members of the phosphoinositide family. The final domain to be mentioned is the Four-point-one-ezrin-radixin-moesin (FERM)-domain. The FERM-domain containing proteins, such as ezrin, radixin and moesin, are known to be of significant importance in the linking of actin filaments to various adhesion proteins. PtdIns(4,5)$P_2$ has been found to be a key regulator of this process although the precise mechanism is still under debate. The current information on FERM-containing proteins suggest that they may contain a PH-related fold as a substructure within the FERM-domain, and some data suggest that Ins(1,4,5)$P_3$ may interact with FERM-domains via binding to a cleft formed by the PH-domain-related structure and a ubiquitin-fold related substructure (Hamada *et al.*, 2000).

At present, nothing is known about the potential function(s) of ENTH-, VHS- or FERM-related domain structures in plant proteins but a recent database search showed that a considerable number of plant (Arabidopsis) proteins contain canonical ENTH-domains and at least 6–8 proteins that contained VHS-like domains were identified. However, no distinct plant FERM-like domains have been identified in the currently available databases (B.K. Drøbak, unpublished). As pointed out above, a strict sequence homology between e.g. mammalian, yeast and plant protein domains may not necessarily be the most reliable way in which to predict potential interactions between proteins and phosphoinositides as such interactions often are determined by a multitude of factors, which may, or may not, be directly related to specific amino acid sequences.

## 9.5 Conclusions

Although this overview only gives a brief summary of some aspects of plant phosphoinositide metabolism and function, it hopefully conveys the fact that this group of lipids acts as a key element in a multitude of cell signalling events as well as having important functions in many other physiological processes. Much has been learned about plant phosphoinositides since their discovery but it is clear that there are still vast areas within this field that remain unexplored. Given their importance in plant and other cells, the study of phosphoinositides provides a fertile, challenging and exciting area of inquiry not only for established researchers but also for the new generation of plant scientists. However, anyone entering this field of study should be prepared for surprises and treat paradigms with respect but also with a healthy degree of scepticism.

## Acknowledgements

The author gratefully acknowledges E.G. Allwood, W.F. Boss, S.K. Dove, M. Durrant, P.J. Hussey and C.J. Staiger for generously sharing ideas and views on many aspects of plant phosphoinositide signalling – some of which are presented in this chapter.

## References

Allwood, E.G., Anthony, R.G., Smertenko, A.P. *et al.* (2002). Regulation of the pollen-specific actin-depolymerizing factor LlADF1. *Plant Cell*, **14**, 2915–2927

Allwood, E.G., Smertenko, A.P. and Hussey, P.J. (2001). Phosphorylation of plant actin-depolymerising factor by calmodulin-like domain protein kinase. *FEBS Letters*, **499**, 97–100.

Balla, T., Downing, G.J., Jaffe, H., Kim, S., Zólyomi, A. and Catt, K.J. (1997). Isolation and molecular cloning of wortmannin-sensitive bovine type II phosphatidylinositol 4-kinases. *Journal of Biological Chemistry*, **272**, 18358–18366

Ballas, L.M. and Bell, R.M. (1981). Topography of glycerolipid synthetic enzymes – synthesis of phosphatidylserine, phosphatidylinositol and glycerolipid intermediates occurs on the cytoplasmic surface of rat-liver microsomal vesicles. *Biochimica et Biophysica Acta*, **665**, 586–595.

Bamburg, J.R. (1999). Proteins of the ADF/cofilin family: essential regulators of actin dynamics. *Annual Review of Cell and Developmental Biology*, **15**, 185–230.

Banerjee, A., Barry, V.A., Dasgupta, B.R. and Martin, T.F.J. (1996). *N*-ethylmaleimide-sensitive factor acts at a profusion ATP-dependent step in $Ca^{2+}$ activated exocytosis. *Journal of Biological Chemistry*, **271**, 20223–20226.

Bonnert, T.P., Garka, K.E., Parnet, P., Sonoda, G., Testa, J.R. and Sims, J.E. (1997). The cloning and characterization of human MyD88: a member of an IL-1 receptor related family. *FEBS Letters*, **402**, 81–84.

Boronenkov, I.V. and Anderson, R.A. (1995) The sequence of phosphatidylinositol-4-phosphate 5-kinase defines a novel family of lipid kinases. *Journal of Biological Chemistry*, **270**, 2881–2884.

Brearley C.A. and Hanke, D.E. (1993) The pathway of synthesis of 3,4- and 4,5-phosphorylated phosphatidylinositols in the duckweed Spirodela polyrhiza L. *Biochemical Journal*, **290**, 145–150.

Bunney, T.D., Watkins, P.A.C., Beven, A.F. *et al.* (2000) Association of phosphatidylinositol 3-kinase with nuclear transcription sites in higher plants.*Plant Cell*, **12**, 1679–1687.

Carlier, M.F. (1998) Control of actin dynamics. *Current Opinion in Cell Biology*, **10**, 45–51.

Carlsson, L., Nyström, L.E., Sundkvist, I., Markey, F. and Lindberg, U. (1977) Actin polymerizability is influenced by profilin, a low molecular weight protein in non-muscle cells. *Journal of Molecular Biology*, **115**, 465–483.

Carpenter, C.L. and Cantley, L.C. (1990) Phosphoinositide kinases. *Biochemistry*, **29**, 11147–11156.

Cheever, M.L., Sato, T.K., de Beer, T., Kutateladze, T.G., Emr, S.D. and Overduin M. (2001) Phox domain interaction with Ptdlns(S)P targets the Vam7 t-SNARE to vacuole membranes *Nature Cell Biology*, **3**, 613–618.

Chen, H., Bernstein, B.W. and Bamburg, J.R. (2000) Regulating actin-filament dynamics *in vivo*. *Trends in Biochemical Sciences*, **25**, 19–23.

Chong, L.D., Traynor-Kaplan, A., Bokoch, G.M. and Schwartz, M.A. (1994). The small GTP-binding protein Rho regulates a phosphatidylinositol 4-phosphate 5-kinase in mammalian cells. *Cell*, **79**, 507–513.

Cooke, F.T. (2002) Phosphatidylinositol 3,5-bisphosphate: metabolism and function. *Archives of Biochemistry and Biophysics*, **407**, 143–151.

Cote, G.G. and Crain, R.C. 1993. Biochemistry of phosphoinositides. *Annual Review of Plant Physiology and Plant Molecular Biology*, **44**, 333–356.

Dasso, M. (1993) RCC1 in the cell cycle: the regulator of chromosome condensation takes on new roles. *Trends in Biochemical Sciences*, **18**, 96–101.

Deak, M., Casamayor, A., Currie, R.A., Downes, C.P. and Alessi, D.R. (1999) Characterisation of a plant 3-phosphoinositide-dependent protein kinase-1 homologue which contains a pleckstrin homology domain. *FEBS Letters*, **451**, 220–226.

Dove, S.K., Cooke, F.T., Douglas, M.R., Sayers, L.G., Parker, P.J. and Michell, R.H. (1997) Osmotic stress activates phosphatidylinositol-3,5-bisphosphate synthesis. *Nature*, **390**, 187–192.

Dove, S.K., Lloyd, C.W. and Drøbak, B.K. (1994) Identification of a phosphatidylinositol 3-hydroxy kinase in plant cells: association with the cytoskeleton. *Biochemical Journal*, **303**, 347–350.

Driscoll, P.C. (2001) Solving the FYVE domain-PtdIns(3)P puzzle. *Nature Structural biology*, **8**, 287–290.

Drøbak, B.K. (1992) The plant phosphoinositide system. *Biochemical Journal*, **288**, 697–712.

Drøbak, B.K. (1996) Metabolism of plant phosphoinositides and other inositol-containing lipids, in *Membranes: Specialized Functions in Plants* (eds M. Smallwood, J.P. Knox and D.J. Bowles), BIOS Scientific Publishers, Oxford, UK, pp. 195–214.

Drøbak, B.K and Heras, B (2002) Nuclear phosphoinositides could bring FYVE alive. *Trends in Plant Sciences*, **7(3)**: 132–138.

Drøbak, B.K., Dewey, R.E. and Boss, W.F. (1999). Phosphoinositide kinases and the synthesis of polyphosphoinositides in higher plant cells. *International Review of Cytology*, **189**, 95–130.

Drøbak, B.K., Watkins, P.A.C., Valenta, R., Dove, S.K., Lloyd, C.W. and Staiger, C.J. (1994) Inhibition of plant plasma membrane phosphoinositide phospholipase C by the actin-binding protein, profilin. *Plant Journal*, **6**, 389–400.

Dudek, H., Datta, S.R., Franke, T.F. *et al.* (1997) Regulation of neuronal survival by the serine-threonine protein kinase Akt. *Science*, **275**, 661–665.

Elge, S., Brearley, C., Xia, H.J., Kehr, J., Xue, H.W. and Mueller-Roeber, B. (2001). An Arabidopsis inositol phospholipid kinase strongly expressed in procambial cells: synthesis of Ptdlns(4,5)P-2 and Ptdlns(3,4,5)P-3 in insect cells by 5-phosphorylation of precursors. *Plant Journal*, **26**, 561–571.

Ellson, C.D. *et al.* (2001) Ptdlns(3)P regulates the neutrophil oxidase complex by binding to the PX domain of p40(phox). *Nature Cell Biology*, **3**, 679–682.

Exton, J.H. (1997). New developments in phospholipase D. *Journal of Biological Chemistry*, **272**, 15579–15582.

Fedorov, A.A., Magnus, K.A., Graupe, M.H., Lattman, E.E., Pollard, T.D., and Almo, S. C. (1994) X-ray structures of isoforms of the actin-binding protein profilin that differ in their affinity for phosphatidylinositol phosphates. *Proceedings of the National Academy of Sciences USA*, **91**, 8636–8640.

Flanagan, C.A. and Thorner, J. (1992) Purification and characterization of a soluble phosphatidylinositol 4-kinase from the yeast *Saccharomyces cerevisiae*. *Journal of Biological Chemistry*, **267**, 24117-24125.

Flanagan, C.A., Schnieders, E.A., Emerick, A.W., Kunisawa, R., Admon, A. and Thorner, J. (1993) Phosphatidylinositol 4-kinase: gene structure and requirement for yeast cell viability. *Science*, **262**, 1444–1448.

Franke, T.F., Kaplan, D.R., Cantley, L.C. and Toker, A. (1997). Direct regulation of the Akt proto-oncogene product by phosphatidylinositol-3,4-bisphosphate. *Science*, **275**, 665–668.

Franke, T.F., Yang, S.I., Chan, T.O. *et al.* (1995) The protein-kinase encoded by the akt protooncogene is a target of the PDGF-activated phosphatidylinositol 3-kinase. *Cell*, **81**, 727–736.

Gehrmann, T., Vereb, G., Schmidt, M. *et al.* (1996). Identification of a 200 kDa polypeptide as type 3 phosphatidylinositol 4-kinase from bovine brain by partial protein and cDNA sequencing. *Biochimica et Biophysica Acta*, **1311**, 53–63.

Goldschmidt-Clermont, P.J., Kim, J.W., Machesky, L.M., Rhee, S.G., and Pollard, T. D. (1991) Regulation of phospholipase C-$\gamma$1 by profilin and tyrosine phosphorylation. *Science*, **251**, 1231–1233.

Goldschmidt-Clermont, P.J., Machesky, L.M., Baldassare, J.J. and Pollard, T.D. (1990) The actin-binding protein profilin binds to $PIP_2$ and inhibits its hydrolysis by phospholipase C. *Science*, **247**, 1575–1578.

Gross, W. and Boss, W.F. (1992) Inositol phospholipids and signal transduction, in *Control of Plant Gene Expression* (ed. D.P.S. Verma) *The Telford Press, Caldwell, New Jersey*, pp. 17–32.

Gungabissoon, R.A., Jiang, C.J., Drobak, B.K., Maciver, S.K. and Hussey, P.J. (1998) Interaction of maize actin-depolymerising factor with actin and phosphoinositides and its inhibition of plant phospholipase C. *Plant Journal*, **16**, 689–696.

Haarer, B.K., Petzold, A.S. and Brown, S.S. (1993). Mutational analysis of yeast profilin. *Molecular and Cellular Biology*, **13**, 7864–7873.

Haarer, B.K., Lillie, S.H., Adams, A.E.M., Magdolen, V., Bandlow, W., and Brown, S.S. (1990) Purification of profilin from *Saccharomyces cerevisiae* and analysis of profilin-deficient cells. *Journal of Cell Biology*, **110**, 105–114.

Hamada, K., Shimizu, T., Matsui, Y., Tsukita, S. and Hakoshima, T. (2000) Structural basis of the membrane-targeting and unmasking mechanisms of the radixin FERM domain. *EMBO Journal*, **19**, 4449–4462.

Hartwig, J.H., Bokoch, G.M., Carpenter, C.L., Janmey, P.A., Taylor, L.A., Toker, A. and Stossel, T.P. (1995) Thrombin receptor ligation and activated Rac uncap actin filament barbed ends through phosphoinositide synthesis in permeabilized human platelets. *Cell*, **82**, 643–653.

Harwood, J.L. (1980). Plant acyl lipids: structure, distribution and analysis, in *The Biochemistry of Plants*, Vol. 4. (ed. P.K. Stumpf), New York, Academic Press, pp. 1–55.

Haugwitz, M., Noegel, A.A., Karakesisoglou, J. and Schleicher, M. (1994) Dictyostelium amoeba that lack G-actin-sequestering profilins show defects in F-actin content, cytokinesis, and development. *Cell*, **79**, 303–314.

Hay, J.C. and Martin, T.F.J. (1993) Phosphatidylinositol transfer protein required for ATP-dependent priming of $Ca^{2+}$-activated secretion. *Nature*, **366**, 572–575.

Hay, J.C., Fisette, P.L., Jenkins, G.H., Fukami, K., Takenawa, T., Anderson, R.A. and Martin, T.F.J. (1995) ATP-dependent inositide phosphorylation required for $Ca^{2+}$-activated secretion. *Nature*, **374**, 173–177.

Heras, B. and Drøbak, B.K. (2002) PARF-1 – an *Arabidopsis thaliana* FYVE-domain protein displaying a novel eukaryotic domain structure and phosphoinositide affinity. *Journal of Experimental Botany*, **53**(368): 565–567.

Hong, Z.G. and Verma, D.P.S. (1994). A phosphatidylinositol 3-kinase is induced during soybean nodule organogenesis and is associated with membrane proliferation. *Proceedings of the National Academy of Sciences USA*, **91**, 9617–9621.

Irvine, R.F., Letcher, A.J., Lander, D.J., Drøbak, B.K., Dawson, A.P. and Musgrave, A. (1989) Phosphatidylinositol(4,5)bisphosphate and Phosphatidylinositol(4)phosphate in plant tissues. *Plant Physiology*, **89**, 888–892.

Itoh, T., Koshiba, S., Kigawa, T., Kikuchi, A., Yokoyama, S. and Takenawa, T. (2001) Role of the ENTH domain in phosphatidylinositol-4,5-bisphosphate binding and endocytosis. *Science*, **291**, 1047–1051.

Jung, J.Y., Kim, Y.W., Kwak, J.M. *et al.* (2002). Phosphatidylinositol 3- and 4-phosphate are required for normal stomatal movements. *Plant Cell*, **14**, 2399–2412.

Kanai, F., Liu, H., Field, S.J. *et al.* (2001). The PX domains of p47phox and p40phox bind to lipid products of PI(3)K. *Nature Cell Biology*, **3**, 675–678.

Kang, F., Purich, D.L. and Southwick, F.S. (1999). Profilin promotes barbed-end actin filament assembly without lowering the critical concentration. *Journal of Biological Chemistry*, **274**, 36963–36972.

Kearns, B.G., McGee, T.P., Mayinger, P., Gedvilaite, A., Phillips, S.E., Kagiwada, S. and Bankaitis, V.A. (1997). Essential role for diacylglycerol in protein transport from the yeast Golgi complex. *Nature*, **387**, 101–105.

Kim, D.H., Eu, Y.-J., Yoo, C.M. *et al.* (2001) Trafficking of phosphatidylinositol 3-phosphate from the *trans*-Golgi network to the lumen of the central vacuole in plant cells. *Plant Cell*, 13, 287–301.

Kutateladze, T. and Overduin, M. (2001) Structural mechanism of endosome docking by the FYVE domain. *Science*, **291**, 1793–1796.

Lambrechts, A., Jonckheere, V., Dewitte, D., Vandekerckhove, J. and Ampe, C. (2002) Mutational analysis of human profilin I reveals a second PI(4,5)-P2 binding site neighbouring the poly(L-proline) binding site. *BMC Biochemistry*, **3**(1): 12.

Lassing, I. and Lindberg, U. (1985). Specific interaction between phosphatidylinositol 4,5-bisphosphate and profilactin. *Nature*, **314**, 472–474.

Lassing, I. and Lindberg, U. (1988) Evidence that the phosphatidylinositol cycle is linked to cell motility. *Experimental Cell Research*, **174**, 1–15.

Liscovitch, M. and Cantley, L.C. (1995). Signal transduction and membrane traffic: the PITP/phosphoinositide connection. *Cell*, **81**, 659–662.

Loijens, J.C. and Anderson, R.A. (1996) Type I phosphatidylinositol-4-phosphate 5-kinases are distinct members of this novel lipid kinase family. *Journal of Biological Chemistry*, **271**, 32937–32943.

Lu, J. and Pollard, T.D. (2001) Profilin binding to poly-*L*-proline and actin monomers along with ability to catalyze actin nucleotide exchange is required for viability of fission yeast. *Molecular Biology of the Cell*, **12**, 1161–1175.

Lu, P.-J., Shieh, W.-R., Rhee, S. G., Yin, H. L. and Chen, C.-S. (1996) Lipid products of phosphoinositide 3-kinase bind human profilin with high affinity. *Biochemistry*, **35**, 14027–14034.

Lundberg, G.A. and Sommarin, M. (1992) Diacylglycerol kinase in plasma membranes from wheat. *Biochimica et Biophysica Acta*,**1123**, 177–183.

Lykidis, A., Jackson, P.D., Rock, C.O. and Jackowski, S. (1997) The role of CDP-diacylglycerol synthetase and phosphatidylinositol synthase activity levels in the regulation of cellular phosphatidylinositol content. *Journal of Biological Chemistry*, **272**, 33402–33409.

Machesky, L.M., Atkinson, S.J., Ampe, C., Vandekerckhove, J. and Pollard, T.D. (1994) Purification of a cortical complex containing two unconventional actins from *Acanthamoeba* by affinity chromatography on profilin–agarose. *Journal of Cell Biology*, **127**, 107–115.

Maciver, S.K. (1998) How ADF/cofilin depolymerizes actin filaments – Commentary. *Current Opinion in Cell Biology*, **10**, 140–144.

Matsuoka, K., Bassham, D.C., Raikhel, N.V. and Nakamura, K. (1995). Different sensitivity to wortmannin of 2 vacuolar sorting signals indicates the presence of distinct sorting machineries in tobacco cells. *Journal of Cell Biology*, **130**, 1307–1318.

McEwen, R.K., Dove, S.K., Cooke, F.T. *et al.* (1999) Complementation analysis in PtdInsP kinase-deficient yeast mutants demonstrates that Schizosaccharomyces pombe and murine Fab1p homologues are phosphatidylinositol 3-phosphate 5-kinases. *Journal of Biological Chemistry*, **274**, 33905–33912.

Meijer, H.J.G., Berrie, C.P., Iurisci, C., Divecha, N., Musgrave, A. and Munnik, T. (2001) Identification of a new polyphosphoinositide in plants, phosphatidylinositol 5-monophosphate (PtIns5P), and its accumulation upon osmotic stress. *Biochemical Journal*, **360**, 491–498.

Memon, A.R. and Boss, W.F. (1990) Rapid light-induced changes in phosphoinositide kinases and $H^+$-ATPase in plasma membrane of sunflower hypocotyls. *Journal of Biological Chemistry*, **265**, 14817–14821.

Mikami, K., Iuchi, S., Yamaguchi-Shinozaki, K. and Shinozaki K. (2000) A novel *Arabidopsis thaliana* dynamin-like protein containing the pleckstrin homology domain. *Journal of Experimental Botany*, **51**, 317–318.

Mikami, K, Takahashi, S., Katagiri, T., Yamaguchi-Shinozaki K. and Shinozaki, K. (1999). Isolation of an *Arabidopsis thaliana* cDNA encoding a pleckstrin homology domain protein, a putative homologue of human pleckstrin. *Journal of Experimental Botany*, **50**, 729–730.

Misra, S. and Hurley, J.H. (1999) Crystal structure of a phosphatidylinositol 3-phosphate-specific membrane targeting motif, the FYVE domain of Vps27p. *Cell*, **97**, 657–666.

Miki, H., Miura, K. and Takenawa, T. (1996) N-WASP, a novel actin-depolymerizing protein, regulates the cortical cytoskeletal arrangement in a PtdIns(4,5)P2-dependent manner downstream of tyrosine kinases. *EMBO Journal*, **15**, 5326–5335.

Moore, J.D. (2001) The Ran-GTPase and cell-cycle control. *BioEssays*, **23**, 77–85.

Moore, T.S., Jr. (1990). Biosynthesis of phosphatidylinositol, in *Inositol Metabolism in Plants* (eds D.J. Morré, W.F. Boss and F.A. Loewus) Wiley-Liss, NY, pp. 107–112.

Mueller-Roeber, B. and Pical, C. (2002). Inositol phospholipid metabolism in Arabidopsis. Characterized and putative isoforms of inositol phospholipid kinase and phosphoinositide-specific phospholipase C. *Plant Physiology*, **130**, 22–46.

Munnik, T., Irvine, R.F. and Musgrave, A. (1994a) Rapid turnover of phosphatidylinositol 3-phosphate in the green alga Chlamydomonas eugametos – signs of a phosphatidylinositide 3-kinase signalling pathway in lower plants. *Biochemical Journal*, **298**, 269–273.

Munnik, T., Irvine, R.F. and Musgrave, A. (1998) Phospholipid signalling in plants. *Biochimica et Biophysica Acta*, **1389**, 222–272.

Munnik, T. Musgrave, A. and Devrije, T. (1994b) Rapid turnover of polyphosphoinositides in carnation flower petals. *Planta*, **193**, 89–98.

Nakagawa, T., Goto, K. and Kondo, H. (1996) Cloning, expression, and localization of 230-kDa phosphatidylinositol 4-kinase. *Journal of Biological Chemistry*,**271**, 12088–12094.

Nikawa, J. and Yamashita, S. (1997) Phosphatidylinositol synthase from yeast. *Biochimica et Biophysica Acta*, **1348**, 173–178.

Palmer, R.H., Dekker, L.V., Woscholski, R., Legood, J.A., Gigg, R and Parker, P.J. (1995) Activation of prk1 by phosphatidylinositol 4,5-bisphosphate and phosphatidylinositol 3,4,5-trisphosphate – a comparison with protein-kinase-c isotypes. *Journal of Biological Chemistry*, **270**, 22412–22416.

Pantaloni, D. and Carlier, M.-F. (1993) How profilin promotes actin filament assembly in the presence of thymosin ß4. *Cell*, **75**, 1007–1014.

Pappan, K., Qin, W., Dyer, J.H., Zheng, L. and Wang, X. (1997) Molecular cloning and functional analysis of polyphosphoinositide-dependent phospholipase D, PLDß, from Arabidopsis. *Journal of Biological Chemistry*, **272**, 7055–7061.

Pike, L.J. (1992) Phosphatidylinositol 4-kinases and the role of polyphosphoinositides in cellular regulation. *Endocrine Reviews*, **13**, 692–706.

Quirk, S., Maciver, S.K., Ampe, C. *et al.* (1993) Primary structure of and studies on Acanthamoeba actophorin. *Biochemistry*, **32**, 8525–8533.

Rameh, L.E., Tolias, K.F., Duckworth, B.C. and Cantley, L.C. (1997a) A new pathway for synthesis of phosphatidylinositol-4,5-bisphosphate. *Nature*, **390**, 192–196.

Rameh, L.E., Arvidsson, A.K., Carraway, K.L. *et al.* (1997b) A comparative analysis of the phosphoinositide binding specificity of pleckstrin homology domains. *Journal of Biological Chemistry*, **272**, 22059–22066.

Randazzo, P.A. (1997) Functional interaction of ADP-ribosylation factor 1 with phosphatidylinositol 4,5-bisphosphate. *Journal of Biological Chemistry*, **272**, 7688–7692.

Roepman, R., vanDuijnhoven, G., Rosenberg, T. *et al.* (1996) Positional cloning of the gene for X-linked retinitis pigmentosa 3: homology with the guanine-nucleotide-exchange factor RCC1. *Human Molecular Genetics*, **5**, 1035–1041.

Sandelius, A.S. and Morré, D.J. (1987) Characteristics of a phosphatidylinositol exchange activity of soybean microsomes. *Plant Physiology*, **84**, 1022–1027.

Sandelius, A.S. and Sommarin, M. (1990) Membrane-localized reactions involved in polyphosphoinositide turnover in plants, in, *Inositol Metabolism in Plants* (eds D.J. Morré, W.F. Boss and F.A. Loewus) Wiley-Liss, NY, pp. 139–161.
Schaller, G.E., Harmon, A. and Sussman, M.R. (1992) Characterization of a calcium- and lipid-dependent protein kinase associated with the plasma membrane in oat. *Biochemistry*, **31**, 1721–1727.
Sexton, J.C. and Moore, T.S. (1981) Phosphatidylinositol synthesis by a $Mn^{2+}$-dependent exchange enzyme in castor bean endosperm. *Plant Physiology*, **68**, 18–22.
Shibasaki, Y., Ishihara, H., Kizuki, N., Asano, T., Oka, Y. and Yazaki, Y. (1997). Massive actin polymerization induced by phosphatidylinositol-4-phosphate 5-kinase in vivo. *Journal of Biological Chemistry*, **272**, 7578–7581.
Skare, P. and Karlsson, R. (2002) Evidence for two interaction regions for phosphatidylinositol(4,5) bisphosphate on mammalian profilin I. *FEBS Letters*, **522**, 119–124.
Smertenko, A.P., Jiang, C.J., Simmons, N.J., Weeds, A.G., Davies, D.R. and Hussey, P.J. (1998) Ser6 in the maize actin-depolymerizing factor, ZmADF3, is phosphorylated by a calcium-stimulated protein kinase and is essential for control of functional activity. *Plant Journal*, **14**, 187–193.
Sohn, R.H., Chen, J., Koblan, K.S., Bray, P.F. and Goldschmidt-Clermont, P.J. (1995) Localization of a binding site for phosphatidylinositol-4,5-bisphosphate on human profilin. *Journal of Biological Chemistry*, **270**, 21114–21120.
Staiger, C.J. (2000) Signaling to the actin cytoskeleton in plants. *Annual Review of Plant Physiology and Plant Molecular Biology*, **51**, 257–288.
Staiger, C.J. and Hussey, P.J. (2003) Actin and actin-modulating proteins, in: *The Plant Cytoskeleton in Cell Differentiation and Development* (ed. P.J. Hussey) Blackwell, Publishers, UK, pp. 31–79.
Staiger, C.J., Goodbody, K.C., Hussey P.J. *et al.* (1993). The profilin multigene family of maize – differential expression of 3 isoforms. *Plant Journal*, **4**, 631–641.
Stenmark, H., Aasland, R., Toh, B.H. and Darrigo, A. (1996) Endosomal localization of the autoantigen EEA1 is mediated by a zinc-binding FYVE finger. *Journal of Biological Chemistry*, **271**, 24048–24054.
Stephens, L.R., Hughes, K.T. and Irvine, R.F. (1991) Pathway of phosphatidylinositol(3,4,5)trisphosphate synthesis in activated neutrophils. *Nature*, **351**, 33–39.
Stevenson, J.M., Perera, I.Y. and Boss, W.F. (1998). A phosphatidylinositol 4-kinase pleckstrin homology domain that binds phosphatidylinositol 4-monophosphate. *Journal of Biological Chemistry*, **273**, 22761–22767.
Sylvia, V., Curtin, G., Norman, J., Stec, J. and Busbee, D. (1988). Activation of a low specific activity form of dna polymerase-alpha by inositol-1,4-bisphosphate. *Cell*, **54**, 651–658.
Takenawa, T. and Itoh, T. (2001) Phosphoinositides, key molecules for regulation of actin cytoskeletal organization and membrane traffic from the plasma membrane. *Biochimica et Biophysica Acta*,**1533**, 190–206.
Tanaka, M. and Shibata, H. (1985) Poly (L-proline)-binding proteins from chick embryos are a profilin and a profilactin. *European Journal of Biochemistry*, **151**, 291–297.
Tolias, K.F., Cantley, L.C. and Carpenter, C.L. (1995) Rho family GTPases bind to phosphoinositide kinases. *Journal of Biological Chemistry*, **270**, 17656–17659.
Valenta, R., Duchêne, M., Pettenburger, K. *et al.* (1991) Identification of profilin as a novel pollen allergen; IgE autoreactivity in sensitized individuals. *Science*, **253**, 557–560.
Van der Biezen, E.A. and Jones, J.D.G. (1998). The NB-ARC domain: a novel signalling motif shared by plant resistance gene products and regulators of cell death in animals. *Current Biology*, **8**, 226–227.
Verheyen, E.M., and Cooley, L. (1994) Profilin mutations disrupt multiple actin-dependent processes during *Drosophila* development. *Development*, **120**, 717–728.
von Witsch, M., Baluska, F., Staiger, C. J. and Volkmann, D. (1998) Profilin is associated with the plasma membrane in microspores and pollen.*European Journal of Cell Biology*, **77**, 303–312.
Wasserman, S. (1998). FH proteins as cytoskeletal organizers. *Trends in Cell Biology*, **8**, 111–115.
Welters, P., Takegawa, K., Emr, S.D. and Chrispeels, M.J. (1994). ATVPS34, A phosphatidylinositol 3-kinase of Arabidopsis thaliana, is an essential protein with homology to

a calcium-dependent lipid-binding domain. *Proceedings of the National Academy of Sciences USA*, **91**, 11398–11402.

Westergren, T., Dove, S.K., Sommarin, M. and Pical, C. (2001). AtPIP5K1, an *Arabidopsis thaliana* phosphatidylinositol phosphate kinase, synthesizes PtdIns(3,4)P-2 and PtdIns(4,5)P-2 *in vitro* and is inhibited by phosphorylation. *Biochemical Journal*, **359**, 583–589.

Whiteford, C.C., Brearley, C.A. and Ulug, E.T. (1997) Phosphatidylinositol 3,5-bisphosphate defines a novel PI 3-kinase pathway in resting mouse fibroblasts. *Biochemical Journal*, **323**, 597–601.

Witke, W., Sutherland, J.D., Sharpe, A., Arai, M. and Kwiatkowski, D.J. (2001) Profilin I is essential for cell survival and cell division in early mouse development. *Proceedings of the National Academy of Sciences USA*, **98**, 3832–3836.

Wong, K. and Cantley, L.C. (1994). Cloning and characterization of a human phosphatidylinositol 4-kinase. *Journal of Biological Chemistry*, **269**, 28878–28884.

Wong, K., Meyers, R. and Cantley, L.C. (1997) Subcellular localizations of phosphatidylinositol 4-kinase isoforms. *Journal of Biological Chemistry*, **272**, 13236–13241.

Xue, H.W., Pical, C., Brearley, C., Elge, S. and Muller-Rober B. (1999). A plant 126-kDa phosphatidylinositol 4-kinase with a novel repeat structure – Cloning and functional expression in baculovirus-infected insect cells. *Journal of Biological Chemistry*, **274**, 5738–5745.

Xu, Y., Hortsman, H., Seet, L., Wong, S.H. and Hong, W. (2001). SNX3 regulates endosomal function through its PX-domain-mediated interaction with PtdIns(3)P. *Nature Cell Biology*, **3**, 658–666.

Yonezawa, N., Homma, Y., Yahara, I., Sakai, H. and Nishida, E. (1991) A short sequence responsible for both phosphoinositide binding and actin binding activities of cofilin. *Journal of Biological Chemistry*, **266**, 17218–17221.

Yonezawa, N., Nishida, E., Iida, K., Yahara, I. and Sakai, H. (1990) Inhibition of the interactions of cofilin, destrin, and deoxyribonuclease-1 with actin by phosphoinositides. *Journal of Biological Chemistry*, **265**, 8382–8386.

Yoshida, S., Ohya, Y., Nakano, A. and Anraku, Y. (1994). Genetic interactions among genes involved in the STT4-PKC1 pathway of *Saccharomyces cerevisiae*. *Molecular & General Genetics*, **242**, 631–640.

Zhang X, Loijens, J.C., Boronenkov, I.V., Parker, G.J., Norris, F.A., Chen, J., Thum, O., Prestwich, G.D., Majerus, P.W. and Anderson, R.A. (1997). Phosphatidylinositol-4-phosphate 5-kinase isozymes catalyze the synthesis of 3-phosphate-containing phosphatidylinositol signaling molecules. *Journal of Biological Chemistry*, **272**, 17756–17761.

Zhao, K., Wang, W., Rando, O.J., Xue, Y., Swiderek, K., Kuo, A., and Crabtree, G. R. (1998) Rapid and phosphoinositol-dependent binding of the SWI/SNF-like BAF complex to chromatin after T lymphocyte receptor signaling. *Cell*, **95**, 625–636.

Zheng, Y., Glaven, J.A., Wu, W.J. and Cerione, R.A. (1996). Phosphatidylinositol 4,5-bisphosphate provides an alternative to guanine nucleotide exchange factors by stimulating the dissociation of GDP from Cdc42Hs. *Journal of Biological Chemistry*, **271**, 23815–23819.

# 10 Oxylipins

Sabine Rosahl and Ivo Feussner

## 10.1 Introduction: synthesis of oxylipins

The composition of lipids is frequently altered during development throughout the whole plant. Apart from the turnover of fatty acids within lipids, formation of oxidised polyunsaturated fatty acids (PUFAs) is an especially important reaction in lipid metabolism. Products of the oxidation of lipids and the resulting metabolites are collectively called oxylipins (Hamberg, 1993; Farmer *et al.*, 2003). This PUFA oxidation is believed to play a regulatory role in plant development and in permanent adjustment to diverse and variable environmental conditions. The last decade has seen a remarkable increase in our understanding of the complexity of the formation of oxylipins (Blée, 1998b). The initial formation of oxylipins may occur either by autoxidation or by the action of enzymes, such as lipoxygenases (LOXs) or α-dioxygenases (α-DOXs). The formation of oxylipins by chemical oxidation (autoxidation) processes is so far regarded as a rather unspecific process leading to complex mixtures of fatty acids harbouring hydroperoxy, hydroxy and epoxy groups in racemic configuration (Frankel, 1998). In addition, substantial amounts of cyclic compounds are found – a group collectively called phytoprostanes being the most prominent (Müller, 1998). The metabolism of PUFAs via the LOX-catalysed step and the subsequent reactions are collectively named the LOX pathway (Feussner and Wasternack, 2002). Recent cloning, expression and functional analysis of genes coding for LOXs and other members of the LOX pathway, as well as metabolite profiling, shed new light on the function of LOXs and the downstream enzymes acting in different branches of the LOX pathway. Consequently, our understanding of the LOX pathway in distinct stress- and developmentally regulated processes has been substantially improved. New aspects of the biochemistry of the LOX pathway will be reviewed in the first section of this chapter. The involvement of oxylipins in signal transduction chains as a consequence of biotic stress is the main focus of the second section and the role of oxylipins in plant pathogen interactions will be summarised in the third section of this chapter.

### *10.1.1 LOX pathway*

The LOX pathway starts with the LOX reaction leading to formation of hydroperoxy PUFAs, which then serve as substrates for at least eight different enzyme families in plants forming the different branches of the LOX pathway (Figure 10.1) (Feussner and Wasternack, 2002). Two additional enzymes are known, which may compete for the PUFA as a substrate: these are α-DOX (Hamberg *et al.*, 1999) and peroxygenase (Blée, 1998a). LOXs (linoleate:oxygen oxidoreductase, EC 1.13.11.12) constitute a large gene family of non-haeme iron containing fatty acid DOXs, which are ubiquitous in plants and animals. LOXs catalyse the regio- and stereospecific insertion of

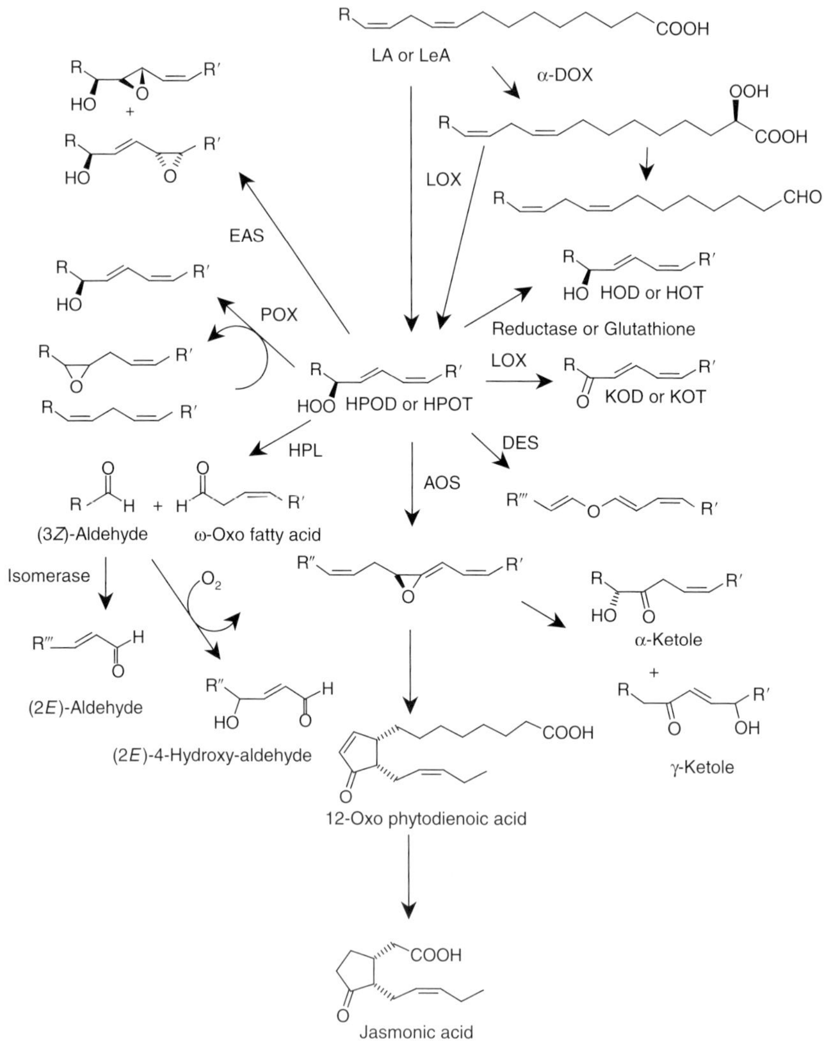


**Figure 10.1** The LOX pathway: metabolic routes of LOX-dependent catabolism of PUFAs in plants.

oxygen to the (1*Z*,4*Z*)-pentadiene system of PUFAs. Therefore, the most prominent substrates in plants are roughanic acid (RA; *all-cis*-7,10,13-hexadecatrienoic acid), linoleic acid (LA; *all-cis*-9,12-octadecadienoic acid) and α-linolenic acid (LeA; *all-cis*-9,12,15-octadecatrienoic acid). Since LA is a major PUFA in the plant kingdom, plant LOXs are classified with respect to their positional specificity of LA oxygenation. LA may be either oxygenated at carbon atom 9 of the hydrocarbon backbone of the fatty acid in case of 9-LOXs, or at C-13 in case of 13-LOXs. This leads to formation of two compounds, the (9*S*)-hydroperoxy- and the (13*S*)-hydroperoxy derivative of LA being the substrate for numerous reactions in the LOX pathway (Feussner and Wasternack, 2002). An alternative classification of plant LOXs has been proposed, which is based on comparison of their primary structure. According to their overall sequence similarity, plant LOXs can be grouped into two subfamilies: those enzymes, which do not possess a transit peptide, have a high sequence similarity (>75%) to one another and are designated *type 1*-LOXs. Within this group, 9- and 13-LOXs can be found. However, another subset of LOXs carries a chloroplast transit peptide sequence. Based on this *N*-terminal extension and the fact that these enzymes show only a moderate overall sequence similarity (~35 %) to one another, they have been classified as *type 2*-LOXs (Shibata *et al.*, 1994). Up to now, these LOX forms all belong to the subfamily of 13-LOXs (Feussner and Wasternack, 2002).

Plant LOXs are versatile catalysts because (1) some of them are found in large amounts in vegetative tissues, (2) they are stable over long periods of time and (3) they are multifunctional enzymes. LOXs catalyse at least three different types of reactions: (1) dioxygenation of lipid substrates (DOX reaction) leading to hydroperoxy lipids, (2) secondary conversion of hydroperoxy lipids (hydroperoxidase reaction) leading to keto lipids and (3) formation of epoxy leukotrienes (leukotriene synthase reaction). However, under physiological conditions, the DOX reaction is most prevalent in plants. The formation of the two different regioisomers of hydroperoxy PUFAs may be explained by the space within the active site and the orientation of the substrate (Feussner and Kühn, 2000). Both factors are important determinants for the positional specificity of plant LOXs, and are modified by additional factors, such as substrate concentration, the physico-chemical state of the substrate, pH or temperature. However, it should be stressed that for other LOXs, regiospecificity may be determined in a more complex manner. For a more detailed discussion on the enzyme mechanism we refer to the following reviews: Gardner (1991), Veldink and Vliegenthart (1991), Feussner and Kühn (2000).

The majority of plant LOXs strongly prefer free fatty acids as substrates. However, as predicted by the model of substrate orientation, 13-LOX activity with PUFAs esterified to more complex lipids has been found in a few cases (Feussner and Kühn, 2000). For example, LOX1 from soybean seeds, LOX H1 from potato leaves and a LOX from cucumber roots showed activity against PUFAs esterified to phospholipids (Matsui *et al.*, 1998; Perez-Gilabert *et al.*, 1998; Leon *et al.*, 2002). This led to the suggestion of an involvement of LOXs in membrane permeabilisation. For another subset of 13-LOXs, e.g. the lipid-body LOX from cucumber, the 13-LOX from barley seedlings and the vegetative LOX VLXD from soybean leaves, activity has been described with PUFAs esterified in neutral lipids such as triacylglycerols (Holtman *et al.*, 1997; Balkenhohl *et al.*, 1998; Fuller *et al.*, 2001). Therefore, an involvement in triacylglycerol catabolism has been suggested for these LOXs (Feussner *et al.*, 2001). Whether there are additional *type 2*-LOXs with specificity towards galactolipids remains to be analysed.

Another aspect in understanding the biochemistry and, thus, the physiological functions of different LOXs is their intracellular localization. This has been analysed in detail in cotyledons, fruits, and leaves of cucumber and soybean. A parallel occurrence of particulate and cytosolic LOX forms was observed. In cotyledons, besides soluble LOXs detected in the cytosol, vacuole and plastid, particulate LOXs were found in microsomal membranes, plasma membranes and lipid bodies. Five different LOXs could be distinguished in etiolated cucumber cotyledons (Feussner and Kindl, 1994), whereas more than six different forms of LOXs were identified in soybean leaves (Grimes *et al.*, 1993). In other organisms such as spinach, barley, tomato, potato or Arabidopsis, a preferential occurrence of LOXs within chloroplasts of leaves has been detected (Feussner *et al.*, 1995; Blée and Joyard, 1996; Heitz *et al.*, 1997; Leon *et al.*, 2002). Here the majority of LOXs seem to appear as soluble forms in the stroma fraction (Bachmann *et al.*, 2002), but at least in case of spinach, substantial LOX activity was detected in the particulate fraction of chloroplasts and could be attributed to the envelope fraction (Blée and Joyard, 1996). The different LOX forms may have a specific location in order to either get access to different substrate pools or to orchestrate the formation of hydroperoxy PUFAs. These might subsequently be channelled into distinct branches of the LOX pathway.

Data from the last 5 years suggest that the majority of accumulating hydroperoxy PUFAs arise from action of LOXs (Figure 10.1). A minority of PUFAs may be converted by α-DOX into α-hydro(pero)xy PUFAs, or by peroxygenase. In addition, autoxidation products, such as phytoprostanes, may be found. Thus, plants predominantly synthesise substances from the (9*S*)-hydroperoxy and the (13*S*)-hydroperoxy types of PUFAs. These are subsequently metabolised via a number of secondary reactions. To date, five major metabolic routes for the metabolism of hydroperoxy fatty acids have been characterised in more detail. For simplicity, all routes are named after the first enzyme starting into the respective pathway. (1) The peroxygenase (POX) or formerly hydroperoxide isomerase pathway: intramolecular oxygen transfer converts fatty acid hydroperoxides to epoxy- or dihydrodiol PUFAs (Blée, 1998a). (2) The AOS pathway: AOS, first named hydroperoxide dehydratase, forms unstable allene oxides, which either undergo non-enzymatic hydrolysis leading to α- and γ-ketols, or may be metabolised to chiral (9*S*,13*S*)-12-oxo phytodienoic acid (OPDA) by an allene oxide cyclase (AOC) (Wasternack and Hause, 2002). (3) The HPL pathway: HPL, first named hydroperoxide isomerase as well, catalyses the oxidative cleavage of the hydrocarbon backbone of fatty acid hydroperoxides. This leads to the formation of short chain aldehydes ($C_6$- or $C_9$-) and the corresponding $C_{12}$- or $C_9$-ω-oxo fatty acids (Noordermeer *et al.*, 2001). (4) The DES pathway: this pathway forms divinyl ethers such as colneleic acid (CA) or colnelenic acid (CnA). (5) The α-DOX pathway: recently, it has been shown that α-DOX is capable of oxygenating (9*S*)-hydroperoxy PUFAs leading to diol PUFAs or odd chain fatty acid derivatives and that 13-LOXs are capable of oxygenating α-DOX-derived (2*S*)-hydro(pero)xy fatty acids (Hamberg *et al.*, 2003). Therefore, the side chain R′ in Figure 10.1 might be different in case of α-DOX-dependent PUFA oxygenation. It should be stressed that, except for POX, cDNAs for all enzymes mentioned above have been isolated over the last years. Moreover, recent cloning of a large number of cDNAs coding for enzymes of the AOS, HPL and DES pathways revealed that all of them belong to one P450-containing enzyme subfamily, named *CYP74* (Feussner and Wasternack, 2002).

In addition to these five pathways, there are other reactions for hydroperoxide metabolism, which are less well-characterised or their physiological significance is less clear. (6) The LOX-catalysed hydroperoxidase reaction (ketodiene forming pathway): under low oxygen pressure, LOXs are capable of catalysing the homolytic cleavage of the O–O bond forming alkoxy radicals that may rearrange to ketodienes (Veldink and Vliegenthart, 1991). (7) The epoxy alcohol synthase (EAS) pathway: epoxy hydroxy fatty acids are formed by intramolecular rearrangement of hydroperoxy fatty acids catalysed by EAS (Hamberg, 1999). Whether this reaction can be put down to the epoxy leukotriene reaction of LOXs or to a yet unidentified enzyme remains to be investigated. In addition, products of the EAS reaction can be regiochemically identical to the POX reaction products, but they differ with respect to their stereochemistry. To date, the EAS pathway has been found only in solanaceous species and these oxylipins may be active in pathogen defence responses (see below). (8) The reductase pathway: Hydroperoxy fatty acids are reduced to their corresponding hydroxy derivatives. This reduction mechanism is yet not established, but may be chemically facilitated by e.g. glutathione, and it is independent from POX (Feussner *et al.*, 1998).

### *10.1.2 The CYP74 family*

Since *CYP74*s are the best characterised members of the LOX pathway except LOXs themselves, they will be described in more detail in the next section. *CYP74*s form an atypical family of P450 monooxygenases and constitute at least three different enzyme subfamilies: AOS, HPL and DES (Figure 10.1). Members of the family do not require either molecular oxygen or NAD(P)H-dependent cytochrome P450-reductase. The new carbon–oxygen bonds are formed by using an acyl hydroperoxide both as the substrate and the oxygen donor. Reduced affinity for CO is a common feature as well of the *CYP74*s, which are similar in catalytic properties to other P450s, such as prostacyclin synthase and thromboxane synthase of the arachidonic acid cascade in animal systems. Characteristic for the AOS, HPL and DES reactions is that they all share an intermediate epoxy allylic carbocation formed from the acyl hydroperoxide but further on, AOS and DES deprotonate the carbocation at different positions leading to stable derivatives, whereas HPLs catalyse a rearrangement of the positive charge leading to a fragmentation of the molecule into two aldehydes (Noordermeer *et al.*, 2001).

The AOS from flax was the first member of the *CYP74* family to be cloned (Song *et al.*, 1993). Therefore, AOS cDNAs are grouped as *CYP74A*. All *CYP74*-cDNAs analysed so far prefer free fatty acids as substrates when expressed as recombinant proteins, whereas fatty acid methyl esters, as well as *N*-acyl(ethanol)amines, are poor substrates (Shrestha *et al.*, 2002). Within the *CYP74A* subfamily, most enzymes are specific for 13-hydroperoxides as substrates and are called 13-AOS. However, enzymes from barley and *Physcomitrella patens* show no substrate specificity for either (9*S*)-hydroperoxides nor (13*S*)-hydroperoxides (Maucher *et al.*, 2000) and are consequently called 9/13-AOS. In addition, cDNAs coding for 9-AOS have been isolated recently from tomato and potato (Feussner and Wasternack, 2002; Itoh *et al.*, 2002). Based on sequence similarity, they belong to the *CYP74C* subfamily. Two other *CYP74* subfamilies contain enzymes with HPL activity with different substrate preference. Thus, *CYP74B* subfamily members accept specifically (13*S*)-hydroperoxides (13-HPLs). HPLs with no substrate preference (9/13-HPLs) are grouped into the subfamily *CYP74C*. A 9-HPL has not been isolated so far (Noordermeer *et al.*, 2001). The DESs belong to the last group of enzymes being recognised as members of the *CYP74* family. So far, only two cDNAs from potato and tomato have been isolated. Both are highly specific for (9*S*)-hydroperoxides (9-DES) and have been grouped into the *CYP74D* subfamily (Itoh and Howe, 2001; Stumpe *et al.*, 2001).

Data on the intracellular localisation of *CYP74*s are scarce and restricted to 13-AOSs, 9/13-AOSs and 13-HPLs, respectively. However, all members appear to be particulate enzymes. A plastidic location seems to be clear in case of those *CYP74* cDNAs that encode chloroplastic transit peptides. This was supported by immunocytochemical analysis in barley leaves and *in vitro* import experiments of cDNAs from tomato (Maucher *et al.*, 2000; Froehlich *et al.*, 2001). In import studies, 13-HPL was localised within the outer envelope, and 13-AOS was found within the inner envelope of spinach chloroplasts. However, the sequence diversity among *CYP74*s suggests that distribution of these enzymes within the cell and even within the plastid envelope should be more complex. Our current knowledge, on the general compartmentation of the LOX pathway, can be summarised as follows (Figure 10.2): since cytosolic 13-LOXs are not found ubiquitously in plants, the cytosol might be mainly a source of (9*S*)-hydroperoxide PUFAs formed by 9-LOXs occurring ubiquitously in plants. This leads to the model that 9-LOX-derived branches of the LOX pathway should be localised outside the plastid, mainly in the cytosol or at the cytosolic side of the outer envelope. Therefore, the formation of products derived from the activity of 9-AOS, 9-DES and 9/13-HPL seem to be localised in the cytosol. Whether 9-LOX-derived hydroperoxy PUFAs may have access to enzymes of the 9/13-AOS type within the plastid remains to be analysed. However, for the chloroplast stroma, the opposite holds true, since, until now, only 13-LOXs were detected

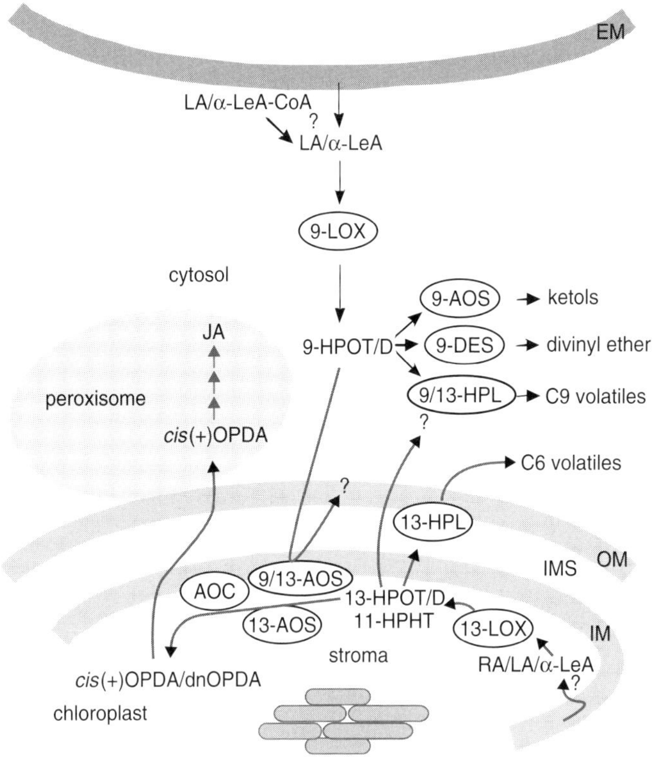


**Figure 10.2** Intracellular localisation of LOX pathway reactions. OM/IM; outer and inner membrane of plastid envelope; IMS, inter membrane space; EM, endomembrane system.

in the group of plastidic *type 2*-LOXs. (13*S*)-hydroperoxide metabolising *CYP74*s as well as other members of the pathway should, therefore, be predominantly localised in the inner envelope or in the stroma. For the 9-LOX pathway, the substrate may then derive from the PUFA-CoA ester pool of the cytosol. Alternatively, since the envelope harbours a substantial amount of phospholipids, PUFAs consumed in the 13-LOX and 9-LOX reactions might be generated by phospholipases presumably located in the envelope thus providing the substrate for both sides of the membranes.

### *10.1.3 Jasmonic acid biosynthesis enzymes*

A main product of the AOS pathway is the phytohormone, jasmonic acid (JA). Since it has numerous functions in signal transduction events discussed in Sections 10.2 and 10.3 of this chapter, its biosynthesis will be described in more detail here (Figure 10.2). As discussed above, the first steps of JA biosynthesis take place in the chloroplast at the level of the free fatty acid derivatives

(13-LOX, 13-AOS- and AOC reaction). Release of the substrate for JA biosynthesis, LeA (which is abundantly esterified in galactolipids and phospholipids) is proposed to occur by the action of phospholipases, such as $PLA_2$ (Narvaez-Vasquez *et al.*, 1999; Dhondt *et al.*, 2000; Göbel *et al.*, 2001). In case of Arabidopsis flowers, the enzyme catalysing this crucial step was identified as a $PLA_1$ localised in the chloroplasts envelope in the mutant *defective in anther dehiscence1* (*dad1*; Ishiguro *et al.*, 2001; Table 9.1). LeA is then oxygenated by *type* 2-13-LOXs to yield (13*S*)-hydroperoxy LeA (13-HPOT), the substrate for AOS (Laudert and Weiler, 1998; Kubigsteltig *et al.*, 1999). Conversion of 13-HPOT by AOS leads to formation of an unstable allene oxide, which is then cyclised to OPDA by the action of AOC, the enzyme catalysing the stereospecificity-determining step of JA biosynthesis (Ziegler *et al.*, 2000). The Arabidopsis transposon insertion mutant *delayed-dehiscence2* (*dde2*) is mutated in the single AOS gene and is male-sterile (von Malek *et al.*, 2002; Table 9.1). Interestingly, overexpression of AOS in Arabidopsis does not lead to constitutively elevated levels of JA, but to earlier and higher JA accumulation after wounding (Laudert *et al.*, 2000). Recently, an alternative lipid-linked

**Table 10.1** Arabidopsis *mutants* affected in oxylipin biosynthesis and/or signalling.

| Mutant | Affected gene/phenotype | Reference |
|---|---|---|
| Biosynthesis *mutants* | | |
| *fad3–2fad7–2fad8* | ER and plastidic fatty acid desaturases | McConn and Browse, 1996 |
| *dad1* | Phospholipase A1 | Ishiguro *et al.*, 2001 |
| *dde2–2* | AOS | von Malek *et al.*, 2002 |
| *dde1* | OPR3 | Sanders *et al.*, 2000 |
| *opr3* | OPR3 | Stintzi and Browse, 2000 |
| *aim1* | β-oxidation | Richmond and Bleecker, 1999 |
| Perception and signalling *mutants* | | |
| *jar1* | Acyl-adenylate-forming enzyme | Staswick *et al.*, 2002 |
| *jin1* | Unknown | Berger *et al.*, 1996 |
| *jin4* | Allelic to jar1 | Berger *et al.*, 1996 |
| *coi1* | F-box protein | Xie *et al.*, 1998 |
| *mpk4* | MAP kinase | Petersen *et al.*, 2000 |
| *axr1* | Ubiquitination | Tiryaki and Staswick, 2002 |
| *jue1 to jue3* | Unknown | Jensen *et al.*, 2002 |
| Constitutive JA response *mutants* | | |
| *cev1* | Cellulose synthase | Ellis *et al.*, 2002 |
| *cet* | Unknown | Hilpert *et al.*, 2001 |
| *cex1* | Unknown | Xu *et al.*, 2001 |
| *joe1, joe2* | Unknown | Jensen *et al.*, 2002 |
| *cir1* | Unknown | Murray *et al.*, 2002 |
| *cas1* | Unknown | Kubigsteltig and Weiler, 2003 |
| *iop1* | Unknown | Penninckx *et al.*, 2003 |

pathway has been proposed, based on the presence of galactolipid-esterified OPDA (Stelmach *et al.*, 2001). In addition, its precursor 13-HPOT has been found esterified to membrane lipids as well (Leverentz *et al.*, 2002; Weichert *et al.*, 2003). Since neither *type 2*-13-LOXs nor a 13-AOS have been identified so far as being able to oxygenate LeA esterified in galactolipids, the occurrence of oxylipins should result from their re-insertion into chloroplast membranes.

The subsequent steps of OPDA modification, i.e. reduction and three β-oxidation steps shortening the side chain and leading to the formation of JA, occur in the peroxisomes (Strassner *et al.*, 2002). It is generally assumed that OPDA is actively exported from the chloroplast and then actively imported to the matrix of peroxisomes. Whether the transfer across two membranes includes an activation of OPDA into OPDA-CoA derivative remains to be analysed. In case of the peroxisomal uptake, the transporter might be the recently identified *pxa1* in Arabidopsis (Zolman *et al.*, 2001). The mutant shows an arrest in early steps of germination, since it seems to be involved in the general uptake of fatty acids in peroxisomes. PXA1 belongs to the family of ATP-binding cassette transporters and it is impaired in auxin metabolism.

Reduction of the cyclopentenone ring to a cyclopentanone derivative is performed by OPDA reductase (Schaller *et al.*, 2000). Arabidopsis *mutants* in the OPR3 gene, *opr3* (Stintzi and Browse, 2000) and *dde1* (Sanders *et al.*, 2000) were identified as male-sterile *mutants*, which can be rescued by exogenous application of JA, but not OPDA. As for the uptake of OPDA, the chain shortening reaction to yield JA is connected to classical fatty acid degradation by β-oxidation. The *abnormal inflorescence meristem1* (*aim1*) mutation again affects inflorescence and floral development in Arabidopsis and the derived amino acid sequence of AIM1 shows extensive similarity to the cucumber multifunctional protein involved in β-oxidation of fatty acids (Richmond and Bleecker, 1999).

## 10.2 Oxylipins as signal molecules

### *10.2.1 Jasmonic acid in wound signalling*

In response to mechanical wounding or herbivore attack, plants activate defence mechanisms, which lead to the synthesis of antinutritional compounds, such as proteinase inhibitors, thereby negatively affecting the growth and development of the attacking pest. Wound signalling and the role of JA in defence against insect attack has been studied in particular detail in solanaceous plants. The discovery of wound-induced expression of proteinase inhibitors in tomato (Green and Ryan, 1972) and potato (Sanchez-Serrano *et al.*, 1986) and the finding that expression of the corresponding genes is inducible by exogenous application of JA (Farmer *et al.*, 1992) led to an initial model of wound signalling in plants (Farmer and Ryan, 1992). According to this, LeA is released from membranes upon wounding by the action of phospholipase $A_2$ and converted to JA, leading to the expression of wound-inducible genes. The current model of wound-induction of defence gene expression includes the 18 amino acid peptide systemin, which is cleaved from the *C*-terminal region of the 200 amino acid precursor prosystemin (Ryan and Pearce, 2003). Thus, after wounding, expression of the prosystemin gene in vascular bundles leads to generation of systemin, which in turn activates a signal transduction cascade involving phospholipase $A_2$, MAP kinases as well as $Ca^{++}$ and calmodulin (Seo *et al.*, 1995; Bögre *et al.*, 1997; Moyen *et al.*, 1998; Bergey and Ryan, 1999; Narvaez-Vasquez *et al.*, 1999). Systemin induces the expression of signalling pathway genes leading to the release of LeA and its conversion to JA. JA biosynthetic enzymes are located in the vein cells (Hause *et al.*, 2000) and they accumulate to higher levels in veins compared to the remaining leaf area (Stenzel *et al.*, 2003). Induction of defence genes, however, occurs in the mesophyll cells (Narvaez Vasquez *et al.*, 1995). $H_2O_2$, which accumulates in cell walls of vascular parenchyma and spongy mesophyll cells in wounded tomato leaves, was shown to be highly mobile, and a concerted action of systemin, JA and $H_2O_2$ was suggested to operate in the local proteinase inhibitor expression (Orozco-Cardenas *et al.*, 2001). Amplification of the wound signal by JA and systemin is assumed to take place in the vascular tissue. This is based on the observation that expression of both prosystemin and JA biosynthetic genes occurs in vascular bundle cells (Hause *et al.*, 2003; Narvaez-Vasquez and Ryan, 2003).

JA does not act alone in inducing gene expression after wounding but does so in concert with other phytohormones e.g. abscisic acid (ABA), ethylene and possibly other compounds (O'Donnell *et al.*, 1996; Dammann *et al.*, 1997). Moreover, not all wound-induced genes are expressed in a JA-dependent manner since both in tomato and in Arabidopsis, JA-independent wound-induced gene expression has been reported (Titarenko *et al.*, 1997; O'Donnell *et al.*, 1998). In Arabidopsis, analyses of JA insensitive *mutants* have provided important insight into distinct components of the JA signal transduction pathway (Table 10.1). Functional evidence for a role of MAP kinases in JA-regulated signalling stems from a mutation in the gene encoding MAP kinase4 (*mpk4*), which is compromised in JA-responsive gene expression (Petersen *et al.*, 2000).

A crucial player in JA signalling has been identified by the isolation of the coronatine-insensitive mutant *coi1* (Feys *et al.*, 1994). The mutation affects a gene coding for an F-box protein, which is part of the SCF complex involved in protein ubiquitination and, possibly, degradation. In general, it seems that plants use the selective protein breakdown by the ubiquitin/26S proteasome pathway as an important strategy to control the level and activity of their constituent proteins. Remarkably, molecular genetic analyses have also connected individual components to almost all aspects of plant biology, including the cell cycle, embryogenesis, photomorphogenesis, circadian rhythms, hormone signalling, homeostasis, disease resistance and senescence. Consequently, it appears that the ubiquitin/26S proteasome pathway competes with transcription complexes and protein kinase cascades as a major player in the regulation of plant cell function (Vierstra, 2003).

Loss of *COI1* results in JA-insensitivity, indicating that removal of a negative factor is important for the activation of JA-dependent responses. A potential substrate for COI1-mediated ubiquitination and thus a possible regulator of JA responses might be a histone deacetylase that was identified as a COI1-interacting protein (Devoto *et al.*, 2002). Interestingly, the auxin resistant mutant *axr1*, which is impaired in a protein required for activation of the ubiquitin-like RUB protein, also displays a JA-insensitive phenotype (Xu *et al.*, 2002). On the other hand, a JA-insensitive mutant isolated in a genetic screen turned out to be an allele of *axr1* (Tiryaki and Staswick, 2002). This mutant exhibits altered responses not only to JA and auxin but also to cytokinin, brassinolide, the ethylene precursor and ABA, suggesting a central role of the AXR1 protein for general hormone signalling. Furthermore, analysis of transgenic plants with reduced COP9 signalosome function displayed a reduced sensitivity to JAME compared to wild-type plants (Feng *et al.*, 2003). Since fertility is not affected in these transgenic plants, signalling through the COP9 signalosome appears not to be of importance for the function of JA in male gametophyte development.

Another jasmonate-insensitive mutant, *jar1*, was isolated based on its resistance to root growth inhibition by the methyl ester of JA (Staswick *et al.*, 1992). The protein encoded by the affected gene belongs to a new class of acyl-adenylate forming enzymes, which activate carboxyl groups for biochemical modifications (Staswick *et al.*, 2002). This suggests that covalent modification of JA is important for distinct functions. Interestingly, jar1 mutant plants are fertile indicating that modification of JA by *JAR1* is not required for male gametophyte development.

Further downstream in JA signalling, numerous promoter analyses of JA-induced genes have led to the identification of JA-response elements (Menke *et al.*, 1999; Brown *et al.*, 2003). The first JA-responsive transcription factors that bind to possess the typical JA-response elements were isolated in a yeast one-hybrid screen as octadecanoid-derivative responsive Catharanthus AP2-domain (ORCA) proteins and by T-DNA activation tagging (Menke *et al.*, 1999; van der Fits and Memelink, 2000). The JA-inducible ORCA transcription factors plant AP2/ERF DNA binding domain (Riechmann and Meyerowitz, 1998) and are able to act as transcriptional activators of JA-responsive gene expression (Menke *et al.*, 1999; van der Fits and Memelink, 2000). Interestingly, the JA-response element contains a GCC box, which has been identified as an ethylene-response element (Ohme-Takagi and Shinshi, 1995) and ERF1 (ethylene response factor1), a transcription factor of ethylene-responsive genes, regulates genes which are responsive to both ethylene and JA (Lorenzo *et al.*, 2003).

### *10.2.2 Systemic wound signalling*

The ability of a plant to activate systemic defence responses after wounding is of importance for the protection of undamaged tissue from insect attack. In search of the systemic signal, research focused first on the oligopeptide systemin. The importance of systemin as a signal for systemic responses in tomato was postulated due to its phloem mobility (Narvaez Vasquez *et al.*, 1995) and based on the observation that systemin-depleted plants are compromised in systemic wound signalling (McGurl *et al.*, 1992; Orozco-Cardenas *et al.*, 1993). On the other hand, constitutive expression of systemin leads to constitutive expression of wound-responsive genes (McGurl *et al.*, 1994). In addition to systemin, hydroxyproline-rich glycopeptides have been identified in tobacco (TobHypSys I and II) and tomato (TomHypSys I to III) which are able to induce defence responses, such as the activation of MAP kinases and the alkalinisation of the cell culture medium (Pearce *et al.*, 2001; Pearce and Ryan, 2003).

Systemin is perceived by the leucine-rich repeat receptor kinase SR160, which, interestingly, appears to be identical to the brassinosteroid receptor kinase *BRI1* (Scheer and Ryan, 2002; Szekeres, 2003). Functional proof that SR160 indeed encodes the systemin receptor stems from experiments in which the transfer of the SR160-encoding gene into tobacco confers systemin-responsiveness to otherwise non-responsive tobacco plants (Scheer *et al.*, 2003).

Systemin has been regarded as the systemically mobile signal responsible for induction of defence responses in distant, unwounded tissue. Recent reports, however, point to JA as the long-distance signal, and systemin is now considered to be required for the production of the systemic wound signal in the wounded leaf but not to travel long distances (Stratmann, 2003).

The function of JA as the long-distance mobile signal was demonstrated by Howe and co-workers. Transgenic tomato plants expressing the prosystemin gene under the control of the 35S promoter show constitutive expression of wound-inducible defence genes (Howe and Ryan, 1999). In a genetic screen, suppressors of this phenotype were isolated, two of which are compromised in systemic wound responses and in the perception of systemin (*spr*, *suppressor of prosystemin-mediated responses*).

One of these suppressor plants, the *spr2* mutant, is impaired in both local and systemic wound responses due to a mutation in the gene encoding a fatty acid desaturase (Li *et al.*, 2003). The loss of LeA and RA fatty acids in the *spr2* mutant correlates with a deficiency in wound-induced JA accumulation, a highly reduced local and an undetectable systemic response to wounding and with a defective defence against insects (Li *et al.*, 2002, 2003).

Using both JA biosynthetic (*spr2*) and JA perception (*jai1*, *jasmonate insensitive1*) *mutants*, grafting experiments revealed that JA biosynthesis is required for the activation of systemic responses only in the wounded leaf, but not in the upper unwounded leaf. In contrast, the ability to perceive JA, which is impaired in the *jai1* mutant, has to exist in the upper, unwounded leaf but not in the wounded leaf. These data show that JA or a derivative of JA (i.e. OPDA) may be the translocated, long-distance systemic signal (Li *et al.*, 2002).

The specific loss of systemic gene expression in response to wounding displayed by the mutant *spr1* is novel and together with the observation that *spr1* plants are insensitive specifically to systemin, renders this mutant particularly interesting for the analysis of systemin-mediated systemic responses (Lee and Howe, 2003). After grafting of *spr1* and wild-type tomato plants, no systemic responses take place if the wounded leaf originates from *spr1*, suggesting that *spr1* is defective in the production of the long-distance signal. Since the upper, unwounded *spr1* leaf shows normal systemic wound responses when grafted onto a wild-type plant, systemin is apparently not the long-distance wound signal (Lee and Howe, 2003; Stratmann, 2003).

### *10.2.3 The role of jasmonic acid in insect resistance*

In addition to *mutants* in oxylipin biosynthesis and signal transduction, transgenic plants with increased or reduced oxylipin biosynthetic capacity have been analysed with respect to their resistance against wounding and herbivore attack. In potato, antisense constructs of a wound- and pathogen-inducible type 2-13-LOX results in increased susceptibility to herbivore feeding (Royo *et al.*, 1999). Despite a lack of detectably reduced JA levels, proteinase inhibitor transcript levels were significantly reduced and weight gain of Colorado potato beetles and larvae of beet armyworm feeding on the transgenic plants was larger than for those feeding on wild-type plants. Thus, JA-inducible proteinase inhibitor expression in wild-type plants is an efficient defence against herbivores.

13-HPL catalyses the formation of aldehydes and ω-oxo-acids. With LA and LEA as substrates, hexanal and 3-hexenal, respectively, as well as 12-oxo-dodecenoic acid are formed. In transgenic potato plants depleted of 13-HPL expression due to antisense inhibition, aldehyde levels are efficiently reduced. Despite unimpaired wound-induced expression of proteinase inhibitor genes, aphids feeding on the transgenic plants exhibit a larger weight gain than those feeding on wild-type plants (Vancanneyt *et al.*, 2001).

The tomato mutant def1 is defective in the wound-induced JA biosynthesis and accumulates less JA (Howe *et al.*, 1996). The mutant plants show decreased resistance against attack by the tobacco hornworm, *Manduca sexta*, indicating that octadecanoid metabolism plays a crucial role for insect defence.

## 10.3 Oxylipins and pathogen defence

Plants respond to pathogen attack with a multicomponent defence response that includes synthesis of antimicrobial compounds, defence gene activation, cell wall reinforcements, synthesis of signalling compounds and, in some cases, hypersensitive cell death (Scheel, 1998). Oxylipins are important for pathogen defence because they can act as antimicrobial substances, as signal molecules or as intermediates in lipid peroxidation processes. Oxylipins from different sources have been implicated in pathogen defence responses. Phytoprostanes, the products of free radical-catalysed oxidation of LeA, which are presumed to be part of an archaic signalling system, accumulate in response to pathogen infection and can induce defence responses such as MAP kinase and defence gene activation (Thoma *et al.*, 2003). Moreover, α-DOX gene expression has been shown to be pathogen-inducible, correlating with hypersensitive cell death, a typical resistance-associated response. Thus, products of the α-DOX reaction are presumed to protect plant tissue from oxidative stress and cell death (Hamberg *et al.*, 2003).

In the following section, we will focus on the importance of JA for pathogen defence and for cross-talk with SA-signalling pathways, mainly in Arabidopsis, and on 9-LOX-derived oxylipins and their impact on defence in solanaceous plants.

### *10.3.1 Jasmonic acid signal transduction mutants and their effects on the pathogen responses*

In addition to the important and well-characterised salicylic acid signalling mechanism for pathogen defence in plants, JA has also been identified as playing a key role in the activation of some defence responses. It is generally assumed that JA-dependent defences are activated by infection with necrotrophic pathogens, which have to kill host cells to obtain nutrients, while biotrophic pathogens, depending on living host cells, induce SA-dependent defence responses. The availability of Arabidopsis *mutants* in JA biosynthesis (*fad3–2 fad7 fad8*, *dad1*, *opr3*, *dde1*, *dde2*; Table 10.1) and in JA perception and signal transduction (*jar1*, *coi1*; Table 10.1) was crucial to the demonstration of JA as an important compound in plant defence. The triple fatty acid desaturase (fad) mutant is defective in the ER-localised $\Delta^{12}$-desaturase and in two plastidic ω3-desaturases and contain less than 0.1% trienoic fatty acids (McConn and Browse, 1996). Thus, the lack of substrate for JA biosynthesis results in negligible levels of JA and renders the plants highly susceptible to insect attack (McConn *et al.*, 1997) and infection by pathogens (Vijayan *et al.*, 1998). Since exogenous application of JA rescues the mutant phenotype, this oxylipin or its derivatives appear to be crucial compounds for the activation of defence responses.

The JA biosynthesis *mutants opr3* and *dde1* carry a mutation in the gene encoding OPDA reductase 3 (Sanders *et al.*, 2000; Stintzi and Browse, 2000). The mutant phenotype of male sterility can be rescued by exogenous application of JA, but not by OPDA, indicating that JA is the signal required for male gametophyte development. In contrast, OPDA is sufficient for successful pathogen defence, since *opr3* plants are fully resistant to insect attack by *Bradysia impatiens* and to infection by the necrotrophic fungus, *Alternaria brassicicola* (Stintzi *et al.*, 2001). Thus, OPDA itself appears to act as a defence-inducing oxylipin, since resistance to herbivores and to fungal infection occurs in the absence of JA.

That OPDA itself can act as a signal molecule has been shown for the system of tendril coiling of *Bryonia dioica*, where application of OPDA induces reactions to a larger scale than JA (Blechert *et al.*, 1999). Moreover, OPDA, but not JA, elicits phytoalexin accumulation in cultured soybean cells (Fliegmann *et al.*, 2003). OPDA, but not JA, accumulates systemically in potato plants induced for systemic resistance by infiltration of *Pseudomonas syringae*, suggesting that the biosynthetic precursor of JA – rather than JA itself – is important for systemic signalling (Landgraf *et al.*, 2002). Microarray analyses indeed have shown that OPDA and JA induce both overlapping and distinct subsets of genes in Arabidopsis (Stintzi *et al.*, 2001). It has been suggested that the electrophilic ring of OPDA is the active structure required for induction (Almeras *et al.*, 2003).

In addition to *mutants* in genes encoding JA biosynthetic enzymes, JA perception and/or signalling *mutants* have been instrumental in elucidating the role of JA in pathogen defence (Table 10.1). In particular, the JA-insensitive mutant *coi1* is more susceptible to the necrotrophic fungi *Alternaria brassicicola* and *Botrytis cinerea* but not to biotrophic fungus such as *Peronospora parasitica* (Thomma *et al.*, 1998). Similarly, *jar1 mutants* are susceptible to the soil fungus *Pythium irregulare,* which does not cause symptoms on wild-type Arabidopsis (Staswick *et al.*, 1998). Moreover, *jar1* plants are also compromised in their ability to mount systemic resistance induced by the non-pathogenic root growth promoting bacteria *Pseudomonas fluorescens* (Pieterse *et al.*, 1998).

Over the last few years, the first *mutants* that show constitutive JA responses have been identified (Table 10.1). In particular, the CEV1 mutant, which was isolated based on constitutive expression of the luciferase reporter gene under the control of the JA-responsive vegetative storage protein promoter, shows constitutive JA-dependent responses, such as stunted roots, high levels of anthocyanins and JA, as well as constitutive expression of JA-responsive defence genes (Ellis and Turner, 2001). In contrast to JA-insensitive *mutants* like coi1, which are susceptible to infection by powdery mildews, cev1 exhibits enhanced resistance to these pathogens (Ellis and Turner, 2001). The cev1 gene encodes a cellulose synthase, suggesting a role of cell wall biosynthesis in JA-dependent defence responses (Ellis *et al.*, 2002). Other *mutants* with constitutively activated JA-dependent responses include the cet, cex, joe, cir, cas and iop *mutants* that show constitutive expression of the JA responses, such as constitutive expression of JA-inducible genes, JA-independent root growth inhibition and resistance to necrotrophic pathogens (Xu *et al.*, 2001; Jensen *et al.*, 2002; Murray *et al.*, 2002; Nibbe *et al.*, 2002; Kubigsteltig and Weiler, 2003; Penninckx *et al.*, 2003). Cloning of the affected genes should identify further components of the JA signal transduction pathway.

### *10.3.2 Cross-talk between salicylic acid and lipid signalling in pathogen defence responses*

Cross-talk between lipid- and SA-dependent signal transduction pathways in pathogen defence responses has been observed in a number of studies, and generally involves antagonistic interactions.

Negative effects of SA on JA were first reported for tomato and potato. Here, acetyl-SA, aspirin, inhibited the synthesis of JA (Pena-Cortes *et al.*, 1993). Since addition of OPDA or JA overcame the inhibitory effect of aspirin, a step in the synthesis of OPDA appears to be affected. Indeed, aspirin was shown to inhibit AOS *in vitro* by acetylation and, thus, irreversible inactivation (Pan *et al.*, 1998). In contrast to these data, JA-induced proteinase inhibitor expression could be significantly inhibited by SA or aspirin, suggesting that SA acts downstream of JA synthesis (Doares *et al.*, 1995).

In Arabidopsis, analysis of *mutants* revealed cross-talk between SA- and JA-dependent signal transduction pathways. Inhibitory effects of SA on JA signalling has been observed in the *eds4* (*enhanced disease susceptibility4*) mutant. Here, enhanced susceptibility to infection by *P. syringae* correlates with reduced levels of SA in response to pathogen attack (Gupta *et al.*, 2000). However, responses to inducers of JA-dependent gene expression are increased, suggesting that in wild-type plants, high SA levels repress JA-dependent responses.

Lack of SA synthesis in the *sid2* (*salicylic acid induction deficient2*) mutant results in enhanced pathogen-inducible PDF1.2 gene expression, suggesting that SA negatively regulates JA-responsive gene expression (Dewdney *et al.*, 2000). Also, in *nahG* plants expressing the gene for a SA hydroxylase, SA is converted to catechol and subsequently, no SA accumulates. Infection of these plants with *P. syringae* leads to large increases in JA levels, which are not observed in wild-type plants. Thus, in the absence of SA, JA-dependent responses become prominent, suggesting that SA negatively regulates JA biosynthesis (Spoel *et al.*, 2003).

Inhibition of SA signalling by JA signalling occurs in the *mpk4* mutant, which exhibits constitutive SA-dependent and reduced JA-dependent defence responses (Petersen *et al.*, 2000). Antagonistic effects have also been revealed by the *coi1* mutant, which is impaired not only in JA signalling but also in resistance to necrotrophic pathogens. However, SA-dependent responses, such as PR gene expression and SA levels are enhanced, suggesting that *COI1* is a negative regulator of SA responses. Moreover, *coi1* plants exhibit increased resistance to the bacterial pathogen *P. syringae*, which might be explained by the increased expression of SA-dependent responses (Kloek *et al.*, 2001).

The central regulator of SA-dependent defence responses, NPR1, appears to be a modulator of the cross-talk between JA and SA signal transduction pathways because inhibition of JA responses does not occur in *npr1* mutant plants (Spoel *et al.*, 2003). Interestingly, suppressors of the the *npr1* phenotype revealed novel interactions of the lipid and SA-dependent signalling pathways. The *npr1*-suppressor mutant *ssi2* (*suppressor of salicylic acid insensitivity2*), in which PR gene expression is constitutively activated, contains increased amounts of SA and exhibits enhanced resistance to virulent and avirulent pathogens (Kachroo *et al.*, 2001). Moreover, *ssi2* plants have decreased oleic acid levels but normal amounts of LeA, possibly due to exchange of PUFAs between the eukaryotic and prokaryotic fatty acid synthesis pathways (see Chapter 5 and Somerville *et al.*, 2000). However, *ssi2 mutants* are unable to activate PDF1.2 expression in response to JA treatment and exhibit enhanced susceptibility to *B. cinerea* (Shah *et al.*, 2001). The *SSI2* gene encodes a stearoyl-ACP-desaturase, suggesting that the loss of SA-dependent responses in *npr1* is rescued by the lack of a fatty acid-derived signalling molecule (Kachroo *et al.*, 2001).

A suppressor of the *ssi2* mutation, *sfd4* (*suppressor of fatty acid (stearoyl) desaturase deficiency4*), again is affected in a gene encoding a 16:1/18:1 ω6-desaturase that catalyses the formation of 16:2 and LA in the plastid (Nandi *et al.*, 2003). The mutant contains significantly less RA in the plastid complex lipid species, which correlates with a loss of *ssi2*-conferred constitutive SA signalling and the regained ability of JA-dependent PDF1.2 gene expression. The data support the hypothesis that lipid-based signals derived from oleic acid and RA are important for the expression of SA-dependent defence responses in Arabidopsis. Moreover, the identification of the *sfd1* mutant, which is impaired in the gene encoding glycerol-3-phosphate dehydrogenase, argues for an importance of fatty acid synthesis for defence signalling (Shah, 2003).

### *10.3.3 9-LOX products – antimicrobial compounds and their impact on lipid peroxidation processes*

9-LOXs are predicted to be localised in the cytoplasm (Figure 10.2) (Feussner and Wasternack, 2002) and their substrates LA and LeA are presumably released from the endomembrane system by phospholipase $A_2$ activities, or might be derived from the cytoplasmic acyl-CoA-pool (Göbel *et al.*, 2001). 9-HPOT and 9-HPOD are substrates for LOX pathway enzymes that catalyse the formation of the divinyl ether containing PUFAs colneleic and colnelenic acid, di- and trihydroxy oxylipins, epoxy PUFAs, as well as hydroxyepoxy PUFAs (Figures 10.1 and 10.2). A role for 9-LOXs in mediating resistance has been postulated based on two observations. First, 9-LOX-derived oxylipins possess antimicrobial activity (Weber *et al.*, 1999; Graner *et al.*, 2003) and, thus, 9-LOXs might restrict pathogen growth by producing toxic or inhibitory compounds. Second, analysis of the lipid peroxidation products in tobacco undergoing an elicitor-induced hypersensitive response revealed that more than 90% of 9-HPOT and 9-HPOD was enzymatically produced, suggesting a crucial role for 9-LOXs in lipid peroxidation and/or membrane disintegration (Rustérucci *et al.*, 1999).

9-LOXs are expressed in response to pathogen infection in a number of pathosystems (Melan *et al.*, 1993; Véronési *et al.*, 1996; Rustérucci *et al.*, 1999; Kolomiets *et al.*, 2000; Göbel *et al.*, 2002). However, most data regarding the elucidation of the role of 9-LOXs and their products for defence responses have been obtained from studies of solanaceous plants. Here, pathogen-induced increases in 9-LOX transcript levels, 9-LOX activity and 9-LOX-derived oxylipins in tobacco and potato (Rustérucci *et al.*, 1999; Weber *et al.*, 1999; Göbel *et al.*, 2001, 2002) suggest a functional significance of 9-LOXs for defence responses. Loss of function experiments, indeed, demonstrated a crucial role of 9-LOXs for pathogen resistance. Antisense-mediated suppression of the elicitor-induced 9-LOX in tobacco converted an incompatible interaction of a resistant tobacco cultivar with *Phytophthora parasitica nicotianae* into a compatible one (Rancé *et al.*, 1998). Thus, 9-LOXs appear to be required for the establishment and/or maintenance of resistance. However, how 9-LOXs are involved in these processes is not known.

Massive lipid peroxidation occurs when plants undergo a hypersensitive response with localised cell death during incompatible or non-host pathogen interactions. These lipid peroxidation processes can occur non-enzymatically by reactive oxygen species or enzymatically by the action of LOXs, peroxidases or α-DOXs. Oxygenation of phospholipids has been demonstrated for some plant LOXs (Matsui *et al.*, 1998; Perez-Gilabert *et al.*, 1998; Leon *et al.*, 2002) supporting the possibility of a role in lipid peroxidation. In animals, direct peroxidation of membrane lipids by LOXs contributes to programmed cell death (Maccarrone *et al.*, 2001). Animal 15-LOX initiates programmed organelle disruption by dioxygenating mitochondrial membranes, resulting in the formation of pore-like structures (van Leyen *et al.*, 1998). Treatment of mitochondria with the soybean 15-LOX leads to membrane disruption and release of cytochrome c, usually a trigger of apoptosis (Maccarrone *et al.*, 2001).

Based on enantiomer analyses, 90% of the lipid hydroperoxides detectable in tobacco during the HR elicited by the *Phytophthora* elicitin, cryptogein, originate from 9-LOX activity (Rustérucci *et al.*, 1999). In potato, the 9-LOX pathway is stimulated upon elicitor treatment and pathogen infection, leading to the formation of hydroxy- and trihydroxy fatty acids, as well as divinyl ether containing PUFAs (Göbel *et al.*, 2001, 2002). Thus, correlative data suggest that lipid peroxidation during the hypersensitive response is catalysed by LOXs. Functional analyses, however, reveal a more complicated scenario. Despite the successful reduction of the pathogen-induced 9-LOX expression, activity and 9-LOX-derived oxylipins by RNA interference in transgenic potato plants, cell death was not inhibited and appeared phenotypically unaltered (Göbel *et al.*, 2003). The increased levels of oxylipins of non-enzymatic origin and of 13-LOX-derived oxylipins suggest that 9-LOX-derived lipid peroxides are apparently not required for the HR in potato (Göbel *et al.*, 2003). Possibly, lipid peroxidation proceeds via enzymatic reactions in the presence of 9-LOXs. Only in the absence of 9-LOXs, autoxidative processes become prominent. It should be kept in mind that enzymatic formation of lipid peroxides by LOXs during hypersensitive response might be of dual advantage to the plant since LOX-derived lipid peroxides are less toxic substances for the plant cell, because they are rapidly metabolised by enzymes of the LOX pathway. Moreover, the antimicrobial activity of 9-LOX-derived oxylipins might contribute to containment of pathogen growth, thus, enhancing the plants' defence capacity.

## 10.4 Conclusions and future prospects

The role of oxylipins in different aspects of plant development and stress responses is now well-established. Oxylipins act as signal molecules, as antimicrobial compounds and they are implicated in lipid peroxidation processes. Cloning of genes encoding a number of oxylipin biosynthetic enzymes has enabled the elucidation of oxylipin synthesis and metabolism as well as functional analyses in transgenic plants. Furthermore, Arabidopsis *mutants* defective in different aspects of oxylipin biosynthesis, metabolism or perception gave important insight into oxylipin and lipid signalling in response to wounding and pathogen attack. Within the last years, the importance of cross-talk between oxylipin signalling and other signalling pathways has been recognised for regulation and fine tuning of the plant responses to stress. With the isolation of Arabidopsis *mutants* defective in cross-talk, novel players are expected to be identified in this crucial area of lipid-based plant signalling molecules.

## References


Almeras, E., Stolz, S., Vollenweider, S., Reymond, P., Mene-Saffrane, L. and Farmer, E.E. (2003) Reactive electrophile species activate defense gene expression in Arabidopsis. *The Plant Journal*, **34**, 205–216.

Bachmann, A., Hause, B., Maucher, H., Garbe, E., Vörös, K., Weichert, H., Wasternack, C. and Feussner, I. (2002) Jasmonate-induced lipid peroxidation in barley leaves initiated by distinct 13-LOX forms of chloroplasts. *Biological Chemistry*, **383**, 1645–1657.

Balkenhohl, T., Kühn, H., Wasternack, C. and Feussner, I. (1998) A lipase specific for esterified oxygenated polyenoic fatty acids in lipid bodies of cucumber cotyledons, in *Advances in Plant Lipid Research* (eds J. Sánchez, E. Cerdá-Olmedo and E. Martínez-Force) Secretariado de Publicaciones de la Universidad de Sevilla, Sevilla, pp. 320–322.

Berger, S., Bell, E. and Mullet, J.E. (1996) Two methyl jasmonate-insensitive mutants show altered expression of *AtVsp* in response to methyl jasmonate and wounding. *Plant Physiology*, **111**, 525–531.

Bergey, D.R. and Ryan, C.A. (1999) Wound- and systemin-inducible calmodulin gene expression in tomato leaves. *Plant Molecular Biology*, **40**, 815–823.

Blechert, S., Bockelmann, C., Fusslein, M., von Schrader, T., Stelmach, B., Niesel, U. and Weiler, E.W. (1999) Structure-activity analyses reveal the existence of two separate groups of active octadecanoids in elicitation of the tendril-coiling response of *Bryonia dioica* Jacq. *Planta*, **207**, 470–479.

Blée, E. (1998a) Biosynthesis of phytooxylipins: the peroxygenase pathway. *Fett-Lipid*, **100**, 121–127.

Blée, E. (1998b) Phytooxylipins and plant defense reactions. *Progress in Lipid Research*, **37**, 33–72.

Blée, E. and Joyard, J. (1996) Envelope membranes from spinach chloroplasts are a site of metabolism of fatty acid hydroperoxides. *Plant Physiology*, **110**, 445–454.

Bögre, L., Ligterink, W., Meskiene, I. *et al.* (1997) Wounding induces the rapid and transient activation of a specific MAP kinase pathway. *Plant Cell*, **9**, 75–83.

Brown, R.L., Kazan, K., McGrath, K.C., Maclean, D.J. and Manners, J.M. (2003) A role for the GCC-box in jasmonate-mediated activation of the *PDF1.2* gene of Arabidopsis. *Plant Physiology*, **132**, 1020–1032.

Dammann, C., Rojo, E. and Sanchez-Serrano, J.J. (1997) Abscisic acid and jasmonic acid activate wound-inducible genes in potato through separate, organ-specific signal transduction pathways. *Plant Journal*, **11**, 773–782.

Devoto, A., Nieto-Rostro, M., Xie, D. *et al.* (2002) COI1 links jasmonate signalling and fertility to the SCF ubiquitin-ligase complex in Arabidopsis. *Plant Journal*, **32**, 457–66.

Dewdney, J., Reuber, T.L., Wildermuth, M.C. *et al.* (2000) Three unique mutants of Arabidopsis identify *eds* loci required for limiting growth of a biotrophic fungal pathogen. *Plant Journal*, **24**, 205–218.

Dhondt, S., Geoffroy, P., Stelmach, B.A., Legrand, M. and Heitz, T. (2000) Soluble phospholipase A2 activity is induced before oxylipin accumulation in tobacco mosaic virus-infected tobacco leaves and is contributed by patatin-like enzymes. *The Plant Journal*, **23**, 431–440.

Doares, S.H., Narvaez-Vasquez, J., Conconi, A. and Ryan, C.A. (1995) Salicylic acid inhibits synthesis of proteinase-inhibitors in tomato leaves induced by systemin and jasmonic acid. *Plant Physiology*, **108**, 1741–1746.

Ellis, C. and Turner, J.G. (2001) The Arabidopsis mutant *cev1* has constitutively active jasmonate and ethylene signal pathways and enhanced resistance to pathogens. *The Plant Cell*, **13**, 1025–1033.

Ellis, C., Karafyllidis, I., Wasternack, C. and Turner, J.G. (2002) The Arabidopsis mutant *cev1* links cell wall signaling to jasmonate and ethylene responses. *The Plant Cell*, **14**, 1557–1566.

Farmer, E.E. and Ryan, C.A. (1992) Octadecanoid precursors of jasmonic acid activate the synthesis of wound-inducible proteinase inhibitors. *The Plant Cell*, **4**, 129–134.

Farmer, E.E., Almeras, E. and Krishnamurthy, V. (2003) Jasmonates and related oxylipins in plant responses to pathogenesis and herbivory. *Current Opinion in Plant Biology*, **6**, 372–378.

Farmer, E.E., Johnson, R.R. and Ryan, C.A. (1992) Regulation of proteinase inhibitor gene expression by methyl jasmonate and jasmonic acid. *Plant Physiology*, **98**, 995–1002.

Feng, S.H., Ma, L.G., Wang, X.P., Xie, D.X., Dinesh-Kumar, S.P., Wei, N. and Deng, X.W. (2003) The COP9 signalosome interacts physically with SCFCOI1 and modulates jasmonate responses. *The Plant Cell*, **15**, 1083–1094.

Feussner, I. and Kindl, H. (1994) Particulate and soluble lipoxygense isozymes – comparison of molecular and enzymatic properties. *Planta*, **194**, 22–28.

Feussner, I. and Kühn, H. (2000) Application of lipoxygenases and related enzymes for the preparation of oxygenated lipids, in *Enzymes in Lipid Modification* (ed. Bornscheuer, U.T.) Wiley-VCH, Weinheim, Germany, pp. 309–336.

Feussner, I. and Wasternack, C. (2002) The lipoxygenase pathway. *Annual Review of Plant Biology*, **53**, 275–297.

Feussner, I., Blée, E., Weichert, H., Rousset, C. and Wasternack, C. (1998) In *Advances in Plant Lipid Research* (eds J. Sánchez, E. Cerdá-Olmedo and E. Martínez-Force) Secretariado de Publicaciones de la Universidad de Sevilla, Sevilla, pp. 311–313.

Feussner, I., Hause, B., Vörös, K., Parthier, B. and Wasternack, C. (1995) Jasmonate-induced lipoxygenase forms are localized in chloroplasts of barley leaves (*Hoerdeum vulgare* cv. Salome). *The Plant Journal*, **7**, 949–957.

Feussner, I., Kühn, H. and Wasternack, C. (2001) Lipoxygenase-dependent degradation of storage lipids. *Trends in Plant Science*, **6**, 268–273.

Feys, B.J.F., Benedetti, C.E., Penfold, C.N. and Turner, J.G. (1994) Arabidopsis mutants selected for resistance to the phytotoxin coronatine are male-sterile, insensitive to methyl jasmonate, and resistant to a bacterial pathogen. *The Plant Cell*, **6**, 751–759.

Fliegmann, J., Schuler, G., Boland, W., Ebel, J. and Mithöfer, A. (2003) The role of octadecanoids and functional mimics in soybean defense responses. *Biological Chemistry*, **384**, 437–446.

Frankel, E.N. (1998) *Lipid Oxidation,* The Oily Press Ltd, Dundee, UK.

Froehlich, J.E., Itoh, A. and Howe, G.A. (2001) Tomato allene oxide synthase and fatty acid hydroperoxide lyase, two cytochrome P450s involved in oxylipin metabolism, are targeted to different membranes of chloroplast envelope. *Plant Physiology*, **125**, 306–317.

Fuller, M.A., Weichert, H., Fischer, A.M., Feussner, I. and Grimes, H.D. (2001) Activity of soybean lipoxygenase isoforms against esterified fatty acids indicates functional specificity. *Archives of Biochemistry and Biophysics*, **388**, 146–154.

Gardner, H.W. (1991) Recent investigations into the lipoxygenase pathway in plants. *Biochimica et Biophysica Acta*, **1084**, 221–239.
Göbel, C., Feussner, I., Hamberg, M. and Rosahl, S. (2002) Oxylipin profiling in pathogen-infected potato leaves. *Biochimica et Biophysica Acta*, **1584**, 55–64.
Göbel, C., Feussner, I. and Rosahl, S. (2003) Lipid peroxidation during the hypersensitive response in potato in the absence of 9-lipoxygenases. *Journal of Biological Chemistry*, **278**, 52834–52840.
Göbel, C., Feussner, I., Schmidt, A., Scheel, D., Sanchez-Serrano, J., Hamberg, M. and Rosahl, S. (2001) Oxylipin profiling reveals the preferential stimulation of the 9- lipoxygenase pathway in elicitor-treated potato cells. *Journal of Biological Chemistry*, **276**, 6267–6273.
Graner, G., Hamberg, M. and Meijer, J. (2003) Screening of oxylipins for control of oilseed rape (*Brassica napus*) fungal pathogens. *Phytochemistry*, **63**, 89–95.
Green, T.R. and Ryan, C.A. (1972) Wound-induced proteinase inhibitors in plant leaves: a possible defense mechanism against insects. *Science*, **175**, 776–777.
Grimes, H.D., Tranberger, T.J. and Franceschi, V.R. (1993) Expression and accumulation patterns of nitrogen-responsive lipoxygenase in soybeans. *Plant Physiology*, **103**, 457–466.
Gupta, V., Willits, M.G. and Glazebrook, J. (2000) *Arabidopsis thaliana* EDS4 contributes to salicylic acid (SA)-dependent expression of defense responses: evidence for inhibition of jasmonic acid signaling by SA. *Molecular Plant Microbe Interactions*, **13**, 503–511.
Hamberg, M. (1993) Pathways in the biosynthesis of oxylipins in plants. *Journal of Lipid Mediators*, **6**, 375–384.
Hamberg, M. (1999) An epoxy alcohol synthase pathway in higher plants: biosynthesis of antifungal trihydroxy oxylipins in leaves of potato. *Lipids*, **34**, 1131–1142.
Hamberg, M., Sanz, A. and Castresana, C. (1999) α-Oxidation of fatty acids in higher plants. Identification of a pathogen-inducible oxygenase (piox) as an α-dioxygenase and biosynthesis of 2-hydroperoxylinolenic acid. *Journal of Biological Chemistry*, **274**, 24503–24513.
Hamberg, M., Sanz, A., Rodriguez, M.J., Calvo, A.P. and Castresana, C. (2003) Activation of the fatty acid α-dioxygenase pathway during bacterial infection of tobacco leaves: formation of oxylipins protecting against cell death. *Journal of Biological Chemistry*, **278**, 51796–51805.
Hause, B., Hause, G., Kutter, C., Miersch, O. and Wasternack, C. (2003) Enzymes of jasmonate biosynthesis occur in tomato sieve elements. *Plant and Cell Physiology*, **44**, 643–8.
Hause, B., Stenzel, I., Miersch, O. *et al.* (2000) Tissue-specific oxylipin signature of tomato flowers: allene oxide cyclase is highly expressed in distinct flower organs and vascular bundles. *The Plant Journal*, **24**, 113–126.
Heitz, T., Bergey, D.R. and Ryan, C.A. (1997) A gene encoding a chloroplast-targeted lipoxygenase in tomato leaves is transiently induced by wounding, systemin, and methyl jasmonate. *Plant Physiology*, **114**, 1085–1093.
Hilpert, B., Bohlmann, H., op den Camp, R. *et al.* (2001) Isolation and characterization of signal transduction mutants of *Arabidopsis thaliana* that constitutively activate the octadecanoid pathway and form necrotic microlesions. *Plant Journal*, **26**, 435–446.
Holtman, W.L., Vredenbregt-Heistek, J.C., Schmitt, N.F. and Feussner, I. (1997) Lipoxygenase-2 oxygenates storage lipids in embryos of germinating barley. *European Journal of Biochemistry*, **248**, 452–458.
Howe, G. and Ryan, C. (1999) Suppressors of systemin signaling identify genes in the tomato wound response pathway. *Genetics*, **153**, 1411–1421.
Howe, G.A., Lightner, J., Browse, J. and Ryan, C.A. (1996) An octadecanoid pathway mutant (JL5) of tomato is compromised in signaling for defense against insect attack. *The Plant Cell*, **8**, 2067–2077.
Ishiguro, S., Kawai-Oda, A., Ueda, J., Nishida, I. and Okada, K. (2001) The DEFECTIVE IN ANTHER DEHISCIENCE gene encodes a novel phospholipase A1 catalyzing the initial step of jasmonic acid biosynthesis, which synchronizes pollen maturation, anther dehiscence, and flower opening in Arabidopsis. *The Plant Cell*, **13**, 2191–2209.
Itoh, A. and Howe, G.A. (2001) Molecular cloning of a divinyl ether synthase – identification as a CYP74 cytochrome P-450. *Journal of Biological Chemistry*, **276**, 3620–3627.
Itoh, A., Schilmiller, A.L., McCaig, B.C. and Howe, G.A. (2002) Identification of a jasmonate-regulated allene oxide synthase that metabolizes 9-hydroperoxides of linoleic and linolenic acids. *Journal of Biological Chemistry*, **277**, 46051–46058.

Jensen, A.B., Raventos, D. and Mundy, J. (2002) Fusion genetic analysis of jasmonate-signalling mutants in Arabidopsis. *The Plant Journal*, **29**, 595–606.

Kachroo, P., Shanklin, J., Shah, J., Whittle, E.J. and Klessig, D.F. (2001) A fatty acid desaturase modulates the activation of defense signaling pathways in plants. *Proceedings of the National Academy of Sciences USA*, **98**, 9448–9453.

Kloek, A.P., Verbsky, M.L., Sharma, S.B. *et al.* (2001) Resistance to *Pseudomonas syringae* conferred by an *Arabidopsis thaliana* coronatine-insensitive (*coi1*) mutation occurs through two distinct mechanisms. *The Plant Journal*, **26**, 509–522.

Kolomiets, M.V., Chen, H., Gladon, R.J., Braun, E.J. and Hannapel, D.J. (2000) A leaf lipoxygenase of potato induced specifically by pathogen infection. *Plant Physiology*, **124**, 1121–1130.

Kubigsteltig, II and Weiler, E.W. (2003) Arabidopsis mutants affected in the transcriptional control of allene oxide synthase, the enzyme catalyzing the entrance step in octadecanoid biosynthesis. *Planta*, **217**, 748–757.

Kubigsteltig, I., Laudert, D. and Weiler, E.W. (1999) Structure and regulation of the *Arabidopsis thaliana* allene oxide synthase gene. *Planta*, **208**, 463–471.

Landgraf, P., Feussner, I., Hunger, A., Scheel, D. and Rosahl, S. (2002) Systemic accumulation of 12-oxo-phytodienoic acid in SAR-induced potato plants. *European Journal of Plant Pathology*, **108**, 279–283.

Laudert, D. and Weiler, E.W. (1998) Allene oxide synthase: a major control point in *Arabidopsis thaliana* octadecanoid signalling. *The Plant Journal*, **15**, 675–84.

Laudert, D., Schaller, F. and Weiler, E.W. (2000) Transgenic *Nicotiana tabacum* and *Arabidopsis thaliana* plants overexpressing allene oxide synthase. *Planta*, **211**, 163–165.

Lee, G.I. and Howe, G.A. (2003) The tomato mutant *spr1* is defective in systemin perception and the production of a systemic wound signal for defense gene expression. *The Plant Journal*, **33**, 567–576.

Leon, J., Royo, J., Vancanneyt, G. *et al.* (2002) Lipoxygenase H1 gene silencing reveals a specific role in supplying fatty acid hydroperoxides for aliphatic aldehyde production. *Journal of Biological Chemistry*, **277**, 416–423.

Leverentz, M.K., Wagstaff, C., Rogers, H.J. *et al.* (2002) Characterization of a novel lipoxygenase-independent senescence mechanism in *Alstroemeria peruviana* floral tissue. *Plant Physiology*, **130**, 273–283.

Li, C., Liu, G., Xu, C. *et al.* (2003) The tomato *suppressor of prosystemin-mediated responses2* gene encodes a fatty acid desaturase required for the biosynthesis of jasmonic acid and the production of a systemic wound signal for defense gene expression. *The Plant Cell*, **15**, 1646–1661.

Li, L., Li, C., Lee, G.I. and Howe, G.A. (2002) Distinct roles for jasmonate synthesis and action in the systemic wound response of tomato. *Proceedings of the National Academy of Sciences USA*, **99**, 6416–6421.

Lorenzo, O., Piqueras, R., Sanchez-Serrano, J.J. and Solano, R. (2003) ETHYLENE RESPONSE FACTOR1 integrates signals from ethylene and jasmonate pathways in plant defense. *The Plant Cell*, **15**, 165–178.

Maccarrone, M., Melino, G. and Finazzi-Agro, A. (2001) Lipoxygenases and their involvement in programmed cell death. *Cell Death and Differentiation*, **8**, 776–784.

Matsui, K., Nishioka, M., Ikeyoshi, M., Matsumura, Y., Mori, T. and Kajiwara, T. (1998) Cucumber root lipoxygenase can act on acyl groups in phosphatidylcholine. *Biochimica et Biophysica Acta*, **1390**, 8–20.

Maucher, H., Hause, B., Feussner, I., Ziegler, J. and Wasternack, C. (2000) Allene oxide synthases of barley (*Hordeum vulgare* cv. Salome): tissue specific regulation in seedling development. *The Plant Journal*, **21**, 199–213.

McConn, M. and Browse, J. (1996) The critical requirement for linolenic acid in pollen development, not photosynthesis, in an Arabidopsis mutant. *The Plant Cell*, **8**, 403–416.

McConn, M., Creelman, R.A., Bell, E., Mullet, J.E. and Browse, J. (1997) Jasmonate is essential for insect defense in *Arabidopsis*. *Proceedings of the National Academy of Sciences USA*, **94**, 5473–5477.

McGurl, B., Orozco-Cardenas, M., Pearce, G. and Ryan, C.A. (1994) Overexpression of the prosystemin gene in transgenic tomato plants generates a systemic signal that constitutively

induces proteinase inhibitor synthesis. *Proceedings of the National Academy of Sciences USA*, **91**, 9799–9802.
McGurl, B., Pearce, G., Orozco-Cardenas, M. and Ryan, C.A. (1992) Structure, expression, and antisense inhibition of the systemin precursor gene. *Science*, **255**, 1570–1573.
Melan, M.A., Dong, X., Endara, M.E., Davis, K.R., Ausubel, F.M. and Peterman, T.K. (1993) An *Arabidopsis thaliana* lipoxygenase gene can be induced by pathogens, abscisic acid, and methyl jasmonate. *Plant Physiology*, **101**, 441–450.
Menke, F.L.H., Champion, A., Kijne, J.W. and Memelink, J. (1999) A novel jasmonate- and elicitor-responsive element in the periwinkle secondary metabolite biosynthetic gene *Str* interacts with a jasmonate- and elicitor-inducible AP2-domain transcription factor, ORCA2. *EMBO Journal*, **18**, 4455–4463.
Moyen, C., Hammond-Kosack, K., Jones, J., Knight, M. and Johannes, E. (1998) Systemin triggers an increase of cytoplasmic calcium in tomato mesophyll cells: $Ca^{2+}$ mobilization from intra- and extracellular compartments. *Plant, Cell and Environment*, **21**, 1101–1111.
Müller, M.J. (1998) Radically novel prostaglandins in animals and plants: the isoprostanes. *Chemistry & Biology*, **5**, R323–R333.
Murray, S.L., Thomson, C., Chini, A., Read, N.D. and Loake, G.J. (2002) Characterization of a novel, defense-related Arabidopsis mutant, *cir1*, isolated by luciferase imaging. *Molecular Plant Microbe Interactions*, **15**, 557–566.
Nandi, A., Krothapalli, K., Buseman, C.M. *et al.* (2003) Arabidopsis *sfd* mutants affect plastidic lipid composition and suppress dwarfing, cell death, and the enhanced disease resistance phenotypes resulting from the deficiency of a fatty acid desaturase. *The Plant Cell*, **15**, 2383–2398.
Narvaez-Vasquez, J. and Ryan, C.A. (2003) The cellular localization of prosystemin: a functional role for phloem parenchyma in systemic wound signaling. *Planta*, **218**, 360–369.
Narvaez-Vasquez, J., Florin-Christensen, J. and Ryan, C.A. (1999) Positional specificity of a phospholipase. A activity induced by wounding, systemin, and oligosaccharide elicitors in tomato leaves. *Plant Cell*, **11**, 2249–2260.
Narvaez Vasquez, J., Pearce, G., Orozco Cardenas, M.L., Franceschi, V.R. and Ryan, C.A. (1995) Autoradiographic and biochemical evidence for the systemic translocation of systemin in tomato plants. *Planta*, **195**, 593–600.
Nibbe, M., Hilpert, B., Wasternack, C., Miersch, O. and Apel, K. (2002) Cell death and salicylate- and jasmonate-dependent stress responses in Arabidopsis are controlled by single *cet* genes. *Planta*, **216**, 120–128.
Noordermeer, M.A., Veldink, G.A. and Vliegenthart, J.F. (2001) Fatty acid hydroperoxide lyase: a plant cytochrome P450 enzyme involved in wound healing and pest resistance. *ChemBioChem*, **2**, 494–504.
O'Donnell, P.J., Calvert, C., Atzorn, R., Wasternack, C., Leyser, H.M.O. and Bowles, D.J. (1996) Ethylene as a signal mediating the wound response of tomato plants. *Science*, **274**, 1914–1917.
O'Donnell, P.J., Truesdale, M.R., Calvert, C.M., Dorans, A., Roberts, M.R. and Bowles, D.J. (1998) A novel tomato gene that rapidly responds to wound- and pathogen-related signals. *The Plant Journal*, **14**, 137–142.
Ohme-Takagi, M. and Shinshi, H. (1995) Ethylene-inducible DNA binding proteins that interact with an ethylene-responsive element. *The Plant Cell*, **7**, 173–182.
Orozco-Cardenas, M., McGurl, B. and Ryan, C.A. (1993) Expression of an antisense prosystemin gene in tomato plants reduces resistance toward *Manduca sexta* larvae. *Proceedings of the National Academy of Sciences USA*, **90**, 8273–8276.
Orozco-Cardenas, M., Narvaez-Vasquez, J. and Ryan, C. (2001) Hydrogen peroxide acts as a second messenger for the induction of defense genes in tomato plants in response to wounding, systemin, and methyl jasmonate. *The Plant Cell*, **13**, 179–191.
Pan, Z.Q., Camara, B., Gardner, H.W. and Backhaus, R.A. (1998) Aspirin inhibition and acetylation of the plant cytochrome P450, allene oxide synthase, resembles that of animal prostaglandin endoperoxide H synthase. *Journal of Biological Chemistry*, **273**, 18139–18145.
Pearce, G. and Ryan, C.A. (2003) Systemic signaling in tomato plants for defense against herbivores. Isolation and characterization of three novel defense-signaling glycopeptide hormones coded in a single precursor gene. *Journal of Biological Chemistry*, **278**, 30044–30050.

Pearce, G., Moura, D.S., Stratmann, J. and Ryan, C.A. (2001) Production of multiple plant hormones from a single polyprotein precursor. *Nature*, **411**, 817–820.

Pena-Cortes, H., Albrecht, T., Prat, S., Weiler, E.W. and Willmitzer, L. (1993) Aspirin prevents wound-induced gene-expression in tomato leaves by blocking jasmonic acid biosynthesis. *Planta*, **191**, 123–128.

Penninckx, I.A., Eggermont, K., Schenk, P.M., Van den Ackerveken, G., Cammue, B.P. and Thomma, B.P. (2003) The Arabidopsis mutant *iop1* exhibits induced over-expression of the plant defensin gene *PDF1.2* and enhanced pathogen resistance. *Molecular Plant Pathology*, **4**, 479–486.

Perez-Gilabert, M., Veldink, G.A. and Vliegenthart, J.F.G. (1998) Oxidation of dilinoleoyl phosphatidylcholine by lipoxygenase 1 from soybeans. *Archives of Biochemistry and Biophysics*, **354**, 18–23.

Petersen, M., Brodersen, P., Naested, H. *et al.* (2000) Arabidopsis MAP kinase 4 negatively regulates systemic acquired resistance. *Cell*, **103**, 1111–1120.

Pieterse, C.M., van Wees, S.C., van Pelt, J.A. *et al.* (1998) A novel signaling pathway controlling induced systemic resistance in Arabidopsis. *The Plant Cell*, **10**, 1571–1580.

Rancé, I., Fournier, J. and Esquerré-Tugayé, M.-T. (1998) The incompatible interaction between *Phytophthora parasitica* var. *nicotianae* race 0 and tobacco is suppressed in transgenic plants expressing antisense lipoxygenase sequences. *Proceedings of the National Academy of Sciences USA*, **95**, 6554–6559.

Richmond, T.A. and Bleecker, A.B. (1999) A defect in beta-oxidation causes abnormal inflorescence development in Arabidopsis. *The Plant Cell*, **11**, 1911–1924.

Riechmann, J.L. and Meyerowitz, E.M. (1998) The AP2/EREBP family of plant transcription factors. *Biological Chemistry*, **379**, 633–646.

Royo, J., Leon, J., Vancanneyt, G. *et al.* (1999) Antisense-mediated depletion of a potato lipoxygenase reduces wound induction of proteinase inhibitors and increases weight gain of insect pests. *Proceedings of the National Academy of Sciences USA*, **96**, 1146–1151.

Rustérucci, C., Montillet, J.L., Agnel, J.P. *et al.* (1999) Involvement of lipoxygenase-dependent production of fatty acid hydroperoxides in the development of the hypersensitive cell death induced by cryptogein on tobacco leaves. *Journal of Biological Chemistry*, **274**, 36446–36455.

Ryan, C.A. and Pearce, G. (2003) Systemins: a functionally defined family of peptide signals that regulate defensive genes in Solanaceae species. *Proceedings of the National Academy of Sciences USA*, **100**, suppl 2, 14577–14580.

Sanchez-Serrano, J., Schmidt, R., Schell, J. and Willmitzer, L. (1986) Nucleotide sequence of proteinase inhibitor II encoding cDNA of potato (*Solanum tuberosum*) and its mode of expression. *Molecular and General Genetics*, **203**, 15–20.

Sanders, P.M., Lee, P.Y., Biesgen, C. *et al.* (2000) The Arabidopsis DELAYED DEHISCENCE1 gene encodes an enzyme in the jasmonic acid synthesis pathway. *The Plant Cell*, **12**, 1041–1061.

Schaller, F., Biesgen, C., Mussig, C., Altmann, T. and Weiler, E.W. (2000) 12-Oxophytodienoate reductase 3 (OPR3) is the isoenzyme involved in jasmonate biosynthesis. *Planta*, **210**, 979–984.

Scheel, D. (1998) Resistance response physiology and signal transduction. *Current Opinion in Plant Biology*, **1**, 305–310.

Scheer, J.M. and Ryan, C.A. (2002) The systemin receptor SR160 from *Lycopersicon peruvianum* is a member of the LRR receptor kinase family. *Proceedings of the National Academy of Sciences USA*, **99**, 9585–9590.

Scheer, J.M., Pearce, G. and Ryan, C.A. (2003) Generation of systemin signaling in tobacco by transformation with the tomato systemin receptor kinase gene. *Proceedings of the National Academy of Sciences USA*, **100**, 10114–10117.

Seo, S., Okamoto, M., Seto, H., Ishizuka, K., Sano, H. and Ohashi, Y. (1995) Tobacco MAP kinase: a possible mediator in wound signal transduction pathways. *Science*, **270**, 1988–1992.

Shah, J. (2003) The salicylic acid loop in plant defense. *Current Opinion in Plant Biology*, **6**, 365–371.

Shah, J., Kachroo, P., Nandi, A. and Klessig, D.F. (2001) A recessive mutation in the Arabidopsis *SSI2* gene confers SA- and NPR1-independent expression of PR genes and resistance against bacterial and oomycete pathogens. *The Plant Journal*, **25**, 563–574.

Shibata, D., Slusarenko, A., Casey, R., Hildebrand, D. and Bell, E. (1994) Lipoxygenases. *Plant Molecular Biology Reporter*, **12**, S41-S42.

Shrestha, R., Noordermeer, M.A., Van der Stelt, M., Veldink, G.A. and Chapman, K.D. (2002) *N*-Acylethanolamines are metabolized by lipoxygenase and amidohydrolase in competing pathways during cottonseed iImbibition. *Plant Physiology*, **130**, 391–401.

Somerville, C., Browse, J., Jaworski, J.G. and Ohlrogge, J. (2000) Lipids, in *Biochemistry & Molecular Biology of Plants* (eds B.B. Buchanan, W. Gruissem and R.L. Jones), American Society of Plant Physiologists, Rockville, Maryland, pp. 456–527.

Song, W.C., Funk, C.D. and Brash, A.R. (1993) Molecular cloning of an allene oxide synthase – a cytochrome-P450 specialized for the metabolism of fatty acid hydroperoxides. *Proceedings of the National Academy of Sciences USA*, **90**, 8519–8523.

Spoel, S.H., Koornneef, A., Claessens, S.M. *et al.* (2003) NPR1 modulates cross-talk between salicylate- and jasmonate-dependent defense pathways through a novel function in the cytosol. *The Plant Cell*, **15**, 760–770.

Staswick, P.E., Su, W. and Howell, S.H. (1992) Methyl jasmonate inhibition of root growth and induction of a leaf protein are decreased in an *Arabidopsis thaliana* mutant. *Proceedings of the National Academy of Sciences USA*, **89**, 6837–6840.

Staswick, P.E., Tiryaki, I. and Rowe, M.L. (2002) Jasmonate response locus *JAR1* and several related Arabidopsis genes encode enzymes of the firefly luciferase superfamily that show activity on jasmonic, salicylic, and indole-3-acetic acids in an assay for adenylation. *The Plant Cell*, **14**, 1405–1415.

Staswick, P.E., Yuen, G.Y. and Lehman, C.C. (1998) Jasmonate signaling mutants of *Arabidopsis* are susceptible to the soil fungus *Pythium irregulare*. *The Plant Journal*, **15**, 747–754.

Stelmach, B.A., Muller, A., Hennig, P., Gebhardt, S., Schubert-Zsilavecz, M. and Weiler, E.W. (2001) A novel class of oxylipins, sn1-O-(12-oxophytodienoyl)-sn2-O-(hexadecatrienoyl)-monogalactosyl diglyceride, from *Arabidopsis thaliana*. *Journal of Biological Chemistry*, **276**, 12832–12838.

Stenzel, I., Hause, B., Maucher, H. *et al.* (2003) Allene oxide cyclase dependence of the wound response and vascular bundle-specific generation of jasmonates in tomato – amplification in wound signalling. *Plant Journal*, **33**, 577–589.

Stintzi, A. and Browse, J. (2000) The Arabidopsis male-sterile mutant, *opr3*, lacks the 12-oxophytodienoic acid reductase required for jasmonate synthesis. *Proceedings of the National Academy of Sciences USA*, **97**, 10625–10630.

Stintzi, A., Weber, H., Reymond, P., Browse, J. and Farmer, E.E. (2001) Plant defense in the absence of jasmonic acid: the role of cyclopentenones. *Proceedings of the National Academy of Sciences USA*, **98**, 12837–12842.

Strassner, J., Schaller, F., Frick, U.B. *et al.* (2002) Characterization and cDNA-microarray expression analysis of 12-oxophytodienoate reductases reveals differential roles for octadecanoid biosynthesis in the local versus the systemic wound response. *The Plant Journal*, **32**, 585–601.

Stratmann, J.W. (2003) Long distance run in the wound response – jasmonic acid is pulling ahead. *Trends in Plant Science*, **8**, 247–250.

Stumpe, M., Kandzia, R., Göbel, C., Rosahl, S. and Feussner, I. (2001) A pathogen-inducible divinyl ether synthase (CYP74D) from elicitor-treated potato suspension cells. *FEBS Letters*, **507**, 371–376.

Szekeres, M. (2003) Brassinosteroid and systemin: two hormones perceived by the same receptor. *Trends in Plant Science*, **8**, 102–104.

Thoma, I., Loeffler, C., Sinha, A.K. *et al.* (2003) Cyclopentenone isoprostanes induced by reactive oxygen species trigger defense gene activation and phytoalexin accumulation in plants. *The Plant Journal*, **34**, 363–375.

Thomma, B., Eggermont, K., Penninckx, I. *et al.* (1998) Separate jasmonate-dependent and salicylate-dependent defense-response pathways in *Arabidopsis* are essential for resistance to distinct microbial pathogens. *Proceedings of the National Academy of Sciences USA*, **95**, 15107–15111.

Tiryaki, I. and Staswick, P.E. (2002) An Arabidopsis mutant defective in jasmonate response is allelic to the auxin-signaling mutant *axr1*. *Plant Physiology*, **130**, 887–894.

Titarenko, E., Rojo, E., Leon, J. and Sanchez-Serrano, J.J. (1997) Jasmonic acid-dependent and -independent signaling pathways control wound-induced gene activation in *Arabidopsis thaliana*. *Plant Physiology*, **115**, 817–826.

van der Fits, L. and Memelink, J. (2000) ORCA3, a jasmonate-responsive transcriptional regulator of plant primary and secondary metabolism. *Science*, **289**, 295–297.

van Leyen, K., Duvoisin, R.M., Engelhardt, H. and Wiedmann, M. (1998) A function for lipoxygenase in programmed organelle degradation. *Nature*, **395**, 392–395.

Vancanneyt, G., Sanz, C., Farmaki, T. *et al.* (2001) Hydroperoxide lyase depletion in transgenic potato plants leads to an increase in aphid performance. *Proceedings of the National Academy of Sciences USA*, **98**, 8139–8144.

Veldink, G.A. and Vliegenthart, J.F.G. (1991) Substrates and products of lipoxygenase catalysis, in *Studies in Natural Products Chemistry*, Vol. 9 (ed. Atta-ur-Rahman), Elsevier Science Publishers B.V., Amsterdam, pp. 559–589.

Véronési, C., Rickauer, M., Fournier, J., Pouénat, M.-L. and Esquerré-Tugayé, M.-T. (1996) Lipoxygenase gene expression in the tobacco-*Phytophthora parasitica nicotianae* interaction. *Plant Physiology*, **112**, 997–1004.

Vierstra, R.D. (2003) The ubiquitin/26S proteasome pathway, the complex last chapter in the life of many plant proteins. *Trends in Plant Science*, **8**, 135–142.

Vijayan, P., Shockey, J., Levesque, C.A., Cook, R.J. and Browse, J. (1998) A role for jasmonate in pathogen defense of *Arabidopsis*. *Proceedings of the National Academy of Sciences USA*, **95**, 7209–7214.

von Malek, B., van der Graaff, E., Schneitz, K. and Keller, B. (2002) The Arabidopsis male-sterile mutant dde2–2 is defective in the ALLENE OXIDE SYNTHASE gene encoding one of the key enzymes of the jasmonic acid biosynthesis pathway. *Planta*, **216**, 187–92.

Wasternack, C. and Hause, B. (2002) Jasmonates and octadecanoids: signals in plant stress responses and development. *Progress in Nucleic Acid Research and Molecular Biology*, **72**, 165–221.

Weber, H., Chetelat, A., Caldelari, D. and Farmer, E.E. (1999) Divinyl ether fatty acid synthesis in late blight-diseased potato leaves. *The Plant Cell*, **11**, 485–493.

Weichert, H., Kolbe, A., Kraus, A., Wasternack, C. and Feussner, I. (2002) Metabolic profiling of oxylipins in germinating cucumber seedlings – lipoxygenase-dependent degradation of triacylglycerols and biosynthesis of volatile aldehydes. *Planta*, **215**, 612–619.

Xie, D.X., Feys, B.F., James, S., Nieto-Rostro, M. and Turner, J.G. (1998) COI1: an Arabidopsis gene required for jasmonate-regulated defense and fertility. *Science*, **280**, 1091–1094.

Xu, L., Liu, F., Lechner, E. *et al.* (2002) The SCF(COI1) ubiquitin-ligase complexes are required for jasmonate response in Arabidopsis. *The Plant Cell*, **14**, 1919–1935.

Xu, L., Liu, F., Wang, Z., Peng, W., Huang, R., Huang, D. and Xie, D. (2001) An Arabidopsis mutant *cex1* exhibits constant accumulation of jasmonate- regulated AtVSP, Thi2.1 and PDF1.2. *FEBS Letters*, **494**, 161–164.

Ziegler, J., Stenzel, I., Hause, B. *et al.* (2000) Molecular cloning of allene oxide cyclase. The enzyme establishing the stereochemistry of octadecanoids and jasmonates. *Journal of Biological Chemistry*, **275**, 19132–19138.

Zolman, B.K., Silva, I.D. and Bartel, B. (2001) The Arabidopsis *pxa1* mutant is defective in an ATP-binding cassette transporter-like protein required for peroxisomal fatty acid b-oxidation. *Plant Physiology*, **127**, 1266–1278.

# 11 Prenyllipids and their derivatives: sterols, prenylquinones, carotenoids and terpenoids

Pierre Benveniste

## 11.1 Introduction

The term ‘prenyllipids’ designates a vast family of compounds originating from isopentenyl diphosphate (IDP), the so-called biogenetically active isoprene unit, which shares its hydrophobic properties with the more common acyl lipids. Because the subject is vast, we shall focus our attention on multicellular plants (bryophytes, pteridophytes, gymnosperms and angiosperms). We shall exclude most microorganisms (bacteria, cyanobacteria, fungi, unicellular algae and protists), higher fungi, algae and animals (nematodes, insects, vertebrates). However, these excluded organisms will be considered occasionally if they are partners in a plant–pathogen interaction.

The localisation of prenyllipids in the plant cell is crucial, because it largely determines their biosynthetic site. At first approach, two main compartments are found. Sterols are generally localised in the cytoplasm, where they are associated with the endomembrane continuum and are concentrated especially in the plasma membrane (Hartmann-Bouillon and Benveniste, 1978). They are considered as absent from the chloroplast (Hartmann *et al.*, 2003), although important exceptions could exist (Hernandez-Pinzon *et al.*, 1999). Tocopherols (tocotrienols), carotenoids and chlorophylls are all found in the plastid compartment. The case of terpenoids is more complex. Monoterpenes and diterpenes are generally synthesised in leucoplasts and in photosynthetic plastids (Bouvier *et al.*, 2000b). In many cases, they diffuse out of plastids and accumulate in specific sink tissues. Sesquiterpenes are usually synthesised in the cytoplasm. These location problems are also important with respect to the sequence of biosynthetic reactions prior to IDP. There are more and more data in higher plants showing that cytoplasmic isoprenoids (sterols) derive from the mevalonic pathway, whereas isoprenoids, which are synthesised in plastids, would derive from the non-mevalonate (MEP) pathway (Hemmerlin *et al.*, 2003). Although this crucial point should be discussed thoroughly for each group of isoprenoids, this topic is beyond the scope of the present chapter and would deserve a whole review on its own. Prenylation of proteins (which plays a very important role in biological events such as signal transduction, membrane dynamics or regulation of cell division) is also beyond the scope of the present review and will not be considered.

In this chapter, we will focus our attention on the support of molecular genetics to elucidate biosynthetic pathways leading to the main classes of prenyllipids. In recent years, the impressive development of molecular biology tools, the sequencing of the *Arabidopsis thaliana* genome, the availability of DNA or transposon tagged *mutants*, multiple possibilities offered by stable transformation with DNA in sense and antisense orientations and gene silencing using appropriate virus-derived vectors have led to many important breakthroughs in the field of isoprenoid biosynthesis, metabolism and function. General schemes concerning biosynthesis of tocopheryl quinones (Hofius *et al.*, 2003), carotenoids (Hirschberg *et al.*, 2001), sterols (Bach and Benveniste, 1997; Hartmann *et al.*, 2003) and terpenoids (Bohlmann *et al.*, 1998) have recently been published.

## 11.2 General considerations

All prenyllipids derive from IDP after several steps of head to tail and/or head to head condensation reactions. The simplest case is that of tocopherols and other prenylquinones. Tocopherol and plastoquinone derive from the condensation of phytyl diphosphate and solanesyl diphosphate, respectively, with homogentisic acid (Cheng *et al.*, 2003). Phytyl diphosphate or solanesyl diphosphate originate from the head to tail condensation of four and nine isoprenic units, respectively. Most carotenoids derive from the head to head condensation of two geranyl–geranyl diphosphate (GGDP) units. This step is followed by a series of desaturation reactions leading to more and more conjugated double bonds that are responsible for the spectrophotometric properties of carotenoids, and the biosynthesis is frequently terminated by cyclisation reactions leading to typical carotenoids such as β-carotene (Hirschberg *et al.*, 2001). Some carotenoids of bacterial origin derive from the condensation of two farnesyl diphosphate (FDP) units (Umeno *et al.*, 2002). Head to head condensation of two FDP units is also involved in the synthesis of triterpenoids and steroids, but this step is not followed by desaturation reactions since the squalene (product of the condensation reaction) and its derivative 2,3-oxidosqualene are cyclised to give tetra- or pentacyclic triterpenes. Triterpenes are then converted aerobically into various classes of compounds (cucurbitacins, limonoids, withanolides, cardiotonic glucosides, steroid alkaloids, steroid sapogenines). Metabolism of triterpenes, such as lanosterol or cycloartenol, leads to sterols and their derivatives (steroid hormones) (Abe *et al.*, 1993).

In the present review, the term ‘terpenoid’ will be restricted (admittedly arbitrarily) to compounds such as monoterpenes, sesquiterpenes and diterpenes deriving from geranyl diphosphate (GDP), FDP and GGDP, respectively, after one or several cyclisation steps (Bohlmann *et al.*, 1998). These compounds share in common the following properties: many of them are volatile and have fragrant odours and often antibiotic properties; some are formed in response to the infection of plants by pathogens and can be considered as phytoalexins (Phillips and Croteau, 1999).

The pathways of monoterpene, sesquiterpene and diterpene biosynthesis are conveniently divided into several parts. The first encompasses the synthesis of IDP, isomerisation to dimethylallyl diphosphate, prenyltransferase-catalysed condensation of these two $C_5$-units to GDP and the subsequent $1'$-4 additions of IDP to generate FDP and GGDP. In the second stage, the prenyl diphosphates undergo a range of cyclisations to produce the parent skeletons of each class. Thus, GDP ($C_{10}$) gives rise to monoterpenes, FDP ($C_{15}$) to sesquiterpenes and GGDP ($C_{20}$) to diterpenes. Finally, the last step consists in a variety of redox modifications of the parent skeletal types to produce thousands of different terpenoids present in resins and essential oils of plant origin (Bohlmann *et al.*, 1998). The elucidation of isoprenoid biosynthesis owes a great deal to the methods of molecular genetics because the enzymes involved in the biosynthesis of these compounds are generally present in very small amounts and they are often bound to membranes. Therefore, they are extremely difficult to purify, which had greatly limited progress in research. Molecular genetic methods have allowed us to overcome these difficulties. Cloning of genes took advantage of the increasing availability of EST cDNA clones from various flowering plants and (for prenylquinones and carotenoids only) of the fact that pathways in plants are sometimes similar to that in cyanobacteria such as *Synechococcus* sp. PCC7942. The cyanobacterial genes then served as molecular probes to isolate the plant orthologues (Hirschberg *et al.*, 2001).

## 11.3 Prenylquinones

Two classes of compounds – tocopherols (tocotrienols) and plastoquinones – share a great popularity in research programmes, for different reasons: tocopherols (tocotrienols) act as antioxidants and free radical scavengers (Cheng *et al.*, 2003); they are essential dietary components for humans. Therefore, many research programmes are aimed at a better understanding of how these compounds are synthesised and what could be the mechanism of their cell protective effects (Ricciarelli *et al.*, 2002). On the other hand, plastoquinones are essential components of the photosynthetic electron transport chain and play a major role in the formation of proton gradients across thylakoid membranes.

### *11.3.1 Properties of prenylquinones*

#### *11.3.1.1 Tocopherols and tocotrienols*

The generic term 'tocopherol' covers a series of hydrophobic compounds ($\alpha$-, $\beta$-, $\gamma$- and $\delta$-tocopherols) where a polar chromanol ring is linked to a phytyl hydrophobic

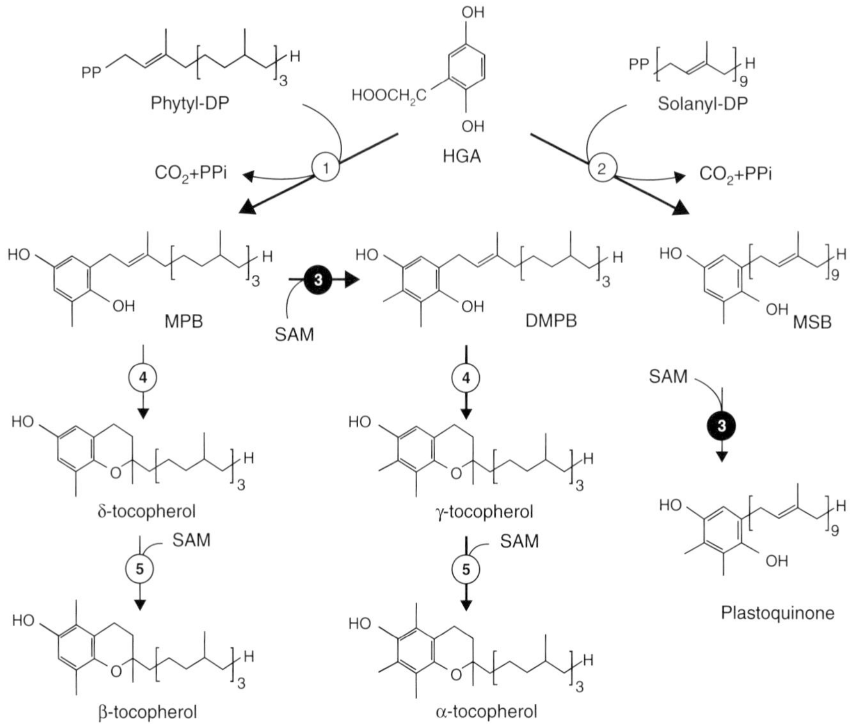


**Figure 11.1** The tocopherol and plastoquinone biosynthetic pathways in plants. Boldface arrows represent the steps leading to α-tocopherol, the most abundant tocopherol produced in wild-type *Arabidopsis thaliana* leaves. **DMPB**, 2,3-dimethyl-5-phytyl-1,4-benzoquinone. **HGA**, homogentisic acid. **MPB**, 2-methyl-6-phytyl-1,4-benzoquinone. **MSB**, 2-methyl-6-solanesyl-1,4-benzoquinone. Enzymes are indicated by circled numbers. **1**, homogentisate phytyltransferase. **2**, homogentisate solanesyltransferase. **3**, **MPB/MSB** methyltransferase. **4**, tocopherol cyclase. **5**, γ-tocopherol methyltransferase. Material copyrighted by the American Society of Plant Biologists, and reprinted from (29), with permission.

chain (Figure 11.1). In the case of tocotrienols, the chromanol ring is tethered to a geranylgeranyl hydrocarbon chain to give α-, β-, γ- and δ-tocotrienols. In plants, a mixture of these eight compounds is generally present, which is collectively referred to as vitamin E. Tocotrienols account for more than 50% of the total vitamin E of the seed endosperm of monocotyledons, whereas tocopherols occur more widely in leaves and seeds of most dicotyledonous plants (Cahoon *et al.*, 2003). The highest biological activity of vitamin E is generally attributed to α-tocopherol, which is often present in only small amounts in the mixture. This biological activity is complex because tocopherols and tocotrienols have been shown to be involved in various cellular processes. The best known effect is the antioxidant one on free radicals of oxygen (hydroxyl and superoxide radicals) and activated oxygen species (singlet oxygen and peroxide) (AOS), which trigger the generation of membrane lipid radicals and alter membrane structure and function (Brigelius-Flohe *et al.*, 2002). Further tocopherols perform other non-antioxidant functions related to signalling and transcriptional regulation (Ricciarelli *et al.*, 2002). Their involvement in the photosynthetic apparatus is still a matter of debate (Grasses *et al.*, 2001). Molecular genetic and genomic approaches allowing one to increase or block tocopherol biosynthesis are essential to a better understanding of functions of these compounds, at least in photosynthetic organisms. One of the prerequisites for such approaches is a better knowledge of the biosynthesis of these compounds.

#### *11.3.1.2 Plastoquinone*

Plastoquinone (PQ) plays a central role in oxygenic photosynthesis; it provides the electronic connection between the two photosystems and generates a transmembrane electrochemical proton gradient used for energy transduction and subsequent synthesis of adenosine triphosphate. PQ receives two electrons from Photosystem II via two plastosemiquinone anions QA and QB stabilised by proteins (de Wijn and van Gorkom, 2001). Then, the doubly reduced dihydroplastoquinone (PQH2) transfers two single electrons to a high-potential electron transfer chain, consisting of the Rieske iron–sulfur protein (which binds an Fe2S2 cluster), a c-type cytochrome (cytochrome f), a b-type cytochrome (cytochrome b6) and subunit IV (Finazzi *et al.*, 2002).

### *11.3.2 Biosynthesis of prenylquinones*

#### *11.3.2.1 Biosynthesis of tocopherol and plastoquinone*

Excellent reviews have been recently published on this topic (Cheng *et al.*, 2003; Hofius and Sonnewald, 2003). Although the plant tocopherol and PQ biosynthetic pathways were elucidated by labelling studies during the 1980s (d'Harlingue and Camara, 1985; Soll *et al.*, 1985), clear dissection of all steps could not be achieved because of difficulties of working with membrane-bound enzymes. Most pathway enzymes have been isolated, purified and characterised only since the 1990s owing to genetic and genomic approaches in *Synechocystis* sp. PCC6803 and *A. thaliana.* These methods allowed cloning of the genes and subsequent expression in bacteria or plants. The tocopherol and PQ biosynthetic pathways in plants are shown in Figure 11.1. In both pathways, the committed step is the condensation of homogentisic acid (HGA) derived from the shikimate pathway with phytyl diphosphate (derived from the MEP) and solanesyl diphosphate to give 2-methyl-6-phytyl-1,4-benzoquinone (MPB) (Collakova and DellaPenna, 2001) and 2-methyl-6-solanesyl-1,4-benzoquinone (MSB) (Norris *et al.*, 1995), respectively. The conversion of MPB to α-tocopherol involves first a methylation step catalysed by a MPB-methyltransferase dependent of S-adenosyl methionine (SAM) giving 2,3-dimethyl-5-phytyl-1,4-benzoquinone (DMPB). The same enzyme was also involved in the methylation of MSB leading to PQ (Cheng *et al.*, 2003). This interesting observation resulted from analysis of a homozygous T-DNA insertional mutant *vte3–2*, which contained neither PQ nor α- + β-tocopherol. MPB methyltransferase has little homology with its *Synechocystis* sp. counterpart. Surprisingly, it has orthologues in archaeobacteria such as *Archaeoglobus fulgidus* (Cheng *et al.*, 2003). The following step during biosynthesis of α-tocopherol is catalysed by a tocopherol cyclase that transforms DMPB into γ-tocopherol. The functional characterisation of this step resulted from an EMS mutant from *A. thaliana* deficient in all tocopherols, lacking tocopherol cyclase activity and called *vte1* (vitamin E deficient). Interestingly, this mutant was viable and the absence of tocopherol did not have a large impact on the photosynthesis, growth and development of the mutant plants (Porfirova *et al.*, 2002). The last step is the conversion of γ-tocopherol into α-tocopherol by the γ-tocopherol methyltransferase (γ-TMT). A γ-TMT gene was isolated from the putative 10-gene tocopherol biosynthetic operon in *Synechocystis* sp. PCC6803 using bioinformatics and gene disruption experiments (Shintani and DellaPenna, 1998). The orthologous gene from *A. thaliana* was then isolated and its overexpression in transgenic *A. thaliana* plants increased strongly the α- to γ-tocopherol ratio, as well as the total vitamin E content compared to wild-type controls (Shintani and DellaPenna, 1998). The same enzyme was capable of converting δ-tocopherol, which lacks a methyl group at the position 7, into β-tocopherol thereby demonstrating that most of enzymes involved in tocopherol biosynthesis have a rather wide substrate specificity. *A. thaliana* γ-TMT bears only little homology with MPB methyltransferases, which are the other methyltransferase involved in tocopherol biosynthesis. An *A. thaliana* mutant (*vte4*) deficient in γ-tocopherol methyltransferase has been characterised and was shown to accumulate γ-tocopherol at the expense of α-tocopherol. Chlorophyll content and photosynthetic quantum yield were very similar in wild-type and mutant, which indicates that γ-tocopherol can be replaced by α-tocopherol to protect the photosynthetic apparatus against oxidative stress (Porfirova *et al.*, 2002).

#### *11.3.2.2 Biosynthesis of tocotrienolquinones*

Tocotrienols represent the primary form of vitamin E in seeds of most monocotyledonous plants, including cereals such as rice and wheat. The biosynthesis of tocotrienols has not been extensively studied in plants. cDNAs encoding homogentisic acid geranylgeranyltransferase (HGGT), which catalyses the committed step of tocotrienol biosynthesis, were isolated from wheat and rice (Cahoon *et al.*, 2003). The deduced polypeptide sequences showed 45–50% identity to homogentisic phytyltransferases (HPTs) reported earlier. HGGT seems to be functionally different from HPT since it displays substrate specificity for GGDP in preference to PDP (Soll *et al.*, 1985). Methylation and cyclisation reactions associated with the tocopherol biosynthesis pathway (Figure 11.1) in spinach chloroplasts are apparently able to convert the HGGT product 2-methyl-6-geranylgeranylbenzoquinol into tocotrienols (Soll and Schultz, 1979). However, this point should be carefully re-examined with the now available purified recombinant methyltransferases that are implicated in tocopherol biosynthesis. Transgenic expression of the barley HGGT in *A. thaliana* leaves resulted in accumulation of tocotrienols, which were absent from leaves of nontransformed plants, and a 10- to 15-fold increase in total vitamin E antioxidants (tocotrienols+ tocopherols). These results demonstrate the ability to enhance the antioxidant content of plants by introduction of an enzyme (HGGT) that redirects metabolic flux (Cahoon *et al.*, 2003).

## 11.4 Carotenoids

We will use this term to design either non-oxygenated compounds such as β-carotene or oxygenated compounds such as zeaxanthin, which are usually called xanthophylls. Plant carotenoids are 40-carbon isoprenoids with a great number (nine and more) of conjugated double bonds. Because of their spectroscopic and physicochemical properties, carotenoids are essential components of photosynthetic organisms. They are present in the light-harvesting antennae of the chloroplasts where they transfer their absorbed energy to chlorophylls. However, the main function of carotenoids in general and of zeaxanthin in particular is to quench the first singlet chlorophyll excited state in a non-radiative manner. Hence, carotenoids have an important role as direct quenchers of excess excitation energy in photosynthesis (Ma *et al.*, 2003). The crucial importance of this function is attested by the fact that *mutants* (*A. thaliana* or *Zea mays*) that are defective in carotenoid biosynthesis or binding have an *albina* phenotype because their chlorophyll is destroyed as soon as it is synthesised. Another very important property of carotenoids present in flowers is to attract pollinator insects or hummingbirds allowing cross-fertilisation (Bradshaw and Schemske, 2003). Finally, carotenoids are precursors of compounds with important physiological activity, such as abscisic acid (ABA) (Rock and Zeevaart, 1991).

### *11.4.1 Biosynthesis of carotenoids*

In higher plants, carotenoids are synthesised inside the plastid by enzymes, which are encoded by the nuclear DNA. The biosynthetic pathway leading from phytoene to carotenoids from higher plants is depicted in Figure 11.2. Like all other isoprenoids, carotenoids are built from isopentenyl diphosphate (IDP). In plastids, IDP is produced via the 1-deoxy-xylulose-5-phosphate (DOXP) pathway from pyruvate and glyceraldehyde-3-phosphate (Rohmer *et al.*, 1999; Schwender *et al.*, 1999). The steps between IDP and phytoene include IDP isomerisation to dimethylallyl diphosphate. There are two IDP isomerase genes in plants and one of them encodes an active polypeptide localised inside plastids (Cunningham *et al.*, 2000). The sequential addition of three IPP molecules to dimethylallyl diphosphate, which

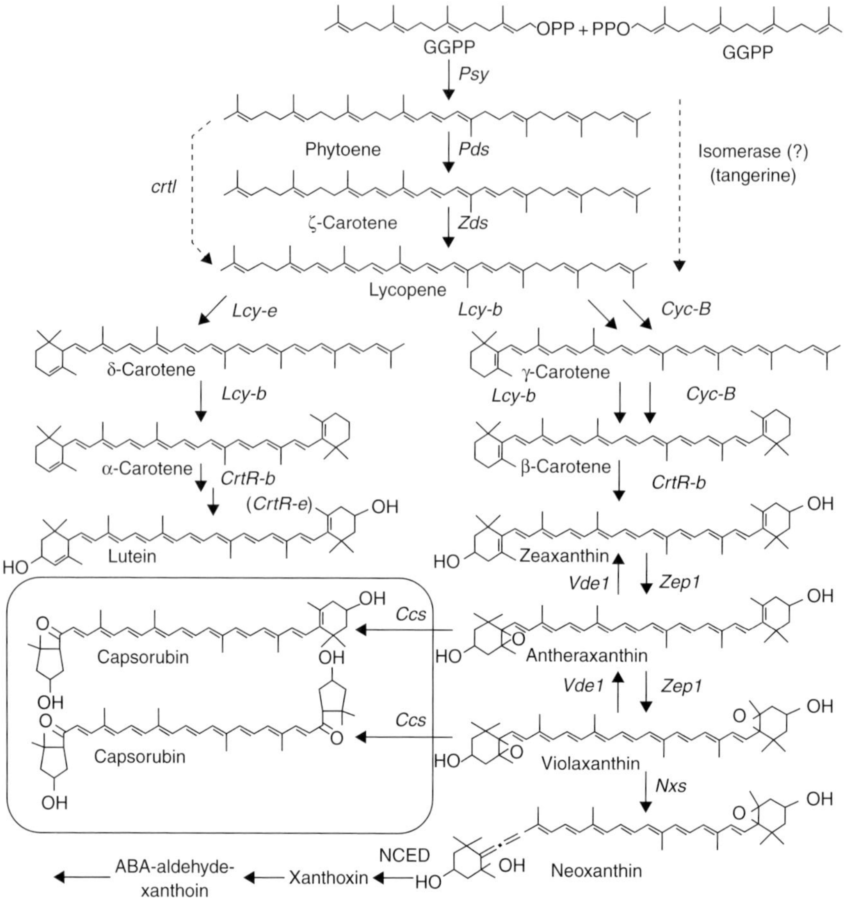


**Figure 11.2** The carotenoid biosynthesis pathway in plants. Enzymes are named according to the designation of their genes. The pathway in the box takes place in chromoplasts of *Capsicum annuum* fruits. ***Psy***, phytoene synthase. ***Pds***, Phytoene desaturase. ***Zds***, ζ-carotene desaturase. ***Lcy***, lycopene cyclase. ***Cyc-B***, Chromoplast-specific lycopene cyclase. ***CrtR-b***, β-ring hydroxylase. ***CrtR-e***, ϵ-ring hydroxylation. ***Zep1***, Zeaxanthin epoxidase. ***Vde1***, Violaxanthin deepoxidase. ***Nxs***, neoxanthin synthase. ***Ccs***, capsoxanthin (capsorubine) synthase. ***Nced***, neoxanthin 9Z-epoxycarotenoid dioxygenase. Material copyrighted by Elsevier and redrawn from (60) with permission.

is catalysed by GGDP synthase (*Ggps*), gives the 20-carbon molecule, GGDP. The genome of *A. thaliana* contains a family of 12 genes homologous to *Ggps*. Among them, 5 *Ggps* genes have been shown to be expressed in different tissues during plant development (Okada *et al*., 2000). The first committed step in the carotenoid pathway is the condensation of two GGDP molecules to produce 15-Z-phytoene, which is catalysed by a membrane-associated enzyme, phytoene synthetase (PSY)

(Camara, 1993). In tomatoes, there are two genes for PSY, *Psy-1*, which encodes a fruit- and flower-specific isoform, and *Psy-2*, which encodes a form that predominates in green tissues (Bartley and Scolnik, 1993). PSY appears to be a rate-limiting enzyme of carotenoid biosynthesis in various plants. An impressive increase of about 50-fold in the carotenoid content of *Brassica oleracea* seeds was achieved by expressing a single gene for PSY (Shewmaker *et al.*, 1999). Two similar enzymes, phytoene desaturase (PDS) and ζcarotene desaturase (ZDS) transform phytoene to lycopene. These FAD-containing enzymes catalyse dehydrogenation reactions resulting in the formation of seven and eleven conjugated double bonds in ζcarotene and lycopene, respectively. These reactions are dependent on PQ (Nievelstein *et al.*, 1995; Norris *et al.*, 1995) and a terminal oxidase (Carol *et al.*, 2001) as electron acceptors. When these two desaturases are expressed in bacterial systems, prolycopene (7,9,7′,9′-tetra-*cis*-lycopene) is obtained in place of all-*trans*-lycopene. Therefore, a carotenoid isomerase should exist in higher plants. Cloning of *tangerine* has revealed a carotenoid isomerase essential for the production of β-carotene and of oxygenated carotenoids in *Lycopersicum esculentum* (Isaacson *et al.*, 2002) and in *A. thaliana* (Park *et al.*, 2002). This enzyme is a flavoprotein, which converts prolycopene into lycopene through isomerisation of four double bonds (Giulano *et al.*, 2002). The function of carotene isomerase in plants is to enable carotenoid biosynthesis to occur in the dark and in non-photosynthetic tissues (Isaacson *et al.*, 2002). Cyclisation of lycopene leads to branching of the biosynthetic pathway; one branch leads to β-carotene and its derivative oxygenated carotenoids, whereas the other leads to α-carotene and lutein.

Lycopene β-cyclase (LCY-B) catalyses a two-step reaction to produce β-carotene, whereas lycopene ϵ-cyclase (LCY-E) creates one ϵ-ring to give δ-carotene. Carotenoids with two ϵ-rings are very rare in nature, however, lettuce (*Lactuca sativa*) contains a bi-cyclase that converts lycopene to lactucaxanthin, a carotenoid possessing two ϵ-rings (Cunningham and Gantt, 2001). It has been shown that the histidine at position 457 acts as a molecular switch determining the number of rings and the position of double bonds. Lycopene cyclases contain a characteristic FAD/NAD(P)-binding sequence motif at the *N*-terminal part of the protein. In tomato there are two lycopene cyclases, LCY-B and CYC-B (chromoplast specific-lycopene cyclase) (Ronen *et al.*, 1999). An enzyme, capsanthine-capsorubine synthase (CCS) has been shown to convert antheraxanthin and violaxanthin to the dark red xanthophylls capsanthin and capsorubin respectively, in *Capsicum annuum* (Bouvier *et al.*, 1994). Interestingly, CYC-B (chromoplast specific-lycopene cyclase) has a stronger sequence identity with CCS than with its paralogue LCY-B.

Hydroxylation of cyclic carotenes at the 3 position in the cyclohexanes is carried out by two types of enzymes – one specific for β-rings (CrtR-b), the other for ϵ-rings (CrtR-e) (Pogson *et al.*, 1996; Sun *et al.*, 1996). Molecular and functional characterisation of β-carotene hydroxylase from *C. annuum* has been performed. The enzyme is membrane-bound, it requires ferredoxin as an electron donor and molecular oxygen, and belongs to a class of monooxygenases, containing iron as well as three typical histidine clusters, which could be involved in $Fe^{2+}$ coordination and subsequent molecular oxygen activation (Bouvier *et al.*, 1998). Carotene hydroxylase converts β-carotene to zeaxanthin via β-cryptoxanthin. Two genes encoding carotene hydroxylase have been found in *A. thaliana* and are probably present in most other higher plants (Hirschberg *et al.*, 2001). Carotenoid β- and ϵ-hydroxylases (*Lut1*) catalyse the formation of lutein from α-carotene. *Lut1* has been shown to encode a cytochrome P450-type monooxygenase, CYP97C1 (Tian *et al.*, 2004). The following step is the epoxidation of zeaxanthin to give antheraxanthin then violaxanthin. This step is catalysed by Zep1 (ABA2), which is a monooxygenase that requires NADPH as electron donor (Bouvier *et al.*, 1996). *Zep1* was cloned from *Nicotiana plumbaginifolia* (Marin *et al.*, 1996) and bell pepper (Bouvier *et al.*, 1996). In leaves, violaxanthin is converted back to zeaxanthin by violaxanthin deepoxidase (VDE) – an enzyme that is activated by low pH and light (Bugos and Yamamoto, 1996). The interconversion of zeaxanthin and violaxanthin is known as the xanthophyll shuttle. Zeaxanthin plays a role in protecting the photosynthetic apparatus from damage by strong light. Lack of this shuttle in the *A. thaliana* mutant *npq1* is due to a loss-of-function mutation in *Vde*. This mutant manifests a phenotype of high susceptibility to intense light (Niyogi *et al.*, 1998). Conversely, transformation of plants with carotene hydroxylase leads to an increase of both zeaxanthin content and high light tolerance (Davison *et al.*, 2002). Neoxanthin is a carotenoid widely present in chloroplasts; it presents the unique particularity to contain an allene group, a chemical motif very rare in nature. Cloning of a gene encoding neoxanthin synthase (*Nxs*) from potato (*Solanum tuberosum*) and tomato has been reported (Al-Babili *et al.*, 2000; Bouvier *et al.*, 2000a).

### *11.4.2 Metabolic engineering of carotenoid biosynthesis*

Transformation of canola (*Brassica napus*) with a gene encoding PSY from *Erwinia*, using the *B. napus* napin gene promoter, leads to an impressive, almost 50-fold, increase in the carotenoid (mainly α- and β-carotene) content of seeds. This resulted in an oil enriched in carotenes (golden oil). Such coloured oils already exist naturally, e.g. in oil palm, but are not commonly found in food crops.

Such vitamin A enriched foods can be useful to combat vitamin A deficiency, which is widespread in developing countries (Shewmaker *et al.*, 1999). This goal was achieved in a system, which already synthesised carotenoids. The next step was to introduce carotenoids in a plant tissue, which did not produce any carotenoids. This was achieved in rice endosperm, which completely lacks carotenoids although it synthesises GGPP. To achieve this goal, the daffodil (*Narcissus pseudonarcissus*) genes phytoene synthase (*Psy*) and lycopene β-cyclase (*Lcyb*) (under control of glutelin gene promoter, an endosperm specific promoter) as well as a bacterial phytoene desaturase gene (*CrtI*) (under control of the cauliflower mosaic virus 35S promoter) were inserted into two plasmids, one carrying *Psy* and *CrtI* and the other providing the sequence coding for *Lcyb*. The combination of both plasmids was thus able to direct β-carotene formation in rice endosperm. The maximum level of carotenoids in the endosperm of plants was 1.6 μg/g. The so-called golden rice did not provide the recommended daily allowance of carotenoids in a diet composed uniquely of rice but, nevertheless, constituted a breakthrough in the metabolic engineering of carotenoids for improved nutritional value of a major crop (Ye *et al.*, 2000). The expression of the algal (*Haematococcus pluvialis*) β-carotene oxygenase gene was directed to seeds of *A. thaliana* by use of the 2S, seed storage protein promoter *napA*. Extracts from seeds of the transgenic plants were red because of accumulation of ketocarotenoids such as 4-keto-lutein (3,3′-dihydroxy-β,ε-carotene-4-one) (Sun *et al.*, 1996). Finally, transformation of plants with β-carotene hydroxylase led to an increase of both zeaxanthin content and high light tolerance confirming that zeaxanthin plays a central role in protection of plants against oxidative damage of membranes (Davison *et al.*, 2002).

### *11.4.3 Catabolism of carotenoids*

Mutants defective in ABA biosynthesis have been found in many plant species. Some of them (mutant *flacca* from *Lycopersicum esculentum* and *vp14* from *Zea mays*) were shown to be defective in enzymes converting neoxanthin (violaxanthin) into ABA (Schwartz *et al.*, 2003). Molecular analysis of *vp14* led to the discovery of a dioxygenase (NCED) capable of cleaving either neoxanthin or violaxanthin at the 9–10 position to give xanthoxin, a precursor of ABA. Neoxanthin was shown to be a better substrate for NCED than violaxanthin. Orthologues of the gene encoding NCED were identified in *Persea americana* (*PaNCED*) and in *A. thaliana* (*AtNCED2*) (Ivchi *et al.*, 2001).

A variety of natural products like ABA are derived from the oxidative cleavage of carotenoids. In 2003, a 7,8(7,8′) carotenoid cleavage dioxygenase (CCD), which specifically catalyses the synthesis of crocetin dialdehyde from zeaxanthin, has been cloned from *Crocus sativus*. Crocetin dialdehyde is the precursor of crocin, the primary pigment of saffron (Bouvier *et al.*, 2003a). A gene encoding a lycopene cleavage dioxygenase has been recently cloned from *Bixa orellana* developing seeds. This enzyme cleaves lycopene to give bixin dialdehyde, which in turn is oxidised by bixin aldehyde dehydrogenase to give norbixin. Finally, norbixin is methylated twice by a norbixin carboxyl methyltransferase to give the stable bixin dimethylester. Bixin is a seed-specific pigment widely used in foods and cosmetics since pre-Columbian times (Bouvier *et al.*, 2003b). Along the same lines, *A. thaliana* carotenoid cleavage dioxygenase 1 (AtCCD1) catalyses the initial step of mycorradicin biosynthesis, which gives the yellow colour of arbuscular mycorrhizal roots (Walter *et al.*, 2000).

## 11.5 Sterols

Several comprehensive reviews are available on this topic (Bach and Benveniste, 1997; Hartmann *et al.*, 2003; Schaller *et al.*, 2003). To avoid redundancy, we will focus our attention on the most recent findings. Sterols are isoprenoid-derived molecules that have essential functions in eukaryotes in general, and in higher plants in particular. In vertebrates, cholesterol is by far the major sterol, whereas a mixture of various sterols is present in higher plants, with sitosterol [(24R)-24-ethyl cholesterol] usually predominating. Higher plants, algae, most fungi and vertebrates synthesise sterols. However, insects, which represent more than 80% of animal species, do not (Nes, 1977). Sterols are present in three forms, free sterols, the major form with a free 3β-hydroxyl, steryl esters and steryl glucosides. Free sterols are integral components of the membrane lipid bilayer and as such they interact with some molecular species of phosphoglycero(sphingo)lipids. The extent of this interaction is thought to play a functional role in the regulation of membrane fluidity and permeability (Demel and de Kruyff, 1976). In addition to this role played by the bulk of sterols, a regulatory function has been assigned to sterols in various animal systems: (1) the covalent binding of cholesterol to the gene, HEDGEHOG is involved in embryonic development of vertebrates (Kip *et al.*, 2000); (2) interaction of cholesterol with caveolin induces the formation of membrane microdomains (caveolae/rafts), which may constitute signalling centres for multiple pathways (Karpen *et al.*, 2001; see also Chapter 7). Finally, sterols are precursors of compounds with a high physiological activity, such as brassinosteroids – an important class of hormones involved in higher plant growth and development (Li *et al.*, 2000)

In spite of this recent knowledge relating to the role of sterols in mediating regulatory processes in animals, an important question linking these findings to plant sterols remains open. In particular, we know very little about the role of typical plant sterols such as sitosterol in signal reception and transduction. In order to address this question, a strategy of gain or loss of function has recently been applied to the sterol biosynthetic pathway in the cases of the following enzymes.

### *11.5.1 2,3(S)-oxidosqualene-cycloartenol cyclase (OSC)*

Whereas 2,3-oxidosqualene (OS) is cyclised into lanosterol in non-photosynthetic eukaryotes, the higher plant OS cyclase (EC 5.4.99.8) catalyses the formation of cycloartenol, precursor of phytosterols in photosynthetic eukaryotes (Figure 11.3) (Nes, 1977). The stereoelectronic factors that underly the catalytic mechanism of these reactions have been discussed elsewhere (Abe *et al.*, 1993). In recent years, an elegant approach led to the isolation of an *A. thaliana* cDNA (*ATCYC*) encoding a cycloartenol synthase (Corey *et al.*, 1993). This was achieved by functional expression of the cDNA in a yeast mutant lacking lanosterol synthase (GL7), followed by a chromatographic screen for successful complementation. This cDNA contained a 2277-bp open reading frame encoding an 86-kD protein, which was about 35% identical to ERG7 from yeast. Single mutations (Y410T or H477N) occurring on the *A. thaliana* cycloartenol synthase have been shown to confer on cycloartenol synthase the ability to form lanosterol as its major product in place of cycloartenol (Segura *et al.*, 2002). Such experiments have far-reaching evolutionary

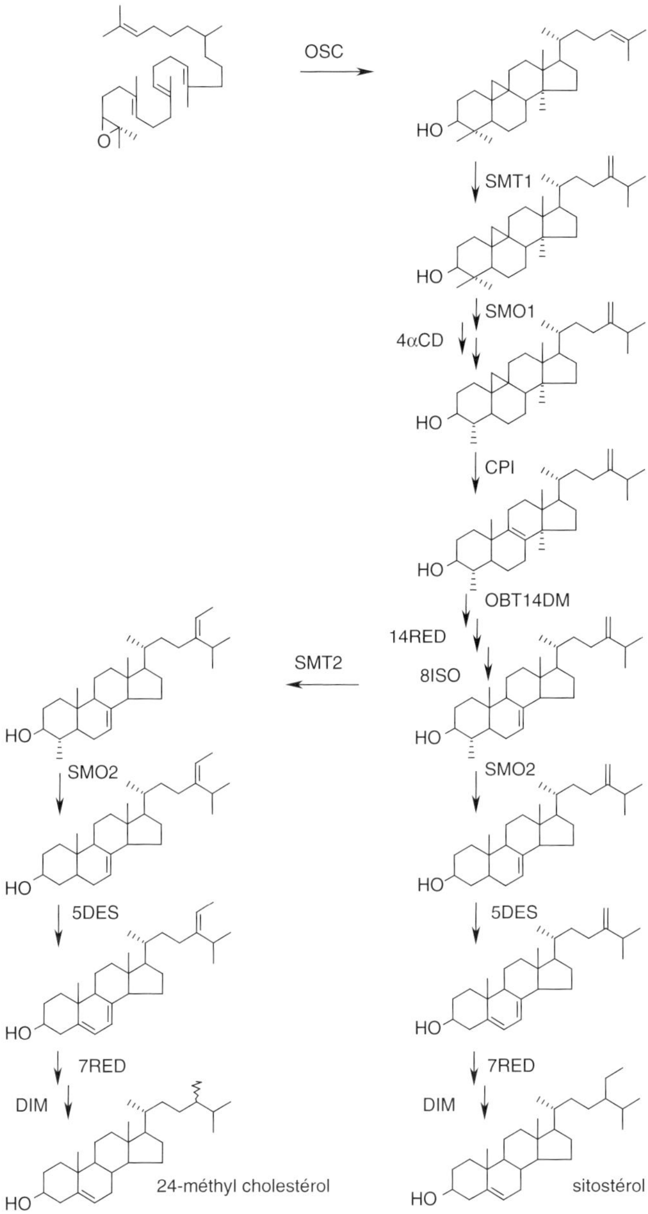


**Figure 11.3** Biosynthesis of phytosterols (segment downstream of 2,3-oxidosqualene). **OSC**, 2,3-oxidosqualene cyclase. **SMT1**, cycloartenol-C24-methyltransferase. **SMT2**, 24-methylene lophenol methyltransferase. **SMO1**, 24-methylene cycloartanol-4-methyl oxidases. **SMO2**, 24-ethylidene lophenol-4-methyl oxidase. **CPI**, cyclopropyl sterol isomerase. **OBT14DM** (**CYP51**), obtusifoliol-14α-demethylase. **14RED**, $\Delta^{14}$-reductase. **8ISO**, $\Delta^{8,7}$-sterol isomerase. **5DES**, $\Delta^{7}$-sterol-C5-desaturase. **7RED**, $\Delta^{5,7}$-sterol reductase. **DIM**, $\Delta^{24}$-sterol-reductase-isomerase.

implications in the context of the dichotomy existing between photosynthetic and non-photosynthetic organisms, which synthesise cycloartenol and lanosterol, respectively. The corresponding single gene in *A. thaliana* (*At2g07050*) consists of 18 exons and 17 introns. It was generally considered that squalene cyclases, such as the squalene-hopane synthase from *Alicyclobacillus acidocaldarius*, were typical of prokaryotes (Wendt *et al.*, 1997), whereas OS cyclases were typical of eukaryotes. However, there are exceptions to this rule since the first prokaryotic OS cyclase gene was found in the myxobacterium, *Stigmatella aurantiaca* (Bode *et al.*, 2003). According to these authors, this gene could encode a cycloartenol synthase however, this exciting claim is still not supported by a clear-cut demonstration of the chemical structure of the cycloartenol produced by the recombinant enzyme.

### *11.5.2 S-Adenosylmethionine-sterol-C-methyltransferases (SMTs)*

Sterols from fungi and higher plants differ from vertebrate sterols by the presence of an extra alkyl group at C-24 (Nes, 2003). This alkylation of the side chain is catalysed by *S*-adenosylmethionine (AdoMet)-sterol-*C*-methyltransferases (SMTs). Chemical and enzymological aspects of these reactions have been detailed elsewhere (Rahier *et al.*, 1984; Nes, 2003).

A *SMT* cDNA of 1411 bp has been cloned from an *A. thaliana* cDNA library (Husselstein *et al.*, 1996). Microsomes from a yeast mutant defective in SMT (*erg6*) expressing the *A. thaliana* SMT were shown to possess AdoMet-dependent sterol-C-methyltransferase activity. Substrate specificity assays provided evidence that the *A. thaliana* cDNA encodes a 24-methylene-lophenol-C-$24^1$-methyltransferase catalysing the second methylation step of plant sterol biosynthesis (Bouvier-Nave *et al.*, 1998). This cDNA was named *ATSMT2-1*. Thorough kinetic studies performed on a recombinant ATSMT2-1 led to essentially identical conclusions (Nes, 2003). Meanwhile another group of cDNAs encoding SMTs has been characterised (Shi *et al.*, 1996; Bouvier-Nave *et al.*, 1998). The deduced proteins are about 80% identical in all possible combinations, but have only 40% identity with SMT2. They, therefore, constitute a group of sterol methyltransferases (SMT1) distinct from the group formed by SMT2. In the presence of AdoMet, delipidated microsomes from *erg6* yeast transformed with tobacco *SMT1* efficiently converted cycloartenol into 24-methylene cycloartanol, but produce little 24-ethylidene lophenol upon incubation with 24-methylene lophenol. Likewise, catalytic constants for cycloartenol and 24-methylene lophenol were 0.01 and 0.001/s, respectively, for cloned *Glycine max* SMT1 (Nes, 2003). This demonstrated that the cDNA *NTSMT1* encoded a cycloartenol-C24 methyltransferase (Bouvier-Nave *et al.*, 1998). As confirmed by the Arabidopsis sequencing project and the above results, the *A. thaliana* genome contains three distinct genes (*At5g13710*, *At1g20330*, *At1g76090*) encoding sterol-C24 methyltransferases (ATSMT1, ATSMT2-1, ATSMT2-2, respectively).

Several *smt1* mutations, which lack SMT1, were isolated from a transposon-tagged Ac transgenic line (Diener *et al.*, 2000). The *smt1* plants have several pleiotropic defects: poor growth and fertility, root sensitivity to $Ca^{2+}$ ions and a loss of proper embryo morphogenesis. *smt1* was shown to have an altered sterol content. It accumulates cholesterol and other 24-dealkylated sterols (Figure 11.3), has much less 24-ethyl sterols, but its campesterol content is similar to that in control plants. *SMT1* was revealed to control the flux of carbon in the sterol biosynthesis pathway in tobacco (Holmberg *et al.*, 2002). The expression of *ATSMT2-1* was modulated in *35S::SMT2-1A. thaliana* in order to study its physiological function. Plants overexpressing the transgene accumulated sitosterol at the expense of campesterol. These plants displayed a reduced stature and growth that could be restored by brassinosteroid treatment. Plants showing co-suppression of *SMT2-1* were characterised by a high campesterol content and a depletion in sitosterol. Pleiotropic effects on development, such as reduced growth, increased branching and low fertility of high campesterol plants were not modified by exogenous brassinosteroids, indicative of specific sterol requirements to promote a normal development. Thus, *ATSMT2-1* has a crucial role in balancing the ratio of campesterol to sitosterol in order to satisfy both growth requirements and membrane integrity (Schaeffer *et al.*, 2001). It was recently shown that in the Arabidopsis *cotyledon vascular pattern* (*cvp1*) mutant, vascular cells were not arranged in parallel lines and were misshapen (Carland *et al.*, 2002). *CPV1* was identified and found to encode SMT2-1. *cvp1 mutants* were strongly enriched in campesterol (7C2) and depleted in sitosterol (7C3) (Carland *et al.*, 2002) resulting in a ratio campesterol to sitosterol very similar to that found in the aforementioned high-campesterol plants. In conclusion, cholesterol, campesterol and sitosterol accumulate (in order of increasing importance) as a result of SMT1 and SMT2 activities.

Cholesterol (7C4) is generally a minor sterol in plants, however, its concentration may increase in Solanaceae, a family of plants often containing steroidal glycoalkaloids. Tubers of potato (*Solanum tuberosum*) contain solanin, a steroidal alkaloid, which is toxic at high doses. Overexpression of *Glycine max* SMT1 in transgenic potato plants resulted in a marked reduction of cholesterol and glycoalkaloid levels, supporting the view that cholesterol is a precursor in steroid alkaloid biosynthesis (Armqvist *et al.*, 2003).

### *11.5.3 4,4-dimethyl sterol and 4α-methyl sterol 4-demethylation (SMOs)*

The passage of 24-methylene cycloartanol to end-pathway sterols involves removal of two methyl groups at position 4 and one methyl at position 14 (Figure 11.2). In contrast to animals and fungi where the two C-4-methyl group are sequentially removed, a series of results has unequivocally established that, in plants, two distinct oxidative systems are involved in the removal of the first and the second C4-methyl of phytosterol precursors (Pascal *et al.*, 1993). In the case of the first enzymatic system, it has been shown that the oxidative conversion of 24-methylene cycloartanol to cycloeucalenol consists of a C-4α-methyl oxidase (SMO1), a C4-decarboxylase/C3-dehydrogenase (Rondet *et al.*, 1999) and a 3-keto reductase (Pascal *et al.*, 1994), and exhibits similar cofactor requirements and inhibitor sensitivity as the corresponding animal systems. This process is repeated with the second methyl group, but the substrate is then 24-methylene or 24-ethylidene lophenol and the product is then episterol or $\Delta^7$-avenasterol, respectively. Cloning and characterisation of *ERG25*, the *S. cerevisiae* gene encoding C4-methyl oxidase has been reported (Bard *et al.*, 1996). This gene encodes a 309-amino acid polypeptide showing a *C*-terminal endoplasmic reticulum (ER) retrieval signal KKXX and three histidine-rich clusters found in eukaryotic membrane bound fatty acid desaturases (Shanklin *et al.*, 1994). The genes encoding subsequent steps (the C4 decarboxylase/C3 dehydrogenase and the 3-ketoreductase) of the C4-demethylation process in yeast have been characterised and designated *ERG26* and *ERG27,* respectively (Gachotte *et al.*, 1998, 1999). The functional identification of two sterol-4α-methyl oxidase (*SMO1* and *SMO2*) cDNAs from *A. thaliana* was achieved by complementation of a yeast *erg25* mutant lacking sterol-4α-methyl oxidation and by a virus-induced gene silencing (VIGS) approach in *Nicotiana benthamiana* (Darnet and Rahier, 2003).

### *11.5.4 Cyclopropyl sterol isomerase (CPI)*

One of the most striking events occurring during sterol biosynthesis is the opening of the cyclopropane ring of cycloeucalenol to give obtusifoliol (Figure 11.1), a step that is restricted to the plant kingdom. This reaction is catalysed by the cyclopropyl sterol isomerase (CPI) (EC 5.5.1.9), the catalytic mechanism, which has been thoroughly studied (Heintz and Benveniste, 1974; Rahier *et al.*, 1989). *A. thaliana* CPI was cloned by functional complementation of a *S. cerevisiae* mutant (Lovato *et al.*, 2000). Expression of an *A. thaliana* cycloartenol synthase cDNA in a *S. cerevisiae* lanosterol mutant (*erg7*) gave rise to a sterol auxotroph because cycloartenol cannot be converted by yeast to ergosterol. This yeast strain was transformed with an *A. thaliana* expression library constructed in a yeast vector (pFL61) and sterol prototrophs were selected. A strain accumulating biosynthetic ergosterol was obtained. The novel phenotype was conferred by an *A. thaliana* cDNA that encoded a 36-kDa protein of 280 amino acids (CPI), which apparently cleaves the cyclopropane ring of a 4 α-methyl cyclopropyl sterol to give a 4 α-methyl-$\Delta^8$-sterol, usable by yeast cells to give ergosterol.

### *11.5.5 Obtusifoliol-14α-demethylase (OBT14DM)*

In animals and fungi, the 14α-methyl group is the first of the C14 and C4 methyls to be removed; however in higher plants, the 14α-methyl is removed after one C-4 methyl has disappeared in the biosynthetic pathway shown in the Figure 11.3. Lanosterol is the substrate of this enzyme in *S. cerevisiae* and in mammals (Aoyama *et al.*, 1994), whereas eburicol is the preferred substrate in filamentous fungi (Delye *et al.*, 1998), and obtusifoliol is the substrate in higher plants and is converted into 4α-methyl-5α-ergosta-8,14,24($24^1$)-trien-3β-ol (Taton and Rahier, 1991). A thorough enzymological study has defined the structural requirements of *Z. mays* obtusifoliol-14α-demethylase (OBT14DM). OBT14DM was shown to be remarkably specific for the obtusifoliol and did not use either lanosterol or eburicol (Taton and Rahier, 1991). cDNAs encoding lanosterol, eburicol and obtusifoliol 14-demethylases have been isolated from mammals (Aoyama *et al.*, 1994), fungi (Kalb *et al.*, 1986) and plants (Bak *et al.*, 1997), respectively. They shared a remarkable amino acid identity ranging from 38% to 65% although they belong to such widely diverging species. Therefore, they were classified in the same family, which is named CYP51 (Nelson *et al.*, 1996). A soluble CYP51 orthologue has been discovered in 2001 in *Mycobacterium tuberculosis*. This enzyme has been shown to demethylate lanosterol and obtusifoliol and to be inhibited by azole antifungals. *Escherichia coli*-expressed MTCYP51 has been recently crystallised in the presence of fluconazole and a crystal structure at 2.2 Å has been reported (Podust *et al.*, 2001). *A. thaliana* genome sequencing has revealed the existence of two genes (At2g17330 and At1g11680) that encode proteins having 65–75% identity with an already characterised OBT14DM from *Sorghum bicolor* (Bak *et al.*, 1997). Both genes possess two exons and an intron. *A. thaliana* plants were transformed with At1g11680 in antisense orientation. The resulting transgenic plants showed a semi-dwarf phenotype in the early growth stage. Their obtusifoliol content increased while campesterol and campestanol levels were unchanged (Kushiro *et al.*, 2001). The endogenous *CYP51* of *Nicotiana benthamiana* was silenced upon inoculation of the plantlets with a *potato virus X* vector harbouring a transcript of a tobacco orthologue of *AtCYP51*. This resulted in a strong accumulation of obtusifoliol, the substrate of CYP51 and other 14α-methyl sterols, at the expense of campesterol and sitosterol and with a concomitant growth reduction phenotype (Burger *et al.*, 2003).

The Antley-Bixler syndrome (ABS), which is a rare multiple anomaly syndrome comprising radiohumeral synostosis, bowed femora, fractures of the long bones and premature fusion of the calvarial sutures has been found in a patient with an intrinsic defect of cholesterol biosynthesis at the level of lanosterol 14α-demethylase. The genetic nature of the defect still remains to be determined (Kelley *et al.*, 2002).

### *11.5.6 $\Delta^{8,14}$-sterol-$\Delta^{14}$-reductase (14RED)*

4α-methyl-5α-ergosta-8,14,24($24^1$)-trien-3β-ol is the product of the C14 demethylation of obtusifoliol in plants. The following step is the hydrogenation by NADPH of the $\Delta^{14}$ double bond to give 4α-methyl-5α-ergosta-8,24($24^1$)-dien-3β-ol (Figure 11.3). *ERG24* is the gene encoding the corresponding enzyme in *S. cerevisiae* (Lorenz *et al.*, 1992). Most interestingly, it has been shown that the human lamin B receptor exhibited sterol C-14-reductase activity in *S. cerevisiae* (Silve *et al.*, 1998). Cloning of the gene (*FACKEL*) encoding this enzyme was performed in 2000 through the isolation of dwarf *mutants*. *FACKEL* (At3g52940) was isolated and was shown to encode a predicted integral membrane protein of 369 amino acids with

8–9 transmembrane segments related to the vertebrate lamin receptor and several sterol C-14 reductases including *S. cerevisiae* sterol C-14 reductase (Jang *et al.*, 2000; Schrick *et al.*, 2000). Functional evidence that *FACKEL* encoded a sterol C-14 reductase was provided by complementation of *erg24*. GC/MS analysis confirmed that *fk* mutations led to accumulation of $\Delta^{8,14}$-sterol intermediates in the biosynthetic pathway preceding the C-14 reductase step. The sterol profile of *fk* calli was similar to that of plant cells treated with A25822B, a very specific inhibitor of $\Delta^{14}$-reductase (Schaller *et al.*, 1994). The *fk* mutation resulted also in reduction of brassinosteroid (BR) content. However, unlike other BR-deficient *mutants*, the defect of hypotyl elongation could not be overcome by exogenous BRs (Jang *et al.*, 2000; Schrick *et al.*, 2001). Thorough microscopical and cytological observations indicated that mutations in the *FACKEL* gene affect body organisation of the *A. thaliana* seedlings and that *FACKEL* was required for cell division and expansion and was involved in proper organisation of the embryo.

### *11.5.7 $\Delta^8-\Delta^7$-sterol isomerase (8ISO)*

When the 14α-methyl group is removed and the 14 double bond is reduced, the resulting $\Delta^8$-sterols are isomerised to $\Delta^7$-sterols in mammals, fungi and higher plants. This process is catalysed by a $\Delta^8-\Delta^7$-isomerase (SI). From enzymatic assays and biogenetic considerations it appears that 4α-methyl-5α-ergosta-8,24($24^1$)-dien-3β-ol (Figure 11.3) is the substrate of this enzyme in higher plants (Taton *et al.*, 1987). A murine SI-encoding cDNA has also been cloned by functional complementation of *erg2*, a *S. cerevisiae* mutant deficient in SI (Bach and Benveniste, 1997). The amino acid sequence deduced from the cDNA open reading frame is highly similar to human emopamil-binding protein (EBP), a protein that is the target for neuroprotective drugs (Silve *et al.*, 1996). Interestingly, the amino acid sequence deduced from the murine (SI) or human (EBP) failed to reveal any noteworthy similarity with *S. cerevisiae* Erg2p (only 13% identity). Mutations in the gene encoding *Homo sapiens* SI were shown to cause X-linked dominant Conradi-Hunermann syndrome (Bravermann *et al.*, 1999). An *A. thaliana* 8ISO cDNA has been isolated by functional complementation of the corresponding *S. cerevisiae* sterol mutant (*erg2*). The full length *A. thaliana* cDNA that complemented the *erg2* mutation contained an open reading frame encoding a protein of 223 amino acids sharing 35% amino acid identity to the *Mus musculus* SI and the *H. sapiens* EBP and very low identity (less than 15%) with Erg2p (Grebenok *et al.*, 1998). The *A. thaliana* protein is encoded by a unique gene (*At1g20050*) possessing 4 exons and 3 introns. In a screen of TDNA and EMS *mutants* of Arabidopsis, a class of recessive seedling-lethal *mutants* defined by two loci, *hydra1* and *hydra2* were identified (Souter *et al.*, 2002). *hydra1 mutants* were shown to be unable to regulate cell size and shape and failed to undergo correct morphogenesis during embryonic and postembryonic development. They were shown to be strongly depleted in campesterol and sitosterol. The *HYDRA1* gene (*At1g20050*) was shown to be the 8ISO mentioned.

### *11.5.8 $\Delta^7$-sterol-C5(6)-desaturase (5DES)*

The $\Delta^7$-sterol C5(6)-desaturase (5DES) from *Z. mays* introduces a double bond at C5 of a Δ7-sterol, it converts 5α-ergosta-7,24-dien-3β-ol and 5α-stigmasta-7,Z-24-dien-3β-ol into ergosta-5,7,24-trien-3β-ol and stigmasta-5,7,Z-24-trien-3β-ol, respectively (Taton and Rahier, 1996). The electron transport system involves NADH, a cytochrome b5 reductase and cytochrome b5 and the desaturase itself (Rahier *et al.*, 1997). A nuclear and recessive mutant of *A. thaliana* affected in sterol biosynthesis (*ste1*) has been isolated and identified. This mutant accumulated $\Delta^7$-sterols at the expense of the normally occurring campesterol and sitosterol (Gatchotte *et al.*, 1995). An *A. thaliana* cDNA encoding a 5DES was isolated and characterised by functional complementation of *erg3*, a yeast mutant with a defective 5DES function (Bach and Benveniste, 1997). The 1141 cDNA contained an open reading frame of 843bp encoding a 281 residue polypeptide (Gatchotte *et al.*, 1996). Three histidine-rich motifs ($HX_3H$, $HX_2HH$ and $HX_2HH$) and three transmembrane segments were found in the *A. thaliana* polypeptide. Alignment of the wild-type with the *ste1* derived protein sequences revealed a single amino acid substitution, i.e. T114 in the wild-type is changed to I114 in Ste1p (Husselstein *et al.*, 1999). The role of 15 residues in the reaction catalysed by *A. thaliana* 5-DES was investigated using site-directed mutagenesis and expression of the mutated enzymes in an *erg3 S. cerevisiae* strain defective in 5-DES. A group of *mutants* was affected in residue 114 based on previous observations indicating that mutant T114I (*ste1*) was deficient in 5-DES. It was shown that the enzyme T114I had an 8-fold higher Km and 10-fold reduced catalytic efficiency. Conversely, the functionally conservative substituted mutant enzyme T114S displayed a 28-fold higher Vmax value and an 8-fold higher Km value than the wild-type enzyme. The data suggested that T114 strongly affected the reactivity of the desaturase (Taton *et al.*, 2000). Deuterium kinetic isotopic effects for the removal of 6α- and 5α-hydrogens were measured. The observed pattern of isotope effects suggested a mechanism for the desaturation of cholest-7-en-3β-ol (5C4) (Rahier *et al.*, 2001). The *A. thaliana* 5-DES considered above is encoded by one gene (*At3g02580*) possessing three exons and two introns. Screening of about 50,000 M2 seeds of an EMS mutant population allowed the isolation of 43 dwarf *mutants*. Two of them (*dwf7-1* and *dwf7-2*) were shown to be null alleles of *STE1* and to be biochemically complemented by brassinolide (BR) and early BR biosynthetic precursors. This work showed for the first time that the reduction of BR biosynthesis in *dwf7* was due to a shortage of substrate sterols (campesterol) and was the direct cause of the dwarf phenotype (Choe *et al.*, 1999). An extremely dwarf mutant (*bul1–1*) of *A. thaliana* was in 2001 isolated and shown to contain as much as 90% of $\Delta^7$-sterols and less than 2% of $\Delta^5$-sterols. It was, therefore, defective in the $\Delta^7$-sterol-C5-desaturation step leading to brassinolide (BR) biosynthesis. Cellular characterisation and rescue experiments with BRs demonstrated the involvement of the 5-DES in BR-dependent plant growth processes (Catterou *et al.*, 2001).

### *11.5.9 $\Delta^{5,7}$-sterol $\Delta^{7}$-reductase (7RED)*

This enzyme catalyses the reduction of the $\Delta^{7}$-double bond of the $\Delta^{5,7}$-sterols into $\Delta^{5}$-sterols and occurs only in vertebrates and higher plants. A microsomal preparation from seedlings of *Z. mays* catalysed the NADPH dependent reduction of the $\Delta^{7}$-bond of $\Delta^{5,7}$-cholestadienol (Rahier and Taton, 1996). The gene encoding the 7RED was cloned by transformation of wild-type *S. cerevisiae* with a cDNA library of *A. thaliana* constructed in the yeast vector pFL61 and the subsequent selection of cells producing $\Delta^{5}$-sterols at the expense of ergosterol (a $\Delta^{5,7}$-sterol) by nystatin selection. This clever strategy allowed recovery of a plasmid bearing a 1290-bp cDNA open reading frame, which encoded a protein of 430 amino acids (Lecain *et al.*, 1996). Screening of a population of dwarf *mutants* led to a series of *mutants* (*dwf5*) presenting a defect in the *7RED* gene (Choe *et al.*, 2000). The *dwf5* plants display the characteristic dwarf phenotype typical of BR *mutants* and were biochemically complemented with exogenous BRs. Sterol analysis showed that $\Delta^{5,7}$-campestenol (a possible substrate of 7RED) accumulated in these *mutants*. The *dwf5* plant could be restored to wild type by ectopic overexpression of the wild-type copy of the gene (Choe *et al.*, 2000). This gene (*At1g50430*) was shown to contain 13 exons and 12 introns. Smith-Lemli-Opitz syndrome (SLOS) is an inborn disorder of sterol metabolism resulting in severe mental retardation. The SLOS gene was revealed as a 7RED (EC 1.3.1.21) required for the *de novo* cholesterol biosynthesis. The protein encoded by the *H. sapiens* 7RED has 35% identity with the *A. thaliana* 7RED (Moebius *et al.*, 1998). Various defects in the 7RED gene (several missense, nonsense and splice site mutations) cause SLOS (Fitzky *et al.*, 1998).

### *11.5.10 $\Delta^{5}$-sterol $\Delta^{24}$-reductase (isomerase) (DIM)*

In animals, the substrate of this enzyme is desmosterol, a $\Delta^{24(25)}$sterol, whereas in higher plants, substrates are 24-methylene cholesterol and isofucosterol (Figure 11.3), two sterols possessing a $C24(24^{1})$ double bond, which are probably isomerised in $\Delta^{24(25)}$-sterols prior to being reduced. The *ERG4* gene provides the enzyme that reduces the double bond at $C24(24^{1})$, the final step in ergosterol biosynthesis. The *diminuto* mutant (*dim*) was initially isolated as a slowly growing dwarf *A. thaliana* mutant with greatly reduced fertility (Klahre *et al.*, 1998). The *DIM* gene (*At3g19820*), also referred to as *DWF1* is composed of two exons and one intron. Both the mutant phenotype and gene expression could be rescued by the addition of exogenous BRs. Analysis of endogenous sterols demonstrated that *dim* was deficient in campesterol, an earlier precursor of BR and in BRs as well. Feeding experiments using deuterium-labelled 24-methylenecholesterol and 24-methyl desmosterol confirmed that *DIM/DWF1* is involved in both the isomerisation and reduction of the $24(24^{1})$ bond and encoded a sterol $C24(24^{1})$ reductase isomerase (Klahre *et al.*, 1998). The peptide sequence of DIM/DWARF1 from *A. thaliana* was shown to have 41% identity with a *H. sapiens* orthologue (*Seladin-1*), but no significant identity with ERG4. Thus, the C24 reduction step seems to be performed by completely different enzymatic systems in higher plants and animals on one hand, and in *S. cerevisiae* (and probably most fungi) on the other hand. The human *DIM/DWARF1* homologue *Seladin-1* has been shown to confer resistance to Alzheimer's disease-associated neurodegeneration and oxidative stress (Greeve *et al.*, 2000). Moreover, *Seladin-1* transcription has been demonstrated to be down regulated in brain regions selectively degenerated in Alzheimer's disease (Iivonen *et al.*, 2002). These exciting results could attribute a novel role of cholesterol in neurodegenerative diseases. The human homologue of *DIM/DWF1* gene (*hDIM*) is a novel ACTH-responsive gene that may be involved in benign adrenocortical adenomas (Sarker *et al.*, 2001).

### *11.5.11 Sterol-$\Delta^{22}$-desaturase*

Ergosterol from *S. cerevisiae*, most fungi and several algae, and stigmasterol, (24S)-24-ethyl-cholesta-5,E-22-dien-3β-ol from most higher plants possess a double bond at C-22. $\Delta^{22}$-sterols are not present in vertebrates. This double bond results from a desaturation step. The gene (*ERG5*) encoding this reaction has been cloned by complementation of an *erg5* mutant (Skaggs *et al.*, 1996). The deduced amino acid sequence (538 amino acid residues) presented general features common to most cytochrome P-450 open reading frames. Surprisingly, the highest homology was not with CYP51A1 (lanosterol-14-demethylase) (Skaggs *et al.*, 1996). Thus, *ERG5* is the first member of the *CYP61* family. Most biosynthetic studies show that sitosterol would be the substrate for a C-22 desaturase leading to stigmasterol (Nes, 1977). Four candidate genes (*At2g28850*, *At2g28860*, *At2g34490*, *At2g34500*) are present in The Arabidopsis Information Resource (TAIR) databases.

As shown above, most genes encoding enzymes of sterol biosynthesis in plants have been cloned isolated and identified. A collection of point-, TDNA- and transposon-mutants has been described in the case of *A. thaliana*. These tools have shown that sterols not only play a structural role in membranes but are also important actors in signal transduction leading to crucial steps of development (Carland *et al.*, 2002; Schaller *et al.*, 2003). In parallel, similar studies have been performed in the animal kingdom, where counterparts of the main plant genes involved in sterol biosynthesis have been cloned in mammals (*Mus musculus*, *H. sapiens*), in birds (*Gallus gallus*) and many *mutants* have been described and characterised. Interestingly, several genetic diseases described earlier have been shown to result from defects in cholesterol biosynthesis (Fitzky *et al.*, 1998; Bravermann *et al.*, 1999; Sarker *et al.*, 2001; Iivonen *et al.*, 2002; Kelley *et al.*, 2002). After isolation of the mutated gene, nonsense or missense mutations leading to loss of enzyme function were identified (Fitzky *et al.*, 1998; Bravermann *et al.*, 1999). These results have demonstrated that cholesterol and/or its derivatives play many more roles than expected. Synaptogenesis in central nervous system has been shown to be promoted by glia-derived cholesterol (Mauch *et al.*, 2001). Unfortunately, in many studies dealing with sterol *mutants* or gene identification, the biochemical characterization is lacking. In all cases, a thorough identification of sterols present in *mutants* should be done. As the products of genes are generally enzymes, a careful enzymological study should be performed in order to find out whether the catalytic properties of enzymes corresponding to the mutated genes are modified, by analysing the kinetic parameters and comparing with those of the wild-type.

## 11.6 Terpenoids, mono-, sesqui- and diterpenes

According to Bohlmann *et al.* (1998), there are in excess of 1000 monoterpenes, more than 7000 sesquiterpenes and more than 3000 diterpenes. The first stage of the biosynthesis involves the synthesis of IDP, followed by isomerisation to dimethylallyl diphosphate, prenyltransferase-catalyzed condensation of these two $C_5$-units to GDP, and the subsequent $1'$-4 additions of IDP to generate FDP and GGDP. The complexity of the biosynthesis of terpenoids is illustrated by the cyclases (second stage), which will convert in one or two steps GDP, FDP and GGDP into the huge array of different carbon skeletons characteristic of the terpenoid family of natural products. The cyclisation steps proceed in a strongly stereo(regio)specific way; the polycyclic hydrocarbons produced by the cyclases possess several asymmetric carbons implying a strict kinetic and stereochemical control of cyclases during the catalytic process. Sometimes the cyclisation step converts the substrate into several products and the cyclase is viewed as multifunctional. Finally, the hydrocarbon resulting from the cyclisation step may be strongly modified by oxidation (third step) to give alcohols, ketones, carboxylic acids functions, which, in turn, may be modified. For instance, alcohols may be acylated by a great variety of acyl donors and carboxylic acids can be methylated. As it is impossible to cover so wide a field of research; we have focused on one particular representative example, namely taxol biosynthesis. This pathway covers many aspects of general terpenoid biosynthesis and raises technical, biosynthetic and economical problems, which are also relevant to most of the other important terpenoids.

### *11.6.1 Biosynthesis of Taxol (paclitaxel)*

The natural product Taxol (a registered trademark of a pharmaceutical company) also called paclitaxel (generic chemical name that will be used below, Figure 11.4) is a structurally complex diterpenoid containing a tricyclic taxan skeleton, which is strongly oxygenated. The hydroxyl groups are either free or esterified with acetic acid at positions 4 and 10, benzoic acid at position 2 and by 2-hydroxyl-3-*N*-benzoylamino-phenylpropanoic acid at position 13. This last substituent has been shown to be especially important for optimal biological activity. Paclitaxel binds to tubulin heterodimers and promotes the assembly and stability of microtubules, disrupting cell division (Li *et al.*, 2000).

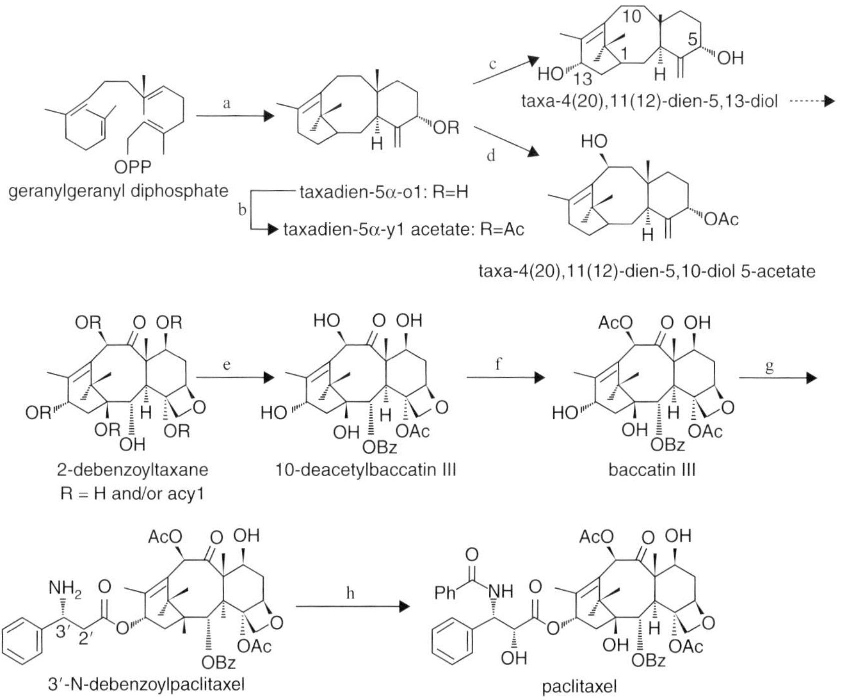


**Figure 11.4** **Paclitaxel** biosynthetic pathway. (a) The cyclisation of **geranylgeranyl diphosphate** to **taxadiene** by taxadiene synthase and the subsequent hydroxylation to **taxadien-5α-ol** by taxadien-5-hydroxylase. (b) The acetylation of **taxadien-5α-ol** by taxadien-5α-ol acetyltransferase. (c) The conversion of **taxadien-5α-ol** to 5,13-diol by a 13α-hydroxylase. (d) The hydroxylation of **taxadien-5α-yl acetate** by a 10β-hydroxylase. (e) The formation of a 2-benzoxy taxoid by a taxane 2α-*O*-benzoyltransferase. (f) The conversion of **10-deacetylbaccatin III** to **baccatin III** by a 10-*O*-acetyltransferase. (g) Side-chain attachment by the phenylpropanoyltransferase. (h) Side-chain benzamidation by **N-debenzoyl-(3′RS)-2′-deoxypaclitaxel** to form **paclitaxel**. The broken arrow indicates multiple convergent steps. Material copyrighted by the National Academy of Sciences, U.S.A. and reprinted from Walker and Croteau, 2000, with permission.

### *11.6.1.1 Cyclisation of geranylgeranyl diphosphate*

The committed step of paclitaxel biosynthesis is the cyclisation of the ubiquitous isoprenoid intermediate GGDP. The conversion of GGDP to taxa-4(5),11(12)-diene (a tricyclic hydrocarbon) is performed by taxadiene synthase. The gene encoding this enzyme was cloned by screening a cDNA library from *Taxus brevifolia* stems with degenerated primers. These primers were synthesised according to consensus sequences of related mono, sesqui- and diterpene cyclases (Wildung and Croteau, 1996). Sequence comparisons with mono-, sesqui- and diterpene cyclases of higher plants show that the taxadiene synthase closely resembles abietane synthase, a diterpene cyclase involved in the synthesis of abietic acid, a diterpene present in many conifers in general and in *Abies grandis* in particular (Vogel *et al.*, 1996).

### *11.6.1.2 Hydroxylations*

The following step is the 5α-hydroxylation of taxadiene. This reaction is performed by a cytochrome P450-dependant hydroxylation, with allylic rearrangement from 4(5) to 4(20) to yield 5α-hydroxy, taxa-4(20), 11(12)-diene (step (a), Figure 11.4), and seems to be a limiting step during paclitaxel biosynthesis (Hefner, 1996). This first oxygenated intermediate is subject to acetyl CoA-dependent transacetylation at C5 (step (b), Figure 11.4) (Walker *et al.*, 2000) before selective hydroxylation at C10 to 5α-acetoxy-10β-hydroxy taxa-4(20), 11(12)-diene (step (d) Figure 11.4) by a different cytochrome P450 monooxygenase (Wheeler *et al.*, 2001). A strategy based on differential display of mRNA-reverse transcription-PCR was devised, using cultured *Taxus cuspidata* cells, induced for paclitaxel production by methyl jasmonate. This strategy revealed 13 full-length closely related cytochrome P450 cDNAs. Functional expression of these enzymes in *S. cerevisiae* was monitored by spectrophotometry coupled to *in vivo* screening of oxygenase activity by feeding taxoid substrates. This strategy yielded a family of taxoid-metabolising enzymes and revealed the taxane 10β-hydroxylase as a 1494-bp cDNA that encodes a 498-residue cytochrome P450 capable of transforming taxadienyl acetate to the 10β-hydroxy derivative (step (d) Figure 11.4) (Schoendorf *et al.*, 2001). Another cDNA, following heterologous expression in yeast, was shown to encode a cytochrome P450-dependent taxoid 14β-hydroxylase, which converted 5α-acetoxy-10β-hydroxy taxa-4(20), 11(12)-diene into 5α-acetoxy-10β,14β-dihydroxy taxadiene (Jennewein *et al.*, 2003). However, this metabolite is not on the route leading to paclitaxel since this latter compound is not oxidised at C-14, which indicates that diverging pathways exist in *T. cuspidata*, some of which are be of potential therapeutic interest.

Other cDNA candidates that did not function in yeast were heterologously expressed by using the *Spodoptera frugiperda* baculovirus-based system and were screened for catalytic capabilities by using taxa-4(20), 11(12)-dien-5α-ol and its acetate ester as test substrates (step (c) Figure 11.4). This approach allowed identification of one of the cytochrome P450 clones as a taxane 13α-hydroxylase by chromatographic and spectrometric characterization of the corresponding recombinant enzyme product (Jennewein *et al.*, 2003). The taxan 13-hydroxylase protein had 63% identity with the aforementioned taxane 10β-hydroxylase. In addition, substrate specificity studies with the available cytochrome P450 hydroxylases indicate that there are several biosynthetic routes to paclitaxel in yew species. One of these routes (c) involved taxa-4(20)11(12)-dien-5,13-diol, the other (d) involved taxa-4(20),11(12)-dien-5,10-diol 5-acetate (Figure 11.4). After several as yet unknown steps, it is suggested that routes converged since 10-deacetylbaccatin III is found in relatively significant amounts in *Taxus baccata*.

*11.6.1.3 Acetylations, benzoylations phenylalanoylation*

A survey of 350 taxoid metabolites characterised to date suggests that, after the acetylation at C5 of taxa-4(20),11(12)-dien-5α-ol (Walker *et al.*, 2000), the second acylation reaction is the benzoylation of the C-2 hydroxyl group of 2-debenzoyltaxane to give 10-deacetylbaccatin III (step (e) Figure 11.4). By employing a homology-based PCR cloning strategy for generating acyltransferase oligonucleotide probes, several DNA fragments were amplified and used to screen a cDNA library constructed from mRNA isolated from methyl jasmonate-induced *Taxus cuspidata* cells. Several full-length acyltransferase cDNAs were obtained and individually expressed in *E. coli*. A cDNA clone encoding a taxane 2α-O-benzoyltransferase has been isolated and characterised after expression in *E. coli*. The recombinant enzyme catalyses the conversion of 2-debenzoyl-7,13-diacetylbaccatin III, a semisynthetic substrate, to 7,13-diacetylbaccatin III and is regiospecific for acylation of the 2-hydroxyl group of the functionalised taxane nucleus (Walker and Croteau, 2000). A second cDNA clone encoding a taxoid C-13 *O*-phenylpropanoyltransferase was isolated, which yielded a recombinant enzyme that catalyses the selective 13-*O*-acylation of baccatin III with β-phenylalanoyl CoA as the acyl donor, to form N-debenzoyl-2′-deoxypaclitaxel (step (g) Figure 11.4). Finally, another cDNA clone encoding the final step in paclitaxel biosynthesis was isolated and expressed in *E. coli*. The recombinant enzyme catalyses the stereoselective coupling of the surrogate substrate *N*-debenzoyl-(3′RS)-2′-deoxypaclitaxel with benzoyl-CoA to form predominantly one 3′-epimer of 2′deoxypaclitaxel (step (h) Figure 11.4) (Walker *et al.*, 2002). In conclusion, application of these transacylase genes in suitable host cells can improve the production yields of paclitaxel and could potentially enable the preparation of second-generation paclitaxel analogues possessing greater bioactivity and improved water solubility.

## 11.7 Conclusions

Isoprenoids play crucial roles in living organisms. First, many of them (sterols, prenylquinones, carotenoids and chlorophylls) belong to so-called primary metabolism because they are constituents of membranes in eukaryotic cells or have essential roles in bioenergetic processes. Cells cannot exist without dolichols, which play a key role in protein glycosylation, or without plasto- and ubiquinones, components of electron transport chains. Activity of several proteins in cell division and signalling is strictly dependent on their farnesylation or geranylgeranylation by specific prenyltransferases. Conversely, higher plants contain myriads of secondary metabolites among them isoprenoids and isoprenoid-derived molecules, which are sources of bioactive compounds that may be transformed into pharmacologically active drugs (Bohlmann *et al.*, 1998). Many of these compounds may also play a role as phytoalexins in plant defence against pathogenic microbes and herbivores (Vogel *et al.*, 1996). Finally, isoprenoids are important in animal and human nutrition. Carotenoids from plants are the natural source of vitamin A. Tocopherols and tocotrienols are constituents of vitamin E. Plant sterols have been shown to have hypocholesterolemic properties. It has been suggested that consumption of tomatoes or lycopene can reduce cancer risk (Erhardt *et al.*, 2003). Links between brain cholesterol content and Alzheimer's disease are repeatedly reported (Iivonen *et al.*, 2002). Pharmacological or nutritional applications of isoprenoids require a better knowledge of their biosynthesis in order to find out the limiting steps. Enzymes catalysing these steps could become targets for genetic engineering experiments in order to optimise production of bioactive terpenoids by crops.

## Acknowledgements

I warmly thank Drs D. DellaPenna (Michigan State University, U.S.A.), R. Croteau (Washington State University, U.S.A.) and J. Hirschberg (Hebrew University of Jerusalem, Israel) for having authorised me to reproduce or redraw Figures 11.1, 11.4 and 11.2 respectively. Many thanks also to Professor T. Bach (University of Strasbourg, France) for having kindly reviewed this article.

## References

Abe, I., Rohmer, M. and Prestwich, G.D. (1993) Enzymatic cyclization of squalene and oxidosqualene to sterols and triterpenes. *Chemical Reviews,* **93**, 2189–2206.

Al-Babili, S., Hugueney, P., Schledz, M. *et al.* (2000) Identification of a novel gene coding for neoxanthin synthase from Solanum tuberosum. *FEBS Letters*, **485**, 168–172.

Aoyama, Y., Funae, Y., Noshiro, M., Horiuchi, T. and Yoshida, Y. (1994) Occurrence of a P450 showing high homology to yeast lanosterol 14-demethylase ($P450_{14DM}$) in the rat liver. *Biochemical and Biophysical Research Communications*, **201**, 1320–1326.

Arnqvist, L., Dutta, P.C., Jonsson, L. and Sitbon, F. (2003) Reducation of cholesterol and glycoalkaloid levels in transgenic potato plants by overexpression of a type 1 sterol methyltransferase cDNA. *Plant Physiology*, **131**, 1792–1799.

Bach, T.J. and Benveniste, P. (1997) Cloning of cDNAs or genes encoding enzymes of sterol biosynthesis from plants and other eukaryotes: heterologous expression and complementation analysis of mutations for functional characterization. *Progress in Lipid Research*, **36**,197–226.

Bak, S., Kahn, R.A., Olsen, C.E. and Halkier, B.A. (1997) Cloning and expression in *Escherichia coli* of the obtusifoliol 14α-demethylase of *Sorghum bicolor* (L.) Moench, a cytochrome P450 orthologous to the sterol 14α-demethylases (CYP51) from fungi and mammals. *Plant Journal*, **11**, 191–201.

Bard, M., Bruner, D.A., Pierson, C.A. *et al.* (1996) Cloning and characterization of *ERG25*, the *Saccharomyces cerevisiae* gene encoding C-4 sterol methyl oxidase. *Proceedings of the National Academy of Sciences USA*, **93**, 186–190.

Bartley, G.E. and Scolnik, P.A. (1993) cDNA cloning, expression during development, and genome mapping of PSY2, a second tomato gene encoding phytoene synthase. *Journal of Biological Chemistry*, **268**, 25718–25721.

Bode, H.B., Zeggel, B., Silakowski, B., Wenzel, S.C., Reichenbach, H. (2003) Steroid biosynthesis in prokaryotes: identification of myxobacterial steroids and cloning of the first bacterial

2.3(S)-oxidosqualene cyclase from the mycobacterium *Stigmatella aurantiaca. Molecular Microbiology,* **47**, 471–481.

Bohlmann, J., Meyer-Gauen, G. and Croteau, R. (1998) Plant terpenoid synthases: molecular biology and phylogenetic analysis. *Proceedings of the National Academy of Sciences USA,* **95**, 4126–4133.

Bouvier, F., Hugueney, P., d'Harlingue, A., Kuntz, M. and Camara, B. (1994) Xanthophyll biosynthesis in chromoplasts: isolation and molecular cloning of an enzyme catalyzing the conversion of 5,6-epoxycarotenoid into ketocarotenoid. *Plant Journal*, **6**, 645–654.

Bouvier, F., d'Harlingue, A., Hugueney, P., Marin, E., Marion-Poll, A. and Camara, B. (1996) Xanthophyll biosynthesis. Cloning, expression, functional reconstitution, and regulation of β-cyclohexenyl carotenoid epoxidase from pepper (*Capsicum annuum*). *Journal of Biological Chemistry*, **271**, 28861–28867.

Bouvier, F., Keller, Y., d'Harlingue, A. and Camara, B. (1998) Xanthophyll biosynthesis: molecular and functional characterization of carotenoid hydroxylases from pepper fruits (*Capsicum annuum* L.). *Biochimica et Biophysica Acta*, **1391**, 320–328.

Bouvier, F. Suire, C. d'Harlingue, A. Backhaus, R.A. and Camara, B. (2000a). Molecular cloning of geranyl diphosphate synthase and compartmentation of monoterpene synthesis in plant cells. *Plant Journal*, **24**, 241–252.

Bouvier, F., d'Harlingue, A., Backhaus, R.A., Kumagai, M.H. and Camara, B. (2000b) Identification of neoxanthin synthase as a carotenoid cyclase paralog. *European Journal of Biochemistry*, **267**, 6346–6352.

Bouvier, F., Suire, C., Mutterer, J. and Camara, B. (2003a) Oxidative remodeling of chromoplast carotenoids: identification of the carotenoid dioxygenase CsCCD and CsZCD genes involved in Crocus secondary metabolite biogenesis. *Plant Cell*, **15**, 47–62.

Bouvier, F., Dogbo, O. and Camara, B. (2003b) Biosynthesis of the food and cosmetic plant pigment bixin (annatto). *Science,* **300**, 2089–2091.

Bouvier-Navé, P., Husselstein, T. and Benveniste, P. (1998) Two families of sterol methyltransferases are involved in the first and the second methylation steps of plant sterol biosynthesis. *European Journal of Biochemistry,* **256**, 88–96.

Bradshaw, H.D. and Schemske, D.W. (2003) Allele substitution at a flower colour locus produces a pollinator shift in monkey flowers. *Nature,* **426**, 176–178.

Braverman, N., Lin, P., Moebius, F.F., Obie, C., Moser, A. (1999) Mutations in the gene encoding 3β-hydroxysteroid-$\Delta^{8,7}$-isomerase cause X-linked dominant Conradi-Hunermann syndrome. *Nature Genetics*, **22**, 291–294.

Brigelius-Flohe, R., Kelly, F.J., Salonen, J.T., Neuzil, J., Zingg, J.M. and Azzi, A. (2002) The European perspective on vitamin E: current knowledge and future research. *American Journal of Clinical Nutrition*, **76**, 703–716.

Bugos, R.C. and Yamamoto, H.Y. (1996) Molecular cloning of violaxanthin de-epoxidase from romaine lettuce and expression in Escherichia coli. *Proceedings of the National Academy of Sciences USA*, **93**, 6320–6325.

Burger, C., Rondet, S., Benveniste, P. and Schaller, H. (2003) Virus-induced silencing of sterol biosynthetic genes: identification of a *Nicotiana tabacum* L. obtusifoliol-14α-demethylase (CYP51) by genetic manipulation of the sterol biosynthetic pathway in *Nicotiana benthamiana* L. *Journal of Experimental Botany*, **54**, 1675–1683.

Cahoon, E.B., Hall, S.E., Ripp, K.G., Ganzke, T.S., Hitz, W.D. and Coughlan, S.J. (2003) Metabolic redesign of vitamin E biosynthesis in plants for tocotrienol production and increased antioxidant content. *Nature Biotechnology*, **21**, 1082–1087.

Camara, B. (1993) Plant phytoene synthase complex: component enzymes, immunology, and biogenesis. *Methods in Enzymology*, **214**, 352–365.

Carland, F.M., Fujioka, S., Takatsuto, S., Yoshida, S. and Nelson, T. (2002) The identification of CVP1 reveals a role for sterols in vascular patterning. *Plant Cell,* **14**, 2045–2058.

Carol, P. and Kuntz, M. (2001) A plastid terminal oxidase comes to light: implications for carotenoid biosynthesis and chlororespiration. *Trends in Plant Science*, **6**, 31–36.

Catterou, M., Dubois, F., Schaller, H., Aubanelle, L., Vilcot, B. (2001) Brassinosteroids, microtubules and cell elongation in *Arabidopsis thaliana*. II. Effects of brassinosteroids on microtubules and cell elongation in the bul1 mutant. *Planta,* **212**, 673–683.

Cheng, Z., Sattler, S., Maeda, H., Sakuragi, Y., Bryant, D.A. and DellaPenna, D. (2003) Highly divergent methyltransferases catalyze a conserved reaction in tocopherol and plastoquinone synthesis in cyanobacteria and photosynthetic eukaryotes. *Plant Cell*, **15**, 2343–2356.

Choe, S., Noguchi, T., Fujioka, S., Takatsuto, S. and Tissier, C.P. (1999) The Arabidopsis *dwf7/ste1* mutant is defective in the $\Delta^7$-sterol C-5 desaturation step leading to brassinosteroid biosynthesis. *Plant Cell*, **11**, 207–221.

Choe, S. Tanaka, A., Noguchi, T., Fujioka, S. and Takatsuto, S. (2000) Lesions in the sterol $\Delta^7$-reductase gene of Arabidopsis cause dwarfism due to a block in brassinosteroid biosynthesis. *Plant Journal*, **21**, 431–443.

Collakova, E. and DellaPenna, D. (2001) Isolation and functional analysis of homogentisate phytyltransferase from Synechocystis sp. PCC 6803 and Arabidopsis. *Plant Physiology*, **127**, 1113–1124.

Corey, E.J., Matsuda, S.P.T. and Bartel, B. (1993) Isolation of an *Arabidopsis thaliana* gene encoding cycloartenol synthase by functional expression in a yeast mutant lacking lanosterol synthase by the use of a chromatographic screen. *Proceedings of the National Academy of Sciences USA*, **90**, 11628–11632.

Cunningham, F.X., Jr. and Gantt, E. (2000) Identification of multi-gene families encoding isopentenyl diphosphate isomerase in plants by heterologous complementation in *Escherichia coli*. *Plant Cell Physiology*, **41**, 119–123.

Cunningham, F.X., Jr. and Gantt, E. (2001) One ring or two? Determination of ring number in carotenoids by lycopene-ϵ-cyclase. *Proceedings of the National Academy of Sciences USA*, **98**, 2905–2910.

Darnet, S. and Rahier, A. (2003) Plant sterol biosynthesis: identification of two distinct families of sterol 4α-methyl-oxidases. *Biochemical Journal*, **378**(Pt 3): 889–898.

Davison, P.A., Hunter, C.N. and Horton, P. (2002) Overexpression of beta-carotene hydroxylase enhances stress tolerance in Arabidopsis. *Nature*, **418**, 203–206.

Delye, C., Bousset, L. and Corio-Costet, M.F. (1998) PCR cloning and detection of point mutations in the eburicol 14α-demethylase (CYP51) gene from *Erysiphe graminis* f. sp. hordei, a 'recalcitrant' fungus. *Current Genetics*, **34**, 399–403.

Demel, R.A. and De Kruyff, B. (1976) The function of sterols in membranes. Biochimica et Biophysica Acta,**457**, 109–113.

Diener, A.C., Li, H., Zhou, W.X., Whoriskey, W.J. and Nes W.D. (2000) *STEROL METHYL TRANSFERASE 1* Controls the level of cholesterol in plants. *Plant Cell*, **12**, 853–870.

de Wijn, R. and van Gorkom, H.J. (2001) Kinetic of electron transfer from Q(a) to Q(b) in photosystem II. *Biochemistry*, **40**, 11912–11922.

d'Harlingue, A. and Camara, B. (1985) Plastid enzymes of terpenoid biosynthesis. Purification and characterization of γ-tocopherol methyltransferase from Capsicum chromoplasts. *Journal of Biological Chemistry*, **260**, 15200–15203.

Erhardt, J.G., Meisner, C., Bode, J.C. and Bode, C. (2003) Lycopene, beta-carotene, and colorectal adenomas. *American Journal of Clinical Nutrition*, **78**,1219–1224.

Finazzi, G. (2002) Redox-coupled proton pumping activity in cytochrome b6f, as evidenced by the pH dependence of electron transfer in whole cells of *Chlamydomonas reinhardtii*. *Biochemistry*, **41**, 7475–7482.

Fitzky, B.U., Witsch-Baumgartner, M., Erdel, M., Lee, J.N. and Paik, Y.K. (1998) Mutations in the $\Delta^7$-sterol reductase gene in patients with the smith-lemli-opitz syndrome. *Proceedings of the National Academy of Sciences USA*, **95**, 8181–8186.

Gachotte, D., Barbuch, R., Gaylor, J., Nickel, E. and Bard, M. (1998) Characterization of the *Saccharomyces cerevisiae* ERG26 gene encoding the C-3 sterol dehydrogenase (C-4 decarboxylase) involved in sterol biosynthesis. *Proceedings of the National Academy of Sciences USA*, **95**, 13794–13799.

Gachotte, D., Husselstein, T., Bard, M., Lacroute, F. and Benveniste, P. (1996) Isolation and characterization of an *Arabidopsis thaliana* cDNA encoding a $\Delta^7$-sterol-C-5-desaturase by functional complementation of a defective yeast mutant. *Plant Journal*, **9**, 391–398.

Gachotte, D., Meens, R. and Benveniste, P. (1995) An Arabidopsis mutant deficient in sterol biosynthesis: heterologous complementation by *ERG3* encoding a $\Delta^7$-sterol-C-5-desaturase from yeast. *Plant Journal*, **8**, 407–416.

Gachotte, D., Sen, S.E., Eckstein, J. Barbuch, R. Krieger, M. (1999) Characterization of the *Saccharomyces cerevisiae ERG27* gene encoding the 3-keto reductase involved in C-4 sterol demethylation. *Proceedings of the National Academy of Sciences USA,* **96**, 12655–12660.

Giuliano, G., Giliberto, L. and Rosati, C. (2002) Carotenoid isomerase: a tale of light and isomers. *Trends in Plant Science*, **7**, 427–429.

Grasses, T., Grimm, B., Koroleva, O. and Jahns, P. (2001) Loss of α-tocopherol in tobacco plants with decreased geranylgeranyl reductase activity does not modify photosynthesis in optimal growth conditions but increases sensitivity to high light stress. *Planta*, **213**, 620–628.

Grebenok, R.J., Ohnmeiss, T.E., Yamamoto, A. Huntley, E. Galbraith, D.W. (1998) Isolation and characterization of an *Arabidopsis thaliana* $\Delta^{8,7}$-sterol isomerase : functional and structural similarities to mammalian $\Delta^{8,7}$-sterol isomerase/emopamil-binding protein. *Plant Molecular Biology,* **3**, 807–815.

Greeve, I., Hermans-Borgmeyer, I., Brellinger, C. Kasper, D. Gomez-Isla, T. (2000) The human DIMINUTO:DWARF1 homolog seladin-1 confers resistance to Alzheimer's disease-associated neurodegeneration and oxidative stress. *Journal of Neuroscience,* **20**, 7345–7352.

Hartmann, M.A. (2003). Sterol metabolism and function in higher plants, in *Lipid Metabolism and Membrane Biogenesis* (ed. G. Daum), *Topics in Current Genetics (Review Series)*, **6**, pp. 183–211.

Hartmann-Bouillon, M.A. and Benveniste, P. (1978). Sterol biosynthetic capability of purified membrane fractions from maize coleoptiles. *Phytochemistry*, **17**, 1037–1042.

Hefner, J., Rubenstein, S.M., Ketchum, R.E., Gibson, D.M., Williams, R.M. and Croteau, R. (1996) Cytochrome P450-catalyzed hydroxylation of taxa-4(5),11(12)-diene to taxa-4(20),11(12)-dien-5α-ol: the first oxygenation step in taxol biosynthesis. *Chemistry & Biology*, **3**, 479–489.

Heintz, R. and Benveniste, P. (1974) Enzymatic cleavage of the 9β,19-cyclopropane ring of cyclopropyl sterols in bramble tissue cultures. *Journal of Biological Chemistry,* **249**, 4267–4274.

Hemmerlin, A., Hoeffler, J.F., Meyer, O. *et al.* (2003) Cross-talk between the cytosolic mevalonate and the plastidial methylerythritol phosphate pathways in tobacco bright yellow-2 cells. *Journal of Biological Chemistry*, **278**, 26666–26676.

Hernandez-Pinzon, I., Ross, J.H., Barnes, K.A., Damant, A.P. and Murphy, D.J. (1999) Composition and role of tapetal lipid bodies in the biogenesis of the pollen coat of *Brassica napus*. *Planta*, **208**, 588–598.

Hirschberg, J. (2001) Carotenoid biosynthesis in flowering plants. *Current Opinion in Plant Biology*, **4**, 210–218.

Hofius, D. and Sonnewald, U. (2003). Vitamin E biosynthesis: biochemistry meets cell biology. *Trends in Plant Science*, **8**, 6–8.

Holmberg, N., Harker, M., Gibbard, C.L., Wallace, A.D., Clayton, J.C. (2002) Sterol C-24 methyltransferase type 1 controls the flux of carbon into sterol biosynthesis in tobacco seed. *Plant Physiology,* **130**, 303–311.

Husselstein, T., Gachotte, D., Desprez, T., Bard, M. and Benveniste, P. (1996) Transformation of *Saccharomyces cerevisiae* with a cDNA encoding a sterol *C*-methyltransferase from *Arabidopsis thaliana* results in the synthesis of 24-ethyl sterols. *FEBS Letters*, **381**, 87–92.

Husselstein, T., Schaller, H., Gachotte, D. and Benveniste, P. (1999) $\Delta^7$-sterol-C5-desaturase molecular characterization and functional expression of wild type and mutant alleles. *Plant Molecular Biology,* **39**, 891–906.

Iivonen, S., Hiltunen, M., Alafuzoff, I. *et al.* (2002) Seladin-1 transcription is linked to neuronal degeneration in Alzheimer's disease. *Neuroscience*, **113**, 301–310.

Isaacson, T., Ronen, G., Zamir, D. and Hirschberg, J. (2002) Cloning of tangerine from tomato reveals a carotenoid isomerase essential for the production of beta-carotene and xanthophylls in plants. *Plant Cell*, **14**, 333–342.

Iuchi, S., Kobayashi, M., Taji, T. *et al.* (2001) Regulation of drought tolerance by gene manipulation of 9-cis-epoxycarotenoid dioxygenase, a key enzyme in abscisic acid biosynthesis in Arabidopsis. *Plant Journal*, **27**, 325–333.

Jang, Y.C., Fujioka, S., Tasaka, M. Seto, H. Takatsuto, S. (2000) A critical role of sterols in embryonic patterning and meristem programming revealed by the *fackel* mutants of *Arabidopsis thaliana. Genes & Development,* **14**, 1485–1497.

Jennewein, S., Rithner, C.D., Williams, R.M. and Croteau, R. (2003) Taxoid metabolism: taxoid 14α-hydroxylase is a cytochrome P450-dependent monooxygenase. *Archives of Biochemistry and Biophysics*, **413**, 262–270.

Kalb, V.F., Loper, J.C., Dey, C.R., Woods, C.W. and Sutter, T.R. (1986) Isolation of a cytochrome P-450 structural gene from *Saccharomyces cerevisiae. Gene,* **45**, 237–245.

Karpen, H.E., Bukowski, J.T., Hughes, T., Gratton, J.P. and Sessa, W.C. (2001) The sonic hedgehog receptor patched associates with caveolin-1 in cholesterol-rich microdomains of the plasma membrane. *Journal of Biological Chemistry,* **276**:19503–19511.

Kelley, R.I., Kratz, L.E., Glaser, R.L., Netzloff, M.L. and Wolf, L.M., (2002) Abnormal sterol metabolism in a patient with Antley-Bixler syndrome and ambiguous genitalia. *American Journal of Medical Genetics,* **110**, 95–102.

Kip Guy, R. (2000) Inhibition of sonic hedgehog autoprocessing in cultured mammalian cells by sterol deprivation. *Proceedings of the National Academy of Sciences USA,* **97**, 7307–7312.

Klahre, U., Noguchi, T., Fujioka, S., Takatsuto, S. and Yokota, T. (1998). The Arabidopsis DIMINUTO/DWARF1 Gene encodes a protein involved in steroid synthesis. *Plant Cell,* **10** 1677–1690.

Kushiro, M., Nakano, T., Sato, K., Yamagishi, K. and Asami, T. (2001) Obtusifoliol 14α-demethylase (CYP51) antisense Arabidopsis shows slow growth and long life. *Biochemical and Biophysical Research Communications,* **285**, 98–104.

Lecain, E., Chenivesse, X., Spagnoli, R. and Pompon, D. (1996) Cloning by metabolic interference in yeast and enzymatic characterization of *Arabidopsis thaliana* sterol $\Delta^7$-reductase. *Journal of Biological Chemistry,* **271**, 10866–10873.

Li,Y., Edsall, R.J., Jagtap, P.G., Kingston, D.G.I. and Bane, S. (2000) Equilibrium studies of a fluorescent paclitaxel derivative binding to microtubules. *Biochemistry*, **39**, 616–623.

Li, J., Nagpal, P., Vitart, V., McMorris, T.C., Chory, J. (1996) A role for brassinosteroids in light-dependent development of Arabidopsis. *Science*, **272**, 398–401.

Lorenz, R.T. and Parks, L.W. (1992) Cloning, sequencing and disruption of the encoding sterol C-14 reductase in *Saccharomyces cerevisiae. DNA and Cell Biology*, **11**, 685–692.

Lovato, M.A. Hart, E.A. Segura, M.J.R. Giner, J.L. and Matsuda, S.P.T. (2000) Functional cloning of an *Arabidopsis thaliana* cDNA encoding cycloeucalenol cycloisomerase. *Journal of Biological Chemistry,* **275**, 13394–13397.

Ma, Y.Z. Holt, N.E., Li, X.P., Niyogi, K.K. and Fleming, G.R. (2003) Evidence for carotenoid involvement in the regulation of photosynthetic light harvesting. *Proceedings of the National Academy of Sciences USA*, **100**(8), 4377–4382.

Marin, E., Nussaume, L., Quesada, A. *et al.* (1996) Molecular identification of zeaxanthin epoxidase of *Nicotiana plumbaginifolia*, a gene involved in abscisic acid biosynthesis and corresponding to the ABA locus of *Arabidopsis thaliana. EMBO Journal*, **15**, 2331–2342.

Mauch, D.H., Nagler, K., Schumacher, S. *et al.* (2001) CNS synaptogenesis promoted by glia-derived cholesterol. *Science*, **294**, 1354–1357.

Moebius, F.F., Fitzky, B.U., Lee, J.N., Paik, Y.K. and Glossmann, H. (1998) Molecular cloning and expression of the human $\Delta^7$-sterol reductase. *Proceedings of the National Academy of Sciences USA,* **95**, 1899–1902.

Nelson, D. Koymans, L., Kamataki, T.Stegeman, J.J. Feyereisen, R. (1996) P450 superfamily: update on new sequences, gene mapping, accession numbers and nomenclature. *Pharmacogenetics*, **6**, 1–42.

Nes, W.R. (1977) The biochemistry of plant sterols. *Advances in Lipid Research*, **15**, 233–324.

Nes, W.D. (2003) Enzyme mechanisms for sterol C-methylations. *Phytochemistry*, **64**, 75–95.

Nievelstein, V., Vandekerchove, J., Tadros, M.H., Lintig, J.V., Nitschke, W. and Beyer, P. (1995) Carotene desaturation is linked to a respiratory redox pathway in *Narcissus pseudonarcissus* chromoplast membranes. Involvement of a 23-kDa oxygen-evolving-complex-like protein. *European Journal of Biochemistry*, **233**, 864–872.

Niyogi, K.K. Grossman, A.R. and Bjorkman, O. (1998) Arabidopsis mutants define a central role for the xanthophyll cycle in the regulation of photosynthetic energy conversion. *Plant Cell*, **10**, 1121–1134.

Norris, S.R., Barrette, T.R. and DellaPenna, D. (1995) Genetic dissection of carotenoid synthesis in Arabidopsis defines plastoquinone as an essential component of phytoene desaturation. *Plant Cell*, **7**, 2139–2149.

Okada, K., Saito, T., Nakagawa, T. Kawamukai, M. and Kamiya, Y. (2000) Five geranylgeranyl diphosphate synthases expressed in different organs are localized into three subcellular compartments in Arabidopsis. *Plant Physiology*, **122**, 1045–1056.

Park, H., Kreunen, S.S., Cuttriss, A.J., DellaPenna, D. and Pogson, B.J. (2002) Identification of the carotenoid isomerase provides insight into carotenoid biosynthesis, prolamellar body formation, and photomorphogenesis. *Plant Cell,* **14**, 321–332.

Pascal, S., Taton, M. and Rahier, A. (1993) Plant sterol biosynthesis: identification and characterization of two distinct microsomal oxidative enzymatic systems involved in sterol C4-demethylation. *Journal of Biological Chemistry,* **268**,11639–11654.

Pascal, S., Taton, M. and Rahier, A. (1994) Plant sterol biosynthesis: identification of a NADPH dependent plant sterone reductase involved in the sterol-4-demethylation. *Archives of Biochemistry and Biophysics,* **312**, 260–271.

Phillips, M.A. and Croteau, R.B. (1999). Resin-based defenses in conifers. *Trends in Plant Science*, **4**, 184–190.

Podust, L.M. Poulos, T.L. and Waterman, M.R. (2001) Crystal structure of cytochrome P450 14α-sterol demethylase (CYP51) from *Mycobacterium tuberculosis* in complex with azole inhibitors. *Proceedings of the National Academy of Sciences USA,* **98**, 3068–3073.

Pogson, B., McDonald, K.A., Truong, M., Britton, G. and DellaPenna, D. (1996) Arabidopsis carotenoid mutants demonstrate that lutein is not essential for photosynthesis in higher plants. *Plant Cell*, **8**, 1627–1639.

Porfirova, S., Bergmuller, E., Tropf, S., Lemke, R. and Dormann, P. (2002) Isolation of an Arabidopsis mutant lacking vitamin E and identification of a cyclase essential for all tocopherol biosynthesis. *Proceedings of the National Academy of Sciences USA*, **99**, 12495–12500.

Rahier, A. (2001) Deuterated $\Delta^7$-cholestenol analogues as mechanistic probes for wild-type and mutated $\Delta^7$-sterol-C5(6)-desaturase. *Biochemistry,* **40**, 256–267.

Rahier, A. and Taton, M. (1996) Sterol biosynthesis : strong inhibition of maize $\Delta^{5,7}$-sterol-$\Delta^7$-reductase by novel 6-aza-*B*-homosteroids and other analogs of a presumptive carbocationic intermediate of the reduction reaction. *Biochemistry,* **35**, 7069–7076.

Rahier, A., Génot, J.C., Schuber, F., Benveniste, P. and Narula, A.S. (1984) Inhibition of (S)-adenosyl-L-methionine-sterol-C24-methyltransferase by analogues of a carbocationic ion high energy intermediate – Structure activity relationships for C-25 heteroatoms (N,As,S) substituted triterpenoid derivatives. *Journal of Biological Chemistry,* **259**, 15215–15223.

Rahier, A. Smith, M. and Taton, M. (1997) The role of cytochrome b5 in 4-methyl-oxidation and C5(6)-desaturation of plant sterol precursors. *Biochemical and Biophysical Research Communications,* **236**, 434–437.

Rahier, A., Taton, M. and Benveniste, P. (1989) Cycloeucalenol-obtusifoliol isomerase. Structural requirements for substrates and inhibitors binding or transformation. *European Journal of Biochemistry*, **181**, 615–626.

Ricciarelli, R., Zingg, J.M. and Azzi, A. (2002) The 80th anniversary of vitamin E: beyond its antioxidant properties. *Biological Chemistry*, **383**, 457–465.

Rock, C.D. and Zeevaart, J.A. (1991) The aba mutant of Arabidopsis thaliana is impaired in epoxy-carotenoid biosynthesis. *Proceedings of the National Academy of Sciences USA*, **88**, 7496–7499.

Rohmer, M. (1999) The discovery of a mevalonate-independent pathway for isoprenoid biosynthesis in bacteria, algae and higher plants. *Natural Product Reports*, **16**, 565–574.

Rondet, S., Taton, M. and Rahier, A. (1999) Identification, characterization, and partial purification of 4α-carboxysterol-C3-dehydrogenase/C4-decarboxylase from *Zea mays. Archives of Biochemistry and Biophysics,* **366**, 249–260.

Ronen, G., Cohen, M., Zamir, D. and Hirschberg, J. (1999) Regulation of carotenoid biosynthesis during tomato fruit development: expression of the gene for lycopene ϵ-cyclase is down-regulated during ripening and is elevated in the mutant Delta. *Plant Journal*, **17**, 341–351.

Sarkar, D. Imai, T. Kambe, F. Shibata, A. Ohmori, S. Siddiq, A. Hayasaka, S. Funahashi, H. and Seo, H. (2001) The human homolog of Diminuto/Dwarf1 gene (hDiminuto): a novel ACTH-responsive gene overexpressed in benign cortisol-producing adrenocortical adenomas. *Journal of Clinical Endocrinology and Metabolism*, **86**, 5130–5137.

Schaeffer, A., Bronner, R., Benveniste, P. and Schaller, H. (2001) The ratio of campesterol to sitosterol with modulates growth in Arabidopsis is controlled by STEROL METHYLTRANSFERASE 2–1. *Plant Journal*, **25**, 605–615.

Schaller, H. (2003) The role of sterol in plant growth and development. *Progress in Lipid Research*, **422**, 163–175.

Schaller, H., Gondet, L., Maillot-Vernier, P. and Benveniste, P. (1994) Sterol overproduction is the biochemical basis of resistance to a triazole in calli from a tobacco mutant. *Planta*, **194**, 295–305.

Schoendorf, A., Rithner, C.D., Williams, R.M. and Croteau, R.B. (2001) Molecular cloning of a cytochrome P450 taxane 10β-hydroxylase cDNA from Taxus and functional expression in yeast. *Peedings of the National Academy of Sciences USA*, **98**,1501–1506.

Schrick, K., Mayer, U., Horrichs, A. Kuhnt, C. and Bellini, C. (2000) FACKEL is a sterol C-14 reductase required for organized cell division and expansion in Arabidopsis embryogenesis. *Genes & Development*, **14**, 1471–1484.

Schwartz, S.H., Qin, X. and Zeevaart, J.A. (2003) Elucidation of the indirect pathway of abscisic acid biosynthesis by mutants, genes, and enzymes. *Plant Physiology*, **131**, 1591–1601.

Schwender, J., Muller, C., Zeidler, J. and Lichtenthaler, H.K. (1999) Cloning and heterologous expression of a cDNA encoding 1-deoxy-D-xylulose-5-phosphate reductoisomerase of Arabidopsis thaliana. *FEBS Letters*, **455**, 140–144.

Segura, M.J.R., Lodeiro, S., Meyer, M.M., Patel, A.J. and Matsuda, S.P.T. (2002) Directed evolution experiments reveal mutations at cycloartenol synthase residue His477 that dramatically alter catalysis. *Organic Letters*, **4**, 4459–4462.

Shanklin, J., Whittle, E. and Fox, B.G. (1994) Eight histidine residues are catalytically essential in a membrane-associated iron enzyme stearoyl-CoA desaturase, and are conserved in alkane hydroxylase and xylene monooxygenase. *Biochemistry,* **33**, 12787–12794.

Shewmaker, C.K., Sheehy, J.A., Daley, M., Colburn, S., and Ke D.Y. (1999) Seed-specific overexpression of phytoene synthase: increase in carotenoids and other metabolic effects. *Plant Journal*, **20**, 401–412.

Shi, J., Gonzales, R.A. and Bhattacharyya, M.K. (1996) Identification and characterization of an S-adenosyl-L-methionine: $\Delta^{24}$-sterol-C-methyltransferase cDNA from soybean. *Journal of Biological Chemistry*, **271**, 9384–9389.

Shintani, D. and DellaPenna, D. (1998) Elevating the vitamin E content of plants through metabolic engineering. *Science*, **282**, 2098–2100.

Silve, S., Dupuy, P.H., Labit-Lebouteiller, C. Kaghad, M. and Chalon, P. (1996) Emopamil-binding protein, a mammalian protein that binds a series of structurally diverse neuroprotective agents, exhibits $\Delta^{8-7}$-sterol isomerase activity in yeast. *Journal of Biological Chemistry*, **271**, 22434–22439.

Silve, S., Dupuy, P.H., Ferrara, P. and Loison, G. (1998) Human lamin B receptor exhibits sterol C14-reductase activity in *Saccharomyces cerevisiae. Biochimica et Biophysica Acta*, **1392**, 233–244.

Skaggs, B.A., Alexander, J.F., Pierson, C.A. Schweitzer, K.S. Chun, K.T. Koegel, C. Barbuch, R. and Bard, M. (1996) Cloning and characterization of the *Saccharomyces cerevisiae* C-22 sterol desaturase gene, encoding a second cytochrome P-450 involved in ergosterol biosynthesis. *Gene*, **169**, 105–109.

Soll, J. and Schultz, G. (1979) Comparison of geranylgeranyl and phytyl substituted methylquinols in the tocopherol synthesis of spinach chloroplasts. *Biochemical and Biophysical Research Communications*, **91**, 715–720.

Soll, J., Schultz, G., Joyard, J., Douce, R. and Block, M.A. (1985) Localization and synthesis of prenylquinones in isolated outer and inner envelope membranes from spinach chloroplasts. *Archives of Biochemistry and Biophysics*, **238**, 290–299.

Souter, M., Topping, J., Pullen, M., Friml, J. and Palme, K. (2002) Hydra mutants of Arabidopsis are defective in sterol profiles and auxin and ethylene signaling. *Plant Cell*, **14**, 1017–1031.

Sun, Z., Gantt, E. and Cunningham, F.X., Jr. (1996) Cloning and functional analysis of the beta-carotene hydroxylase of *Arabidopsis thaliana*. *Journal of Biological Chemistry*, **271**, 24349–24352.

Taton, M. and Rahier, A. (1991) Properties and structural requirements for substrate specificity of cytochrome P450-dependent obtusifoliol 14α-demethylase from maize (*Zea mays*) seedlings. *Biochemical Journal*, **277**:483–492.

Taton, M. and Rahier, A. (1996) Plant sterol biosynthesis: identification and characterization of higher plant ?7-sterol-C5(6)-desaturase. *Archives of Biochemistry and Biophysics*, **325**, 279–288.

Taton, M., Benveniste, P. and Rahier, A. (1987) Comparative study of the inhibition of sterol biosynthesis in *Rubus fruticosus* suspension cultures and *Zea mays* seedlings by [N-(1,5,9-trimethyldecyl]-4α,10-dimethyl-8-aza-trans-decal-3β-ol and derivatives. *Phytochemistry*, **26**, 385–392.

Taton, M., Husselstein, T., Benveniste, P. and Rahier, A. (2000) Role of highly conserved residues in the reaction catalyzed by recombinant $\Delta^7$-sterol-C5(6)-desaturase studied by site-directed mutagenesis. *Biochemistry*, **39**, 701–711.

Tian, L., Musetti, V. Kim, J., Magallanes-Lundback, M. and DellaPenna, D. (2004) The Arabidopsis LUT1 locus encodes a member of the cytochrome P450 family that is required for carotenoid ϵ-ring hydroxylation activity. *Proceedings of the National Academy of Sciences USA*, **101**, 402–407.

Umeno, D. Tobias, A.V. and Arnold, F.H. (2002) Evolution of the C30 carotenoid synthase CrtM for function in a C40 pathway. *Journal of Bacteriology*, **184**, 6690–6699.

Vogel, B.S., Wildung, M.R., Vogel, G. and Croteau, R. (1996) Abietadiene synthase from grand fir (*Abies grandis*). cDNA isolation, characterization, and bacterial expression of a bifunctional diterpene cyclase involved in resin acid biosynthesis. *Journal of Biological Chemistry*, **271**, 23262–23268.

Walker, K. and Croteau, R. (2000) Taxol biosynthesis: molecular cloning of a benzoyl-CoA:taxane 2alpha-O-benzoyltransferase cDNA from taxus and functional expression in *Escherichia coli*. *Proceedings of the National Academy of Sciences USA*, **97**, 13591–13596.

Walker, K., Long, R. and Croteau, R. (2002) The final acylation step in taxol biosynthesis: cloning of the taxoid C13-side-chain *N*-benzoyltransferase from Taxus. *Proceedings of the National Academy of Sciences USA*, **99**, 9166–9171.

Walker, K., Schoendorf, A. and Croteau, R. (2000) Molecular cloning of a taxa-4(20),11(12)-dien-5alpha-ol-O-acetyl transferase cDNA from Taxus and functional expression in *Escherichia coli*. *Archives of Biochemistry and Biophysics*, **374**, 371–380.

Walter, M.H., Fester, T. and Strack, D. (2000) Arbuscular mycorrhizal fungi induce the non-mevalonate methylerythritol phosphate pathway of isoprenoid biosynthesis correlated with accumulation of the 'yellow pigment' and other apocarotenoids. *Plant Journal*, **21**, 571–578.

Wendt, K.U., Poralla, K. and Schulz, G.E. (1997) Structure and function of a squalene cyclase. *Science*, **277**, 1811–1814.

Wheeler, A.L., Long, R.M., Ketchum, R.E., Rithner, C.D., Williams, R.M. and Croteau, R. (2001) Taxol biosynthesis: differential transformations of taxadien-5α-ol and its acetate ester by cytochrome P450 hydroxylases from Taxus suspension cells. *Archives of Biochemistry and Biophysics*, **390**, 265–278.

Wildung, M.R. and Croteau, R. (1996) A cDNA clone for taxadiene synthase, the diterpene cyclase that catalyzes the committed step of taxol biosynthesis. *Journal of Biological Chemistry*, **271**, 9201–9204.

Ye, X., Al-Babili, S., Kloti, A. *et al.* (2000) Engineering the provitamin A (β-carotene) biosynthetic pathway into (carotenoid-free) rice endosperm. *Science*, **287**, 303–305.

# List of Abbreviations

| | |
|---|---|
| 10:0 | Capric acid |
| 12:0 | Lauric acid |
| 14RED | $\Delta^{8,14}$-sterol-$\Delta^{14}$-reductase |
| 16:0 | Palmitic acid |
| 18:0 | Stearic acid |
| $18{:}1 cis\Delta^{6}$ | Petroselinic acid |
| $18{:}1 cis\Delta^{9}$ | Oleic acid |
| $18{:}2 cis\Delta^{9,12}$ | Linoleic acid |
| 20:1 | Eicosenoic acid |
| $22{:}1 cis\Delta^{13}$ | Erucic acid |
| 24:1 | Nervonic acid |
| 5DES | $\Delta^{7}$-sterol-5-desaturase |
| 7RED | $\Delta^{5,7}$-sterol-$\Delta^{7}$-reductase |
| 8ISO | $\Delta^{8}$–$\Delta^{7}$-sterol isomerase |
| *A. thaliana* | *Arabidopsis thaliana* |
| AAPT | Aminoalcoholphosphotransferase |
| ABA | Abscisic acid |
| ABC | ATP-binding cassette |
| ABS | Antley-Bixler syndrome |
| ACAT | Acetyl-CoA: ACP acyltransferase |
| ACBP | Acyl-CoA binding proteins |
| ACCase | Acetyl-CoA carboxylase |
| ACP | Acyl carrier protein |
| ACS | Acyl-CoA synthetase |
| ACX | Acyl-CoA oxidase |
| ADRP | Adipose differentiation-related protein |
| AIP1 | Actin-interacting protein 1 |
| AOC | Allene oxide cyclase |
| AOPPs | Aryloxyphenoxypropionic acids |
| AOS | Allene oxide synthase |
| Apo-B | Apolipoprotein B |
| ASP | Acylation stimulating protein |
| ASTM | American Society for Testing and Materials |
| *B. napus* | *Brassica napus* |
| BCCP | Biotin carboxyl carrier protein |

| | |
|---|---|
| BEE | Base-exchange enzyme |
| BHK | Baby hamster kidney |
| BKR | β-Ketoacyl-ACP reductase |
| BR | Brassinosteroid |
| BSA | Bovine serum albumin |
| CaMV | Cauliflower mosaic virus |
| CCD | Carotenoid cleavage enzyme |
| CCS | Capsanthin-capsorubine synthase |
| CCT | Phosphocholine cytidylyltransferase |
| CD | Circular dichroism |
| CDP | Cytidine diphosphate |
| CDP-DAG | CDP-diacylglycerol |
| CDP-EA | CDP-ethanolamine |
| CDS | CDP-diacylglycerol synthase |
| *CER* | *eeceriferum*; gene involved in wax production in Arabidopsis |
| CHAPS | 3-[(3-cholamidopropyl)dimethylammonio]-1-propanesulfonate |
| CHDs | Cyclohexanediones |
| CISK | Cytokine-independent survival kinase |
| CK | Choline kinase |
| CoA | Coenzyme A |
| CPI | Cyclopropyl sterol isomerase |
| CPT | CDP-choline:1,2-diacylglycerol cholinephosphotransferase |
| CT | Carboxyltransferase |
| CTP | Cholinephosphate acyltransferase |
| $C_{X:Y}$ | A fatty acyl group containing X carbon atoms and Y double bonds |
| CYC | Chromoplast specific-lycopene cyclase |
| DAG | Diacylglycerol |
| DAGPP | Diacylglycerol-pyrophosphate |
| DAGTA | Diacylglycerol transacylase |
| DES | Divinyl ether synthase |
| DGAT | Diacylglycerol acyltransferase |
| DGDG | Digalactosyldiacylglycerol |
| DGTA | Diacylglycerol transacylase |
| DHAP | Dihydroxyacetone-phosphate |
| DMPB | 2,3-Dimethyl-5-phytyl-1,4-benzoquinone |
| DOX | Dioxygenase |
| DPG | Diphosphatidylglycerol (cardiolipin) |
| DPGS | Diphosphatidylglycerol synthase |
| DSC | Differential scanning calorimetry |
| *E. coli* | *Escherichia coli* |
| *E. lagascae* | *Euphorbia lagascae* |
| EA | Ethanolamine |
| EACT | Phosphoethanolamine cytidylyltransferase |

| | |
|---|---|
| EAK | Ethanolamine kinase |
| EAS | Epoxy alcohol synthase |
| EBP | Emopamil-binding protein |
| EFA | Essential fatty acid |
| ENR | Enoyl-ACP reductase |
| ER | Endoplasmic reticulum |
| EST | Expressed sequence tag |
| FA | Fatty acid |
| FAE | Fatty acid elongase |
| FAR | Fatty acyl-CoA reductase |
| FAS | Fatty acid synthase |
| FAT | Fatty acyl-ACP thioesterase |
| Fd | Ferredoxin |
| FDP | Farnesyl diphosphate |
| FERM | Four-point-one-ezrin-radixin-moesin |
| FFA | Free fatty acid |
| FNR | NADPH/ferredoxin $NADP^+$ reductase |
| FTIR | Fourier-transform infrared spectroscopy |
| G3P | *sn*-glycerol-3-phosphate |
| Gal | D-galactose |
| GATA | GATA applies to a family of transcriptional activating proteins that bind to DNA sites with the consensus sequence [A/T][GATA][A/G]. |
| GC | Gas chromatography |
| GC/MS | Gas chromatography/mass spectrometry |
| GDP | Geranyl diphosphate |
| GEF | Guanine nucleotide exchange factor |
| GFP | Green fluorescent protein |
| GGDP | Geranylgeranyl diphosphate |
| GGGT | Ggalactolipid:galactolipid galactosyltransferase |
| GLA | γ-Linolenic acid |
| GLC | Gas–liquid chromatography |
| Glc | D-glucose |
| GPAT | Glycerol-3-phosphate acyltransferase |
| GPDH | Glycerol-3-phosphate dehydrogenase |
| GPI | Glycosylated phosphatidylinositol |
| Gro | Glycerol |
| Gro-3-P | Glycerol-3-phosphate |
| GRP | Glycine rich proteins |
| GUS | β-Glucuronidase |
| HCV | Hepatitis C core protein (or virus) |
| HDL | High density lipoprotein |
| HGA | Homogentisic acid |
| HGGT | Homogentisic acid geranylgeranyltransferase |
| HO-18:1 | Ricinoleic acid |
| HPL | Hydroperoxide lyase |
| HPOD | Hydroperoxy octadecadienoic acid |
| HPOT | Hydroperoxy octadecatrienoic acid |

| | |
|---|---|
| HPT | Homogentisic phytyltransferase |
| ICL | Isocitrate lyase |
| IDP | Isopentenyl diphosphate |
| IV | Iodine value |
| JA | Jasmonic acid |
| JAME | Jasmonic acid methyl ester |
| KAS | β-Ketoacyl-ACP synthase |
| KAT | 3-Ketoacyl-CoA thiolase |
| KCS | β-Ketoacyl-CoA synthase |
| *L. alba* | *Limnanthes alba* |
| *L. douglassii* | *Limnanthes douglasii* |
| LA | Linoleic acid |
| LACS | Long-chain acyl-CoA synthetase |
| LCY | Lycopene cyclase |
| LeA | Linolenic acid |
| LOX | Lipoxygenase |
| LPA | Lysophosphatidic acid |
| LPAAT | Lysophosphatidic acid acyltransferase |
| LPC | Lysophosphatidylcholine |
| LPCAT | Lysophosphatidylcholine acyltransferase |
| LPE | Lysophosphatidylethanolamine |
| LPL | Lysophospholipase |
| LTP | Lipid transfer protein |
| *M. ramanniana* | *Mortierella ramanniana* |
| MAG | Monoacylglycerol |
| MD | Microspore-derived |
| MEP | 2-Methyl-erythritol-4-phosphate |
| MFP | Multifunctional protein |
| MGAT | Acyl-CoA:monoacylglycerol acyltransferase |
| MGDAG | Monogalactosyldiacylglycerol |
| MPB | 2-Methyl-6-phytyl-1,4-benzoquinone |
| MS | Malate synthase |
| MSB | 2-Methyl-6-solanesyl-1,4-benzoquinone |
| MTP | Microsomal transport protein |
| *N. tabacum* | *Nictotinia tabacum* |
| NADH | Nicotinamide adenine dinucleotide |
| NADPH | Nicotinamide adenine dinucleotide phosphate |
| NAPE | *N-acyl*-phosphatidylethanolamine |
| NCED | Neoxanthin cleavage enzyme dioxygenase |
| NEM | *N*-ethylmaleimide |
| NMR | Nuclear magnetic resonance |
| OBT14DM | Obtusifoliol-14α-demethylase (CYP51) |
| OPDA | 12-Oxo-phytodienoic acid |
| OPR | 12-Oxo-phytodienoate reductase |
| ORF | Open reading frame |
| OSC | 2,3-Oxidosqualene-cycloartenol cyclase |
| P | Phosphate |
| *P. americana* | *Persea americana* |

| | |
|---|---|
| PA | Phosphatidate |
| PA | Phosphatidic acid |
| PAP | Phosphatidic acid phosphatase |
| PAT | Perlipin |
| PC | Phosphatidylcholine |
| PDAT | Phospholipid:diacylglycerol acyltransferase |
| PDK1 | 3-Phosphoinositide-dependent protein kinase-1 |
| PDS | Phytoene desaturase |
| PE | Phosphatidylethanolamine |
| PEA | Phosphoethanolamine |
| PEAMT | Phosphoethanolamine methyltransferase |
| PEPCK | Phosphoenolpyruvate carboxykinase |
| PG | Phosphatidylglycerol |
| PGP | Phosphatidylglycerol-3-phosphate |
| PGP1, PGP2 | Phosphatidylglycerol-3-phosphate synthase 1 or 2 |
| PGPP | Phosphatidylglycerol-3-phosphate phosphatase |
| PGT | Processive galactosyltransferase |
| PHA | Polyhydroxyalkanoate |
| PI | Phosphatidylinositol |
| PIP2 | Phosphatidylinositol-bisphosphate |
| PIS | Phosphatidylinositol synthase |
| PKB | Protein kinase B alpha |
| PL | Phospholipase |
| PLA, PLB, PLC, PLD | Phospholipase A, B, C or D |
| $PLA_2$ | Phospholipase $A_2$ |
| PLAP | Plastid lipid-associated protein |
| PLC | Phospholipase C |
| PLD | Phospholipase D |
| PM | Plasma membrane |
| POX | Peroxygenase |
| PP17 | Placental proteins 17 |
| PQ | Plastoquinone |
| PQH2 | Dihydroplastoquinone |
| PS | Phosphatidylserine |
| PSD1, PSD2, PSD3 | Phosphatidylserine decarboxylase 2, 3 |
| PSS | Phosphatidylserine synthase |
| PSY | Phytoene synthetase |
| PUFA | Polyunsaturated fatty acid |
| RA | Roughanic acid |
| RCC1 | Regulator of Chromosome Condensation 1 |
| RNAi | RNA interference |
| ROS | Reactive oxygen species |
| rTMT | r-Tocopherol methyltransferase |
| SA | Salicylic acid |
| SAG | Senescence-associated gene |
| SAM | S-adenosylmethionine |
| SDC | Serine decarboxylase |

| | |
|---|---|
| *S. cerevisiae* | *Saccharomyces cerevisiae* |
| SDR | Short-chain alcohol dehydrogenase |
| SEM | Scanning electron microscopy |
| Ser | L-serine |
| SER | Smooth endoplasmic reticulum |
| SMO | Sterol-C4$\alpha$-methyl oxidase |
| SMT | *S*-adenosylmethionine-sterol-*C*-methyltransferase |
| sn | Stereospecific numbering |
| SQD2 | Sulfoquinovosyldiacylglycerol synthase |
| SQDG | Sulfoquinovosyldiacylglycerol |
| SRP | Signal recognition particle |
| *T. majus* | *Tropaeolum majus* |
| TAG | Triacylglycerol |
| TCA | Tricarboxylic acid |
| TEM | Transmission electron microscopy |
| TIP47 | Tail-interacting 47 kDa protein |
| TLC | Thin-layer chromatography |
| UDP-Gal | UDP-D-galactose |
| UDP-Glc | UDP-D-glucose |
| UDP-SQ | UDP-sulfoquinovose |
| UGE | UDP-D-glucose 4-epimerase |
| UV | Ultra violet |
| VDE | Violaxanthin deepoxidase |
| VIGS | Virus-induced gene silencing |
| VLCFA | Very-long-chain fatty acid |
| VLDL | Very-low-density lipoprotein |
| VOC | Volatile organic chemicals |
| WS | Wax synthase |
| *Z. mays* | *Zea mays* |
| ZDS | $\zeta$-carotene desaturase |
| ZEP | Zeatin epoxidase |

# Index